Achieving sustainable cultivation of grain legumes

Volume 1: Advances in breeding and cultivation techniques

It is widely recognised that agriculture is a significant contributor to global warming and climate change. Agriculture needs to reduce its environmental impact and adapt to current climate change whilst still feeding a growing population, i.e. become more 'climate-smart'. Burleigh Dodds Science Publishing is playing its part in achieving this by bringing together key research on making the production of the world's most important crops and livestock products more sustainable.

Based on extensive research, our publications specifically target the challenge of climate-smart agriculture. In this way we are using 'smart publishing' to help achieve climate-smart agriculture.

Burleigh Dodds Science Publishing is an independent and innovative publisher delivering high quality customer-focused agricultural science content in both print and online formats for the academic and research communities. Our aim is to build a foundation of knowledge on which researchers can build to meet the challenge of climate-smart agriculture.

For more information about Burleigh Dodds Science Publishing simply call us on +44 (0) 1223 839365, email info@bdspublishing.com or alternatively please visit our website at www.bdspublishing.com.

Related titles:

Achieving sustainable cultivation of grain legumes Volume 2:
Improving cultivation of particular grain legumes
Print (ISBN 978-1-78676-140-8); Online (ISBN 978-1-78676-142-2, 978-1-78676-143-9)

Achieving sustainable cultivation of cassava Volume 1: Cultivation techniques
Print (ISBN 978-1-78676-000-5); Online (ISBN 978-1-78676-002-9, 978-1-78676-003-6)

Achieving sustainable cultivation of cassava Volume 2: Genetics, breeding, pests and diseases
Print (ISBN 978-1-78676-004-3); Online (ISBN 978-1-78676-006-7, 978-1-78676-007-4)

Managing soil health for sustainable agriculture Volume 1: Fundamentals
Print (ISBN 978-1-78676-188-0); Online (ISBN 978-1-78676-190-3, 978-1-78676-191-0)

Managing soil health for sustainable agriculture Volume 2: Monitoring and management
Print (ISBN 978-1-78676-192-7); Online (ISBN 978-1-78676-194-1, 978-1-78676-195-8)

Chapters are available individually from our online bookshop: https://shop.bdspublishing.com

climate•SMART•publishing
IN AGRICULTURAL SCIENCE

BURLEIGH DODDS SERIES IN AGRICULTURAL SCIENCE

NUMBER 35

Achieving sustainable cultivation of grain legumes

Volume 1: Advances in breeding and cultivation techniques

Edited by Dr Shoba Sivasankar, formerly International Crops Research Institute for the Semi-Arid Tropics (ICRISAT), India; Dr David Bergvinson and Dr Pooran Gaur, International Crops Research Institute for the Semi-Arid Tropics (ICRISAT), India; Dr Shiv Kumar Agrawal, International Center for Agricultural Research in the Dry Areas (ICARDA), Morocco; Dr Steve Beebe, International Center for Tropical Agriculture (CIAT), Colombia; and Dr Manuele Tamò, International Institute of Tropical Agriculture (IITA), Benin

burleigh dodds
SCIENCE PUBLISHING

Published by Burleigh Dodds Science Publishing Limited
82 High Street, Sawston, Cambridge CB22 3HJ, UK
www.bdspublishing.com

Burleigh Dodds Science Publishing, 1518 Walnut Street, Suite 900, Philadelphia, PA 19102-3406, USA

First published 2018 by Burleigh Dodds Science Publishing Limited

Library of Congress Control Number: 2017938497

British Library Cataloguing in Publication Data
A catalogue record for this book is available from the British Library

ISBN 978-1-78676-136-1 (print)
ISBN 978-1-78676-138-5 (online)
ISBN 978-1-78676-139-2 (online)
ISSN 2059-6936 (print)
ISSN 2059-6944 (online)

Typeset by Deanta Global Publishing Services, Chennai, India
Printed by Lightning Source

Contents

Series list

Title	Series number
Achieving sustainable cultivation of maize - Vol 1 From improved varieties to local applications Edited by: Dr Dave Watson, CGIAR Maize Research Program Manager, CIMMYT, Mexico	001
Achieving sustainable cultivation of maize - Vol 2 Cultivation techniques, pest and disease control Edited by: Dr Dave Watson, CGIAR Maize Research Program Manager, CIMMYT, Mexico	002
Achieving sustainable cultivation of rice - Vol 1 Breeding for higher yield and quality Edited by: Prof. Takuji Sasaki, Tokyo University of Agriculture, Japan	003
Achieving sustainable cultivation of rice - Vol 2 Cultivation, pest and disease management Edited by: Prof. Takuji Sasaki, Tokyo University of Agriculture, Japan	004
Achieving sustainable cultivation of wheat - Vol 1 Breeding, quality traits, pests and diseases Edited by: Prof. Peter Langridge, The University of Adelaide, Australia	005
Achieving sustainable cultivation of wheat - Vol 2 Cultivation techniques Edited by: Prof. Peter Langridge, The University of Adelaide, Australia	006
Achieving sustainable cultivation of tomatoes Edited by: Dr Autar Mattoo, USDA-ARS, USA & Prof. Avtar Handa, Purdue University, USA	007
Achieving sustainable production of milk - Vol 1 Milk composition, genetics and breeding Edited by: Dr Nico van Belzen, International Dairy Federation (IDF), Belgium	008
Achieving sustainable production of milk - Vol 2 Safety, quality and sustainability Edited by: Dr Nico van Belzen, International Dairy Federation (IDF), Belgium	009
Achieving sustainable production of milk - Vol 3 Dairy herd management and welfare Edited by: Prof. John Webster, University of Bristol, UK	010
Ensuring safety and quality in the production of beef - Vol 1 Safety Edited by: Prof. Gary Acuff, Texas A&M University, USA & Prof.James Dickson, Iowa State University, USA	011
Ensuring safety and quality in the production of beef - Vol 2 Quality Edited by: Prof. Michael Dikeman, Kansas State University, USA	012
Achieving sustainable production of poultry meat - Vol 1 Safety, quality and sustainability Edited by: Prof. Steven C. Ricke, University of Arkansas, USA	013
Achieving sustainable production of poultry meat - Vol 2 Breeding and nutrition Edited by: Prof. Todd Applegate, University of Georgia, USA	014
Achieving sustainable production of poultry meat - Vol 3 Health and welfare Edited by: Prof. Todd Applegate, University of Georgia, USA	015

Acknowledgements

We wish to acknowledge the following for their help in reviewing particular chapters:

- Chapter 5: Dr Alemayehu Teressa Negawo, International Livestock Research Institute (ILRI), Ethiopia
- Chapter 8: Dr Mike Grusak, USDA-ARS, USA
- Chapter 19: Professor Ravindra Chibbar, University of Saskatchewan, Canada; Dr Jenny Wood, Department of Primary Industries – New South Wales, Australia

Introduction

Grain legumes are widely cultivated, particularly for their dry seeds (known as pulses). They include common beans and lentil which are grown globally, as well as regionally-specific types of pulses such as cowpea, faba beans and pigeonpea. The FAO defines pulses as crop plant members of the *Leguminosae* family (commonly known as the pea family) that produce edible seeds. In this definition, only legumes harvested for dry grain are classified as pulses, excluding species such as soybean and groundnut. However, broader definitions of grain legumes include groundnut and soybean as well.

Grain legumes are important in the developing world for a number of reasons. They are a rich source of protein and fibre, minerals and vitamins. In addition, their rapid growth and ability to fix nitrogen and improve soil health makes them a key rotation crop in the sustainable intensification and diversification of smallholder farming. This makes grain legumes a key food security crop. However, yields in developing countries are low as a result of such factors as the need for improved varieties of seed, poor seed distribution, the impact of pests and diseases, as well as vulnerability to poor soils, drought and other effects of climate change. There is now a rich body of research addressing these challenges.

The challenges facing grain legumes are addressed in the two volumes of *Achieving sustainable cultivation of grain legumes*:

- Volume 1 Advances in breeding and cultivation techniques
- Volume 2 Improving cultivation of particular grain legumes

This volume (Volume 1) focussing on breeding and cultivation. The volume reviews key developments in understanding crop physiology and genetic diversity as well as breeding. The book also covers advances in grain legume cultivation, from variety selection to post-harvest storage, and discusses the latest trends in disease, insect pest and weed management.

Part 1 Plant physiology and breeding

The first part of the volume summarises advances in understanding crop physiology and genetic diversity, and examines how this understanding has informed the development of new varieties. Chapter 1 discusses advances in understanding grain legume physiology, with a particular focus on stomatal behaviour and the plant's response to abiotic stress. Grain legumes can fulfill several roles. A potential constraint on their use is their response to drought in the more arid conditions predicted in some parts of the world due to climate change. The chapter reviews the responses of various grain legumes to water deficit conditions, and then discusses how grain legumes can be bred for the stomatal characteristics most appropriate for water-scarce environments.

Chapter 2 continues the previous chapter's focus on advances in understanding grain legume physiology, focusing on how root system architecture (RSA) influences root foraging and resource uptake from the soil. RSA thus determines plant growth and productivity in grain legumes. Understanding RSA, nutrient uptake and its response to drought and other abiotic stresses is vital for breeding super genotypes for efficient water and nutrient acquisition and with enhanced adaptation to abiotic stresses. The chapter describes the role of root architecture in plant health, available approaches for measuring root architecture, variability of RSA traits across genotypes and its implications for breeding drought-resistant varieties. The chapter explains how combined root phenotyping, non-destructive imaging, root model simulations and molecular techniques can serve as tools in breeding legume genotypes with optimised root system for enhanced adaptation to target environments.

Chapter 3 moves from the physiology of grain legumes to their genetics, and concentrates on the challenge of exploiting the genetic diversity of grain legumes. An increase in the production and marketing of grain legumes would help to diversify diets and improve human nutrition, as well as contributing to the sustainability of cropping systems and long-term soil health. However, the projected impacts of climate change, as well as existing constraints in cultivation, pose challenges to increasing production. One of the key resources for meeting these challenges is conserving genetic diversity for crop improvement, whether *ex situ* in genebanks and *in situ* with farmers or in protected areas. The chapter reviews global strategies to assess the status of the major grain legume genetic resources held in *ex situ* collections, develop a vision of a more secure global system of conservation, and identify priorities to facilitate the establishment of this global system. As an example, the chapter focuses on the characterization and exploitation of the genetic resources of chickpea and pigeonpea.

Complementing the previous chapter's insights into genetics, Chapter 4 looks at advanced breeding techniques for grain legumes. The challenge of grain legume production is to continue increasing productivity while reducing the significant seed yield gap between developed and developing countries/regions. Advanced breeding techniques have an important role to play in the era of genomics. The chapter describes the main grain legume breeding programs, including breeding targets such as stressors and phenotypes. The chapter examines grain legume reference genome sequences, legume common lineages and synteny, and describes the use of whole-genome and reduced representation resequencing and SNP chips to identify target traits.

Going beyond the advanced breeding techniques discussed in the preceding chapter, Chapter 5 addresses the genetic modification of grain legumes. Most grain legumes are grown predominantly in the low-input production systems in the developing countries of Asia and Africa and are subject to various production constraints. The chapter addresses progress in and prospects for transgenic interventions in the improvement of grain legumes, concentrating on chickpea, pigeonpea, cowpea, lentils, peas, peanuts and other grain legumes. The chapter examines the key challenges in the commercialization of genetically-engineered grain legumes, and shows how these technological interventions provide opportunities for enhancing efficiency and effectiveness of breeding programs for complex traits, as well as increasing the rate of genetic gain in grain legumes.

The focus of Chapter 6 is specifically on using breeding and genetic engineering techniques to develop drought- and heat-tolerant varieties of grain legume. Despite the complexity imposed by the polygenic nature of plant tolerance to drought and to heat, significant strides have been made in understanding the underlying response and adaptation of plants to these stresses. Together with advanced and accessible technologies such as next-generation sequencing and precision phenotyping, this has enabled the assembly of molecular and physiological tools in grain legumes that support the development of tolerance through traditional and molecular breeding. The chapter provides an overview of physiological, biochemical and molecular responses and adaptation mechanisms to drought, and to heat, with specific examples from grain legumes. It examines traditional breeding and the utilization of natural genetic diversity, developments in molecular breeding including the identification of genetic associations and quantitative trait loci (QTL) and functional genomics studies on stress-relevant candidate genes or gene families. Finally, the chapter discusses critical contemporary considerations to strengthen research on developing tolerance to drought and heat.

Complementing the previous chapter's focus on drought- and heat-tolerance, Chapter 7 examines another specific application of breeding and genetic engineering techniques in developing pest- and disease-resistant varieties of grain legume. Average yields of most grain legumes are still relatively low due in part to susceptibility to pests and diseases. The chapter explores current advances in developing pest- and disease-resistant cultivars of a variety of grain legume crops, including faba beans, peas, lentils, chickpeas, grass peas and common beans. The chapter focuses, in particular, on breeding cultivars less susceptible to foliar diseases caused by biotrophic or necrotrophic pathogens, soil borne diseases, as well as nematodes, bacteria, insect pests and parasitic weeds.

The concluding chapter of this section, Chapter 8, considers the bio-fortification of grain legumes. Micronutrient malnutrition (MNM) is one of the world's major health threats. Iron (Fe) and zinc (Zn) deficiencies affect more than a third of the world's population, particularly women and children. Biofortified legume crops are being developed with higher levels of bioavailable Fe and Zn to alleviate MNM and improve the nutritional status of affected populations. The chapter reviews the genetic variability of Fe and Zn content in many legume crops, and shows how this is being used to guide breeding as well as improve agronomic management.

Part 2 Cultivation

The second part of the volume reviews improvements in cultivation techniques to make the most of these new varieties, from variety selection and seed quality management, through pest and disease management to storage and quality assessment. The focus of Chapter 9 is on variety selection and seed quality management in grain legume cultivation. The chapter examines the key issues and challenges in variety selection and quality control. The chapter also includes some detailed case studies on how research has been used to improve legume cultivation in practice.

Chapter 10 examines grain legumes in integrated crop management systems. The chapter looks at how integrated crop management strategies can be used to increase grain legume production in rainfed, resource-poor farming systems. Generally, grain legumes fail to reach even half of their potential yields in these systems. For rainfed grain legumes the major contributor to the yield gap is sub-optimal soil moisture, along with a suite of nutrient, pest and disease constraints. The challenge is to identify remedial action within the means of resource-poor farmers. This requires greater emphasis on farmer-participatory research to identify local constraints, and engaging farmers in testing locally feasible solutions. Examples of this approach are documented in the chapter. Particular areas in need of intensive on-farm research include adapting grain legume farming to conservation agriculture and exploring means to increase cropping intensity of grain legumes in cereal-dominated cropping systems. The chapter suggests that a concerted shift in international and national efforts to support farmer-participatory approaches is needed.

Complementing Chapter 10, Chapter 11 focuses on the challenge of ensuring sustainable grain legume-cereal cropping systems. There is currently renewed interest in intercropping in Europe in order to achieve sustainable, ecological or eco-functional intensification of agricultural production, particularly in organic farming. The chapter summarises data from over 50 field experiments undertaken since 2001 on cereal-grain legume intercropping in 13 sites in southern and western France as well as in Denmark using spring and winter cereal-grain legume intercrops. The chapter also addresses the effects of intercropping on yields and quality, the agronomical performance of intercropping and cultivation practices in intercropping.

Continuing the theme of integrated cultivation strategies, Chapter 12 goes on to consider soil and nutrient management in grain legume cultivation. Due to their role in improving soil fertility, grain legumes can be integrated into farming systems as part of soil fertility management. This offers a potential pathway towards sustainable intensification. The chapter focuses on how to maximize biological nitrogen fixation and yields of grain legumes through soil and nutrient management practices. These techniques are illustrated in practice through a case study looking at biological nitrogen fixation in the Guinea Savanna zone of Ghana.

Chapter 13 considers a major challenge to integrated management, namely diseases affecting grain legumes and their management. The chapter reviews the main grain legume diseases before considering the relative benefits and disadvantages of both traditional and integrated disease management techniques. The chapter reviews the components of integrated disease management and practical developments in the field, with a particular focus on modelling, sampling and identification. Finally, the chapter considers advanced and rapid analysis techniques in integrated disease management.

Moving from integrated disease management to integrated pest management (IPM), Chapter 14 addresses insect pests and integrated pest management techniques in grain legume cultivation. As an example, it focuses on cowpea, a staple legume food crop grown and consumed in the dry savanna regions of sub-Saharan West Africa. The chapter describes the pests that attack cowpea at every stage of its development, including aphids, thrips, pod sucking bugs, and lepidopteran pod borers. The chapter reviews current control measures and their limitations. It then discusses ways of developing IPM strategies by exploiting knowledge of pest biology, host plant resistance (including *Bt* cowpea) and biocontrol. The chapter also emphasises the importance of disseminating new information effectively to farmers using cell phone technology, discussing the example of Scientific Animations Without Borders (SAWBO) which uses cell phones to distribute downloadable video content.

Complementing the focus of the previous chapters on diseases and pests, Chapter 15 moves on to look at weed management in grain legume cultivation. Managing weeds in a grain legume crop is critical for obtaining optimum yield. All grain legumes are relatively poor competitors with weeds and thus are prone to yield reductions when grown in the presence of weeds. The chapter outlines the nature and challenges of weed interference with crop growth and methods of weed control, as well as offering two detailed case studies. The chapter demonstrates that a combination of practices, such as cultivar selection, planting date, tillage system, in-crop tillage or cultivation, and/or herbicides is needed to increase the likelihood of successful weed control.

The theme of Chapter 16 is the challenge of grain legume storage in developing countries. Grain stored after harvest is often subject to major quantitative and qualitative losses. The major cause of loss is storage insects and mycotoxin-producing organisms, which reduce the nutritional and monetary value of the grain. The chapter describes various methods of storage-insect control that have the potential to help farmers with limited resources.

Complementing the previous chapter, Chapter 17 concentrates on the drying, handling, storing and quality monitoring of pulses. The chapter discusses how drying, storage and handling can adversely affect the quality of pulses. It reviews how pulse quality is assessed, and discusses strategies for pulse drying, handling and storage in order to maintain quality. The chapter describes techniques for grain monitoring and automated aeration control, as well as non-destructive techniques for quality assessment of pulses, including colour and hyperspectral imaging.

Continuing the focus on quality, Chapter 18 reviews the dietary health benefits and anti-nutritional factors in grain legumes. Legumes are a significant source of essential nutrients including dietary fiber, protein, dietary minerals, and carbohydrates. Abundant in both soluble and insoluble fiber, legumes support colonic and overall health. The chapter describes the effect of grain legume consumption in combatting gut inflammation, the health benefits of dietary fiber, and possible connections between dry grain legume consumption and carcinogenesis. The chapter also explores possible chemical toxicant exposure from grain legume consumption.

The final chapter in the book, Chapter 19, also assesses the nutritional quality of grain legumes, but from an economic perspective. It discusses the potential economic value that can potentially be derived from exploiting the nutritional potential of grain legumes. It assesses current constraints on production, including constraints in the availability of credit, inputs and information on best practice. It also reviews ways of changing consumer behaviour to boost demand to allow the expansion of grain legume cultivation.

Part 1

Plant physiology and breeding

Chapter 1

Advances in understanding grain legume physiology: stomatal behaviour and response to abiotic stress

E. Troyo–Diéguez and A. Nieto-Garibay, Centro de Investigaciones Biológicas del Noroeste, México; J. L. García-Hernández, Universidad Juárez del Estado de Durango, México; P. Preciado-Rangel, Instituto Tecnológico de Torreón, México; F. A. Beltrán-Morales and F. H. Ruiz-Espinoza, Universidad Autónoma de Baja California Sur, México; and B. Murillo-Amador, Centro de Investigaciones Biológicas del Noroeste, México

1 Introduction

Grain legumes will have an increasing role in agriculture in both less developed and developed regions of the world. A major reason for this prediction is that grain legumes can fulfil several roles: they can be an immediate seed crop for consumption or marketing, a means to increase soil fertility as a result of its nutrient-rich residues, and a key component of livestock feed. In many cropping systems, grain legumes are likely to be needed to fulfil all three roles of increased seed yield, enhanced soil fertility, and cattle feed (Sinclair and Vadez, 2012). A potential constraint on their use is their response to drought in the more arid conditions predicted in some parts of the world due to climate change (Chaves et al., 2002; Hulme, 2005).

Grain legumes vary widely in water requirements (Berger et al., 2002). One study found total water use by peanut the highest among some cultivated legumes, followed by cowpea, soybean and mungbean (Pandey et al., 1984a,b). Crop water requirements

http://dx.doi.org/10.19103/AS.2017.0023.01

also depend on the stage of growth with, for example, four stages for common bean (*Phaseolus vulgaris*), each with its own water requirements (Simonne et al., 2007). Water availability affects nutrients absorption and, in consequence, plant growth by influencing physiological processes and conditions (Hungria and Vargas, 2000). Due to current environmental issues concerning the use of water for irrigation, the improvement of crop water-use efficiency (WUE) has become a priority (Wullschleger et al., 1998; Turner, 2001). As noted, in the face of climate change there is a growing need to develop plants that are able to withstand more prolonged periods of drought (Dracup et al, 1980; Passioura, 1996). To optimise water use, knowledge of above- and below-ground plant physiological responses is necessary (Vandeleur et al., 2005).

2 Grain legume response to water deficit

Water status in plants is affected by factors such as air temperature, solar radiation, air humidity, wind intensity and soil humidity (Katerji and Rana, 2011). Plant responses to conditions of water scarcity are complex (Condon and Hall, 1997; Porch et al., 2004). Alternatives to maintain productivity include reducing transpiration or obtaining a more efficient use of the transpired water during photosynthesis. Responses usually involve a mix of stress avoidance and tolerance strategies (Subbarao et al., 1995; Chaves et al., 2002). Both responses have, for example, been shown in cowpea (*Vigna unguiculata*). In the first case, plants produced seed before reaching the permanent wilting stage. Although the plants did not recover after the stress period, viable seeds were produced, thus ensuring the next generation. In other conditions, plants were able to both withstand the period of drought stress by reducing water loss and recover to such an extent that they managed to produce seed after being re-watered (de Ronde and Spreeth, 2007). A key feature of drought adaptation strategies is to maintain turgidity at a relatively positive value (osmotic adjustment) (Turner et al., 2001; Erre et al., 2009). Osmotic adjustment is an important mechanism of drought resistance in legumes such as chickpea and common bean (Amede and Schubert, 2003a).

Stomatal closure has been identified as one of the first plant responses to water deficit, with the reduction in gas exchange, reducing water loss although sacrificing carbon assimilation in some extent (Chaves et al., 2002). A plant under drought adaptation is closely dependent on transpiration and photosynthesis. In the case of a single leaf, the rate of transpiration is strongly affected by changes in stomatal conductance (Yang et al., 2004). It is less well known whether leaf-level responses translate into changes in canopy evapotranspiration (Bernacchi, 2007). Negative feedback between evaporative demand and transpiration tends to produce approximately constant rates of evapotranspiration for a wide range of vegetation types.

It has been suggested that stomatal closure may be induced by changes in chemical signalling and/or the hydraulic status of the plant (Chaves et al., 2002). According to Heath and Mansfield (1971), the major factors influencing stomata to open or close are:

- Light: Stomatal aperture is inversely proportional to light intensity. There is a clear spectral sensitivity; only blue and, to a lesser extent, red lights cause opening.
- Carbon dioxide: Increasing CO_2 concentration causes stomatal closure, and decreasing its concentration causes opening.

- Temperature: Increases in temperature cause stomatal opening up to an optimum degree, above which additional increase in temperature causes stomatal closure.
- Available moisture: water stress causes stomatal closure.

Stomatal closure is a key first step in response to drought since it is more rapid and flexible than alternative mechanisms such as changes in leaf area, root growth or osmotic adjustment (Chaves, 1991; Amede and Schubert, 2003b).

The size of stomata changes markedly in response to a number of physiological factors and environmental conditions (Plaut and Moreshet, 1973). Stomatal aperture or conductance depends on several variables, namely, bulk leaf water content or water potential, which may itself be a function of transpiration rate (Jarvis, 1975). In many species, stomatal conductance depends also on the intercellular space carbon dioxide concentration, which in turn depends upon the rate of photosynthesis (Jarvis and Morison, 1981). Results obtained by Yang et al. (2004) indicate that stomatal aperture may be more closely related to photosynthetic characteristics and seed chemical composition in grass pea, and water deficit may enhance the correlations, as well as the vapour pressure deficit (VPD).

The number and distribution of stomata in the unit leaf area have an important role in these processes by adjusting CO_2, O_2 and moisture exchange between the leaves and the atmosphere (Brownlee, 2001). The number and distribution can vary by genotype. In a study of ten faba bean genotypes, for example, Peksen et al. (2006) found that stomatal numbers per mm^2 on the lower and upper epidermis of the leaflets ranged from 260 to 306 and 237 to 288, respectively. Genotypes differed only for the number of stomata per mm^2 on upper epidermis of the leaflet ($p < 0.05$) and leaflet number $leaf^{-1}$ ($p < 0.01$), although the number of stomata per mm^2, stoma width and length on the lower epidermis of the leaflet were greater ($p < 0.01$). A stomatal density comparison among three legumes – cowpea (*Vigna unguiculata*) (Fig. 1), common bean (*Phaseolus vulgaris*) (Fig. 2) and chick pea (*Cicer arietinum*) (Fig. 3) – was undertaken by the Electron Microscope Lab at CIBNOR using a scanning electron microscopy (SEM) Hitachi S-3000N. Stomatal density was calculated from 180 to 200, 160 to 180 and 220 to 225 stomata/mm^2, respectively. Stomatal size was 18–23, 20–25 and 18–22 μm, in the same order. The study concluded that chickpea has the most efficient stomatal density for water conservation. Physiological traits in faba bean related to drought tolerance are also discussed in Khan et al. (2007).

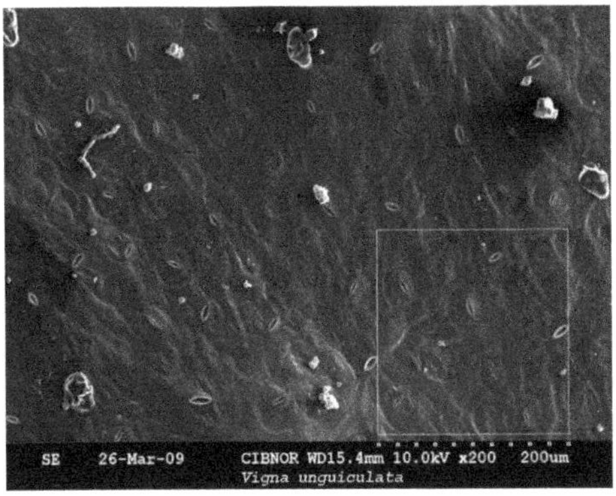

Figure 1 Electron microscope photograph of cowpea (*Vigna unguiculata*) abaxial leaf tissue.

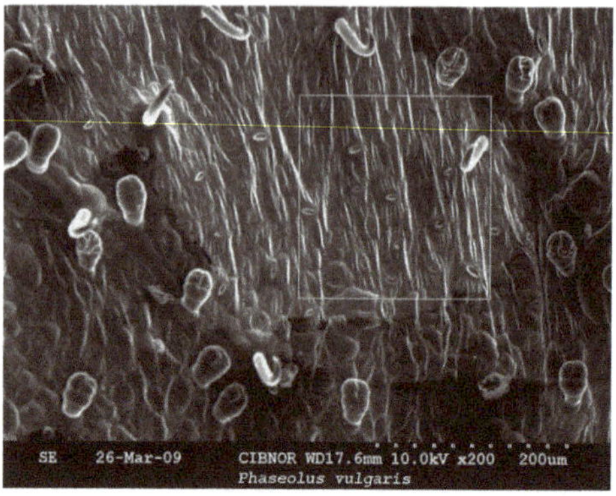

Figure 2 Electron microscope photograph of common bean (*Phaseolus vulgaris*) abaxial leaf tissue.

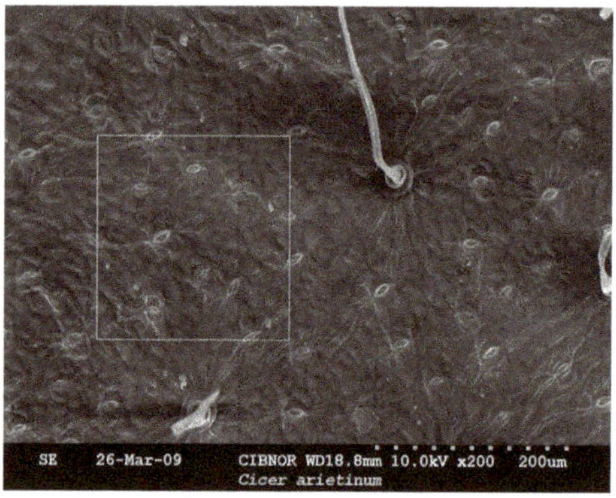

Figure 3 Electron microscope photograph of chick pea (*Cicer arietinum*) abaxial leaf tissue.

Figure 4 Assessing stomatal characteristics: *Dolichos lablab* grown on arid yermosols in Baja California Sur, Mexico.

3 Breeding for stomatal characteristics

Stomatal density and pore area are important for the control of stomatal transpiration in drought-stressed plants (Simova-Stoilova et al., 2016). It has been suggested that stomatal frequency may be modified through breeding to improve water use (Tanzarella et al., 1984; Sharma-Natu and Ghildiyal, 2005). The main focus must be on a balance between the conservation of water, which could be achieved by searching for fewer or smaller stomata, or for those species or varieties with stomata able to partially close during most of the photoperiod, and productivity (varieties with higher stomata density or larger ones, opened along the photoperiod) (Fig. 4). However, the optimal combination depends upon a series of complex interrelations, certain plant characteristics such as the size and number of leaves and the environment in which the plant grows (Wright et al., 1993; Subbarao et al., 1995). The aim is to increase the photosynthetic rate, through selection of appropriate stomatal characteristics, so as to achieve an optimal balance between the production of high performances of cultivars and the conservation of water (Long and Bernacchi, 2003).

There are two particular reasons for focusing on research trying to manipulate the stomatal apparatus. The first is to maximize plant productivity, and consequently the yield, by increasing the rates of assimilation. From an increase in stomatal conductance, a parallel increase in the partial CO_2 pressure would be expected within the leaf and, therefore, an increase in the photosynthesis in those plants (particularly the C3 type) in which the photosynthetic apparatus is not saturated with CO_2 (Turner et al., 2001; Vara et al., 2005). Having stomata that close in response to a severe 'pressure' (water demand) can be more useful for a crop, than stomata that are able to maintain low conductance. Stomatal closing in response to an increasing water deficit would tend to improve water use efficiency (WUE) and drought tolerance (Bozoglu and Karayel, 2006).

Because of the complexity of common legumes, research about their genomics first began on *Medicago truncatula* (Barker et al., 1990) and *Lotus japonicus* (Handberg and Stougaard, 1992). Although legumes like chickpea, common bean and pigeonpea constitute an important dietary component for the majority of populations in many developing countries, they did not immediately benefit from the genomic revolution (Varshney et al., 2010). However, as a result of community efforts, draft genome sequences have become available recently for pigeonpea (Varshney et al., 2012) and chickpea (Varshney et al., 2013). The use of marker-assisted selection and other molecular breeding systems for tracking single gene traits and quantitatively inherited characteristics will likely increase the overall efficiency and effectiveness of legumes improvement programmes in the foreseeable future and provide new opportunities for their development as a food staple and economic resource, as suggested for cowpea (Timko and Singh, 2008). Another important element is pre-breeding programmes for diversification of the primary gene pool (Sharma et al., 2013).

Whole genome sequences of both model and crop legumes are expected to offer new opportunities in legume breeding (Varshney et al., 2015). Large-scale analyses by using different genomics technologies are providing extensive data sets that will help identify potential candidate genes for an increase in intrinsic resistance and/or tolerance levels in important legume crops. Identification of these candidate genes may allow their direct application in crop improvement. However, in most cases, the roles of these candidate genes remain unknown and it will be important to carry out functional studies as a preliminary step towards their use in genetic improvement (Reddy et al., 2012).

4 Where to look for further information

Further information may be found at the following web sites: http://www.fao.org/in-action/inpho/crop-compendium/legumes/en/, http://www.arccjournals.com/journals/legume-research-an-international-journal, http://www.plantphysiol.org/, https://www.journals.elsevier.com/journal-of-plant-physiology, and related prestigious journals.

Legume production data for Mexico can be found in the Agriculture, Livestock, Fisheries and Food Ministry (SAGARPA) web page: http://www.sagarpa.gob.mx/quienesomos/datosabiertos/siap/Paginas/default.aspx.

For legume information in México and Latin-America contact Dr. Enrique Troyo (etroyo04@cibnor.mx) and/or Dr. Bernardo Murillo (bmurillo04@cibnor.mx).

5 Acknowledgements

This work was financed by Centro de Investigaciones Biológicas del Noroeste (CIBNOR, S.C.) and Consejo Nacional de Ciencia y Tecnología (CONACyT) through the Project CB-2009-01-0134460. Technician Ariel Cruz-Villacorta helped to process electron microscope photographs at the SEM Lab of CIBNOR. Many thanks are due to the field technicians Ariel Cruz, Alvaro Gonzalez-Michel, Carmen Mercado, Lidia Hirales, Saúl Briseño, Pedro Luna, Adrián Jordán and José Raymundo Ceseña for their helpful assistance in the maintenance of the experimental field plots.

6 References

Amede, T. and Schubert, S. (2003a). Mechanisms of drought resistance in grain legumes I: Osmotic adjustment. *Ethiop. J. Sci.*, 26(1), 37–46.

Amede, T. and Schubert, S. (2003b). Mechanisms of drought resistance in grain legumes II: Stomatal regulation and root growth. *Ethiop. J. Sci.*, 26(2), 137–44.

Barker, D., Bianchi, S., Blondon, F., Dattee, Y., Duc, G., Essad, S., Flament, T., Gallusci, T., Genier, G., Guy, P., Muel, X., Tourner, J., Denarie, J. and Huguet, T. 1990. *Medicago truncatula*, a model plant for studying the molecular genetics of the Rhizobium-legume symbiosis. *Plant Mol. Biol. Rep.*, 8, 40–9.

Berger, J. D., Robertson, L. D. and Cocks, P. S. 2002. Agricultural potential of Mediterranean grain and forage legumes: Key differences between and within *Vicia* species in terms of phenology, yield, and agronomy give insight into plant adaptation to semi-arid environments. *Genet. Resour. Crop Evol.*, 49, 313–25.

Bernacchi, C. 2007. Canopy Evapotranspiration Measurements on Soybean and Maize in An Open-Air Controlled Experiment. ASA-CSSA-SSSA International Annual Meetings. 4–8 November 2007. New Orleans, Louisiana, USA.

Bozoglu, H. and Karayel, R. 2006. Investigation of stomata densities in pea (*Pisum sativum* L.) Lines/Cultivars. *Online J. Biol. Sci.*, 6(2): 56–61.

Brownlee, C. 2001. The long and short of stomata density signals. *Trends Plant Sci.*, 6: 441–2.

Chaves, M. 1991. Effects of water deficits on carbon assimilation. *J. Exp. Bot.*, 234, 1–16.

Chaves, M., Pereira, J., Maroco, J., Rodrigues, M. and Picardo, C. 2002. How plants cope with water stress in the field: Photosynthesis and growth. *Ann. Bot.*, 89, 907–16.

Condon, A. G. and Hall, A. E. 1997. Adaptation to diverse environments: Genotypic variation in water-use efficiency within crop species. In Jackson, L. E. (Ed.), *Agricultural Ecology*. Academic Press: San Diego, CA, pp. 79–116.

de Ronde, J. A. and Spreeth, M. H. 2007. Development and evaluation of drought resistant mutant germ-plasm of *Vigna unguiculata*. *Water SA*, 33(3): 381–6.

Dracup, J. A., Lee, K. S. and Paulson Jr., E. G. 1980. On the definition of droughts. *Water Resour. Res.*, 16(2), 297–302.

Erre, P., Chessa, I., Nieddu, G. and Jones, P. G. 2009. Diversity and spatial distribution of *Opuntia* spp. in the Mediterranean Basin. *J. Arid. Environ.*, 73(12): 1058–66.

Handberg, K. and Stougaard, J. 1992. *Lotus japonicus*, an autogamous, diploid legume species for classical and molecular genetics. *Plant J.*, 2, 487–96.

Heath, O. V. S. and Mansfield, T. A. 1971. The movement of stomata. In Wikins, M. B. (Ed.), *The Physiology of Plant Growth and Development*. McGraw-Hill: New York, pp. 302–32.

Hulme, P. E. 2005. Adapting to climate change: Is there scope for ecological management in the face of a global threat? *J. Applied Ecol.*, 42, 784–94.

Hungria, M. and Vargas, M. A. T. 2000. Environmental factors affecting N2 fixation in grain legumes in the tropics, with an emphasis on Brazil. *Field Crops Res.*, 65(2–3), 151–64.

Jarvis, P. G. 1975. Water transfer in plants. In deVries, D. A. and Afgan, N. H. (Eds), *Heat and Mass Transfer in the Biosphere. 1. Transfer Processes in Plant Environment*. Scripta Book Co.: Washington, D.C., pp. 369–94.

Jarvis, P. G. and Morison, J. I. L. 1981. The control of transpiration and photosynthesis by the stomata. In Jarvis, P. G., Mansfield, T. A. (Eds), *Stomatal Physiology*. Cambridge Univ. Press: Cambridge, Great Britain, p. 295.

Katerji, N. and Rana, G. 2011. Crop reference evapotranspiration: A discussion of the concept, analysis of the process and validation. *Water Resour. Manage.*, 25, 1581–600.

Khan, H. R., Link, W., Hocking, T. J. and Stoddard, F. L. 2007. Evaluation of physiological traits for improving drought tolerance in faba bean (*Vicia faba* L.). *Plant Soil*, 292(1–2): 205–17.

Laranjo, M., Alexandrea, A. and Oliveira, S. 2014. Legume growth-promoting rhizobia: An overview on the Mesorhizobium genus. *Microbiol. Res.*, 169(1), 2–17.

Long, S. P. and Bernacchi, C. J. 2003. Gas exchange measurements, what can they tell us about the underlying limitations to photosynthesis? Procedures and sources of error. *J. Exp. Bot.*, 54: 2393–401.

Pandey, R. K., Herrera, W. A. T. and Pendleton, J. W. 1984a. Drought response of grain legumes under irrigation gradient: I. Yield and yield components. *Agron. J.*, 76, 549–53.

Pandey, R. K., Herrera, W. A. T. and Pendleton, J. W. 1984b. Drought response of grain legumes under irrigation gradient: II. Plant water status and canopy temperature. *Agron. J.*, 76, 553–7.

Passioura, J. B. 1996. Drought and drought tolerance. *Plant Growth Regul.*, 20(2):79–83.

Peksen, E., Peksen, A. and Artik, C. 2006. Comparison of leaf and stomatal characteristics in faba bean (*Vicia faba* L.). *J. Biol. Sci.*, 6(2): 360–4.

Plaut, Z. and Moreshet, S. 1973. Transport of water in plant – atmosphere system. In: Yaron, B., Danfors, E. and Vaadia, Y. (Eds), *Arid Zone Irrigation. Ecological Studies, Vol. 5*. Springer Verlag: Berlin, p. 434.

Porch, T. G., Dickson, M. H., Long, M. C., Viands, D. R. and John, M. 2004. General combining ability effects for reproductive heat tolerance in snap bean. *J. Agr. Univ. P. R.*, 88(3–4), 161–4.

Reddy, D. S., Bhatnagar-Mathur, P., Vadez, V. and Sharma, K. K. 2012. Grain legumes (soybean, chickpea, and peanut): Omics approaches to enhance abiotic stress tolerance. In Tuteja, N., Gill, S. S., Tiburcio, A. F. and Tuteja, R. (Eds), *Improving Crop Resistance to Abiotic Stress*. First Edition. Wiley-VCH Verlag GmbH & Co. KGaA: Berlin, Germany.

Sharma-Natu, P. and Ghildiyal, M. C. 2005. Potential targets for improving photosynthesis and crop yield. *Curr. Sci.*, 88(12): 1918–28.

Sharma, S., Upadhyaya, H. D., Varshney, R. K. and Gowda, C. L. L. 2013. Pre-breeding for diversification of primary gene pool and genetic enhancement of grain legumes. *Front. Plant Sci.*, 4(309), 1–14. doi: 10.3389/fpls.2013.00309.

Simova-Stoilova, L., Vassileva, V. and Feller, U. 2016. Selection and breeding of suitable crop genotypes for drought in a changing climate: Which morphological and physiological properties should be considered? *Agriculture*. 6, 1–19

Simonne, E. H., Dukes, M. D. and Haman, D. Z. 2007. Principles and practices of irrigation management for vegetables. Chapter 8. Document AE260, Horticultural Sciences Dept., UF/IFAS, Fla. Coop. Ext. Serv. USA, pp. 33–9.

Sinclair, T. R. and Vadez, V. 2012. The future of grain legumes in cropping systems. *Crop & Pasture Sci.*, 63, 501–12.

Subbarao, G. V., Johansen, C., Slinkard, A. E., Nageswara Rao, R. C., Saxena, N. P. and Chauhan, Y. S. 1995. Strategies for improving drought resistance in grain legumes. *Crit. Rev. Plant Sci.*, 14(6): 469–523.

Tanzarella, O. A., de Pace, C. and Filippetti, A. 1984. Stomatal frequency and size in *Vicia faba* L. *Crop Sci.*, 24, 1070–6.

Timko, M. P. and Singh, B. B. 2008. Cowpea, a Multifunctional Legume, pp. 227–58. In Moore, P. H. and Ming, R. (Eds), *Genomics of Tropical Crop Plants*. Springer: New York, p. 582.

Turner, N. C. 2001. Optimizing water use. In Nosberger, J., Geiger, H. H. and Struik, P. C. (Eds), *Crop Science: Progress and Prospects*. CABI Publishing. Wallingford Oxfordshire, UK. pp. 119–35.

Turner, N. C., Wright, G. C. and Siddique, K. H. M. 2001. Adaptation of grain legumes (pulses) to water-limited environments. *Adv. Agron.*, 71, 193–231.

Vandeleur, R., Niemietz, C., Tilbrook, J. and Tyerman, S. D. 2005. Roles of aquaporins in root responses to irrigation. *Plant Soil*, 274:141–61.

Vara, P. P. V., Allen Jr., L. H. and Boote, K. J. 2005. Crop responses to elevated carbon dioxide and interaction with temperature: Grain legumes. *J. Crop Improv.* 13(1/2), 113–55.

Varshney, R. K., Chen, W., Li, Y., Bharthi A. K., Saxena, R. K., Schlueter, J. A., Donoghue, M. A., Azam, S., Fan, G., Whaley, A. M., Farmer, A. D., Sheridan, J., Iwata, A., Tuteja, R., Penmetsa, R. V., Wu, W., Upadhyaya, H. D., Yang, S., Shah, T., Saxena, K. B., Michael, T., McCombie, W. R., Yang, B., Zhang, G., Yang, H., Wang, J., Spillane, C., Cook, D. R., May, G. D., Xu, X. and Jackson, S. A. 2012. Draft genome sequence of pigeonpea (Cajanus cajan), an orphan legume crop of resource-poor farmers. *Nat. Biotechnol.* 30, 83–9.

Varshney, R. K., Kudapa, H., Pazhamala, L., Chitikineni, A., Thudi, M., Bohra, A., Gaur, P. M., Janila, P., Fikre, A., Kimurto, P. and Ellis, N. 2015. Translational genomics in agriculture: Some examples in grain legumes. *Crit. Rev. Plant Sci.* 34:169–94.

Varshney, R. K., Song, C., Saxena, R. K., Azam, S., Yu, S., Sharpe, A. G., Cannon, S., Baek, J., Rosen, B. D., Tar'an, B., Millan, T., Zhang, X., Ramsay, L. D., Iwata, A., Wang, Y., Nelson, W., Farmer, A. D., Gaur, P. M., Soderlund, C., Penmetsa, R. V., Xu, C., Bharti, A. K., He, W., Winter, P., Zhao, S., Hane, J. K., Carrasquilla-Garcia, N., Condie, J. A., Upadhyaya, H. D., Luo, M., Thudi, M., Gowda, C. L. L., Singh, N. P., Lichtenzveig, J., Gali, K. K., Rubio, J., Nadarajan, N., Dolezel, J., Bansal, K. C., Xu, X., Edwards, D., Zhang, G., Kahl, G., Gil, G., Singh, K. B., Datta, S. K., Jackson, S. A., Wang, J. and Cook, D. R. 2013. Draft genome sequence of chickpea (*Cicer arietinum*) provides a resource for trait improvement. *Nat. Biotechnol.* 31: 240–6.

Varshney, R. K., Thudi, M., May, G. D. and Jackson, S. A. 2010. Legume genomics and breeding. *Plant Breed. Rev.*, 33: 257–304.

Wright, G. C., Hubick, K. T., Farquhar, G. D., Nageswara Rao, R. C. 1993. Genetic and environmental variation in transpiration efficiency and its correlation with carbon isotope discrimination and specific leaf area in peanut. In: Ehleringer, J. R., Hall, A. E., Farquhar, G. J. (Eds), *Stable Isotopes and Plant Carbon-water Relations*. Academic Press: San Diego, CA, USA, pp: 245–67.

Wullschleger, S. D., Meinzer, F. C. and Vertessy, R. A. 1998. A review of whole-plant water use studies in trees. *Tree Physiol.*, 18, 499–512.

Yang, H. M., X. Y. Zhang and G. X. Wang. 2004. Relationships between stomatal carácter, photosynthetic character and seed chemical composition in grass pea at different water availabilities. *J. Agr. Sci.*, 142(6), 675–81.

Chapter 2

Advances in understanding grain legume physiology: understanding root architecture, nutrient uptake and response to abiotic stress

Yinglong Chen, The University of Western Australia, Australia; Northwest A&F University, China; Ivica Djalovic, Institute of Field and Vegetable Crops, Serbia; and Kadambot H. M. Siddique, The University of Western Australia, Australia

1 Introduction

Grain legumes are an important and cheap source of protein with a total growth area of some 190 million ha globally (Reeves et al., 2016). Grain legumes play a distinct role in agricultural ecosystems due to their ability to fix nitrogen (N) symbiotically (Siddique et al., 1999; Rubiales and Mikic, 2015). It is estimated that the crop legume–rhizobia symbioses fix some 21 Mt nitrogen annually with up to 1/3 nitrogen returning to soils (Herridge et al., 2008; Foyer et al., 2016). The most important grain legumes are soybean (*Glycine max*), chickpea (*Cicer arietinum*), narrow-leafed lupin (*Lupinus angustifolius*), common bean (*Phaseolus vulgaris*), mash bean (*Vigna mungo*), mung bean (*Vigna radiata*), lentil (*Lens culinaris*), etc.

Grain legumes are mostly grown rainfed and often experience drought stress from terminal drought to intermittent drought at various growth stages (Purushothaman et al., 2013). Root system architecture (RSA), and anatomical and physiological traits may contribute to drought tolerance and adaptation to other environmental stresses. Crop genotypes with particular RSA traits, such as deep roots, increased root density in subsoil, increased root hair length

http://dx.doi.org/10.19103/AS.2017.0023.37

and density and/or xylem diameters, are often associated with efficient water and nutrient uptake (Wang et al., 2006; Watt et al., 2008; Chen et al., 2015a). Under drought conditions, roots extend their length, increase their surface area and deplete immobile nutrients (Lynch and Brown, 2012). Reduction in root growth under drought conditions reduces the nutrient acquisition capacity of root systems which eventually affects growth, development, yield and yield quality traits (Marchner, 1995). It is essential to have a better understanding of root traits and their functions for improving the legume selection strategy to sustain productivity for mitigating climate change, particularly drought stress.

2 The role of root system architecture in plant health

The root system which links the soil, where water and nutrients reside, to the above-ground part of the plant, where these resources are mainly used, is complex. Root systems perform many essential adaptive functions including water and nutrient uptake, anchorage to the soil and the establishment of biotic interactions at the rhizosphere. Root systems of major agricultural crops are poorly adapted to soils that mostly have poor water-holding capacity and nutrient deficiencies (Siddique et al., 2001). Root responses to the availability of water and nutrients, as well as to chemical properties of the soil, in turn, affect the uptake of soil water and nutrients (Feddes and Raats, 2004).

RSA refers to *in situ* spatial distribution of the root system within the rooting volume (Hinsinger et al., 2011; Lynch, 1995, 2007; Manschadi and Manske, 2013). RSA plays a vital role in the exploration of soil zones and acquisition of soil water and nutrients (Chen et al., 2009; Gregory et al., 2009; Hammond et al., 2009; Lynch and Brown, 2012). It is plastic and dynamic, allowing plants to respond to their environments in order to optimise acquisition of important soil resources (Zhu et al., 2011). The root and its architecture, the spatial configuration of the root system over time (Lynch, 2007), play an important role in soil exploration; nutrient acquisition; interaction with symbiotic soil organisms, for example, rhizobia and arbuscular mycorrhizal fungi (Hetrick 1991; Parniske, 2008); pathogen defence and soil stabilisation (den Herder et al., 2010). Roots also serve as host for many soil microbes whose proliferation in or outside of the surface of roots is inadvertently catalysed by the release of C from root cells to the rhizosphere (Lambers et al., 2009).

Most soils are inherently low in nutrients, in particular P and N, which consequently leads to poor crop yield (Lynch and Brown, 2008). Legume plants have high P requirement to stimulate root development and plant growth, initiate nodule formation as well as promote the efficiency of the rhizobium–legume symbiosis (Vance, 2001). Lateral roots play an essential role in searching for nutrient availability in the soil and are especially important for P acquisition since P is an immobile nutrient element in the soil surface layers (Lynch and Brown, 2001). Legume roots can be adapted to low soil P condition by increasing the growth of roots such as basal and adventitious roots, modifying root architecture and the ability to extract P from the soil (Lynch, 1995). Low soil P availability improves root architecture traits including the primary root length, root branching, number and length of lateral roots and enhancement of root hairs and cluster root formation (Jin et al., 2012). According to Zhu and Lynch (2004), low P availability in soils affects lateral root branching, root density and length of root hairs and also parenchyma formation (Fan et al., 2003).

Studies have also shown RSA alterations in relation to nutrient placements and deficiency. Two narrow-leafed lupin genotypes with contrasting RSA traits exhibited genotype-dependent root growth plasticity in response to P supply, with the greatest response to banded P 10 cm below the soil surface compared to top-dressed P supply (Chen et al., 2013b). Lupin plants supplied with banded P had the largest root system and the greatest P-uptake efficiency. This study also showed that addition of P fertiliser significantly stimulated root branching in the topsoil, whereas plants without P supply had relatively deeper roots. These findings demonstrate that selecting and breeding cultivars with root systems that use water and nutrients efficiently is an important strategy to increase crop adaptation to edaphic stress (Siddique et al., 2001; Chen et al., 2016).

3 Measuring root system architecture

Quantifying the architecture of root systems is important as a means of determining the availability and accessibility of water and nutrients and hence improving crops and managing soils to improve crop yields. Measurements of RSA traits are often achieved from the laboratory other than the field to facilitate detailed, replicated and standardised quantification and allow ready comparison in trait variation between species and between genotypes of the same species (Gregory, 2006). Many studies have been carried out on root growth in controlled conditions using young seedlings grown in homogenous media such as gels, hydroponics or sand because seedling roots can be screened rapidly in such media, and root length and elongation rates of individual roots can be determined relatively easily.

Measuring the architecture of root systems growing in soil is difficult and often involves the destruction of the soil habitat (Chen et al., 2014). Traditional methods for measuring roots in the soil such as excavation of the root system (Jackson, 1995; Trachsel et al., 2011), coring methods and trench profiles (Kucke et al., 1995) are destructive, tedious, time-consuming and the accuracy of their results is often not great (Smit et al., 2000). Excavation of roots from the soil and washing them in order to measure their length and architecture remains the most common and simplest method of studying roots in the field. Trench profiles can also be used to measure root spatial distribution and have potential to provide quantitative information on the root system. But unfortunately the trench method involves a major disturbance of the soil and provides measurements only for a small fraction of the root system (Pierret et al., 2003). Classical techniques are not often suitable in investigating root developmental processes (Liedgens et al., 2000) and may lead to loss of information on fine root parts such as root hairs (Smit et al., 2000).

In situ imaging of roots refers to imaging techniques, carried out either in the field or in controlled environments in which the rooting medium for the plant is soil (Rodgers et al., 2004). Popular *in situ* root observations and measurements employ rhizotrons, rhizoboxes or minirhizotrons. These methods are often used for the direct observation of the root growth with time, and for the characterisation of rhizosphere biochemical properties including hydration, pH and carboxylate concentration at spatial and temporal scales (den Herder et al., 2010; Wenzel et al., 2001; Chen et al., 2015b; Fig. 1). The comparatively easier access and more favourable geometry provided by a flat viewing plane of rhizotrons offer greater scope for quantitative measurements to be made (Schmidt et al., 2011). Though the technique allows the application of demographic methods to roots in the field environment, facilitating the tabulation of developmental stages from initial growth to senescence, it has proven difficult to convert root counts into more useful data such as root length densities or biomass.

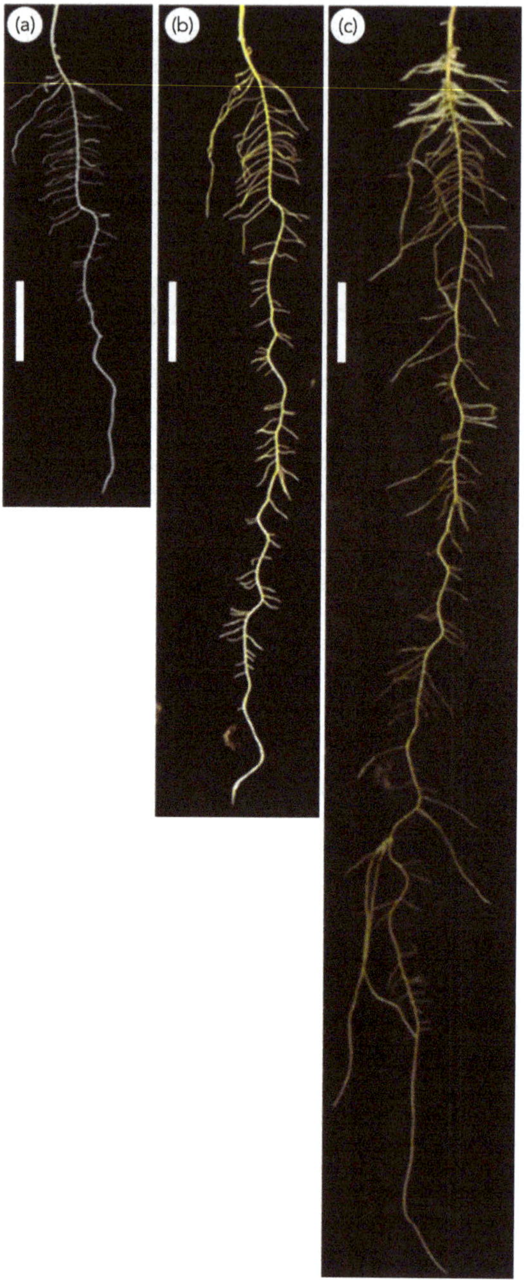

Figure 1 Example images of root morphology of a wild *Lupinus angustifolius* genotype grown in the semi-hydroponic phenotyping system (Chen et al., 2011a). Root images (a)–(c) were taken at 2, 4 and 6 weeks after planting respectively, showing the actual two-dimensional root architecture (Bar = 5 cm).

Recent advances in imaging using neutron or X-ray imaging, for example, are providing new ways to measure root characteristics in soil (Pierret et al., 2003). However, many imaging techniques employ two-dimensional imaging and influence root growth since it is forced to grow at the surface of the imaging window. With developments in radiation tomography, such as X-ray computed tomography, neutron tomography and magnetic resonance imaging, it is now possible to image the 3D architecture of soil and root structures non-destructively with some limitations on pot size, resolutions and scanning duration, reviewed in Chen et al. (2015a).

In the past several years, the interest in plant high-throughput phenotyping techniques for characterising RSA traits has grown tremendously. A novel semi-hydroponic phenotyping system was initially designed for characterising root trait variability in a large set of collections in narrow-leafed lupin (Chen et al., 2011a,b, 2012). The system was constructed using inexpensive and easily obtained materials including 240-L wheelie bins, clear acrylic panels covered with black calico cloth and a controlled water-spraying system. Root architecture, root elongation rate and branching patterns can be monitored for up to 6-week growing periods of crop species (Fig. 2). The system allows for rapid

Figure 2 Exposed root system of a narrow-leafed lupin plant grown (a) in a rhizobox with the glass panel removed, or (b) in a PVC column with rhizosphere soil attached, for collecting rhizosphere carboxylates from individual root tips using anion exchange members. Selected root tips for sampling are shown with (a) map pins or number (b). Bar = 8 cm.

Table 1 Descriptive statistics of 30 root-related traits in 270 chickpea genotypes grown in a semi-hydroponic phenotyping platform. Traits with CVs (coefficients of variation) ≥0.30 appear in bold type. Probability values (*P*) were based upon a GLM multivariate analysis and appear in bold if <0.01 and italic if <0.05 (extracted from Chen et al., 2017)

Trait	Abbreviation	Description	Unit	Minimum	Maximum	Mean	CV*	*P* value*
Taproot length zone1	TRL_z1	Taproot length at branching zone (z1)	cm	17.0	80.0	55.7	0.21	**<0.001**
Taproot length zone2	TRL_z2	Taproot length at non-branching zone (z2)	cm	2.67	36.3	16.4	**0.37**	*0.042*
Taproot length	TRL	Taproot length (i.e. root depth, z1+z2)	cm	38.3	105	72.1	0.17	**0.001**
Total root length	RL	Total root length per plant	cm	305	3824	1175	**0.42**	**<0.001**
Total branch length	BL	Total root branch length	cm	267	3765	1103	**0.44**	**<0.001**
Total branch number	BN	Total branch number	number per root	23.1	764	188	**0.66**	**<0.001**
Average branch length	ABL	Branch length divided by branch number	cm per branch	3.95	12.1	6.40	0.21	**<0.001**
Total root area	RA	Root surface area	cm^2	94.8	1027	319	**0.42**	**<0.001**
Total root volume	RV	Root volume	cm^3	2.36	22.03	6.95	**0.43**	**<0.001**
Root diameter	RD	Average root diameter	Mm	0.68	1.07	0.87	0.07	**<0.001**
Specific root length	SRL	Root length over root dry mass	mg^{-1} dry mass	32.5	264	65.7	**0.46**	**<0.001**
Branch length over taproot	BLR_tap	Branch length divided by taproot length	mm^{-1} taproot	6.62	77.8	15.6	**0.46**	**<0.001**
Branch density	BD	Branch number divided by taproot length	cm^{-1} taproot	0.57	15.7	2.61	**0.67**	**<0.001**
Branch intensity	BI	Branch number divided by total root length	cm^{-1} root	0.08	0.24	0.16	0.20	**<0.001**
Root tissue density	RTD	Root dry mass divided by root volume	mg cm^{-3}	1.27	9.87	3.05	0.27	**<0.001**

Topsoil root length	RL_top	Root length in section 1 (s1, 0–20 cm)	Cm	140	923	440	0.29	**<0.001**
Topsoil branch length	BL_top	Branch length in section 1 (s1, 0–20 cm)	Cm	120	903	420	**0.30**	**<0.001**
Topsoil root diameter	RD_top	Average root diameter in section 1 (s1, 0–20 cm)	Mm	0.77	1.09	0.92	0.06	**<0.001**
Root length s2	RL_s2	Root length in section 2 (s2, 20–40 cm)	Cm	71.6	868	426	0.28	**<0.001**
Root diameter s2	RD_s2	Average root diameter in section 2 (s2, 20–40 cm)	Mm	0.64	1.00	0.82	0.08	**<0.001**
Root length s3	RL_s3	Root length in section 3 (s3, 40 cm and beyond)	Cm	21.1	2262	309	**0.63**	**<0.001**
Root diameter s3	RD_s3	Root diameter in section 3 (s3, 40 cm and beyond)	Mm	0.62	1.05	0.86	0.09	**<0.001**
Subsoil root length	RL_sub	Root length in subsoil (s2 and s3, 20 cm and beyond)	Cm	95.8	3112	735	**0.58**	**<0.001**
Subsoil branch length	BL_sub	Branch length in subsoil (s2 and s3, 20 cm and beyond)	Cm	77.5	3073	683	**0.62**	**<0.001**
Subsoil root diameter	RD_sub	Root diameter in subsoil (s2 and s3, 20 cm and beyond)	Mm	0.64	1.03	0.84	0.08	**<0.001**
Root length ratio	RLR_top/sub	Topsoil root length divided by subsoil root length		0.21	2.20	0.70	**0.46**	**<0.001**
Branch length ratio	BLR_top/sub	Topsoil root length divided by subsoil root length		0.19	2.44	0.73	**0.49**	**<0.001**
Root growth rate	RGR	Average taproot growth rate	cm d^{-1}	1.10	3.00	2.06	0.17	**0.001**
Root mass	RM	Root dry mass	mg	69.3	370	198	**0.32**	**<0.001**
Root mass ratio	RMR	Root-to-shoot dry mass ratio		0.19	1.55	0.68	**0.31**	0.275

measurement of 2D root architecture over time with minimal disturbance to plant growth and without destructive root sampling. The semi-hydroponic phenotyping system was efficient in screening root traits in core collections of narrow-leafed lupin (Chen et al., 2011b, 2012), chickpea (Chen et al., 2017), barley, wheat and maize (in preparation). Some 30 root-related traits are characterised in narrow-leafed lupin (Chen et al., 2011a, 2016) and chickpea (Table 1; Chen et al., 2017). Phenotyping studies conducted at various institutions on several crops involving both tap-rooted legume species and fibrous-root-system wheat and barley demonstrate that the semi-hydroponic system is a desirable tool for examining root architecture of deep root systems and large sets of plants in a relatively small space.

Increasing number of laboratory-based methods are being used to screen large numbers of plants and to measure selected architectural traits. Phenotyping systems have focused on automated solutions for data acquisition using platforms that combine a vehicle and perhaps some robotics with imaging systems and sensors, such as Automated Growscreen-Rhizo system with rhizoboxes developed by Forschungszentrum Jülich, Germany (Nagel et al., 2012). These techniques use sensor systems and automated computer algorithms to extract phenotypic traits for large genetic mapping populations using non-destructive and non-invasive sampling methods (Fiorani and Schurr, 2013). With such new techniques, plant phenotypes can be linked with functional genomics and environmental factors, thus facilitating plant breeding and leading towards improved crop production and yield stability. High-throughput phenotyping procedures capable of phenotyping large numbers of lines or individual plants accurately with a fraction of the time, cost and labour of current techniques are thus needed (White et al., 2012).

4 Variability of root system architecture traits across genotypes

Different crop species and different genotypes of the same species may vary in RSA, resulting in differences in root functioning. Genotypic variations in RSA have a major impact on the efficient acquisition of resources from soils. For example, genotypic variation in common bean cultivars were observed in branching density of tap and lateral roots under drought stress (Fenta, 2014).

Numerous studies have examined genotypic variation of RSA traits in grain legumes in relation to P efficiency. Root density in the upper soil layers is the most important trait associated with improved acquisition of P. Genetic variability exists in the soybean germplasm for root architectural traits, such as root angle (Zhao et al., 2004), root diameter and rooting depth (Ao et al., 2010), and morphology traits such as root length (Zhao et al., 2004), root volume, area and length (Ao et al., 2010). In common bean, P availability in the soil has been shown to control various characteristics of root architecture such as branching, basal root length and adventitious roots in the topsoil (Miguel et al., 2013). Common bean genotypes grown under P-deficit environments showed differential stimulation of shallower basal root growth angles (Liao et al., 2001). Increases in production of adventitious roots in common bean help in P acquisition by improving plant foraging in the most P-rich soil environment, and genotypes with shallower root systems were more competitive for topsoil P acquisition than those having deep root systems (Miller et al., 2003). Results from both experimental observations and geometric model simulations indicate that genotypes with shallow basal roots are more effective in P acquisition than genotypes with deeper roots, especially in stratified soils with heterogeneous P distribution (Ge et al., 2000; Lynch and Brown, 2001). The existence of genetic variation for P efficiency offers the possibility to develop bean genotypes with superior adaptation to low-P soils.

Using the semi-hydroponic phenotyping system (Chen et al., 2011a), a wide range of variability in RSA traits across 270 chickpea genotypes was observed, with 17 measured root traits (out of 30) having coefficients of variation greater than 0.30 (Table 1; Chen et al., 2017). This study revealed that root morphological and architectural traits differed significantly among the tested genotypes – all of the measured root-related traits differed significantly among genotypes ($P \leq 0.05$) except for the root–shoot mass ratio ($P = 0.275$) (Table 1). Most chickpea genotypes grew vigorously with the average root growth rate ranging from 1.1 to 3.0 cm d^{-1}. Rooting depth ranged from 38.3 to 105 cm and total root length ranged from 305 to 3824 cm. Similarly, significant differences in the rooting pattern and branching type were observed among 111 wild genotypes of narrow-leafed lupin using the semi-hydroponic phenotyping system (Chen et al., 2011a,b, 2012; 2016). Root architectural and morphological traits, including taproot length (i.e. root depth (Fig. 3), root length, branch length and root area and volume displayed large variations among genotypes ($P < 0.01$). This was the first study in characterising root trait variations in chickpea germplasm, providing useful data for further breeding programmes ultimately aimed at developing new cultivars with root traits for improved adaptation to specific environmental conditions, such as low water availability.

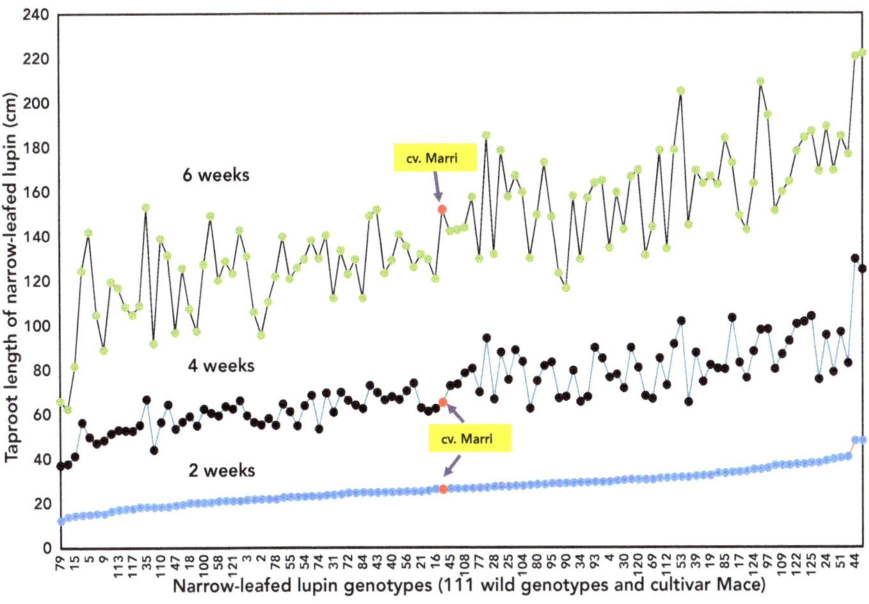

Figure 3 Genotypic variation in taproot length (i.e. root depth) among 111 wild genotypes of narrow-leafed lupin 2, 4 and 6 weeks after growing in the novel semi-hydroponic phenotyping system. A local cultivar 'Marri' was included in the experiment as the reference/control. Genotypes are in order of taproot root length at two weeks. A list of narrow-leafed lupin genotypes included in this study can be found in Chen et al. (2016).

Plant root systems are astonishingly plastic in their architecture, which allows for optimal exploitation of diverse soil structures and conditions. Therefore, some caution should be exercised in translating results of screening RSA traits from non-soil media to what might happen in field conditions, and in verifying the screening system adopted (Gregory et al., 2009; Chen et al., 2015a). Considerably consistent rankings of ten selected narrow-leafed lupin genotypes for several important RSA traits between experiments conducted under controlled environments using either the semi-hydroponic phenotyping system (Chen et al., 2011a), or soil media, confirmed the capacity of the semi-hydroponic system to represent growth conditions in simple, uniform soil environments (Chen et al., 2011b, 2012).

5 Implications for breeding drought-resistant varieties

Adaptation to drought in legume crops may be enhanced by breeding genotypes with deeper rooting for a target production zone where the crop is grown under limited rainfall with substantial water available deep in the soil. The size of a root system, root branching and its spatial distribution in soil, and hydraulic conductance influence water uptake to meet the transpiration requirements of the shoot (Passioura, 1982). Crop plants often maintain higher root length densities than are required in the surface layer to facilitate rapid uptake of rainfall, extract relatively immobile nutrients like P, and compete with weeds or other neighbouring plants for both water and nutrients (Subbarao et al., 1995). Rooting depth, root length density, root hair and specific root length are the most important RSA traits in relation to water-uptake efficiency. Studies on root anatomical traits showed that grain legumes are capable of regulating the necessary tissue development for appropriate hydraulic conductance during secondary thickening of the root system depending on the soil moisture status (Purushothaman et al., 2013). Example root traits associated with drought tolerance in grain legume species are summarised in Table 2. Root traits that overcome abiotic constraints are critical to maintain structural and functional properties, and are considered first-order targets in breeding programmes for rainfed environments. Depending on the target environment, certain root traits may be more important than others, and targeting of genetic improvement depends on the type of traits required relative to those present in current cultivars.

Genetic gains in forage and grain production are important targets for plant breeding programmes and these gains could be enhanced by understanding the root traits that contribute to improved plant performance. For example, well-developed root systems have been shown to be linked to drought tolerance as an avoidance mechanism guaranteeing plant productivity under water-limited conditions (Asfaw and Blair, 2012; Comas et al., 2013; Idrissi et al., 2015). Developing cultivars with enhanced drought tolerance by conventional breeding often has limited success due to the complexity of this trait and the difficulties with finding reliable and suitable phenotyping methods.

Breeding efforts that select for and modify specific root traits are limited despite the importance of the root systems for enhancing the acquisition of water and nutrients (Zhu et al., 2011). However, it is difficult to screen large numbers of accessions for these root traits using conventional methods. A method using herbicide injected deep in the soil was used to identify cowpea (Robertson et al., 1985) and peanut (Khalfaoui and Havard, 1993) genotypes for differences in rate of rooting – genotypes with earlier herbicide symptoms extracted more soil moisture deep in the profile by the roots. Applying a marker-assisted

Table 2 Example root traits identified showing associations with drought stress in some important grain legume species

Legume species	Root traits	References
Chickpea (*Cicer arietinum*)	Prolific root system; root length density, root dry weight, rooting depth	Gaur et al. (2008); Kashiwagi et al. (2005, 2006); Zaman-Allah et al. (2011)
Common bean (*Phaseolus vulgaris*)	Rooting depth; primary root length	Beebe (2012); Beebe et al. (2013); Polania et al. (2009); White and Castillo (1992)
Cowpea (*Vigna unguiculata*)	Deep rooting	Hall et al. (2012)
Faba bean (*Vicia faba*)	Root traits	Khan et al. (2010)
Common vetch (*Vicia sativa*)	Deep rooting	Lynch and Wojciechowski (2015)
Pea (*Pisum sativum*)	Lateral roots	Rewald et al. (2011)
Lentil (*Lens culinaris*)	Root dry weight; root length; taproot length, lateral root number	Gahoonia et al. (2006); Kumar et al. (2012); Sarker et al. (2005)
Soybean (*Glycine max*)	Root length; root surface area; root volume; root angle; tap and lateral root diameter; tap and lateral root branching densities	Fenta et al. (2014)

selection for these traits would offer an interesting alternative in breeding programmes targeting drought tolerance. As such, identifying and mapping DNA markers linked to genes controlling rooting patterns associated with drought tolerance will assist in reliable and efficient identification and development of tolerant cultivars. Mapping quantitative trait loci (QTLs) for root traits and their use in marker-assisted breeding (e.g. moving a favourable QTL allele present in exotic germplasm into elite cultivars) is an alternative method for selecting root traits that are difficult to phenotype. During the last 10–15 years, the increasing availability of molecular markers has allowed researchers and breeders to track segments of the genome linked to specific phenotypes of interest in QTL-mapping and genome-wide association studies (see Korte and Farlow, 2013 for a review). SSRs or SNPs mapped to the genome have thus been (and are) extensively used for Marker (or Gene)-Assisted Selection in plant breeding programmes (Xu and Crouch, 2008). Several studies have shown that root traits are polygenically controlled, whereas they also identified related QTL) for different species such as common bean (Asfaw and Blair, 2012), soybean (Brensha et al., 2012) and chickpea (Kashiwagi et al., 2014).

6 Conclusion

Characterising the variability in RSA traits among a large number of genotypes of the same crop species using novel high-throughput phenotyping platforms is considered as the first step in crop breeding. Knowledge on RSA, water and nutrient uptake and root responses to drought and other abiotic stresses is vital for breeding super genotypes for efficient water and nutrient acquisition and with enhanced adaptation to edaphic stresses. Breeding for legume cultivars with enhanced drought tolerance by conventional breeding methods often has limited success due to the complexity of this trait and the difficulties with finding reliable and suitable phenotyping methods. Recent developments of high-throughput phenotyping platforms and non-destructive imaging technologies shed light on characterising genotypic variability in RSA. Genotypes with contrasting RSA traits identified under controlled environments can be validated in the field under various environmental conditions and the combination of phenotyping and molecular approaches identify candidate genotypes with suitable root traits for potential breeding for efficient water and nutrient capture in stressful or poor soil environments. In addition to root trait phenotyping and genotyping, root structural–functional models are promising for selecting genotypes of grain legume species with optimised root system for adaptation to target environments.

7 Where to look for further information

Plant Roots: The Hidden Half (Fourth Edition) edited by Amarm Eshel and Tom Beeckman and the early editions of this volume are good introductions to understanding plant root system and functions, and methods in root study.

A number of special issues of journals with a focus on root research have been published, such as:

- *Annals of Botany* 'Matching roots to their environment' (2013)
- *Field Crops Research* 'Root Systems Form and Function in Relation to Yield' (2014)
- *Plant Physiology* 'Focus Issue on Plant Roots' (2015)
- *Journal of Experimental Botany* 'Roots Down Under – ISRR-9 Special Issue' (2016)
- *Annals of Botany* 'Roots Down Under – ISRR-9 Special Issue' (2016)
- *Plant and Soil* 'Roots Down Under – ISRR-9 Special Issue' (2016)
- *Agriculture* 'Dynamics of Root–Soil–Microbial Interactions' (2017).

The International Society of Root Research (ISRR) (https://www.rootresearch.org/) can be contacted for cooperation and communication between root researchers around the world. One of its main activities is to hold regular scientific meetings for root researchers to share their latest research findings and to promote collaborations. The next meeting (ISRR-10) will be held in Israel from 8 to 12 July, 2018 with a theme of Exposing the Hidden Half – Root Research at the Forefront of Science (http://www.ortra.com/events/isrr10/Home.aspx).

Rhizosphere 5 (The Fifth International Conferences on Rhizosphere), devoted to the study of the rhizosphere, will take place from 7 to 11 July, 2019 in Saskatoon, Saskatchewan Canada (http://www.rhizo5.org/home/index.html). The general theme for the conference is 'Shining light on the plant root–soil interface'. Past Rhizosphere conferences were held in Munich (2004), Montpellier (2007), Perth (2011) and Maastricht (2015).

Most recently, The International Symposium on Crop Roots and Rhizosphere Interactions was successfully held in Yangling, China, from 9 to 13 October, 2017, co-organised by Northwest A&F University (China) and The University of Western Australia (Australia) (http://121.195.186.141/~YChen?language=en). The second symposium is proposed to be held in July 2019 in Lanzhou, China.

8 Acknowledgements

The Australian Research Council (DP0988193, DP130104825), CGIAR (CRP-Grain Legumes Competitive Grants), National Natural Science Foundation of China (31471946) and Chinese Academy of Sciences ('100 Talents' Program) are acknowledged.

9 References

Ao, J., Fu, J., Tian, J., Yan, X. and Liao, H. (2010). Genetic variability for root morph – architecture traits and root growth dynamics as related to phosphorus efficiency in soybean. *Funct. Plant Biol.* 37(4): 304–12.

Asfaw, A. and Blair, M. (2012). Quantitative trait loci for rooting pattern traits of common beans grown under drought stress versus non-stress conditions. *Mol. Breed.* 30: 681–95.

Beebe, S. (2012). Common bean breeding in the tropics. *Plant Breed. Rev.* 36: 357–426.

Beebe, S. E., Rao, I. M., Blair, M. W. and Acosta-Gallegos J. A. (2013). Phenotyping common beans for adaptation to drought. *Front. Physiol.* 4: 35.

Brensha, W., Kantartzi, S. K., Meksem, K., Grier, R. L., Barakat, A. and Lightfoot, D. A.and Kassem, M. A. (2012). Genetic analysis of root and shoot traits in the 'Essex' by 'Forrest' recombinant inbred line (RIL) population of soybean (*Glycine max* (L.) Merr.). *J. Plant Genome Sci.* 1: 1–9.

Chen, W. R., He, Z. L., Yang, X. E. and Feng, Y. (2009). Zinc efficiency is correlated with root morphology, ultrastructure, and antioxidative enzymes in rice. *J. Plant Nutr.* 32: 287–305.

Chen, Y., Dunbabin, V. M., Diggle, A. J., Siddique, K. H. M. and Rengel, Z. (2011a). Development of a novel semi-hydroponic phenotyping system for studying root architecture. *Funct. Plant Biol.* 38: 355–63.

Chen, Y., Dunbabin, V. M., Postma, J., Diggle, A. J., Palta, J. A., Lynch, J. P., Siddique, K. and Rengal, Z. (2011b). Phenotypic variability and modelling of root structure of wild *Lupinus angustifolius* genotypes. *Plant Soil* 348: 345–64.

Chen, Y., Dunbabin, V. M., Diggle, A. J., Siddique, K. H. M. and Rengel, Z. (2012). Assessing variability in root parameters of wild *Lupinus angustifolius* germplasm: basis for modelling root system structure. *Plant Soil* 354: 141–55.

Chen, Y., Dunbabin, V. M., Diggle, A. J., Siddique, K. H. M. and Rengel, Z. (2013a). Phosphorus starvation boosts carboxylate secretion in P-deficient genotypes of *Lupinus angustifolius* with contrasting root structure. *Crop Pasture Sci.* 64: 588–99.

Chen, Y., Dunbabin, V. M., Postma, J. A., Diggle, A. J., Siddique, K. H. M. and Rengel, Z. (2013b). Modelling root plasticity and response of narrow-leafed lupin to heterogeneous phosphorus supply. *Plant Soil* 372: 319–37.

Chen, Y., Palta, J., Clements, J., Buirchell, B., Siddique, K. H. M. and Rengel, Z (2014). Root architecture alteration of narrow-leafed lupin and wheat in response to soil compaction. *Field Crops Res.* 165: 61–70.

Chen, Y., Djalovic, I. and Rengel, Z. (2015a). Phenotyping for root traits. In: Kumar, J., Pratap, A. and Kumar, S. (Eds), *Phenomics of Crop Plants: Trends, Options and Limitations* (pp. 102–28). Springer-Verlag, Berlin Heidelberg.

Chen, Y., Siddique, K. H. M. and Rengel, Z. (2015b). Temporal and spatial dynamics of rhizosphere chemistry among cluster-root-forming white lupin and yellow lupin and non-cluster-root narrow-leafed lupin. In: *International Society of Root Research 9th International Symposium*, Canberra, Australia, 6–9 October 2015.

Chen, Y., Shan, F., Nelson, M. N., Siddique, K. H. M. and Rengel, Z. (2016). Root trait diversity, molecular marker diversity and trait-marker associations in a core collection of *Lupinus angustifolius*. *J. Exp. Bot.* 67: 3683–97.

Chen, Y., Ghanem, M. E. and Siddique K. H. M. (2017). Characterising root trait variability in chickpea (*Cicer arietinum* L.) germplasm. *J. Exp. Bot.* 68: 1987–99.

Comas, L. H., Becker, S. R., Cruz, V. M. V., Byrne, P. F. and Dierig, D. A. (2013). Root traits contributing to plant productivity under drought. *Front. Plant Sci.* 4: 442.

den Herder, G., van Isterdael, G., Beeckman, T. and de Smet, I. (2010). The roots of a new green revolution. *Trends Plant Sci.* 15: 600–7.

Fan, M., Zhu, J., Richard, C., Brown, K. M. and Lynch, J. P. (2003). Physiological roles for aerenchyma in phosphorus-stressed roots. *Funct. Plant Biol.* 30: 493–506.

Feddes, R. A. and Raats, P. A. C. (2004). Parameterizing the soil-water-plant root system, p. 95–141. In: Feddes, R. A., Rooji, G. H. de and Dam, J. C. van (Eds), *Unsaturated-zone Modeling: Progress, Challenges and Applications*. Wageningen, The Netherlands, pp. 95–141.

Fenta, B. A., Beebe, E. S., Kunert, J. K., Burridge, D. J., Barlow, M. K., Lynch, P. J. and Foyer, H. C. (2014). Field phenotyping of soybean roots for drought stress tolerance. *Agronomy* 4: 418–35.

Fiorani, F. and Schurr, U. (2013). Future scenarios for plant phenotyping. *Ann. Rev. Plant Biol.* 64: 267–91.

Foyer, C. H., Lam, H. M., Nguyen, H. T., Siddique, K. H., Varshney, R. K., Colmer, T. D., Cowling, W., Bramley, H., Mori, T. A., Hodgson J. M. and Cooper J. W. (2016). Neglecting legumes has compromised human health and sustainable food production. *Nat. Plants* 2: 16112.

Gahoonia T. S., Ali O, Sarker A, Nielsen N. E. and Rahman M. M. (2006). Genetic variation in root traits and nutrient acquisition of lentil (*Lens culinaris* Medikus.) genotypes. *J. Plant Nutr.* 29: 643–55.

Gaur, P. M., Krishnamurthy, L. and Kashiwagi, J. (2008). Improving drought-avoidance root traits in chickpea (*Cicer arietinum* L.)-current status of research at ICRISAT. *Plant Prod. Sci.* 11(1): 3–11.

Ge, Z., Rubio, G. and Lynch, J. P. (2000). The importance of the root gravitropism for interroot competition and phosphorus acquisition efficiency: results from a geometric simulation model. *Plant Sci.* 218: 159–71.

Gregory, P. J. (2006). *Plant Roots: Their Growth, Activity and Interactions with Soil*. Blackwell Scientific Publications, Oxford, UK, p. 318.

Gregory, P. J., Bengough, A. G., Grinev, D., Schmidt, S., Thomas, W. T. B., Wojciechowski, T. and Young, M. (2009). Root phenomics of crops: opportunities and challenges. *Funct. Plant Biol.* 36: 922–9.

Hall, A. E. (2012). Phenotyping cowpeas for adaptation to drought. *Front. Physiol.* 3: 155.

Hammond, J. P., Broadley, M. R., White, P. J., King, G. J., Bowen, H. C., Hayden, R., Meacham, M. C., Mead, A., Overs, T., Spracklen, W. P. and Greenwood, D. J. (2009). Shoot yield drives phosphorus use efficiency in *Brassica oleracea* and correlates with root architecture traits. *J. Exp. Bot.* 60: 1953–68.

Herridge, D. F., Peoples, M. B. and Boddey, R. M. (2008). Global inputs of biological nitrogen fixation in agricultural systems. *Plant Soil* 311, 1–18.

Hetrick, B. A. D. (1991). Mycorrhizas and root architecture. *Experientia* 47: 355–62.

Hinsinger, P., Brauman, A., Devau, N., Gérard, F., Jourdan, C., Laclau, J. P., Le Cadre, E., Jaillard, B. and Plassard, C. (2011). Acquisition of phosphorus and other poorly mobile nutrients by roots: Where do plant nutrition models fail? *Plant Soil* 348: 29–61.

Idrissi, O., Udupa, S. M., DeKeyser, E., VanDamme, P. and DeRiek, J. (2015). Functional genetic diversity analysis and identification of associated simple sequence repeats and amplified fragment length polymorphism markers to drought tolerance in lentil (*Lens culinaris* ssp. *Culinaris Medicus)* Landraces. *Plant Mol. Biol. Rep.* 34: 659–80.

Jackson, L. E. (1995). Root architecture in cultivated and wild lettuce (*Lactuca* spp.). *Plant Cell Environ.* 18: 885–94.

Jin, J., Tang, C. X., Armstrong, R. and Sale, P. (2012). Phosphorus supply enhances the response of legumes to elevated CO_2 (FACE) in a phosphorus-deficient Vertisol. *Plant Soil* 358: 86–99.

Kashiwagi, J., Krishnamurthy, L., Upadhyaya, H. D., Krishna, H., Chandra, S., Vincent, V. and Serraj, R. (2005). Genetic variability of drought-avoidance root traits in the mini-core germplasm collection of chickpea (*Cicer arietinum* L.). *Euphytica* 146: 213–22.

Kashiwagi, J., Krishnamurthy, L., Crouch, J. H. and Serraj, R. (2006). Variability of root characteristics and their contribution to seed yield in chickpea (*Cicer arietinum* L.) under terminal drought stress. *Field Crop Res.* 95: 171–81.

Kashiwagi, J., Krishnamurthy, L., Gaur, P. M., Shandra, S. and Upadhyaya, H. D. (2014). Estimation of gene effects of the drought avoidance root characteristics in chickpea (*C. arietinum* L.). *Field Crop Res.* 105: 64–9.

Khalfaoui, J.-L. B. and Havard, M. (1993). Screening peanut cultivars in the field for root growth: A test by herbicide injection in the soil. *Field Crops Res.* 32: 173–9.

Khan H. R., Paull, J. G., Siddique, K. H. M. and Stoddard F. L. (2010). Faba bean breeding for drought-affected environments: A physiological and agronomic perspective. *Field Crops Res.* 115: 279–86.

Korte, A. and Farlow, A. (2013). The advantages and limitations of trait analysis with GWAS: A review. *Plant Methods* 9(1): 29.

Kücke, M., Schmid, H. and Spiess, A. (1995). A comparison of four methods for measuring roots of field crops in three contrasting soils. *Plant Soil* 172: 63–71.

Kumar, J., Basu, P. S., Srivastava, E., Chaturvedi, S. K., Nadarajan, N. and Kumar, S. (2012). Phenotyping of traits imparting drought tolerance in lentil. *Crop Pasture Sci.* 63: 547–54.

Lambers, H., Mougel, C., Jaillard, B., Hinsinger and P. (2009). Plant-microbe-soil interactions in the rhizosphere: An evolutionary perspective. *Plant Soil* 321: 83–115.

Liao, H., Yan, X., Rubio, G., Beebe, S. E., Blair, M. W. and Lynch, J. P. (2001). Genetic mapping of basal root gravitropism and phosphorus acquisition efficiency in common bean. *Funct. Plant Biol.* 31: 959–70.

Liedgens, M., Richner, W., Stamp, P. and Soldati, A. (2000). A rhizolysimeter facility for studying the dynamics of crop and soil processes: description and evaluation. *Plant Soil* 223: 87–97.

Lynch, J. (1995). Root architecture and plant productivity. *Plant Physiol.* 109: 7–13.

Lynch, J. P. (2007). Roots of the second green revolution. *Aust. J. Bot.*, 55: 493–512.

Lynch, J. P. and Brown, K. M. (2001). Top soil foraging-an architectural adaptation of plants to low phosphorus availability. *Plant Soil* 237: 225–37.

Lynch, J. P. and Brown, K. M. (2008). Root strategies for phosphorus acquisition. *Plant Ecophysiol.* 7: 83–116.

Lynch, J. P. and Brown, K. M. (2012). New roots for agriculture: Exploiting the root phenome. *Philos. Trans. R Soc. B* 367: 1598–604.

Lynch, J. P. and Wojciechowski, T. (2015). Opportunities and challenges in the subsoil: Pathways to deeper rooted crops. *J. Exp. Bot.* 66(8): 2199–210.

Manschadi, A. M. and Manske, G. G. B. (2013). Root architecture and resource acquisition: Wheat as a model plant, Ch. 22, pp. 1–18. In: Eshel, A. and Beeckman, T. (Eds), *Plant Roots–the Hidden Half* (Fourth Edition). CRC Press, London.

Marchner, H. (1995). *Mineral Nutrition of Higher Plants*. Academic, San Diego.

Miguel, M. A., Widrig, A., Vieira, R. F., Brown, K. M. and Lynch, J. P. (2013). Basal root whorl number: A modulator of phosphorus acquisition in common bean (*Phaseolus vulgaris*). *Ann. Bot.* 112(6): 973–82.

Miller, C. R., Ochoa, I., Nielsen, K., Beck, D. and Lynch, J. L. (2003). Genetic variation for adventitious rooting in response to low phosphorus availability: Potential utility for phosphorus acquisition from stratified soils. *Func. Plant Biol.* 30: 973–85.

Nagel, K. A., Putz, A., Gilmer, F., Heinz, K., Fischbach. A., Pfeifer, J., Faget, M., Blossfeld, S., Ernst, M., Dimaki, C. and Kastenholz, B. (2012). GROWSCREEN-Rhizo is a novel phenotyping robot enabling simultaneous measurements of root and shoot growth for plants grown in soil-filled rhizotrons. *Funct. Plant Biol.* 39: 891–904.

Parniske, M. (2008). Arbuscular mycorrhiza: The mother of plant root endosymbioses. *Nat. Rev. Microbiol.* 6: 763–75.

Passioura, J. B. (1982). The role of root system characteristics in the drought resistance of crop plants. In: *Drought Resistance in Crops with Emphasis on Rice* (pp. 71–82). IRRI, Philippines.

Pierret, A., Kirby, M. and Moran, C. (2003). Simultaneous X-ray imaging of plant root growth and water uptake in thin-slab systems. *Plant Soil* 255: 361–73.
Polania, J., Rao, I. M., Beebe, S. and Garcia, R. (2009). Root development and distribution under drought stress in common bean (*Phaseolus vulgaris* L.) in a soil tube system. *Agron. Columbiana* 27: 25–32.
Purushothaman, R., Zaman-Allah, M., Mallikarjuna, N., Pannirselvam, R., Krishnamurthy, L. and Gowda, C. L. (2013). Root anatomical traits and their possible contribution to drought tolerance in grain legumes. *Plant Prod. Sci.* 16(1): 1–8.
Reeves, T. G., Thomas, G. and Ramsay, G. (2016). *Save and Grow in Practice: Maize, Rice, Wheat. A Guide to Sustainable Cereal Production*. FAO, UN.
Rewald, B., Ephrath, J. E. and Rachmilevitch, S. (2011). A root is a root is a root? Water uptake rates of *Citrus* root orders. *Plant Cell Environ.* 34: 33–42.
Robertson, B. M., Hall, A. E. and Foster, K. W. (1985). A field technique for screening for genotypic differences in root growth. *Crop Sci.* 25: 1084–90.
Rodgers, H. L., Day, F. P. and Atkinson, R. (2004). Root dynamics in restored and naturally regenerated Atlantic white cedar wetlands. *Restoration Ecol.* 12: 401–11.
Rubiales, D. and Mikić, A. (2015). Introduction: Legumes in sustainable agriculture. *Crit. Rev. Plant Sci.* 34: 2–3.
Sarker, A., Erskine, W. and Singh, M. (2005). Variation in shoot and root characteristics and their association with drought tolerance in lentil landraces. *Gen. Res. Crop. Evol.* 52: 89–97.
Schmidt, H., Eickhorst, T. and Tippkotter, R. (2011). Monitoring of root growth and redox conditions in paddy soil rhizotrons by redox electrodes and image analysis. *Plant Soil* 341: 221–32.
Siddique, K. H. M., Loss, S. P., Regan, K. L. and Jettner, R. (1999). Adaptation of cool season grain legumes in Mediterranean type environments of south-western Australia. *Aust. J. Agric. Res.* 50: 375–87.
Siddique, K. H. M., Regan, K. L., Tennant, D. and Thomson, B. D. (2001). Water use and water use efficiency of cool season grain legumes in low rainfall Mediterranean-type environments. *Eur. J. Agron.* 15: 267–80.
Smit, A. L., Bengough, A. G., Engels, C., van Noordwijk, M., Pellerin, S., van de Geijn, S. C. (Eds) (2000). *Root Methods: A Handbook*. Springer, Heidelberg.
Subbarao, G. V., Johansen, C., Slinkard, A. E., Nageswara Rao, R. C., Saxena, N. P., Chauhan, Y. S. and Lawn R. J. (1995). Strategies for improving drought resistance in grain legumes. *Crit. Rev. Plant Sci.* 4(6): 469–523.
Trachsel, S., Kaeppler, S. M., Brown, K. M., Lynch, J. P. (2011). Shovelomics: Hhigh throughput phenotyping of maize (*Zea mays* L.) root architecture in the field. *Plant Soil* 341: 75–87.
Vance, C. P. (2001). Symbiotic nitrogen fixation and phosphorus acquisition. Plant nutrition in a world of declining renewable resources. *Plant Physiol.* 127: 390–7.
Wang, H., Inukai, Y. and Yamauchi, A. (2006). Root development and nutrient uptake. *Crit. Rev. Plant Sci.* 25: 279–301.
Watt, M., Magee, L. J. and McCully M. E. (2008). Types, structure and potential for axial water flow in the deepest roots of field-grown cereals. *New Phytol.* 178: 135–46.
Wenzel, W. W., Wieshammer, G., Fitz, W. J. and Puschenreiter, M. (2001). Novel rhizobox design to assess rhizosphere characteristics at high spatial resolution. *Plant Soil* 237: 37–45.
White, J. W., Andrade-Sanchez, P., Gore, M. A., Bronson, K. F., Coffelt, T. A., Conley, M. M., Feldmann, K. A., French, A. N., Heun, J. T., Hunsaker, D. J., Jenks, M. A., Kimball, B. A., Roth, R. L., Strand, R. J., Thorp, K. R., Wall, G. W. and Wang, G. (2012). Field-based phenomics for plant genetics research. *Field Crop Res.* 133: 101–12.
White J. W. and Castillo J. A. (1992). Evaluation of diverse shoot genotypes on selected root genotypes of common bean under soil water deficits. *Crop Sci.* 32: 762–5.
Xu, Y. and Crouch, J. H. (2008). Marker-assisted selection in plant breeding: From publications to practice. *Crop Sci.* 48(2): 391–407.

Zaman-Allah, M., Jenkinson, D. M. and Vadez, V. (2011). Chickpea genotypes contrasting for seed yield under terminal drought stress in the field differ for traits related to the control of water use. *Funct. Plant Biol.* 38: 270–81.

Zhao, J., Fu, J., Liao, H., He, Y., Nian, H., Hu, Y., Qiu, L., Dong, Y. and Yan, X. (2004). Characterization of root architecture in an applied core collection for phosphorus efficiency of soybean germplasm. *Chinese Sci. Bull.* 49(15): 1611–20.

Zhu, J. M. and Lynch, J. P. (2004). The contribution of lateral rooting to phosphorus acquisition efficiency in maize (*Zea mays* L) seedlings. *Funct. Plant Biol.* 31: 949–58.

Zhu, J. M., Ingram, P. A., Benfey, P. N. and Elich, T. (2011). From lab to field, new approaches to phenotyping root system architecture. *Curr. Opin. Plant Biol.* 14: 310–17.

Chapter 3

Conserving and characterizing the genetic diversity of grain legumes

P. J. Bramel and H. D. Upadhyaya, Global Crop Diversity, Germany and International Crops Research Institute for the Semi-Arid Tropics (ICRISAT), India

1 Introduction

Pulses or grain legumes are crops that belong to species in the *Fabaceae* family (Smykal et al., 2015). These include globally important pulses such as beans (*Phaseolus vulgaris*), chickpeas (*Cicer arietinum*), lentils (*Lens culinaris*), cowpeas (*Vigna unguiculata*), peas (*Pisum sativum*), faba beans (*Vicia faba*) and pigeon peas (*Cajanus cajan*). There are also more regional or local pulses such as grasspea (*Lathyrus sativus*), tepary beans (*Phaseolus acutifolius*), scarlet runner beans (*Phaseolus coccineus*), lima beans (*Phaseolus lunatus*), mung beans (*Vigna radiata*), adzuki beans (*Vigna angularis*), Bambara groundnuts (*Vigna subterranea*), black gram (*Vigna mungo*), moth bean (*Vigna acontifolia*), creole bean (*Vigna reflexopiloxa*) and rice bean (*Vigna umbellata*). All of these crops are mainly used in local diets for their high protein content. When they are grown, they contribute to soil health with their nitrogen-fixation properties due to their symbiotic relationship with nitrogen-fixing bacteria (Foyer et al., 2016). Groundnuts (*Arachis hypogaea*) and soybeans (*Glycine max*) are also grain legumes but not pulses, due to their high oil content as well as protein content. All of these grain legumes are critical components of sustainable cropping systems (Siddique et al., 2001), diversified protein balanced diets and the health of humans and soils (Foyer et al., 2016).

http://dx.doi.org/10.19103/AS.2017.0023.02

The year 2016 was declared as the 'International year of pulses' to bring needed attention to this undervalued commodity (FAO, 2014b). These crops are prevalent in subsistent farming systems globally but are viewed as specialty crops in the broader agricultural and market context. FAO (2014b) concluded that from 1961 to 2012, there were massive gains in yield and production for maize, wheat, rice and soybean, from 200 to 800% growth versus only 54% for pulses. The pulses are made up of a variety of species that are grown in a wide range of environments globally. There is much less investment in international and national research on crop improvement, agronomic practices, as well as market or product development for pulses. Both the production and consumption have increased over time since 1961 but recently production and consumption have been slowing down due to competition with meat and milk as income increases. This slowdown does not predict a surplus of pulses in the future or a drop in demand since, in many of the consuming countries, the population is growing at a rate that exceeds farming output. Approximately 84% of the pulse production in the world is used to meet domestic consumption and only 16% is traded internationally. Countries like India and China are being forced to import pulses; thus international trade in pulses is expected to continue to grow much more rapidly than production.

Generally there is a need to increase the production and marketing of these crops, not only for dietary diversification and improved nutritional and health benefits but also for the sustainability of the future cropping systems and long-term soil health (Foyer et al., 2016). This will be a challenge given the impact of change in the climate in many of the current production areas, especially given the more local adaptation and use of these crops. One of the key resources for adapting to these challenges through crop improvement will be the genetic diversity conserved *ex situ* in gene banks and *in situ* with farmers or in protected areas. Thus, Foyer et al. (2016), Upadahyaya et al. (2011a) and Smykal et al. (2015) reviewed the status of the genetic resources of the grain legume crops currently conserved in some of the key gene banks. From these reviews it is clear that there is a need to secure the conservation of these genetic resources for use now and for future generations. Thus efforts being made to increase our characterization and exploitation of these genetic resources will be reviewed for chickpeas and pigeonpeas, as an example. The new approaches being taken for enhancing the use of these resources will also be considered.

2 Conservation of grain legume genetic resources: overview and global assessment of germ plasm collections

2.1 Overview

The global nature of grain legume production and consumption means that there is a global interdependency for food supply and genetic resources. This interdependency was recognized in the First and Second Global Plan of Actions for Plant Genetic Resources for Food and Agriculture (FAO, 1996, 2011a) and was the basis for the International Treaty for Plant Genetic Resources for Food and Agriculture. Khoury et al. (2015) assessed this interdependency based upon the nutritional value, productivity, food supplies and gross domestic product to determine the degree to which international exchange of plant genetic resources underpins national food supplies, production and economies. The assessment demonstrated that national food supplies and production systems are highly interdependent in relation to plant genetic resources. The result was consistent across all regions and continents, even in countries located in areas of high indigenous crop diversity. This interdependence has increased in the past 50 years due to globalization of food systems. Thus, a global system for the conservation of this diversity as a global public good is critical to ensure availability of and access to the resource for all future generations.

Gepts (2006) concluded that a number of technical and financial issues are key challenges in the future to *ex situ* conservation but also a clear long-term political agenda and commitment to these collections in the future. To maintain these commitments, a much greater effort is needed to raise public awareness of the value of these conserved resources and the needs for sustained conservation and use of these global public goods. The Global Crop Diversity Trust was established to secure the resources needed to ensure sustained support for the long-term conservation of *ex situ* collections forever, https://www.croptrust.org, and facilitate the development of a global system for *ex situ* conservation of the crop diversity the world depends upon for food security. In 2006, the Global Crop Diversity Trust facilitated the development of a global *ex situ* conservation strategy with the conservation and use community. The Second Global Plan of Action for PGRFA (FAO, 2011a) served as a framework for the actions needed to promote the development of an efficient system for the conservation and use of plant genetic resources for food and agriculture through better management, cooperation and coordination. The eighteen priority actions of the Global Plan of Action for PGRFA formed the elements of the global strategies for the key grain legumes.

The Global Conservation Strategy for *ex situ* conservation of chickpeas, lentils, lathyrus (grass pea) and faba bean was finalized in 2008 with the leadership of Dr Geoffrey Hawtin and the support of the GCDT and GRDC in Australia (Hawtin, 2008a,b,c,d). The Global Conservation Strategy for *ex situ* conservation of cowpeas was finalized in 2010 with the leadership of IITA and the support of the GCDT and GRDC in Australia (Dumet and Fatokun, 2011). A draft Global Conservation Strategy for *ex situ* conservation of pigeonpeas was finalized in 2011 with the leadership of ICRISAT and the support of the GCDT (Singh et al., 2011). A draft global conservation strategy has been developed for Phaseolus (bean) genetic resources with the leadership of Dr Daniel Debouck of CIAT, mainly based upon expert consultations and was supported by GCDT and GRDC in Australia (Debouck, 2014). All of these grain legume crops are listed in Annex I of International Treaty on Plant Genetic Resources for Food and Agriculture (ITPGRFA) (FAO, 2009).

All of these strategies were the result of a survey of genebank curators and a consultation process with experts. Most of these global strategies are nearly ten years old, but when developed were seen as a dynamic process with a need for monitoring and updating. These strategies have been used as a guidance for the global community to ensure actions are taken to address the future needs for conservation and use in these crops. Global *ex situ* conservation strategies have not been developed for soybeans and groundnuts as well as for minor grain legumes, such as green gram, black gram, adzuki beans, Bambara groundnuts and among others. Some of these are not explicitly included in the multilateral system (FAO, 2009), so their global access and benefit sharing for exchange are not clear.

Frankel (1977) categorized crop genetic resources into six broad groups that include the wild relatives of crop species, landraces, genetic stocks, obsolete cultivars, modern cultivars and breeding lines. These genetic resources represent the gene pool potentially available to breeders and other users of collections. This broad pool can be further subdivided into primary, secondary and tertiary gene pools (Harlan and de Wet, 1971). The primary pool consists of the biological species, including cultivated, wild and weedy forms of the crop, and gene transfer within this group is considered to be easy. In the secondary gene pool are the species from which gene transfer is possible but difficult, while the tertiary gene pool is composed of species from which gene transfer is possible only with great difficulty. Clearly the boundaries on these groups are fuzzy and also change with technology. Consequently, Smartt (1984) have suggested the gene pools concept of Harlan and De Wet (1971) be modified to increase the number of gene pools from three to four to coincide with, respectively, populations, species, genera and tribes.

2.2 Global assessment of grain legume crop germ plasm collections

Smykal et al. (2015) did an extensive review of the phylogeny, crop evolution, domestication and status of conserved germ plasm of economical important species of the *Fabaceae* family. The review included an inventory of the key genetic resources conserved *ex situ*, cultivated as well as wild, for the grain and forage legumes. This inventory included the collections that are held in *ex situ* gene banks as well as in botanical gardens and herbariums. They also reviewed the status of *in situ* reserves, especially for the wild relatives.

Smykal et al. (2015) reported that there were an estimated 985 452 accessions of the major grain legumes held in *ex situ* collections globally due to the importance of grain legumes to many countries. One of the major sources of information on genebank holdings in Smykal et al. (2015) was public databases such as FAO-WIEWS (http://www.fao.org/wiews-archive/wiews.jsp) and Genesys (https://www.genesys-pgr.org). Foyer et al. (2016) also presented a systematic inventory of grain legume collections that had published accession-level information in publically available databases such as Genesys, the Chinese Crop Germ Plasm Information System, the National Institute of Agrobiological Sciences database in Japan and the National Bureau of Plant Genetic Resources database in India. In their assessment they also listed the main holding institutes with the percentage of the global accessions held. Both of these assessments are given in Table 1, although the number of accessions given in Foyer et al. (2016) was much less for the main grain legumes since they involved a smaller sample of the gene banks globally.

The most comprehensive inventory is given in FAO-WIEWS (2011b) but this is based on reporting from 2009. So for the assessment in Table 1, the inventory was updated for specific gene banks from up-to-date accession-level information given in Genesys. In Table 1, institutes who reported holding more than 100 accessions in FAO-WIEWS and Genesys were designated as holders of significant collections for eighteen major and minor legume crops. The total number of accessions held in significant collections was 970 064 or 94.2% of all the accessions reported in these two databases. The largest collections held were for crops such as soybeans, common beans, groundnuts and chickpeas. Overall these four crops account for 61% of all the accessions conserved globally *ex situ* for the major and minor legumes reported in Table 1. The smallest numbers of institutes and accessions were found for the local legumes crops such as Grasspea, tepary beans, scarlet runner beans, lima beans, Bambara groundnuts, black gram, adzuki beans and moth beans.

Major *ex situ* collections were defined as institutes that held more than 500 accessions (Table 1). The number of institutes who hold major collections ranged from 1 for Bambara groundnuts to 61 for common beans, while the proportion of the accessions held by

Table 1 The number of accessions cited in Smykal et al. (2015) and Foyer et al. (2016), the number of institutes and number or proportion of accessions held that reported in FAO-WIEWS and Genesys as holding significant major or the largest collections for eighteen major and minor legume crops

		FAO-WIEWS and Genesys					
	Cited in literature	Significant (>100 accessions)		Major (>500 accessions)		Largest (>1000 accessions)	
Crop	No. of accessions	No. of institutes	No. of accessions	No. of institutes	% of accessions	No. of institutes	% of accessions
Pea	98947[1]	56	71357	31	93.0%	24	85.8%
Chickpeas	98313[1]	41	96029	22	94.7%	11	86.3%
Faba beans	38000[1]	45	40514	21	85.6%	9	65.4%
Lentils	58407[1]	44	56188	21	88.5%	11	77.3%
Grasspea	38360[1]	17	15962	8	86.4%	5	70.7%
Common beans	261968[1]	103	190671	61	94.4%	41	87.9%
Tepary beans	1257[2]	4	809	0	0.0%	0	0.0%
Scarlet runner beans		14	4677	3	52.2%	0	0.0%
Lima beans	6420[2]	15	12961	7	89.3%	0	0.0%
Soybeans	156849[1]	92	206993	35	89.4%	20	79.2%
Cowpeas	65323[1]	52	72461	20	89.7%	13	82.6%
Bambara groundnuts	2183[2]	10	3878	1	52.4%	1	52.4%
Mung bean	23658[2]	29	42091	14	93.4%	9	84.9%
Black gram	1668[2]	11	7032	5	81.5%	2	52.2%
Adzuki beans	9978[2]	7	9500	3	90.4%	3	91.5%
Moth bean		3	3517	2	93.1%	2	93.1%
Pigeonpea	40820[1]	22	37828	7	92.3%	6	90.5%
Groundnuts	128435[1]	61	98136	27	87.4%	18	81.3%
Total	**1030586**		**970604**		**90.6%**		**80.8%**

[1] Smykal et al. (2015).
[2] Table 2 of Foyer et al. (2016).

the major collections ranged from 52.2% for scarlet runner beans to 94.7% for chickpeas. Overall crops, about 91% of the accessions are conserved in major collections. Five International gene banks (ICARDA, ICRISAT, CIAT, IITA and ILRI) and 72 national institutions conserve these major collections.

A large collection holder was defined as an institute that held more than 1000 accessions. Overall, about 81% of the accessions are conserved by the largest collections. There are six National Institutes who hold major collections of most of the grain legumes assessed in Table 1. These are the Australia Grains Genebank in Australia, CENARGEN in Brazil, ICS-CAAS in China, NBPGR in India, VIR in Russia and USDA-ARS in the United States. Institutes in Bulgaria, Chile, Ecuador, Ethiopia, France, Germany, Hungary, Indonesia, Iran, Japan, Mexico, Philippines, Ukraine and Zambia hold large collections of at least four of these grain legumes. Foyer et al. (2016) also assessed the proportion of the global accessions held in the main institutions that are included here.

The assessment in Table 1 is based solely on the size of the collections. This illustrates that the global system for these crops is made up of a few large collections that conserve the vast majority of the accessions in *ex situ* conservation. This does not infer that these larger collections hold the most important genetic diversity since many of the collections have accessions in common and are held outside the centre of origin for the crop. There may be a number of important small collections in terms of the unique nature of the accessions held. Foyer et al. (2016) concluded that several accessions are duplicated across collections but determining this based upon the inventory list is not easy. This assessment of the exact nature of the accessions held in the collections was done as part of the global conservation strategies and will be discussed later.

A key component of the global *ex situ* system will be a global linkage of all these substantial national, regional and international collections, especially those that contain significant numbers of unique accessions not duplicated elsewhere, such as the genetic stock and mutant collections described in Upadhyaya et al. (2011a). These unique genetic resources are held outside of the national gene banks by individual researchers, so one goal of any global system would be to locate those collections, determine their longer-term value, assess their risk status and, if necessary, safety duplicate these into more publically held collections.

3 Conservation of grain legume genetic resources: status of *ex situ* conservation of grain legume genetic resources

As part of the various global grain legume conservation strategy developments, a survey was distributed to curators of collections for chickpea, lentil, grasspea, faba bean, and cowpea to gather basic information on the numbers and types of accessions held, the conditions under which they were stored, the routine operations for maintenance of the accession, the accession-level documentation, safety duplication and their accessibility (Hawtin, 2008a,b,c,d; Dumet and Fatokun, 2011). No survey was done for the Phaseolus strategy or for any of the other crops given in Table 1. Upadhyaya et al. (2011a) also described these global strategies and concluded that the surveys provided valuable information on who holds what, how it is maintained and shared, the gaps in collections and the common issues faced by these collection holders. They described the baseline assessment that was made by the crop community of actions that needed to be taken forward.

In relation to the adequacy of storage facilities and conditions, gene banks generally have adequate storage facilities for medium- and long-term storage. In general, the larger seed size of many of the legumes results in the need for more space for long-term storage and this can significantly affect the cost for conservation. For cowpeas in gene banks in Africa, there are significant constraints to storage due to the lack of consistent electricity supplies.

Grain legumes have evolved various physiological and physical processes to manage the timing of germination in their natural environments and many of these have gone through modification from selection during domestication (Kigel et al., 2015). A review of the nature of physiological and physical dormancy in legumes is given in Kigel et al. (2015). In general, one of the key issues for conservation is the impact of seed coat impermeability or seed hardness on seed viability monitoring. For some legumes, such as lentils, the low seed moisture that is recommended for long-term storage can result in greater seed hardness. The seed coat can be fragile in some of the crops and this will increase the possibility of damage during post-harvest seed handling. This damage reduces viability and also increases the risk of pathogen infestation. The high oil content of crops like groundnuts and soybeans can impact the long-term viability of these seeds. Seed health is a greater challenge for grain legumes than cereals (Girish et al., 2001), and not just from the risk of pathogen attack with the seed coat damage. Grain legumes suffer more from seed-borne pathogens and viruses than cereals. Thus, the longer-term impact of these challenges to conservation is the difficulty in determining the timing and methodology for the routine monitoring of seed viability; the larger packaging and greater space requirement for seed storage; the lower seed longevity that requires more frequent regeneration and the increased risk of seed health issues for regeneration, distribution and seed longevity. The only crop for which the survey asked for information on the seed viability or germ plasm health monitoring was cowpeas. Many of the gene banks surveyed for cowpeas did not do an initial germination test after harvest, nor did they do routine monitoring of germination of the stored seeds. There was also very little routine monitoring of seed health.

There are established international standards for the conservation of orthodox seeds such as grain legumes (FAO, 2014a). In the future, to assure the security of the conservation of grain legume accessions, it will be necessary to implement quality-management systems for some of these key routine processes such as seed viability testing, germ plasm health testing, regeneration and accession-level documentation. The respondents for the cowpea survey concluded the main constraints to *ex situ* conservation in Africa collections were the unreliability of electricity supply, the lack of appropriate storage containers, high humidity storage conditions, prevalence of biotic stresses and sustainability of funding for routine operations of the gene banks. Over all gene banks, problems in maintaining standards arise because of lack of sufficient and consistent annual operating budgets to meet the costs of routine operations such as regeneration, characterization and documentation, especially in developing countries. This lack of funding can lead to a number of outstanding issues including:

- Backlogs in regeneration, which is threatening the viability and safety of a significant number of genebank accessions (summary of status of respondents given in Table 2)
- Formal arrangements for safety duplication of accessions which is also contributing to the threat to the safety of unique accessions (summary of respondents given in Table 2)

Table 2 The number of respondents and the key results of the surveys completed as part of the development of the global *ex situ* conservation strategies for chickpea, lentil, grasspea, faba bean and cowpea

Crop	No. of respondents	Quality of collections	Regeneration backlog	Safety duplication
Chickpea	21	12 institutions with more than 50% landraces. 9 institutions have 1–4% wild species while 1 has more than 20% wild species. 6 institutions have more than 50% of the accession collected from within the country	Significant regeneration needs for 8 countries who responded	14 collections indicate more than 40% safety duplicated but not all to site outside country. Some very important collections like NBPGR, Turkey, Nepal and Ethiopia have no or very little safety duplication
Lentil	19	13 institutes with more than 50% landraces, 8 institutes with wild accessions. 7 institutes with accession mainly collected from within the country	Important collections in Iran, Russia, Chile, Turkey, Syria, Hungary, China, Pakistan, Bangladesh, Ukraine, Nepal, Morocco, Slovakia and Mexico have significant backlogs	10 institutions with adequate safety duplication but many important collection like those in NBPGR, Turkey, Spain and Ethiopia have no safety duplication
Grasspea	13	NBPGR, VIR, Ethiopia, Nepal, Portugal, Spain and Turkey hold majority of their accessions of national origin	7 institutes have more than 50% regeneration needs	The strategy came up with action plan to securely safety duplicate the unique accessions
Faba bean	29	20 institutions have majority landraces, 15 institutions have wild relatives, 17 collections with majority from original country, large number of collections with unique locally adapted germ plasm	Adequate information on regeneration backlog missing for many collections, 12 institutions have more than 50% regeneration needs	10 institute with more than 50% safety duplicated. Many important collections are not safety duplicated and very few with adequate formal arrangements
Cowpea	15	Quality was assessed based upon age of collection and 13 holding diverse species and overall 60% landraces and 2% wild species	Backlog not assessed	8 institutes with no safety duplication

- Inability to provide seed and information of accessions on request
- Lack of passport, characterization and other documentation of accessions in collections.

The long-term conservation and distribution of accessions requires that seeds are periodically screened for viability, and when viability falls below an internationally accepted level, the accession is regenerated and fresh seed is placed in the genebank. Regeneration is a relatively expensive procedure and is often one of the first activities postponed when funds supporting genebank activities decline. It is clear that regeneration of accessions is probably the single greatest threat to the safety of grain legume accessions held in globally important gene banks. Regeneration needs are generally greater in developing countries but there are even backlogs in some major developed countries. In the various surveys given in Table 2, there were significant regeneration backlogs, and some of the main collections with unique, local accessions had not responded to the survey question, such as for lentils. From 2008 to 2012, the Global Crop Diversity Trust managed a large-scale effort to address the urgent regeneration need for about 80 000 accessions of crops held in 77 national collections (https://www.croptrust.org/what-we-do/supporting-the-global-system/rescuing-diversity). The accessions to be regenerated were identified with the finding of the crop strategies.

The faba bean strategy concluded that there is a need to conduct a survey of regeneration protocols globally to ensure security. Upadhyaya et al. (2011a) reviewed the few studies that had been done on allelic changes in an accession that had undergone periodic regeneration. They concluded that these studies did imply that there is a need to review and improve regeneration protocols to accommodate larger population sizes as well as monitor the genetic integrity of the accession over time to reduce the risk of erosion.

Another major deficiency identified in this survey was the lack of safety duplication of many important collections (Table 2). Safety duplication refers to the need for duplicate samples of accessions in a collection to be maintained in more than one genebank as a form of insurance against disastrous loss (for example, due to fire, earthquake or war). 'Black box' duplication – whereby the genebank of origin is responsible for the quality of the stored samples and their regeneration, and the recipient genebank merely holds the samples in long-term storage – is considered the most convenient and cost-effective method of safety duplication. Duplication of accessions for safety is a requirement of agreed international standards for genebank management. Yet, there are many globally important collections that do not have unique accessions duplicated elsewhere for safety, for example, in key collections for chickpeas and lentils. In the strategy for grasspea, an action plan was developed to safety duplicate unique accessions in some of the largest gene banks. Clearly, this is a major area of concern, and full safety duplication, of at least unique accessions, must be the goal of an efficient rationalized global system of collections. The lack of safety duplication is linked, in part, to the need for urgent regeneration of samples in many collections. Curators sensibly see no point in sending seed of accessions that have low viability to duplicate black box storage. Consequently, these two safety issues, the regeneration of accessions and the safety duplication, need to be considered simultaneously with samples of newly regenerated seed also used for safety duplication.

The inability of collections to distribute accessions upon request can be due to the lack of availability of the seed. Availability of an accession for distribution is related to the viability of the seed, its health and its legal status. National governmental phytosanitary regulations can significantly hamper germ plasm exchange. This is related to the plant and seed health. The legal status of an accession is related to accessibility under either the internationally agreed terms of access and benefit sharing provided for in the multilateral system as set out in the SMTA by the International Treaty on Plant Genetic Resources for Food and Agriculture (FAO, 2009), by national or local regulations, or term-limited material transfer agreements (MTA) or patents. The conditions for distribution, use and benefit sharing should be made available to potential users. As the multilateral system expands, the access to many major collections should improve, but realistically, some collections or accessions may continue to be governed by local regulations, or by term-limited MTA or patents, and not by the SMTA of the ITPGRFA. Collection holders should make more of the accessions available in the multilateral system as the implementation of the ITPGRFA expands. In the survey on faba beans and cowpeas, the respondents were asked to give more details on distributions; domestic versus international users, type of user, farmer versus researcher; and type of agreement used, SMTA versus institutional MTA. ICARDA, IITA, USDA and European gene banks did most of the international distributions. Farmers were significant recipients of accession from the national gene banks.

It is clear from the surveys of collections that there is a great variation in the systems used to record, store and distribute information. A major step in developing an integrated global accession-level information system is to ensure adequate information storage and retrieval infrastructure and to rectify the deficiencies in the documentation of accession in collections. This will involve, at a minimum, to improve the accession-level documentation of collections to enhance accessibility and as a tool to assist in the management and rationalization of collections, using GRIN-Global or other genebank management systems; utilize globally agreed protocols for data ontology and quality; and publish all key accession-level data, such as passport, characterization, evaluation and other relevant data from individual collections to global portals, such as Genesys, to allow for easy access by all collection holders and the collections' user clientele.

A global portal to publish and share accession-level information exist through Genesys (https://www.genesys-pgr.org). Currently, the main contributors of accession-level information in Genesys are the CGIAR international gene banks, USDA-ARS in the United States, the Australian Grains Genebank, and the national institutes in European countries who make their data available through EURISCO (http://eurisco.ipk-gatersleben.de). For example, this global portal currently host 100 239 accessions of common beans that are held mainly by CIAT, USDA-ARS in the United States, IPK in Germany and VIR in Russia. These three institutions hold about 60% of the accessions reported in Genesys. The accessions in Genesys are 66% landraces or farmers varieties. The information on the accession includes collection management such as safety duplication, availability and germ plasm storage type. It also includes passport and characterization data but the quality of this information varies. The quality of the passport information is scored for each holding institute using the passport data completeness index (PDCI) as described by van Hintum et al. (2011). For example, the collection held by the CIAT has an average PDCI of 4.51.

The development of genebank management software packages, like GRIN-Global, will facilitate the secure conservation and use of the accession held globally (http://www.grin-global.org/index.php/Main_Page). This system can be used by an individual genebank to capture a diversity of accession-level information as well as enhance the management of the collection. It allows for automation of data collection and of monitoring. It allows for greater sharing of this information internally and externally. It would facilitate the publishing of accession-level information from individual gene banks to an integrated web-based system like Genesys that allows the searching of the databases of all cooperating gene banks. This has clearly been one of the major challenges facing the development of an integrated global conservation system for grain legume genetic resources.

4 Conservation of grain legume genetic resources: *ex situ* collections of grain legumes

4.1 Composition of accession held in *ex situ* collections of grain legumes

In most of the surveys and in the consultation workshop, an assessment was made for the quality of the collections based on the type of accessions conserved (wild species, landraces, breeding material or others) as well as the source of the accessions (collected within country, outside country, introduced from others or other sources). A summary of the quality of the collections, based on the type and source of accessions, is also given in Table 2. Cowpeas used a different approach to assessing the quality of collections. This analysis identified key collections that hold a high percentage of unique local accessions.

For example in chickpeas (Hawtin, 2008a), NBPGR in India reported that their accessions were 1% wild species and 65% landraces or farmer's varieties and 89% were collected from within India. PGRI in Pakistan had 1% wild species and 98% landraces with 70% of the accession collected from within Pakistan. AARI in Turkey had 1% wild relatives and 98% landraces but 100% of the accession in their collections was originally collected from within Turkey. ABD in Nepal has a smaller collection but 100% is landraces and 97% collected from Nepal. BCRI in Ethiopia has 75% landraces with 71% of the accessions collected from Ethiopia. Even CRF-INIA in Spain had 1% wild species, and 60% landrace with 58% of the accession collected from Spain. Thus for chickpeas, these collections would be considered as important in terms of the conservation of unique local diversity. The same Institutions in India, Turkey, Ethiopia, Nepal and Spain held key collections of lentils (Hawtin, 2008d). In both crops, VIR in Russia held a collection with more than 90% landraces but almost 95% of these were collected from outside Russia. In their case, given the age of the collections, there is a high probability that these are unique accessions.

For grasspea and faba beans, the local nature of these fairly minor grain legume can be seen in the high proportion of collections that held mainly landraces collected within the country (Hawtin, 2008b,c). There has been little breeding effort in this crop and not that much germ plasm exchange so the national origin of most collections would have been expected. The ICRISAT, ICARDA and IITA collections for all of these crops have a high percentage of landraces but these have been assembled and collected from many countries. They also hold a higher percentage of breeding material than the national gene banks.

4.2 Redundancies in *ex situ* collections of grain legumes

The task of rationalizing collections and identifying duplicates is difficult and time consuming. Over time, a variety of names and numbering systems have been used for the same accessions in different gene banks. Single plant selections can be made from an accession and given new accession numbers. Landraces may have been collected from several geographically distinct sites and represent genetically different samples but may have the same name. Many accessions have limited passport data available that makes the determination of their origin and uniqueness difficult. Hazenkamp et al. (2014) discussed the use of a central crop registry that was developed using the main wheat and rice collections. They compiled accession-level information on passport data for the collections, established standardized data fields for comparison, assessed similarity of accession within and amongst collections, and feedback the results to the institutes to improve data quality. They described a methodology to assess similarities in collections using this crop registry of accessions. For wheat in the CIMMYT, USDA-ARS and ICARDA collections, they found over 50% of the similarities were direct donor–recipient relationships or due to the subdivision of an accession into further new accession based on single plant selection.

The crop registries could be expanded to include other crops to gain a much better understanding of the duplicates within and between collections globally. This is a key first step in rationalization and it will empower curators to identify unique and duplicated accessions that are held in other collections. Curators can make decision about reducing redundancy and negotiate with other gene banks to insure safety duplication. It will allow them to focus on the conservation of the unique accessions and establish safety duplication for these. Upadhyaya et al. (2011a) concluded that the development of crop registries was a recurrent recommendation in the grain legumes conservation strategies. They indicated that ICRISAT was developing a registry for chickpeas.

The European Cooperative Programme for Plant Genetic Resources (ECPGR) involves 43 countries that aim to facilitate long-term conservation on a cooperative basis and to increase utilization (http://www.ecpgr.cgiar.org). AEGIS was established by ECPGR to create a European-wide integrated system to securely conserve unique and important accessions for Europe and make them available (Engels and Maggioni, 2011). This effort would lead to increased security of long-term conservation with a more systematic safety duplication, facilitated access, improved quality standards for conservation, cost-efficient conservation, reduced redundancies and improved information sharing from the greater collaboration of European collection holders. So far 34 countries have become member of AEGIS (http://www.ecpgr.cgiar.org/aegis/aegis-homepage). For example, within the European collection, there are currently 124 accessions from *Phaseolus* and 527 accessions from *Pisum*.

It must be noted, however, that the identification of genetic duplicates within and between collections is not a trivial exercise. Genetic drift within accessions due to out-crossing, founder effect from restricted sampling, genotypes by environment interactions, unrecognized selection pressures and mechanical mixtures have been observed. Upadhaya et al. (2011a), Smykal et al. (2015) and Foyer et al. (2016) concluded that genotyping collections with greater genomic saturation and precision and with reduced per-sample costs will aid in the characterization of genetic diversity, both unique and common. It can be used to detect redundancies within and between accessions in collections. It will also help in the monitoring and management of genetic integrity of the accessions during regeneration. It will also be a tool to enhance the use of accessions.

4.3 Gaps in *ex situ* collections of grain legumes

An important aim for the global grain legume conservation strategy was to identify gaps in existing collections. For most of the crops, further collection of landraces that were likely to still exist in the field in areas that are poorly represented in gene banks were identified as priority areas. For example, the chickpea global conservation strategy identified key under-represented areas for landraces in the Hindu Kush-Himalayan region, the west and north of China, Ethiopia for desi chickpeas, Uzbekistan, Armenia and Georgia (Hawtin, 2008a). Collecting priorities should be decided in conjunction with national genebank and experts who have knowledge of the local material.

Though the collections of grain legumes at ICRISAT genebank are large, they cannot be considered as complete. Critical assessment of grain legume collections revealed several potential gaps. The genebank at ICRISAT conserves more than 20 600 accessions of chickpea from 59 countries. Major chickpea-growing countries were explored during collection missions launched by ICRISAT for its mandate crops during 1974–1993. Russia and CIS, Colombia, Peru, Western Nepal, Eritrea and Myanmar were considered as the geographical gaps in cultivated chickpea collection. Singh et al. (1997) identified geographical gaps for wild species in the ICRISAT collection from Russia and CIS, Turkey, Afghanistan, Iraq and Ethiopia. The Mediterranean region, which is one of the two primary centres of diversity, contributed only 1240 accessions (7.4%) and was under-represented. Ethiopia, which is the secondary centre of diversity, was also under-represented with only 928 accessions (Upadhyaya, 2003).

The pigeonpea germ plasm collection at ICRISAT genebank consists of more than 13 700 accessions from 74 countries. Germ plasm collection has covered the major geographic areas of importance. However, West Africa and SouthEast Asia (Philippines, Indonesia, Thailand, Malaysia and China) and to some extent Latin America are not yet covered by collections (Singh et al., 2006). The areas that are yet to be surveyed for pigeonpea collection in India include Uttar Pradesh and Uttaranchal (districts of Mirzapur, Bundelkhand region, Tarai), Madhya Pradesh (Rewa, Sidhi, Gwalior, Morena, Bhind region), Maharashtra (Nasik, Dhule) and adjoining North Karnataka, southeastern districts of Bihar, parts of Orissa, Rajasthan, Gujarat, Nilgiri and Ragan hills and the northeast region. In Uganda, emphasis would be in eastern and northern parts of the country. The focus for the future should be more on trait-based collections and wild species (Singh et al., 2006).

Results of studies using passport and characterization data and GIS software such as FloraMap and ArcGIS revealed several geographical and trait-diversity gaps in the pigeonpea germ plasm from East and Southern Africa (Upadhyaya et al., 2015). A total of 84 districts located in 35 provinces of four East African countries and 54 districts located in 18 provinces of three Southern African countries were found as the important geographical gaps. Five districts of three provinces in Ethiopia, 12 districts of three provinces in Kenya, 37 districts of 14 provinces in Tanzania, 30 districts of 15 provinces in Uganda, six districts of three provinces in Malawi, 28 districts of eight provinces in Mozambique and 20 districts of seven provinces in Zambia were identified as the geographical gaps in the collections from ESA countries. Eight districts in Eastern, Central and Coast provinces in Kenya were identified as gaps in phenotypic diversity for all traits. A total of 22 districts located in five countries were identified as gaps in diversity for one or the other trait (Upadhyaya et al., 2015).

Collection of wild species has been given less emphasis than the collection of land races in the past, as they were considered to be under less threat given their unaided survival and reproduction.However, there is a need to substantially increase the genetic diversity of the wild species conserved *ex situ*. More and more wild species are being threatened by changing patterns of land use and climate change. There is also an increased demand for readily available samples of wild species for researchers and breeders. Table 2 indicates that few institutions hold a significant number of wild accessions. Singh et al. (2014) reviewed the collections of the wild *Lens* held globally. From their assessment, they identified key gaps where there is an urgent need to collect and protect the populations *in situ*.

Upadhyaya et al. (2011b) studied the occurrence of *C. scarabaeoides* near the sites of origin for pigeonpea. They reported a total of 790 districts located in 118 provinces in Bangladesh, Cambodia, India, Indonesia, Laos, Malaysia, Myanmar, Nepal, Papua New Guinea, Philippines, Thailand and Vietnam as geographical gaps in the collection. Among wild species, those endemic to Australia are inadequately represented in the global collection. The species *C. cinerius, C. confertiflorus, C. lanceolatus, C. marmoratus* and *C. latisepalus*, which were considered to be important for their drought tolerance, are yet to be collected. Therefore, under- and unexplored districts for cultivated and wild species of pigeonpea should be the target of future collection missions for pigeonpea germ plasm.

The conservation of genetic and mutant stocks is important to ensure that this historic research legacy is not lost and that this significant gap in collections is addressed (Upadhyaya et al., 2011a). Obsolete and new cultivars are often more rigorously characterized and broadly evaluated due to release practices and regulations, resulting in a higher short-term 'value' for immediate, direct use in grain legume breeding programmes. In many cases these are not conserved for the long term, but as more data, such as phenotyping, genotyping and nutritional profiles, are generated from this germ plasm, there is a need to fill this conservation gap for the future users.

5 Global *ex situ* conservation strategy for grain legumes

The global conservation strategies concluded that the current system for *ex situ* conservation of grain legumes is not global but consists of poorly linked international and national collections which hold some unknown degree of common and unique accessions that are conserved with limited sustained support and are not very well known or available to users. Therefore, this global system should ensure availability of these genetic resources for future generation users so that they will be able to resolve new challenges of production. However, since the current system does not meet the need of users, nor is it able to take advantage of new technologies and tools to enhance its use, a new model for managing global resources in a secure, cost-effective, sustainable global system will need to be adopted to benefit research, production and consumption (Bramel, 2017).

Bramel (2017) recognized some basic principles for this global conservation and use system. There is a need to ensure secure conservation with quality-management system and safety backups. Viable, healthy, fully documented accessions, including the information, should be available to users in a timely manner upon request. A rational, cost-effective global system should have centralization of some specialized activities, such as international distribution or safety duplication, and decentralization of a few other activities at the regional or national level, such as local regeneration, distributions, evaluations, plant breeding and outreach to farmers. A sustainable global system should have multiple types of collections such as international or global collections, national or local collections, specialized collections for wild relatives or genetic stocks, and research collection from breeders. Collections should be linked to better meet user needs to ensure that the genetic diversity conserved is appropriate with minimum redundancies to reduce collection size and complexity. The documentation on accessions that are shared in an integrated system should include passport, characterization, evaluation and genomic data. Users should be able to access this information from a global platform that integrates accession-level data from individual collections into a single portal with tools to enhance search and selection. Finally, a legal framework that facilitates access and the sharing of benefits should be used, such as a fully functioning multilateral system under the ITPGRFA.

Each institution holding a collection in this global system should have primary responsibility for the long-term conservation of the germ plasm that is unique to their collection. The institution may choose to outsource some of the functions associated with the long-term conservation of their unique accessions (e.g. regeneration), but those decisions remain their own responsibility. All the unique accessions would be conserved and managed in accordance with agreed international scientific and technical standards utilizing a quality-management system. All the unique accessions would also be 'black box' safety duplicated in at least one other gene bank to avoid loss due to mishap or catastrophe. Complete passport and characterization data would be available for all the accessions and accessible online. The databases of all the collections in the network would be linked through an integrated information platform that would allow public web access.

The global conservation strategies for chickpea, lentil, grasspea and faba bean describe a global approach to conserving the genepool. They recognized that gene banks differed in the degree of genetic variation they conserved for the global community (quality and quantity of their accessions), the status of the routine operations in terms of meeting international standards, the international availability of the accession to users and the completeness of accession-level information that is shared globally. Based on these criteria they classified gene banks into three groups that would take on different roles in the global systems for *ex situ* conservation. These groups consisted of gene banks that meet international standards and make their accession available globally; gene banks that need to upgrade their operations to better meet the conservation and user needs locally; and gene banks that should be supported to safety duplicate their unique accession within the gene banks of the first two groups. Many of the gene banks in the last group were found in countries where the crop had limited production or where the information on the collections was not well known. In all cases, there was a recognition of the need for all gene banks to share accession-level information at the global level in order to facilitate the global system collaboration.

There is an example of an integrated regional genebank system for cooperation and collaboration in the conservation and use of grain legume genetic resources. ECPGR is a collaborative programme among most European countries that aims at contributing to a rational and effective conservation of plant genetic resources and facilitates their utilization in Europe (http://www.ecpgr.cgiar.org). In Europe, an ECPGR Working Group on Grain Legumes was established in 1991 (http://www.ecpgr.cgiar.org/working-groups/grain-legumes/). Since its inception, the Working Group has covered a range of species (*Arachis*, *Cicer*, *Glycine max*, *Lens*, *Lupinus*, *Phaseolus*, *Pisum*, *Vicia faba* and *Vigna*). Despite the large number of crops, countries and collections involved, ECPGR provides a very useful model and source of experience for global grain legume conservation.

The secure conservation of grain legume genetic resources will also depend upon secure annual resources. The mission of the Global Crop Diversity Trust is to both facilitate the development of this global system for wheat and support the long-term conservation of the most important global collections (https://www.croptrust.org). Consistent support of the routine conservation needs for these global collections is too important to be left to the vagaries of the annual funding cycle of international donors or national governments. Thus a Crop Diversity Trust fund has been established not only to generate annual income to maintain these collections at agreed standards but also to ensure the long-term availability, security and use of these collections. Currently, the Crop Trust supports part of the routine operating cost for the long-term conservation of grain legumes at CIAT, ICRISAT, ICARDA and IITA.

6 Characterizing the genetic diversity of chickpeas and pigeonpeas

The use of PGR in crop improvement is one of the most sustainable ways to conserve valuable genetic resources for the present and future use in crop improvement programmes (McCouch et al., 2013). Extensive use of fewer and closely related parents in crop improvement programmes could result in new cultivars with increased vulnerability to pests and diseases. Breeders prefer to use their own working collections including mostly improved materials (Nass and Paterniani, 2000). They are often reluctant to use wild species and landraces available in gene banks, thereby increasing the gap between available genetic resources and their use in breeding programmes (Marshall, 1989). The reasons for the low use of germ plasm include i) lack of accurate and precise large-scale multilocation evaluation data of germ plasm; ii) lack of rational systematic entry points into the vast international collections and iii) lack of robust, cost-effective tools to facilitate the efficient utilization of germ plasm in crop breeding (Dwivedi et al., 2009; Upadhyaya et al., 2011a). The identification of useful germ plasm for breeding programmes is a basic need to enhance the use of conserved accessions. This could be achieved by two approaches: (i) identifying regions from where germ plasm that has a higher mean for desired traits as well as genetic diversity and (ii) developing and systematically evaluating representative core collections of legumes.

6.1 Diversity in *ex situ* collections

Studies on phenotypic diversity in germ plasm collections are very limited. Studies at ICRISAT revealed a wide variation in the germ plasm collections of chickpeas and pigeonpeas for morphological, agronomic and nutritional traits as well as stress resistance. Narayan and Macefield (1976) assessed the diversity of a large collection of 5466 accessions. Moreno and Cubero (1978) studied 150 accessions from five geographical regions. Murthy (1975) used 459 accessions from 16 countries. At ICRISAT, an assessment was made for the entire collection of chickpea, covering 16 820 accessions from 43 countries that had been characterized for seven morphological and 13 agronomic traits and wilt reaction during 1974–1998 (IBPGR, ICRISAT and ICARDA, 1993; Upadhyaya, 2003). The ICRISAT collection has a high representation from India in South Asia with (7174 accessions) and Iran (4838 accessions) in West Asia. The Mediterranean region, which is one of the two primary centres of diversity, contributed only 1240 accessions (7.4%) and was under-represented in the collection at ICRISAT. Ethiopia, which is the secondary centre of diversity, was also under-represented (928 accessions) although this was 83% of the accession from the Africa region.

None of the morphological descriptors was monomorphic and most showed at least two relatively frequent phenotypic classes. The kabuli types that are characterized by no anthocyanin, white flower, owl's head seed shape, a smooth seed surface and beige seeds were more frequent in the collection from East Asia, the Mediterranean region and Europe (Upadhyaya, 2003). Desi-type accessions with pink flowers, brown or yellow-brown seeds, angular seed shape and rough seed surface were abundant in South Asia, Southeast Asia and Africa. Erect and spreading growth habits were found in high frequencies in East Asia. Semi-erect and semi-spreading growth habits were evenly distributed in South Asia, whereas in the rest of the regions, except South-East Asia, semi-erect accessions were predominant. Significant differences were observed among regions for all 13 agronomic traits (Upadhyaya, 2003). Accessions from Africa were early to flower (54.4 days), while those from South-East Asia had a later maturity (109.7 days). On the other hand, collections from East Asia were late to flower (70.1 days) and those from Europe were late to mature (124.6 days), produced the highest number of pods per plant (53.2) and larger seeds (25.7 g 100 seeds^{-1}), and had the highest plot yield (1469 kg ha^{-1}).

Significant variability for most of the traits in all the 25 years was maintained to the ICRISAT chickpea collection (Upadhyaya, 2003). The variances among regions were heterogeneous for all the 13 agronomic traits. East Asia had the highest variances for days to 50% flowering, plant height, plant width and days to maturity; Africa for apical primary branches, pods per plant and seeds per pod and Europe for basal primary branches and basal secondary branches, apical secondary branches, and plot yield. West Asia region had the highest pooled Shannon-Weaver diversity index (H`) for morphological (0.39 ± 0.070) and agronomic traits (0.57 ± 0.012) (Upadhyaya, 2003). Seed colour (0.62 ± 0.083) among morphological traits and days to 50% flowering (0.58 ± 0.025) among agronomic traits had the highest pooled H`. Over all the 21 traits, West Asia (0.50 ± 0.033) had the highest H` and South-East Asia the least pooled H` (0.31 ± 0.035). This may be because the southwest Asia, which is one of the primary centres of diversity, was classified as West Asia.

A hierarchical cluster analysis conducted on the first three principal component scores (total variation accounted for was 91%) resulted in two clusters (Upadhyaya, 2003). Africa, South Asia and South-East Asia grouped together to form Cluster 1 and the Americas, Europe, West Asia, Mediterranean region, and East Asia formed Cluster 2. The accessions from Cluster 1 were predominantly of desi type with a low 100-seed weight, whereas those from Cluster 2 were predominantly of kabuli type with a high 100-seed weight. Accessions in Cluster 2 took more days to flower and mature with taller plants, more tertiary branches and higher 100-seed weight than the accessions in Cluster 1. Accessions in Cluster 1 had wider plants, more basal primary branches, apical primary branches, basal secondary branches, apical secondary branches, pods per plant, seeds per pod and higher plot yield than in Cluster 2. Accessions resistant to Fusarium wilt were available from all regions but their proportion was slightly higher amongst the accessions from South Asia, South-East Asia and West Asia. This may be due to the fact that while Fusarium wilt is a widespread disease in large chickpea-growing countries (Nene et al., 1991), it is a more serious disease in these three regions (Nene and Reddy, 1987).

The reports on pigeonpea germ plasm characterization are not exhaustive and mainly focused on smaller germ plasm sets. For example, 320 accessions were characterized for five traits by Dhari et al. (1997), 168 accessions for 20 traits by Satpute et al. (1994), 75 accessions for seven traits by Singh et al. (2002), 67 accessions for 10 traits by Hazarika et al. (1986) and 40 accessions for 11 traits by Kyi et al. (2001). The ICRISAT genebank conserves 13 778 pigeonpea germ plasm accessions from 74 countries. Studies using

11 402 accessions characterized during 1976–2001 revealed a large variation in the collection for 14 qualitative and 12 quantitative traits (Upadhyaya et al., 2007). Pigeonpea is an often cross-pollinated crop with an average out-crossing of 13.1% and range of 9.7–24.1% (Githiri et al., 1991). Therefore a lot of heterogeneity (more than one descriptor state occurring in an accession) exists in the accession that have been collected from the farmers' fields. This genetic variability within accessions has been maintained, but the variance component amongst accessions is still significant for most of the traits in all the 22 years of germ plasm characterization, indicating that over the years significant diversity has been maintained in the germ plasm collection.

The accessions were grouped based on geographical proximity and similarity of the climate, and this grouping revealed a large variation for the region of South India, Maldives and Sri Lanka but minimum variation for Europe and Oceania (Reddy et al., 2005; Upadhyaya et al., 2005). Accessions from Oceania were earliest to flower (121 days) and mature (179 days) with short stature (138 cm), whereas those from Africa took the maximum number of days for flowering (162 days) and maturity (226 days) and were tall (203 cm). The number of primary branches was the highest in accessions from America (14.84) and Europe (14.70), while the accessions from South India, Maldives and Sri Lanka produced more secondary branches and racemes and had high harvest index, shelling percentage and seed yield (76.87 g per plant).

Shannon-Weaver diversity index (H`) (Shannon and Weaver, 1949) indicated that the accessions from Indonesia, Philippines and Thailand had the highest pooled H` for qualitative traits (0.349 + 0.059) and accessions from Africa for quantitative traits (0.613 + 0.006) (Upadhyaya et al., 2005). African accessions also had the highest pooled H` (0.464 + 0.039) over all the traits, while those from Oceania had the lowest pooled H` (0.337 + 0.037).

A hierarchical cluster analysis conducted on the first three PC scores (92.28% variation) resulted in three clusters (Upadhyaya et al., 2005). Semi-spreading growth habit, green stem colour, indeterminate flowering pattern and yellow flower were predominant among the qualitative traits. Primary seed colour had maximum variability, with orange colour followed by cream being two of the most frequent secondary colours in the collection.

6.2 Developing core and mini core collections

Reduced subsets, such as core collections (Frankel, 1984), that represent a high proportion of the diversity of the entire collection of a given species have been suggested as an approach to enhance the utilization of germ plasm in crop improvement programmes. Reddy et al. (2005) and Upadhyaya et al. (2001) have developed core collections of the ICRISAT crops, including chickpea and pigeonpea. When the entire collection is very large, even a 10% core collection is still unwieldy for meaningful evaluation and convenient exploitation. Upadhyaya and Ortiz (2001) postulated the concept of the mini core collection and described a two-stage strategy to develop the mini core collection. The first stage involves developing a representative core collection (about 10%) from the entire collection using all the available information, such as origin, geographical distribution and characterization data of accessions. The second stage involves an evaluation of the core collection for various morphological, agronomic and quality traits, and selection of a further subset of about 10% accessions of the original core collection. Thus the mini core collection will contain 10% of the core, or ~1% of entire collection, that represents the diversity of the entire collection (Upadhyaya and Ortiz, 2001). Upadhyaya and Ortiz (2001), Upadhyaya et al. (2006c, 2010) and Upadhyaya (2015) reported on mini core collections of chickpea and pigeonpea. A summary of these efforts is given in Table 3.

Mini core collections developed for the ICRISAT genebank serve as a gateway for accessing the genetic diversity of the entire collection by the global research community. The mini core collections are a resource for in-depth evaluation of the diversity within a collection for increased use of accessions in crop improvement programmes. So far, the genebank has provided 296 sets of mini core collections including those of chickpea and pigeonpea to scientists in 36 countries and 115 sets to scientists at ICRISAT for the evaluation and identification of promising accessions. Extensive multi-discipline evaluation of mini core collections identified new promising sources of resistance to biotic stresses, abiotic stresses, nutritional traits and grain yield in chickpea (Gaur et al., 2013; Upadhyaya et al., 2013a) and pigeonpea (Sharma et al., 2012; Upadhyaya et al., 2006a; Upadhyaya, 2015) for use in crop improvement programmes.

Some of the chickpea mini core accessions showed multiple resistances to abiotic and biotic stresses but only a few had good agronomic and seed-quality traits (Upadhyaya et al., 2013a). Early maturity and seed size are agronomically important traits, and accessions with early maturity and large seed size have been reported in chickpea (Upadhyaya et al., 2006b). In contrast, a number of accessions had no resistance to abiotic or biotic stresses but were agronomically superior and adapted to diverse environments. Both of these accession types would be a good parent in breeding programmes (Upadhyaya et al., 2013a).

More recently, large variability was observed among chickpea mini core accessions for oligosaccharides such as sucrose (3.6–54.1 mg g^{-1}), raffinose (0.2–15.1 mg g^{-1}), stachyose (2.8–59.4 mg g^{-1}) and ciceritol (4.8–90.7 mg g^{-1}). Clearly, there is a need to screen mini core collections to identify germ plasm with nutraceuticals properties for use as ingredients in functional foods for improving human health (Upadhyaya et al., 2006b).

6.3 Composite collections and reference sets for assessing molecular diversity

Assessing the range of diversity and understanding the genetic structure of gene pools is critical for the effective use of genetic resources. Past studies indicated abundant diversity among wild *cicer*, but limited variation in cultivated chickpea (Kazan and Muehlbauer, 1991; Shan et al., 2005; Rao et al., 2007) and pigeonpea (Burns et al., 2001; Odeny et al., 2007). As part of Generation Challenge Program (GCP), ICRISAT had developed global composite collections of chickpea (3000 accessions) and pigeonpea (1000 accessions)

Table 3 Summary of core and mini core collections of legumes developed at ICRISAT genebank, India

	Entire collection	No. of accessions used	No. of accessions in core	No. of accessions in mini core	Reference
Chickpea	20 764	16 991	1956	211	Upadhyaya and Ortiz, 2001
Pigeonpea	13 778	12 153	1256	146	Upadhyaya et al., 2006c

that include mini core collections (Upadhyaya et al., 2006b, 2011c,d). The GCP aimed at exploiting the genetic diversity housed in existing germ plasm collections with a focus on improving the drought tolerance in crops (www.generationcp.org). The reduced subsets (both conventional mini core collections and reference sets) are an ideal as diversity panels for studying population structure, population diversity, and association genetics for agronomically beneficial traits.

Molecular characterization of composite sets using 50 SSRs in chickpea and 20 SSRs in pigeonpea has helped in the understanding of genetic diversity and population structures in each species. In chickpea, the 48 SSR markers detected 1683 alleles in 2915 accessions, of which 935 were considered rare, 720 common and 28 most frequent. The alleles per locus ranged from 14 to 67and the polymorphic information content ranged from 0.467 to 0.974. Marker polymorphism varied between groups of accessions in the composite collection and reference set. A number of group-specific alleles were detected: 104 in Kabuli, 297 in desi, and 69 in wild *Cicer*; 114 each in Mediterranean and West Asia, 117 in South and South-East Asia and 10 in African region accessions. Desi and kabuli shared 436 alleles, while wild *Cicer* shared 17 and 16 alleles with desi and kabuli, respectively. Desi chickpea contained a higher proportion of rare alleles (53%) than kabuli (46%), while wild *Cicer* accessions were devoid of rare alleles. A genotype-based reference set captured 1315 (78%) of the 1683 composite collection alleles of which 463 were rare, 826 common and 26 the most frequent alleles (Upadhyaya et al., 2008).

In pigeonpea, analysis using 20 SSR markers data on 952 accessions detected 197 alleles, of which 115 were rare and 82 were common alleles. Gene diversity varied from 0.002 to 0.726. Sixty group-specific alleles were detected in 45 wild accessions and 64 unique alleles in 907 cultivated accessions. Geographically, 32 unique alleles were found in Southern Indian provinces, Maldives and Sri Lanka, seven in Indonesia, Philippines and Thailand, four in Northwestern Indian provinces, Iran and Pakistan, and two in Africa. Wild types as a group were genetically more diverse than the cultivated types (Upadhyaya et al., 2011c). A genotype-based reference set of 300 genetically diverse accessions have been established. The reference set captured 95% of the 197 alleles found in the composite collection. The genotype-based reference sets are ideal for allele mining, association genetics, mapping and cloning gene(s), and in applied breeding for the development of broad-based elite breeding lines/cultivars with superior yield and enhanced adaptation to diverse environments. The genetically diverse reference set offer breeders the opportunity to combine multiple resistances into an agronomically improved genetic background for genetic mapping populations for trait mapping and cultivar development.

6.4 Germ plasm enhancement through wide crosses

The narrow genetic diversity in cultivated germ plasm has hampered the effectiveness of conventional breeding as well as the development and utilization of genomic tools. Wild relatives of grain legumes are reported to harbour beneficial alleles and genes for resistance/tolerance to biotic and abiotic stresses and for increased grain yield and quality. For example some of the wild relatives of chickpea have also been reported to possess high seed protein ((Dwivedi et al., 2005; Nguyen et al., 2005; Sharma et al., 2007). Singh et al. (2005) and Dwivedi et al. (2007) found that some of the interspecific derivatives with *C. reticulatum* or *C. echinospermum* in their pedigree showed high degree of resistance to wilt, foot rot and root rot diseases and produced greater biomass and/or seed yield.

A number of the wild *Cajanus* species, especially from the secondary gene pool, that are cross compatible with cultivated pigeonpea have been used for the genetic improvement of pigeonpea (Upadhyaya et al. 2013b). The most significant achievement is the development of a unique cytoplasmic nuclear male sterility system (CMS). Saxena et al. (2010) described eight unique CMS systems with cytoplasm derived from wild *Cajanus* species. A4 cytoplasm derived from *C. cajanifolius* (Saxena et al., 2005) is stable across environments with a good fertility restoration system and has been used to develop the world's first commercial pigeonpea hybrid, ICPH 2671 (Saxena et al., 2010).

Evaluation of wild *Cajanus* species has identified accessions having resistance to alternaria blight (Sharma et al., 1987), Phytophthora blight (Rao et al., 2003), sterility mosaic virus (SMV) (Kulkarni et al., 2003), pod fly (Sharma et al., 2003), root-knot nematodes (Sharma et al., 1993; Sharma 1995), tolerance to salinity (Rao et al., 2003; Srivastava et al., 2006), drought (Rao et al., 2003) and photoperiod insensitivity (Rao et al., 2003). Wild *Cajanus* species, especially, *C. scarabaeoides, C. acutifolius, C. platycarpus, C. reticulates, C. sericeus* and *C. albicans* have been reported to have resistance to pod borer, *Helicoverpa armigera* (Mallikarjuna et al., 2007; Sujana et al., 2008; Sharma et al., 2009). Some wild *Cajanus* species namely *C. scarabaeoides, C. sericeus, C. albicans, C. crassus, C. platycarpus* and *C. cajanifolius* have higher seed protein content (average 28.3%) compared to pigeonpea cultivars (24.6%) (Singh and Jambunathan, 1981).

Utilization of wild *Cajanus* species has not only led to the development of the CMS systems and to cultivars with resistant/tolerant to biotic/abiotic stresses but also contributed significantly towards the improvement of agronomic performance and nutritional quality of cultivated pigeonpea (Upadhyaya et al., 2013b). Breeding lines with high protein content have been developed from crosses with *C. sericeus, C. albicans*, and *C. scarabaeoides*. A high protein line, ICPL 87162 with a 30–34% protein content, was developed from the cross *C. cajan* x *C. scarabaeoides* (Reddy et al., 1997).

6.5 Past use and impact of accessions conserved in *ex situ* collections

The collections of two important grain legumes, chickpea and pigeonpea, conserved at the ICRISAT genebank are a major source of variability for researchers globally. Seeds of the conserved accessions are available free of cost at the ICRISAT genebank under the Standard Material Transfer Agreement (SMTA) of International Treaty on Plant Genetic Resources for Food and Agriculture (ITPGRFA), for use in research and for training purposes. To date, the ICRISAT genebank provided 151 939 samples of chickpea to researchers in 88 countries and 74 783 samples of pigeonpea to researchers in 113 countries. Within ICRISAT, 190 636 samples of chickpea and 86 900 samples of pigeonpea were provided to researchers.

Globally the collections in gene banks are large but only a few accessions (<1%) have been used in crop improvement programmes. A large gap exists between the availability and actual utilization of the conserved accessions. This is true for CGIAR institutes as well as for the national institutes. At ICRISAT, during 35 years, only eight pigeonpea accessions were supplied more than 200 times. In chickpea, only 91 accessions were used in developing breeding lines (Upadhyaya et al., 2006a).

Despite the constraints to the use of the conserved accessions, samples of grain legumes supplied from ICRISAT genebank have been used as raw materials in crop improvement programmes and academic studies globally. Overall crops that ICRISAT conserves, national institutes have released more than 1000 varieties in 81 countries utilizing germ plasm and breeding lines from ICRISAT that are contributing to global food security. Some of the elite accessions have been released directly as cultivars, 22 chickpea accessions in 23 countries and 11 pigeonpea accessions in 7 countries. ICC 11879, a chickpea landrace, was released as a variety in eight Mediterranean countries and ICC 13816 was released as a variety in seven countries. A vegetable pigeonpea landrace from India (ICP 7035) has been released as cultivar in India, Fiji, Nepal, China and Philippines. Wilt-resistant pigeonpea landrace ICP 8863 that was released as Maruti in India generated benefits of US$ 75 million by 1996 with a 73% internal rate of return (Bantilan and Joshi 1996). In addition, the grain legume collections at ICRISAT genebank have served as an insurance against the genetic erosion globally. The genebank was able to restore chickpea accessions to Ethiopia (931) and India (7488) and pigeonpea accessions to India (5977) and Sri Lanka (71), when national gene banks have lost their local collections due to natural calamities, strife, etc.

7 Conclusion

To ensure that future users will have this diversity available for grain legumes, such as chickpeas and pigeonpeas as well as all the other crops, there are urgent needs to better secure their conservation. The crop communities for beans, chickpeas, cowpeas, pigeonpeas, grasspeas, and faba beans have developed global *ex situ* conservation strategies. The global conservation strategies concluded that the current system for *ex situ* conservation of grain legumes is not global but consist of poorly linked international and national collections which hold some unknown degree of common and unique accession, that are conserved with limited sustained support and not very well known or available to users. Some of the key needs for the global conservation system related to the regeneration of accessions, safety duplication of collections, a global information system to enhance the use of conserved germ plasm, global rationalization of accessions, and collections to fill significant gaps. A review of the characterization and utilization of diversity within chickpeas and pigeonpeas found that *ex situ* collections are currently underutilized but with the identification of subsets, more intense evaluation, and the application of the new tools from genomics, the exploitation of these collections is projected to increase. The conserved collections of grain legumes are critical resources for future food systems. Thus, there is a need to consider a new model for managing global resources in a secure, cost-effective, sustainable global system.

8 References

Bantilan, M. C. S. and P. K. Joshi. 1996. *Returns to Research and Diffusion Investments on Wilt Resistance in Pigeonpea. (In En. Summaries in En, Fr.)*. Impact series no. 1. International Crops Research Institute for the Semi-Arid Tropics, Patancheru 502 324, Andhra Pradesh, India, 36p.

Bramel, P. J. 2017. Wheat genetic resources: Global conservation and use for the future. In Langridge, P. (Ed.), *Achieving Sustainable Cultivation of Wheat Vol I*. Burleigh Dodds Science Publishing, Cambridge, UK, pp. 3–24

Burns, M. J., K. J. Edwards, H. J. Newbury, B. V. Ford-Lloyd and C. D. Baggot. 2001. Development of simple sequence repeat (SSR) markers for the assessment of gene flow and genetic diversity in pigeonpea (*Cajanus cajan* L.). *Molecular Ecology Notes* 1: 283–5.

Debouck, D. G. 2014 (Draft). Conservation of Phaseolus beans genetic resources: A strategy. *Global Crop Diversity Trust*. https://www.croptrust.org/our-work/supporting-crop-conservation/conservation-strategies/.

Dhari, R., K. Kumar, Y. S. Tomer and R. P. S. Kharb. 1997. Evaluation of pigeonpea germplasm for yield and its component traits. *Annals of Biology* 13: 221–3.

Dumet, D. and C. Fatokun. 2011. Global strategy for the conservation of cowpea (Vigna unguiculata subsp. unguiculata). *Global Crop Diversity Trust*. https://www.croptrust.org/our-work/supporting-crop-conservation/conservation-strategies/.

Dwivedi, S. L., M. W. Blair, H. D. Upadhyaya, R. Serraj, J. Balaji, H. K. Buhariwalla, R. Ortiz and J. H. Crouch. 2005. Using pigeonpea genomics to exploit grain legumes biodiversity in crop improvement. *Plant Breeding Reviews* 26: 171–357.

Dwivedi, S. L., H. D. Upadhyaya and C. L. L. Gowda. 2009. Approaches to enhance the value of genetic resources in crop improvement. The FAO Regional Project 'Capacity building and enhanced regional collaboration for the conservation and sustainable use of plant genetic resources in Asia'. (http://www.fao.org/docrep/014/am265e00.pdf).

Dwivedi, S. L., H. D. Upadhyaya, H. T. Stalker, M. W. Blair, D. J. Bertioli, S. Nielen and R. Ortiz. 2007. Enhancing crop genepools with beneficial traits using wild relatives. *Plant Breeding Reviews* 30: 179–230.

Engels, J. M. M. and L. Maggioni. 2011. *AEGIS: A Regionally Based Approach to PGR Conservation*. CAB International, Wallingford, UK.

FAO. 1996. *The Global Plan of Action for the Conservation and Sustainable Utilization of Plant Genetic Resources for Food and Agriculture*. FAO, Rome, Italy. http://www.fao.org/agriculture/crops/thematic-sitemap/theme/seeds-pgr/gpa/gpa-background/en.

FAO. 2009. *The International Treaty for Plant Genetic Resources for Food and Agriculture*. FAO, Rome, Italy. ftp://ftp.fao.org/docrep/fao/011/i0510e/i0510e.pdf.

FAO. 2011a. *Second Global Plan of action for Plant Genetic Resources for Food and Agriculture*. Commission on Genetic Resources for Food and Agriculture Food and Agriculture Organization of the United Nations (FAO), Rome. http://www.fao.org/agriculture/crops/thematic-sitemap/theme/seeds-pgr/gpa/en.

FAO. 2011b. FAO-World Information and Early Warning on PGRFA. http://www.fao.org/wiews-archive/germplasm_query.htm?i_l=EN.

FAO. 2014a. *Genebank Standards for Plant Genetic Resources for Food and Agriculture*. Rev ed. FAO, Rome.

FAO. 2014b. *Pulses in Food Outlook: Biannual Report on Global Food Markets*, FAO, Rome, pp. 63–7.

Foyer, C. H., H.-M. Lam, H. T. Nguyen, K. H. M. Siddique, R. K. Vershney, T. D. Colmer, W. Cowling, H. Bramley, T. A. Mori, J. M. Hodgson, J. W. Cooper, A. J. Miller, K. Kunert, J. Vorster, C. Cullis, J. A. Ozga, M. L. Wahlqvist, Y. Liang, H. Shou, K. Shi, J. Yu, N. Fodor, B. N. Kaiser, F-L Wong, B. Villiyodan and M. J. Considine. 2016. Neglected legumes has compromised human health and sustainable food production. *Nature Plants* 2:Article 16112.

Frankel, O. H. 1977. Natural variation and its conservation, pp. 21–4. In A. Muhammed and R. C. von Botstel (Eds), *Genetic Diversity of Plants*. Plenum Press, New York.

Frankel, O. H. 1984. Genetic perspective of germplasm collection. In W. Arber, K. Limensee, W. J. Peacock and P. Stralinger (Eds), *Genetic Manipulations: Impact on Man and Society*. Cambridge University Press, Cambridge, pp. 61–170.

Gaur, P. M., A. K. Jukanti, S. Samineni, S. K. Chaturvedi, S. Singh, S. Tripathi, I. Singh, G. Singh, T. K. Das, M. Aski, N. Mishra, N. Nadarajan and C. L. L. Gowda. 2013. Large variability in chickpea for tolerance to herbicides imazethapyr and metribuzin. *Agronomy* 3: 524–36.

Gepts, P. 2006. Plant genetic resource conservation and utilization: The accomplishment and future of a societal insurance policy. *Crop Science* 46: 2278–92.

Girish, A. G., Singh, S. D., Chakraborty, S. K., Prasada Rao, R. D. V. J., Surender, A., Varaprasad, K. S. and Bramel-Cox, P. J. 2001. Seed-associated microflora of ICRISAT mandate crops. *Seed Science and Technology* 29: 429–43.

Githiri, S. M., P. M. Kimani and K. N. Saxena. 1991. Natural out-crossing in dwarf pigeonpea. *Euphytica* 53(1): 37–9.

Harlan, J. R. and J. M. J. De Wet. 1971. Toward a rational classification of cultivated plants. *Taxon* 20(4): 509–17.

Hawtin, G. 2008a. Global strategy for the ex situ conservation of chickpeas (Cicer L.). *Global Crop Diversity Trust.* https://www.croptrust.org/our-work/supporting-crop-conservation/conservation-strategies/.

Hawtin, G. 2008b. Global strategy for the ex situ conservation of faba beans (Vicia faba L.). *Global Crop Diversity Trust.* https://www.croptrust.org/our-work/supporting-crop-conservation/conservation-strategies/.

Hawtin, G. 2008c. Global strategy for the ex situ conservation of Lathyrus (grasspea), with special reference to Lathyrus sativus, L. cicero, L. ochrus. *Global Crop Diversity Trust.* https://www.croptrust.org/our-work/supporting-crop-conservation/conservation-strategies/.

Hawtin, G. 2008d. Global strategy for the ex situ conservation of lentil (Lenns Miller). *Global Crop Diversity Trust.* https://www.croptrust.org/our-work/supporting-crop-conservation/conservation-strategies/.

Hazarika, G. N., V. P. Singh and B. P. S. Malik. 1986. Divergence analysis in pigeonpea. *Crop Improvement* 13: 113–16.

Hazenkamp, Th., T. S. Payne and N. R. Sackville Hamilton. 2014. Assessing rice and wheat germplasm collections using similarity groups. *Genetic Resource Crop Evaluation* 61: 841–51.

IBPGR, ICRISAT and ICARDA, 1993. *Descriptors for Chickpea (Cicer arietinum L.).* International Board for Plant Genetic Resources, Rome, Italy; International Crops Research Institute for the Semi-Arid Tropics, Patancheru; India and International Center for Agriculture Research in the Dry Areas, Aleppo, Syria.

Kazan, K. and F. J. Muehlbauer. 1991. Allozyme variation and phylogeny in annual species of *Cicer* (leguminosae). *Plant Systematics and Evolution* 175: 11–21.

Khory, C. K., H. A. Achicanoy, A. D. Bjorkman, C. Navarro-Racines, L. Guarino, X. Flores-Palacios, J. M. M. Engels, J. H. Wiersema, H. Dempewolf, J. Ramírez-Villegas, N. P. Castañeda-Álvarez, C. Fowler, A. Jarvis, L. H. Rieseberg and P. C. Struik. 2015. Estimation of countries' interdependence in plant genetic resources provisioning national food supplies and production systems. *Research Study* 8, ITPGRFA. http://www.planttreaty.org/sites/default/files/files/ITPGRFA_Interdependence_text_2015_9_18_0_Main%20text.pdf.

Kigel, J., L. Rosental and A. Fait. 2015. Seed physiology and germination of grain legumes. *Handbook of Plant Breeding* 10(11): 327–63.

Kulkarni, N. K., A. S. Reddy, P. Lava Kumar, J. Vijaynarasimha, K. T. Rangaswamy, V. Muniappa, L. J. Reddy, K. B. Saxena, A. T. Jones and D. V. R. Reddy. 2003. Broad-based resistance to pigeonpea sterility mosaic disease in accessions of *Cajanus scarabaeoides* (L.) Benth. *Indian Journal of Plant Protection* 31: 6–11.

Kyi, K. M., T. M. Than, W. Twat, Y. Y. Myint, M. Ito and K. Irie. 2001. Characterization and preliminary evaluation of pigeonpea germplasm in Myanmar. *International Chickpea and Pigeonpea Newsletter* 8: 27–8.

Mallikarjuna, N., H. C. Sharma and H. D. Upadhyaya. 2007. Exploitation of wild relatives of pigeonpea and chickpea for resistance to *Helicoverpa armigera*. *SAT eJournal* 3(1): 4.

Marshall, D. R. 1989. Limitations to the use of germplasm collections. In A. H. D. Brown, O. H. Frankel, D. R. Marshall and J. T. Williams (Eds) *The Use of Plant Genetic Resources*. Cambridge University Press, New York, pp. 105–20.

McCouch, S., G. J. Baute, J. Bradeen, P. Bramel, P. K. Bretting, E. Buckler, J. M. Burke, D. Charest, S. Cloutier, G. Cole, H. Dempewolf, M. Dingkuhn, C. Feuillet, P. Gepts, D. Grattapaglia, L. Guarino, S. Jackson, S. Knapp, P. Langridge, A. Lawton-Rauh, Q. Lijua, C. Lusty, T. Michael, S. Myles, K. Naito, R. L. Nelson, R. Pontarollo, C. M. Richards, L. Rieseberg, J. Ross-Ibarra, S. Rounsley, R. S. Hamilton, U. Schurr, N. Stein, N. Tomooka, E. van der Knaap, D. van Tassel, J. Toll, J. Valls, R. K. Varshney, J. Ward, R. Waugh, P. Wenzl and D. Zamir. 2013. Agriculture: Feeding the future. *Nature* 499: 23–4.

Moreno, M. T. and J. I. Cubero. 1978. Variation in *Cicer arietinum* L. *Euphytica* 27: 465–85.

Murty, B. R. 1975. Biology of adaptation in chickpea. In *International Workshop on Grain Legumes (Pigeonpeas and Chickpeas)*. International Crops Research Institute for the Semi-Arid Tropics (ICRISAT), Hyderabad, India, pp. 239–51.

Narayan, R. K. J. and A. J. Macefield. 1976. Adaptive responses and genetic divergence in world germplasm collection of chick pea (*Cicer arietinum* L.). *Theoretical Applied Genetics* 47: 179–87.

Nass, L. L. and E. Paterniani. 2000. Pre-breeding: A link between genetic resources and maize breeding. *Scientia Agricola* 57: 581–7.

Nene, Y. L., M. V. Reddy, M. P. Haware, A. M. Ghanekar and K. S. Amin. 1991. Field diagnosis of chickpea diseases and their control. Information Bulletin no. 28. International Crops Research Institute for the Semi-Arid Tropics, Patancheru, India.

Nene, Y. L. and M. V. Reddy. 1987. Chickpea diseases and their control. In M. C. Saxena and K. B. Singh (Eds), *The Chickpea*. CAB International, Wallingford, UK, pp. 233–70.

Nguyen, T. T., P. W. J. Taylor, R. J. Redden and R. Ford. 2005. Resistance to Ascochyta rabiei (Pass.) Lab in a wild *Cicer* germplasm collection. *Australian Journal of Experimental Agriculture* 45: 1291–6.

Odeny, D. A., B. Jayasree, M. Ferguson, D. Hoisington, J. Crouch and C. Gebhardt. 2007. Development, characterization and utilization of microsatellite markers in pigeonpea. *Plant Breeding* 126: 130–6.

Rao, L. S., P. U. Rani, P. S. Deshmukh, P. A. Kumar and S. K. Panguluri. 2007. RAPD and ISSR finger printing in cultivated chickpea (*Cicer arietinum* L.) and its wild progenitor *Cicer reticulatum* Ladizinsky. *Genetic Resources and Crop Evolution* 54:1235–44.

Rao, N. K., L. J. Reddy, and P. J. Bramel. 2003. Potential of wild species for genetic enhancement of some semi-arid food crops. *Genetic Resources and Crop Evolution*, 50, 707–21.

Reddy, L. J., K. B. Saxena, K. C. Jain, U. Singh, J. M. Green, D. Sharma, D. J. Faris, A. N. Rao, R. V. Kumar, and Y. L. Nene. 1997. Registration of high-protein pigeonpea elite germplasm ICPL 87162. *Crop Science*, 37: 294.

Reddy, L. J., H. D. Upadhyaya, C. L. L. Gowda and S. Singh. 2005. Development of core collection in pigeonpea [*Cajanuscajan* (L.) Millsp.] using geographical and qualitative morphological descriptors. *Genetic Resources and Crop Evolution*, 52: 1049–56.

Satpute, R. G., D. Khare and P. Das. 1994. Collection, evaluation, and maintenance of pigeonpea germplasm from Madhya Pradesh, India. *International Chickpea and Pigeonpea Newsletter* 1: 26–8.

Saxena, K. B., R. V. Kumar, V. A. Dalvi, N. Mallikarjuna, C. L. L. Gowda, B. B. Singh, S. B. S. Tikka, K. B. Wanjari, L. B. Pandet, L. M. Paralkar, M. K. Patel, B. Shiying and Z. Xuxiao. 2005. Hybrid breeding in grain legumes – a success story of pigeonpea. In M. C. Khairwal and H. K. Jain (Eds), *Proceedings of the International Food Legumes Research Conference*, New Delhi, India.

Saxena, K. B., R. Sultana, N. Mallikarjuna, R. K. Saxena, R. V. Kumar, S. L. Sawargaonkar, and R. K. Varshney. 2010. Male-sterility systems in pigeonpea and their role in enhancing yield. *Plant Breeding* 129, 125–34.

Shan, F., H. C. Clarke, J. A. Plummer, G. Yan and K. H. M. Siddique. 2005. Geographical patterns of genetic variation in the world collection of wild annual *Cicer* characterized by amplified fragment length polymorphisms. *Theoretical Applied Genetics* 110: 381–91.

Shannon, C. E. and W. Weaver. 1949. *The Mathematical Theory of Communication*. Univ Illinois Press, Urbana.

Sharma, H. C., C. L. L. Gowda, P. C. Stevenson, T. J. Ridsdill-Smith, S. L. Clement, G. V. R. Rao, J. Romeus, M. Miles and M. El-Bouhssini. 2007. Host plant resistance and pest management in chickpea. In S. S. Yadav, R. J. Redden, W. Chen and B. Sharma (Eds) *Chickpea Breeding and Management*. CABI, Nosworthy Way, Wallingford, Oxfordshire, UK, pp. 520–37.

Sharma, D., J. Kannaiyan and K. B. Saxena. 1987. Sources of resistance to alternaria blight in pigeonpea. *SABRAO Journal of Breeding and Genetics* 19(2), 109–14.

Sharma, H. C., G. Pampapathy and L. J. Reddy. 2003. Wild relatives of pigeonpea as a source of resistance to the podfly (*Melalagromyza* obtuse Malloch) and pod wasp (*Tranaostigmodes cajaninae* La Salle). *Genetics Resource and Crop Evolution* 50: 817–24.

Sharma, H. C., G. Sujana and D. M. Rao. 2009. Morphological and chemical components of resistance to pod borer, *Helicoverpa armigera* in wild relatives of pigeonpea. *Arthropod – Plant Interactions* 3(3): 151–61. Springer SBM, Dordrecht, Netherlands.

Sharma, M., A. Rathore, U. N. Mangala, R. Ghosh, S. Sharma, H. D. Upadhyaya and S. Pande. 2012. New sources of resistance to fusarium wilt and sterility mosaic in a mini core collection of pigeonpea germplasm. *Europian Journal of Plant Pathology* 133: 707–14.

Sharma, S. B. 1995. Resistance to *Rotylenchulus reniformis, Heterodera cajani* and *Meloidogyne javanica* in accessions of *Cajanus platycarpus. Plant Disease* 79: 1033–5.

Sharma, S. B., P. Remanandan and K. C. Jain. 1993. Resistance to cyst nematode (*Heterodera cajani*) in pigeonpea cultivars and wild relatives of *Cajanus. Annals of Applied Biology* 123: 75–81.

Siddique, K. H. M., K. L. Regen, D. Tennant and B. D. Thomson. 2001. Water use and water use efficiency of cool season legumes in low rainfall Mediterranean-type environment. *Europian Journal of Agronomy* 15: 267–80.

Singh, A. K., N. Singh, S. P. Singh, N. B. Singh and J. Smartt. 2006. Pigeonpea. In B. S. Dhillon et al. (Eds) *Plant Genetic Resources: Food Grains*. Narosa Publishing House, New Delhi, 2005, pp. 222–37.

Singh, K. B., R. P. S. Pundir, L. D. Robertson, H. A. van Rheenen, U. Singh, T. J. Kelley, P. Parthasarathy Rao, C. Jhoanson and N. P. Saxena. 1997. Chapter 8: Chickpea. In D. Fuccillo, L. Sears and P. Stapleton (Eds), *Biodiversity in Trust*. Cambridge University Press, The Edinburgh Building, Cambridge CB 2 2RU, United Kingdom, pp. 100–13.

Singh, L., C. L. L. Gowda and H. D. Upadhyaya. 2011. Strategy for the global conservation of pigeonpea genetic resources. Global Crop Diversity Trust and ICRISAT.

Singh, M., I. S. Bisht, S. Kumar, M. Dutta, K. C. Bansal, M. Karale, A. Sankar, A. Amri, S. Kumar and S. K. Datta. 2014. Global wild annual Lens collections: A potential resources for lentil base broadening and yield enhancement. *PLoS ONE* 9(9): e107781. doi:10.1371/journal.pone.0107781.

Singh, S., R. K. Cumber, N. Joshi and K. Singh. 2005. Introgression from wild *Cicer reticulatum* to cultivated chickpea for productivity and disease resistance. *Plant Breeding* 124: 477–80.

Singh, S. P., U. K. Tripathi, D. N. Ram, S. Kumar and S. Kumar. 2002. Evaluation of germplasm in pigeonpea [*Cajanus cajan* (L.) Millsp]. *Progressive Agriculture* 2(2): 163–5.

Singh, U. and R. Jambunathan. 1981. Protease inhibitors and *in vitro* protein digestibility of pigeonpea [*Cajanus cajan* (L.) Millsp.] and its wild relatives. *Journal of Food Science and Technology* 18: 246–7.

Smartt, J. 1984. Gene pools in grain legumes. *Journal of Economic Botany* 38(1): 24–35.

Smykal, P., C. J. Coyne, M. J. Ambrose, N. Maxted, H. Schaefer, M. W. Blair, J. Berger, S. L. Greene, M. N. Nelsen, N. Besharat, T. Vymyslicky, C. Toker, R. K. Saxena, M. Roorkiwal, M. K. Pandey, J. Hu, Y. H. Li, L. X. Wang, Y. Guo, L. J. Qiu, R. J. Redden and R. K. Varshney. 2015. Legume crops phylogeny and genetic diversity for science and breeding. *Critical Reviews in Plant Sciences* 34: 1–3, 43–104. doi:10.1080/07352689.2014.897904.

Srivastava, N., V. Vadez, H. D. Upadhyaya and K. B. Saxena. 2006. Screening for intra and inter specific variability for salinity tolerance in Pigeonpea (*Cajanus cajan*) and its related wild species. *Journal of SAT Agricultural Research*. http: icrtest:8080/Journal/crop improvementv2il/v2ilscreening for.Pdf.

Sujana, G., H. C. Sharma and D. M. Rao. 2008. Antixenosis and antibiosis components of resistance to pod borer *Helicoverpa armigera* in wild relatives of pigeonpea. *International Journal of Tropical Insect Science* 28(4): 191–200.

Upadhyaya, H. D. 2003. Geographical patterns of variation for morphological and agronomic characteristics in the chickpea germplasm collection. *Euphytica* 132: 343–52.

Upadhyaya, H. D. 2015. Establishing core collections for enhanced use of germplasm in crop improvement. *Journal of Crop Breeding and Genetics* 1–1:1–12.

Upadhyaya, H. D. and R. Oritz. 2001. A mini core subset for capturing diversity and promoting utilization of chickpea genetic resources in crop improvement. *Theoretical Applied Genetics* 102: 1292–98.

Upadhyaya, H. D., P. J. Bramel and S. Singh. 2001. Development of a chickpea core subset using geographic distribution and quantitative traits. *Crop Science* 41: 206–10.

Upadhyaya, H. D., C. L. L. Gowda, H. K. Buhariwalla and J. H. Crouch. 2006a. Efficient use of crop germplasm resources: Identifying useful germplasm for crop improvement through core and min-core collections and molecular marker approaches. *Plant Genetic Resources* 4(1): 25–35.

Upadhyaha, H. D., C. L. L. Gowda, R. P. S. Punir and B. R. Ntare. 2007. Use of core and mini-core collections in preservation and utilization of genetic resources in crop improvement. In Vodouhe, R., Atta-Krah, K., Achigan-Dako, G. E., Eyog-Matig, O. and Avohou, H. (Eds) *Plant genetic resources and food security in West and Central Africa Regional Conference*, 26–30 April 2004. Bioversity, Rome, 365p.

Upadhyaya, H. D., Dronavalli, N., S. L. Dwivedi, J. Kashiwagi, L. Krishnamurthy, S. Pande, H. C. Sharma, V. Vadez, S. Singh, R. K. Varshney and C. L. L. Gowda. 2013a. Mini core collection as a resource to identify new sources of variation. *Crop Science* 53: 2506–17. doi:10.2135/cropsci2013.04.0259.

Upadhyaya, H. D., S. L. Dwivedi, M. Ambrose, N. Ellis, J. Berger, P. Smykal, D. Debouck, G. Duc, D. Dumet, A. Flavell, S. K. Sharma, N. Mullikarjuna, and C. L. L. Gowda. 2011a. Legume genetic resources: Management, diversity assessment, and utilization in crop improvement. *Euphytica* 180: 27–47.

Upadhyaya, H. D., S. L. Dwivedi, M. Baum, R. K. Varshney, S. M. Udupa, C. L. L. Gowda, D. Hoisington and S. Singh. 2008. Genetic structure, diversity and allelic richness in composite collection and reference set in chickpea (*Cicer arietinum* L.). *BMC Plant Biology*. doi:10.1186/1471-2229-8-106.

Upadhyaya, H. D., B. J. Furman, S. L. Dwivedi, S. M. Udupa, C. L. L. Gowda, M. Baum, J. H. Crouch, H. K. Buhariwalla and S. Singh. 2006b. Development of a composite collection for mining germplasm possessing allelic variation for beneficial traits in chickpea. *Plant Genetic Resources* 4: 13–19. doi:10.1079/PGR2005101.

Upadhyaya, H. D., R. P. S. Pundir, C. L. L. Gowda, K. N. Reddy and S. Singh. 2005. Geographical patterns of diversity for qualitative and quantitative traits in the pigeonpea germplasm collection. *Plant Genetic Resources* 3(3): 331–52.

Upadhyaya, H. D., K. N. Reddy, M. Irshad Ahmed, C. L. L. Gowda, M. Thimma Reddy and S. Ramachandran. 2015. Identification of gaps in pigeonpea germplasm from East and Southern Africa conserved at the ICRISAT genebank. *Indian Journal of Plant Genetic Resources* 28(2): 180–8.

Upadhyaya, H. D., K. N. Reddy, R. P. S. Pundir, S. Singh, C. L. L. Gowda and M. Irshad Ahmed. 2011b. Diversity and geographical gaps in *Cajanus scarabaeoides* (L.) Thou. germplasm conserved at the ICRISAT genebank. *Plant Genetic Resources: Characterization and Utilization* 11(1): 3–14. doi:10.1017/S1479262111000736.

Upadhyaya, H. D., K. N. Reddy, S. Shivali, R. K. Varshney, R. Bhattacharjee, S. Singh and C. L. L. Gowda. 2011c. Pigeonpea composite collection and identification of germplasm in crop improvement programmes. *Plant Genetic Resources: Characterization and Utilization* 9(1): 97–108.

Upadhyaya, H. D., K. N. Reddy, S. Singh and C. L. L. Gowda. 2013b. Phenotypic diversity in *Cajanus* species and identification of promising sources for agronomic traits and seed protein content. *Genetic Resource and Crop Evolution* 60: 639–59. doi:10.1007/s10722-012-9864-0.

Upadhyaya, H. D., L. J. Reddy, C. L. L. Gowda, K. N. Reddy and S. Singh. 2006c. Development of a mini core subset for enhanced and diversified utilization of pigeonpea germplasm resources. *Crop Science* 46: 2127–32.

Upadhyaya, H. D., M. Thudi, N. Dronavali, N. Gujaria, S. Singh, S. Sharma and R. K. Varshney. 2011d. Genomic tools and germplasm diversity for chickpea improvement. *Plant Genetic Resources* 9: 45–58.

Upadhyaya, H. D., D. Yadav, N. Dronavalli, C. L. L. Gowda and S. Singh. 2010. Mini core germplasm collection for infusing genetic diversity in plant breeding programs. *Electronic Journal of Plant Breeding* 1(4): 1294–309.

van Hintum, Th., F. Menting and E. van Strien. 2011. Quality indicators for passport data in ex situ genebanks. *Plant Genetic Resources* 9: 478–85.

Chapter 4

Advanced breeding techniques for grain legumes in the genomics era

Juan M. Osorno and Phillip E. McClean, North Dakota State University, USA; and Timothy Close, University of California (Riverside), USA

1 Introduction

Fabaceae (formerly Leguminosae) is the second most economically important botanical family for human consumption. However, historically grain legumes (a.k.a. pulses) have received less attention, resources and scientific efforts towards crop improvement, especially when compared to cereal crops (Foyer et al., 2016; Sivasankar et al., 2016). This is also reflected in the fact that the grain legume scientific community is significantly smaller at the global scale. Despite these limitations, modest increases in seed yield and productivity have been accomplished for most grain legumes (CGIAR, 2012; Curran et al., 2017; Vandemark et al., 2014). The challenge is to continue increasing productivity while reducing the significant seed yield gap between developed and developing countries/ regions. In addition, the economic and nutritional relevance of grain legumes has become a more prominent topic in recent years (Joshi and Rao, 2017; McDermott and Wyatt, 2017). The Food and Agriculture Organization (FAO) declared 2016 the International Year of Pulses (http://iyp2016.org/; http://www.fao.org/pulses-2016/en/). The reader may refer to other chapters within this book for more extensive information on the economic and social relevance of grain legumes today.

http://dx.doi.org/10.19103/AS.2017.0023.03

Most grain legumes belong to the Faboideae (Syn. Papilionoideae) subfamily which includes more than 500 genera and 14 000 species distributed across diverse geographical regions and ecosystems worldwide (Smartt, 1990). For example, over 77 species have been described for *Phaseolus* spp. (Freytag and Debouck, 2002), and more than 50 market classes or seed types have been reported within *P. vulgaris*, which is the most important grain legume worldwide (Voysest, 2000). In terms of economic importance and use, FAO recognizes at least 10 pulses or grain legume groups (FAO, 1994, 2016):

1 Dry beans (*Phaseolus* spp. including several species now in *Vigna*); common bean, kidney bean, navy bean, pinto bean, haricot bean (*Phaseolus vulgaris*); lima bean, butter bean (*Phaseolus lunatus*); adzuki bean (a.k.a. azuki bean) (*Vigna angularis*), mung bean, golden gram, green gram (*Vigna radiata*); black gram, urd bean (*Vigna mungo*); scarlet runner bean (*Phaseolus coccineus*); rice bean (*Vigna umbellata*); moth bean (*Vigna aconitifolia*); tepary bean (*Phaseolus acutifolius*); and cowpea (a.k.a. niébé, black-eyed pea, blackeye, southern pea, blackeye pea, crowder pea, lubia, coupe, frijole, long bean) (*Vigna unguiculata* L. *Walp*). Cowpea is reported separately in countries where only cowpea is produced.
2 Dry broad beans (*Vicia faba*), horse bean (*Vicia faba equina*), faba bean or broad bean (*Vicia faba*), field bean (*Vicia faba*).
3 Dry peas (*Pisum* spp.), garden pea (*Pisum sativum* var. *sativum*), protein pea (*Pisum sativum* var. *arvense*).
4 Chickpea, garbanzo, bengal gram (*Cicer arietinum*).
5 Pigeon pea, arhar/toor, cajan pea, Congo bean, gandules (*Cajanus cajan*).
6 Lentil (*Lens culinaris*).
7 Bambara groundnut, earth pea (*Vigna subterranea*).
8 Vetch, common vetch (*Vicia sativa*).
9 Lupins (*Lupinus* spp.).
10 Minor pulses (a.k.a. pulses nes – not elsewhere specified), including lablab, hyacinth bean (*Lablab purpureus*), Jack bean (*Canavalia ensiformis*), sword bean (*Canavalia gladiata*), winged bean (*Psophocarpus tetragonolobus*), velvet bean, cowitch (*Mucuna pruriens* var. *utilis*), yam bean (*Pachyrhizus erosus*).

Other minor grain legumes not included within the statistics compiled by the FAO may include horse gram (*Macrotyloma uniflorum* (Lam.) Verdc.), fenugreek (*Trigonella* spp.) and gemsbok bean (*Tylosema esculentum* (Burch.) A. Schreib.), among others (Belamkar et al., 2015; Bennett et al., 2001; Singh et al., 2014). In addition, there are several legume tree species in which seeds are also consumed in a similar way as grain legumes.

Given the fact that FAO statistical databases pool together several grain legume species within the same class, it is sometimes hard to come up with accurate estimates by specific regions or countries since there may be different preferred grain legumes depending on the geographical area. For example, both *Phaseolus* beans and species now belonging to the *Vigna* genera (e.g. golden, green and black gram, mung bean, adzuki bean, cowpea, among others) are grouped within the same category (dry beans) because of their suggested common evolutionary past (Goel et al., 2002; Verdcourt, 1970, 1980). Therefore, it is difficult to obtain reliable estimates of production of

Phaseolus and *Vigna* separately (Joshi and Rao, 2017; Beebe, 2012). One of the best attempts to obtain reliable estimates is the one made by Akibode and Maredia (2011). Similar situations can be found in other groups of grain legumes created by FAO and therefore, caution should be exercised when reporting production statistics of these species.

Grain legumes have a long, interesting evolution and domestication history that spans across many regions and ethnic groups globally (Albala, 2007; Muñoz et al., 2017). Grain legumes show incredible phenotypic plasticity and adaptation to diverse environments worldwide as they can be found across almost all latitudes and between zero and 3500 masl (Pratap and Kumar, 2011). The first evidence of grain legumes comes from ~11 000 years ago in the Fertile Crescent, the Mediterranean Basin and the Middle East regions that were home to some of the earliest human civilizations (Muehlbauer and Kaiser, 1992; Perez de la Vega et al., 2012). In the same way, common bean was one of the early crops adopted and domesticated in the Americas ~8 000 years ago (Bitocchi et al., 2012, 2017; Schmutz et al., 2014).

Another major contrast found among grain legumes is the difference between cool versus warm season legume species. This is mostly related to the ability of some legume species to tolerate some light frost and grow/thrive under temperate climates (Muehlbauer, 1993). Cool season legumes grow vegetatively during the cool season and flower and produce seeds as daylengths become progressively longer. Pea, chickpea, lentil, faba bean and grass pea are typical examples of cool season legumes. On the other hand, warm season legumes are much more sensitive to frost and they also require a larger amount of heat units in order to grow normally. Warm season pulses are characterized by epigeal germination, a period of rapid vegetative growth, followed by flowering when daylengths become progressively shorter during the growing season. Common bean, mung bean, cowpea and pigeon pea are good examples of warm season legumes.

Regardless of their environmental preference, grain legume breeding tools and methods are somewhat common among them and can be used across species in many cases. With few exceptions, most grain legume species are predominantly self-pollinated. Consequently, most breeding techniques are geared towards inbred line development. With the exception of few legume species such as soybean (*Glycine max* L.) and peanut (*Arachis hypogea* L.), which are mostly used for oil and animal feeding, and legumes used as horticultural crop (e.g. snap bean, green peas, etc.), most breeding efforts still remain within the public institutions and governments. A combination of factors may explain the lack of interest by the private sector. These include the lack of economic resources of grain legume farmers in developing countries, lower commodity prices, the lack of adequate seed systems, lack of a culture that values superior quality seed, among others. However, there are examples of successful private enterprises that provide certified seed at various quality levels. For example, some farmer associations in developing countries are well organized and have started their own quality-declared seed production systems in which their best growers are hired to produce high quality seed for the rest of the associates. Similarly, there are many successful examples of private seed operations in developing countries providing seed of the highest quality standards.

2 Main grain legume breeding programmes

As mentioned before, most of the breeding efforts for grain legumes occur within the public sector. Breeding programmes are diverse and vary by region and target area. Grain legume breeding efforts within the CGIAR system are mostly split between ICRISAT, ICARDA, IITA and CIAT (http://grainlegumes.cgiar.org/). ICRISAT focuses on chickpea, groundnut and pigeon pea; ICARDA on kabuli chickpea, lentil, faba bean and grass pea and is exploring expansion into field pea; IITA focuses on cowpea, soybean (mostly for human consumption) and more recently bambara groundnut; and CIAT focuses on common bean. These programmes rely heavily on the network of national programmes already existent within each country as an outlet for the improved varieties. This system has proven successful in many cases, but an unintended consequence of this system is that it creates high dependence of the national programmes for new genetics coming from these centres. Several national programmes in Latin America, Africa and Asia make their own crosses and develop their own locally adapted populations and varieties using both local and imported germplasm. With very few exceptions, it is hard to find a single cultivar with wide geographical adaptation and therefore, this combination of local+CGIAR material approach helps to accelerate increases in local/regional productivity.

In many cases, research programmes at universities are conducting most of the breeding efforts for grain legumes. This is the typical case in North America (the United States, Canada), Europe and Oceania. Private breeding companies also play a major role by releasing improved cultivars in these countries. In some regions, cultivars coming from private companies are even more popular than the public ones, proving that private investments in grain legume breeding and seed systems are profitable if reliable seed systems are in place and if there is a recognized value of quality seed. In Africa, a combination of both national programmes and universities are leading the breeding efforts for grain legumes and it varies by country. However, the national programmes appear to have a stronger role/presence.

3 Breeding targets: stressors and phenotypes

Breeding methods such as pedigree, single seed descent, backcross, recurrent selection, among others are very commonly used in grain legumes (Fehr, 1987; Beaver and Osorno, 2009). One important feature almost unique to grain legume breeding is the need to conserve specific market characteristics of seed size, shape and colour (Beaver and Osorno, 2009; de Ron, 2015; Smartt, 1990). This is mostly dictated by cultural preferences that vary even among regions within the same country (Voysest, 2000). However, these sometimes-stringent requirements of visual seed traits for market/consumer acceptance consequently makes genetic progress slower compared to other crop species. Therefore, grain legume breeders must deal with a suite of target traits that combine breeding aspects of both agronomic and horticultural crops. More recently, the increased interest in genetically improving crops for nutritional aspects has added another layer of complexity for a successful increase of overall productivity.

Similar to breeding methods, the production challenges and seed yield-limiting factors across grain legumes are very similar, with a combination of biotic and abiotic stresses. In some cases, the same pathogen or insect species can attack multiple grain legume species (Agrios, 2004; Schwartz et al., 2005). Therefore, it is important for grain legume breeders to clearly define their breeding targets or phenotypes. An attempt to summarize/compile these target traits is available in Table 1. Most of these targets are tied either directly or indirectly to seed yield and hence, increased productivity, and several of them are unique to specific regions and/or production systems.

Table 1 Target breeding traits most commonly used in grain legume breeding programmes. Information compiled from Araujo et al. (2015); Beaver and Osorno (2009); Bohra et al. (2014a,b); Jacob et al. (2016); Kumar et al. (2011); Miklas et al. (2006); Muñoz et al. (2017); Osorno and McClean (2014); Pandey et al. (2017); Rubiales et al. (2015); Varshney (2016); Zargar et al. (2015)

Target trait	Importance	Geographical distribution	Mode of inheritance†	Number of available markers
Seed yield components				
Pods per plant	Very high	Worldwide	Multi-genic	Low
Seeds per pod	Very high	Worldwide	Multi-genic	Low
Seed size	Very high	Worldwide	Oligo-genic	High
Biotic stressors				
Fungi	Very high	Worldwide	Single and multi-genic	Very high
Bacteria	High	Worldwide	Oligo- and multi-genic	Medium
Virus	Very high	Worldwide	Single gene	Very high
Nematodes	Moderate	Regional	Oligo-genic	Low
Insects	High	Regional	Single and multi-genic	Low
Parasitic weeds	Low	Regional	Single and oligo-genic	Low
Abiotic stressors				
Drought	Very high	Worldwide	Multi-genic	Very high
Heat	Very high	Worldwide	Multi-genic	Medium
Flooding	Very high	Worldwide	Multi-genic	Low
Low fertility	Very high	Regional	Oligo- and multi-genic	Medium
High/low pH	Very high	Worldwide	Multi-genic	Low
Cold/frost	Moderate	Regional	Oligo- and multi-genic	Low
Other traits				
Days to maturity	Very high	Regional	Oligo- and multi-genic	Medium
Days to flowering	Moderate	Regional	Oligo- and multi-genic	Low
Plant architecture	High	Regional	Multi-genic	Low
Seed nutritional content	Very high	Regional	Oligo- and multi-genic	Medium

† Single = controlled by one gene; Oligo-genic = two or three genes with major effect; Multi-genic = multiple genes with minor additive effects.

Another challenge in breeding programmes is the need to evaluate/screen thousands of genotypes for multiple target traits in a timely, efficient and accurate manner. High-throughput phenotypic tools (a.k.a. phenomics) have the potential in some circumstances to screen thousands of lines in a timely fashion, which allows faster discovery of new traits and combinations (Araus and Cairns, 2014; Sankaran et al., 2015).

The three-tiered breeding strategy proposed by Kelly et al. (1998) for common bean remains relevant 20 years later for many crops. However, with the recent advances in genotyping methods and tools, a floodgate of new information is now available not only to geneticists and other scientists working on the understanding of evolutionary patterns, but also to plant breeders trying to map/locate genes of economic importance (Osorno and McClean, 2014). The possibility of using new markers in grain legumes that could aid in the selection process in a more accurate way is a reality as shown by many examples (see reviews from Kumar et al., 2011; Pandey et al., 2016). Newer molecular markers being reported appear to be more reliable across different genetic backgrounds thanks to the screening of numerous genotypes with a vast number of markers increasing rapidly. This enables recognition of multiple recombination events as well as genomic blocks that are highly conserved. Then both phenotypic and genotypic information can be correlated by taking advantage of linkage disequilibrium (LD) and using it in association mapping approaches (Zargar et al., 2015).

There are multiple opportunities to apply these modern tools along the breeding process. At the earliest stages, the simplest application would be confirming a successful cross between two parents by screening the F_1 progenies. This would require in theory just a single marker that is polymorphic between two parents. Later during the selection process across early generations, a breeder may simply wish to follow a few traits that are essential for the target geographical region. Most often, this will be a marker for essential disease resistance genes or other simply inherited traits (Miklas et al., 2006; Pandey et al., 2016). At the end of the breeding process, a line near commercial release might be fingerprinted with a suite of markers just for identity preservation purposes. These marker assays are generally straightforward, and in most cases, they are affordable for most small- to medium-sized breeding programmes.

Breeders are now being asked to not only monitor their programmes, but also use their programmes for marker/trait association studies. Given that many of the simple traits have abundant markers available (even if they are not used extensively), the traits of interest now are complex or quantitative in nature with heritability estimates that range from low to high. Considerations about the genotype by environment interaction are very important when trying to study these complex multi-genic traits.

More recently, efforts are underway to improve the accuracy and prediction rate of genomic selection in crops. In this case, a suite of random markers well distributed across the genome are used to obtain estimated breeding values (GEBV) based on the performance of a training population. Those parameters are then used to select specific lines within the breeding programme during early generations. In this way, target population sizes are reduced significantly for multi-location yield testing, which is a major task in any breeding programme. Theoretically, it is very logical and intuitive, and it has shown its effectiveness in other breeding systems such as in animal breeding, in addition to reducing costs mostly in phenotyping. However, most recent efforts in GS in grain legumes are resulting in prediction rates that are not higher than 60–70% (Zargar et al., 2015). A range of factors may be affecting these low prediction rates, including the lack of better prediction algorithms available, low marker density/coverage, small training population size, among others. Some studies argue that GS is well suited for crops with a high heterotic expression, but not for self-pollinated crops, which is the case of most grain legumes.

Alternative populations besides the typical F_2 or bi-parental RILs are being used more recently for gene mapping. One example of those is the Multi-parent Advanced Generation Inter-Cross population or MAGIC populations (Cavanagh et al., 2008; Kover et al., 2009). Instead of crossing two lines as in any bi-parental population, a MAGIC population is established by crossing together multiple founder lines or genotypes. Such populations are more genetically diverse than those established from just two parents. In addition, rather than searching for associations immediately after crossing, the population is first cycled through several additional generations of crossing. Each extra generation mills the genetic contribution from the founder lines in a finer way given the multiple recombination events. As a result, associations are located with greater accuracy and are more useful, especially in chromosomal regions with low recombination rates. Thus, MAGIC populations could be considered an intermediate stage between naturally occurring accessions and existing synthetic populations. However, it is important to maximize the genetic diversity to improve results (Huang et al., 2013). One disadvantage of MAGIC populations is that they take time to develop, but results have shown that mapping accuracy and detection is much improved over bi-parental RIL populations. Current efforts are focused on either producing or already using these MAGIC populations in many crop species including some grain legumes such as common bean, chickpea, pigeon pea, cowpea and faba bean, among others (Huang et al., 2015; Huynh et al., 2017).

The development and exploitation of hybrid crops and its tremendous impact in increasing seed yield in cross-pollinated crops is well documented. A different story is found for self-pollinated crops such as most grain legumes, where modest attempts to produce hybrids have produced limited results. In most grain legumes, the lack of natural cross-fertilization and low heterotic expression (hybrid vigour) are the major bottlenecks in exploiting hybrid vigour at a commercial scale because their flower structure favours self-pollination (Saxena et al., 2011; Suso et al., 2015). Early efforts on developing hybrid soybean were made during the 1970s and 1980s in the United States with not much success or impact (Palmer et al., 2001). The only exceptional case of exploitation of hybrid vigour in grain legumes worth mentioning is pigeon pea, which has obtained seed yield increases in India between 30% and 50% by exploiting hybrid vigour and cytoplasmic male sterility (Saxena et al., 2013). Until 2016, three commercial pigeon pea hybrids have been released. The first report of grain legume hybrid was a pigeon pea (ICPH 2671), which was commercially released in 2010 (Sawargaonkar et al., 2016). Current efforts are underway to explore/exploit heterosis in chickpea, pea, faba bean, cowpea, black gram, mung bean, lablab and soybean, among others.

Another area of critical importance for legume improvement that has been successful in many other crops is the transgenic and gene editing technologies. The obvious examples in legumes are glyphosate and insect resistance in soybean, which have been very successful in countries such as the United States, Canada, Brazil and Argentina, among others. Most soybean varieties in these countries are using transgenic technology for both herbicide and insect resistance. Besides glyphosate, a more recent herbicide-tolerant variety that has been commercially deployed is Dicamba-resistant soybeans in which two different events have been released by two companies. It is important to note that virtually all transgenic soybean varieties are used for oil or feed production instead of human/food consumption. There is an ongoing discussion of the impact and adoption of transgenic technologies in feed versus food for direct human consumption because of potential negative consumer reaction to these products.

Among the factors that have contributed to the lack of progress in developing transgenic grain legumes are the difficulty in obtaining reproducible/repeatable results due to the lack of competent totipotent cells for transformation, the long period of time required to develop transgenics, lack of long-term funding, the absence of a coordinated research by the scientific community, and as mentioned before, an unfavourable public perception towards the use of transgenic crops (Eapen, 2008). In many cases, genetic transformation occurs, but regeneration of differentiated tissues becomes the limiting factor, so it is a tissue culture problem rather than the transformation protocol *per se*.

Even though genetic transformation has been reported for several grain legumes, transgenic cultivars have not yet been commercially released outside of soybean (Eapen, 2008), except for the virus-resistant common bean in Brazil (Bonfin et al., 2007). The scientific group at the Brazilian Institute of Agricultural Research (EMBRAPA) successfully obtained resistance to bean golden mosaic virus in common bean via particle bombardment. This viral disease is the most important yield-limiting factor for common bean in Brazil. Virus replication is blocked inside the plant's vascular system by silencing expression of the *repAC1* viral gene, which encodes the only protein strictly essential for viral genome replication (Aragão and Faria, 2009; Aragão et al., 2013). The event was approved by the National Technical Commission on Biosafety (CTNBio), but the first successful event was obtained using a pinto bean cultivar (Olathe), which is not a preferred commercial market class in Brazil. Since the resistance is inherited in a Mendelian fashion (Faria et al., 2014), intensive backcrossing was made during the last seven years to produce commercial cultivars of the preferred carioca market class. At least two transgenic lines will be commercially released in the near future (Faria et al., 2016) and therefore, it will be an interesting case to follow closely regarding farmer adoption and consumer/industry acceptance of a transgenic grain legume.

Gene editing tools such as the recent Zinc Finger Nucleases (ZFNs), Transcriptional Activator-Like Effector Nucleases (TALENs) and Clustered Regularly Interspaced Short Palindromic Repeats (CRISPR/Cas9) technologies offer an interesting option for grain legume improvement by making point changes/edits in the DNA sequence or its promoters that will alter or block the expression of a specific gene. One of the main arguments of this technology is that it would be considered as non-GMO since there is no physical insertion of foreign/alien genetic material. However, this is still under debate in many countries and it does not seem to have a consensus regarding its classification.

The only example in crop species of gene editing technology that has been deployed commercially can be found in canola (*Brassica* spp.) with herbicide (sulphonylurea) resistance (https://www.cibus.com/products.php). Potentially, it could be commercially available very soon in maize (*Zea mays* L.), tomato (*Solanum lycopersicum* L.), cotton (*Gossypium hirsutum* L.), rice (*Oryza sativa* L.) and mushrooms, among few others (Bomgardner, 2017; van de Wield et al., 2017; Waltz, 2016; Weeks, 2017). In the case of legumes, not many examples could be found in the literature, except for allergen reduction in peanuts and oil composition in soybean (Demorest et al., 2016; Scheben et al., 2017; Weeks, 2017). No examples in grain legumes were found except some 'proof-of-concept' articles with no target traits of economic importance yet to be reported as successful.

A more futuristic example of potential solutions to many production problems is the use of RNA interference (RNAi) mechanisms by directly spraying nanoparticles embedded with specific RNAi sequences that would block/reduce or increase gene expression either in the host or the pathogen/insect (Mitter et al., 2017; Wang and Jin, 2017).

Current efforts are also underway for creating reference genomes for pathogens. This is a great complement for a better understanding of host–pathogen interactions and how to design the best strategies for genetic resistance against diseases. Several examples exist in cereal diseases but information is more limited in the case of grain legumes. Some of these pathogens have a broad range of host species including legumes, making the genome sequence useful for multiple crop species simultaneously. For example, there is a draft sequence for white mould (*Sclerotinia sclerotiorum* (Lib.) de Bary), which has been reported in more than 400 species (Amselem et al., 2011). Complete genomes are available for *Xanthomonas fuscans* subsp. *fuscans* (Darrasse et al., 2013), *Pseudomonas syringae* pv. *syringae* (Fail et al., 2005), *Pythium ultimum* (Lévesque et al., 2010), *Fusarium graminearum* (Ma et al., 2010) and *F. oxysporum* (Williams et al., 2016). Partial/draft genomes have been determined for *Xanthomonas axonopodis* pv. *phaseoli* (Aritua et al., 2015), *Pseudomonas syringae* pv. *phaseoli*, *Curtobacterium flaccumfaciens*, *Rhizoctonia solani* (Hane et al., 2014), *Alternaria alternata* and *Colletotrichum lindemuthianum*. In the same way, some sequencing and transcriptomics data are available for *Uromyces appendiculatus* (Link et al., 2014) and *Ascochyta rabiei* (Fondevilla et al., 2015), among others (this list of pathogen sequences was provided in part by J. Pasche and K. Simons, Department of Plant Pathology, North Dakota State University).

4 The grain legume reference-genome sequences

The development of a reference- or draft-genome sequence is nearly a prerequisite for modern genome-assisted breeding. The genome sequence project begins with a random fractionation of the genome into short fragments that are either partially or fully sequenced to obtain a collection of reads. These sequence reads are first assembled into overlapping reads that are termed contigs. One indicator of the quality of a genome is the L50 contig. The L50 is the length of contig among all contigs that sums to 50% of the predicted genome size. Longer L50 lengths indicate higher quality genomes. Table 2 summarizes the major statistics from the ten legume crop species that have been sequenced to date, and for all legumes except common bean the L50 ranges from 18 to 183 kb. The common bean and cowpea genomes are exceptions, primarily because they were developed from long-read technologies. Using sequencing reads derived from fragments of longer length (~2.0–~10.0 kb), contigs derived from very short reads can be assembled into scaffolds. Again, a longer scaffold L50 indicates a higher quality genome. Finally, scaffolds are ordered into pseudochromosomes using a high-density, genetically derived molecular marker map. Reference genomes have long contiguity in which contigs and scaffolds are assembled into pseudochromosomes.

The most complete reference genomes consist of pseudochromosomes whose cumulative sizes nearly equal the genome size as estimated by flow cytometry or *k*-mer procedures. Draft genomes also gather all the read information, but the sum of pseudochromosome sizes falls far below that of draft genomes. Here, if the total length of the pseudochromosomes is ≥85% of the estimated genome size, it is considered a reference genome, whereas those <85% are considered as draft genomes. This distinction, however, does not limit the utility of draft genomes for many of the molecular breeding applications such as single-nucleotide polymorphisms (SNP) discovery, molecular genetic mapping and variety identification.

Table 2 Characteristics of genome sequences currently available in legumes

Species name	Common name	Genotype	Year	Publication	Technical method	# Chrom	Est. genome size/ assembled scaffold size (Mb)
Glycine max	Soybean	Williams 82	2010	*Nature* 463:178	Sanger	20	1115/978
Cajanus cajan	Pigeon pea	ICPL 87119	2012	*Nat Biotechnol* 30:83	Sanger, Illumina	11	833/606
Cicer arientinum	Chickpea	CDC Frontier	2013	*Nat Biotechnol* 31:240	Sanger, Illumina	8	738/533
Medicago truncatula	Medicago	A17	2014	*Nature* 480:520; *BMC Genomics* 15:312	Sanger, 454, Illumina	8	454/384
Phaseolus vulgaris	Common bean	G19833	2014	*Nat Genet* 46:707	Sanger, Illumina, PacBio	11	587/521
Vigna radiata	Mung bean	VC1973A	2014	*Nat Commun* 5:5443	454, Illumina	11	543/431
Vigna angularis	Adzuki bean	Gyeongwor	2015	*Scientific Reports* 5:8069	454, Illumina	11	612/399
Phaseolus vulgaris	Common bean	BAT93	2016	*Genome Biol* 17:32	454, Sanger, Illumina, Solid	11	556/495
Arachis duranensis	Peanut A genome	V14167	2016	*Nat Genet* 48:438	Illumina, Sanger	10	1076/1211
Arachis ipaensis	Peanut B genome	K30076	2016	*Nat Genet* 48:438	Illumina, Sanger	10	1361/1512
Vigna unguiculata	Cowpea	IT97K-499-35	2017	www.phytozome.net	PacBio, Bionano	11	622/519

% Genome coverage based on total assembled length	Pseudo-chromo-somes length	% Genome coverage based on pseudo-chromo-somes length	Repeat content (%)	Chrom size range (Mb)	# Genes/ transcripts	Contig N50/ L50 (#/ kb)	Scaffold N50/L50 (#/Mb)	Genome duplication history	# NBS-LRR gene models
88	951	85	57	37–62	56 044/ 88 647	1548/ 183	10/48.6	Eudicot 3x + Legume 2x + (2x)	465[A]
73	247	30	52	10–48	40 071/na	7815/22	380/0.38	Eudicot 3x + Legume 2x	289[A]
72	347	36	49	na	28 255/na	na/24	na/0.65	Eudicot 3x + Legume 2x	121
85	366	80	31	32–55	31 661/ 41 159	na	53/1.3	Eudicot 3x + Legume 2x	571[A]
89	508	87	45	32–60	27 197/ 31 688	73/1900	5/49.7	Eudicot 3x + Legume 2x	337[A]
79	314	56	50	na	22 427/na	na/42	na/1.5	Eudicot 3x + Legume 2x	73
65	210	34	43	na	26 857/na	na/22	3,883/ 0.70	Eudicot 3x + Legume 2x	87
89	451	81	33	33–51	30 491/ 66 634	na/18	267/0.45	Eudicot 3x + Legume 2x	234
113	1025	95	62	78–133	36 734/ 88 643	12 992/ 22	334/0.9	Eudicot 3x + Legume 2x	345
111	1338	98	69	109–149	41 840/ 94 419	15 898/ 23	86/5.3	Eudicot 3x + Legume 2x	397
83	473	76	na	34–65	29 773/ 42 287	16 418/ 12	6/41.7	Eudicot 3x + Legume 2x	na

For legumes, progress towards these genome-wide scale sequences has spanned the past ten years. Over that time, a significant switch has occurred in the technologies used to generate the raw reads that are assembled into the genome sequence. Soybean was the first legume for which a reference-genome sequence was developed (Schmutz et al., 2010). The soybean (*Glycine max* L.) reference genome has the distinction of being one of the last released sequences based on Sanger sequencing reads, a technology that allows for a higher degree of assembly accuracy, but is quite expensive to generate. All subsequent genome sequences were based almost exclusively on short-read technologies developed by Illumina (sequencing-by-synthesis; www.illumina.com). Essentially discontinued technologies from Roche (454) and ABI (SOLiD; www.thermofisher.com) were used to develop a few genome sequences. Over a decade of data collection, the *Medicago truncatula* reference-genome sequence project evolved as an international collaborative effort that initially focused on sequence reads of BACs from low copy regions of the genome to a whole-genome shotgun using Illumina and Roche 454 technologies (Young et al., 2011; Tang et al., 2014). This resulted in a reference-genome sequence that utilized sequencing reads from all technologies. The first common bean (*P. vulgaris*) reference-genome sequence (Schmutz et al., 2014) was a collaboration among several US institutions that was funded by the US Department of Agriculture, the Department of Energy Joint Genome Institute and HudsonAlpha Institute of Biotechnology. The reference genotype for this project was Chaucha Chuga (CIAT identifier: G19833), which belongs to the Andean gene pool. This genome was initially sequenced primarily using Roche 454 technology (Schmutz et al., 2014), and then again sequenced using the PacBio (Pacific Biosciences, Inc; www.pacb.com) long-read method. The PacBio reads were the primary source of reads used to develop the version 2.0 assembly of the reference common bean genome (phytozome.jgi.doe.gov/pz/portal.html#!info?alias=Org_Pvulgaris). The latest version has the longest scaffold L50 of any legume species (Table 2). A second common bean genotype was sequenced by a Spanish, Mexican and Brazilian consortium (Vlasova et al., 2016). Here the genotype was the Middle American BAT93 genotype and the project resulted in a draft genome sequence. Cultivated peanut is a tetraploid AB species, and the two parents of that tetraploid (A genome, *Arachis duranensis*; B genome, *Arachis ipanensis*) were individually sequenced (Bertioli et al., 2016). Draft and reference genomes are also available for four other legume crop species: pigeon pea (Varshney et al., 2012), chickpea (Varshney et al., 2013), mung bean (Kang et al., 2014), cowpea (Lonardi et al., 2017; Muñoz-Amatriaín et al., 2017) and two versions of adzuki bean (Kang et al., 2015; Yang et al., 2015).

5 Legume common lineages and synteny

What has been learned collectively and individually from these genome-sequencing projects? Prior to this era of intense whole-genome sequencing, there was partial sequence data, most often from sequencing the expressed portion of the genome (ESTs). Even with these pieces of sequence there were hints that all legumes shared a common duplication event. This was first observed with a Ks analysis that compared EST-derived contigs of soybean and *Medicago* (Schlueter et al., 2004). Genome-wide Ks analyses develop a distribution of synonymous nucleotide substitution values among paralogous duplicates within a species. The distribution generally has several peaks related to different duplication events in the species history, and the peak values are indicative of a major divergence event. All individual legume genome projects identified a similar Ks peak with a value of ~0.60–0.65. When the

Ks distributions for all crop legumes were corrected relative to soybean, the shared pan-legume duplication was dated at ~59 million years ago (Mya) (Wang et al., 2017). Subsequent analyses (Wang et al., 2016) demonstrated that the peanut lineage diverged 49.1 Mya from the main legume lineage, and the cool season *Medicago* (*Medicago*, chickpea) and warm season common bean (common bean, soybean, pigeon pea, mung bean and adzuki bean) lineages diverged 48.1 Mya. Within the warm season legume lineage, Ks analyses also supported the concept that soybean was a paleopolyploid (Schmutz et al., 2010) that diverged from the other warm season legumes (common bean, mung bean and adzuki bean) ~13 Mya. Genome evolution is also associated with chromosome breakage and reunion to develop a different chromosome number than the ancestral species. The ancestral eudicot species is estimated to have contained 21 chromosomes that resulted from an ancient triplication event ~130 Mya (Bowers et al., 2003; Jaillon et al., 2007). That number appears to have been reduced to a base number of 11 chromosomes in the common bean lineage and seven chromosomes in the *Medicago* lineage.

One of the best examples of close synteny can be observed within the Phaseolinae sub-tribe between *Vigna* and *Phaseolus* (Lucas et al., 2011; Muñoz-Amatriaín et al., 2017). It is important to note that early botanists and taxonomists originally considered these two species as the same and it was not until the last century when the *Phaseolus/Vigna* complex was finally split into two separate genera. Three *Vigna* species and *P. vulgaris* each have a basic chromosome number of 11, but the numbering of chromosomes has been designated independently by each research group. Among these species, the *P. vulgaris* genome was the first to be published, thus establishing a precedent and rational baseline for a more uniform chromosome numbering system, at least among *Phaseolus* and *Vigna* species. However, cowpea serves as a convenient nodal species for this cross-genus comparison now that the cowpea genome assembly has become the most complete of any *Vigna* species (Lonardi et al., 2017, www.phytozome.net). As summarized in Table 3, six of the 11 cowpea chromosomes are fully syntenic with six of the 11 chromosomes of *P. vulgaris* and each of the remaining five cowpea chromosomes is related to parts of two *P. vulgaris* chromosomes. From Table 3, it can be surmised that, among warm season legumes with a basic chromosome number of 11, *Vigna radiata* (mung bean) is the most syntenic with cowpea, having nine fully syntenic chromosomes, followed by *Vigna angularis* (adzuki bean) with eight and *P. vulgaris* with six fully syntenic chromosomes. Synteny clearly extends to soybean as well, but due to the polyploid-like genetic composition of soybean, and a basic chromosome number that is not an integer multiple of 11, the relationships between cowpea and soybean are much more complicated.

It should be noted that as sequence assemblies continue to be refined, it seems likely that the extent of synteny will appear to be more extensive than presently thought. For example, in adzuki bean, the less developed genome assembly of Kang et al. (2015), which is not included in Table 3, appears to be significantly less syntenic with the cowpea genome than does the more fully developed adzuki bean genome assembly of Yang et al. (2015). In addition, some apparent inversions relative to cowpea in the initially released sequence assembly of *P. vulgaris* no longer appear to be inversions in the newest *P. vulgaris* assembly. Another reason which may affect synteny estimates is that a substantial number of segments of the cowpea genome are yet of unknown orientation and were therefore arbitrarily given an orientation in the current cowpea pseudomolecules, causing segments to falsely appear as inversions in synteny analyses.

The breakage/reunion events associated within the two lineages can be visualized as chromosome-based synteny blocks among any two legumes that can be exploited for genetic mapping purposes. From a plant breeding perspective, the synteny relationship

Table 3 Synteny between cowpea chromosomes (pseudomolecules) and those of *Phaseolus vulgaris, Vigna radiata, Vigna angularis* and *Glycine max*

New cowpea LG	Old cowpea LG	*P. vulgaris*	*V. radiata*	*V. angularis*	*G. max*
Vu01	4	5, 1	3, 5	4	3, 11, 18, 19
Vu02	7	2,3	11	7	1, 2, 9, 11
Vu03	3	2,3	7	10, 11	5, 7, 8, 13, 16, 17
Vu04	11	4	1	2	9, 13, 16, 19
Vu05	1	5, 8	4, 5	4	9, 12, 13, 18
Vu06	6	6	10	3	8, 9, 11, 15, 18
Vu07	2	7	8	6	10, 20
Vu08	5	1,8	6	1	2, 14, 17
Vu09	8	9	5	5, 9	4, 6, 15
Vu10	10	10	9	8	1, 3, 7, 16
Vu11	9	11	2	2, 5	6, 11, 12
Total simple	11	6	9	8	0

Adapted from Lonardi et al. (2017).

is useful when the location of a gene associated with a phenotype of economic value in one species can be used to map that gene in a second species. For example, legumes in the common bean lineage can display either a determinate or indeterminate growth habit. For soybeans grown in the northern US states, the indeterminate growth habit is preferred because it performs and yields better during a rather shorter growing season. Yet, in the southern states, indeterminacy will lead to lodging, disease susceptibility and harvesting issues which consequently will result in reduced seed yield. Therefore, determinate varieties are preferred in these growing regions. Similarly, for common bean, there is also a yield drag associated with determinate growth habit. The major determinacy gene in soybean, *Dt1*, was cloned and located to chromosome Gm19. The Gm19 region in which *Dt1* is located is syntenic to chromosome Pv01 in common bean. Genetic and functional analyses have conclusively shown that the common bean determinacy gene, *Fin*, is located on a chromosomal Pv01 block that is syntenic to the *Dt1* containing block on Gm19 (Repinski et al., 2012). Further analysis of the sequence of *Fin* among a large genotype collection demonstrated that it was a frequent mutational target (Kwak et al., 2012). Similar examples can be found for disease resistance genes shared among legume species. It is this type of comparative analysis of legume species within a shared lineage that will prove useful to expand the molecular genetic tools that can aid legume crop improvement.

For all crop species, disease is a major factor that prevents the plant from reaching its full yield potential. Compared with other legume species, common bean seems to be susceptible to more major biotic stressors than other legumes. Diseases such as common bacterial blight (*Xanthomonas axonopodis*), bean common mosaic virus, bean rust (*Uromyces appendiculatus* (Pers.) Unger), anthracnose (*Colletotrichum lindemuthianum* (Sacc. & Magn.) Lams. -Scrib) and angular leaf spot (*Pseudocercospora griseola* Syn. *Phaeoisariopsis griseola* (Sacc.) Ferraris). For this purpose, disease resistance has been the primary research focus as breeders and geneticists strive to provide the right combination of genes to maximize yield. Resistances to diseases, especially those that are controlled genetically by a major factor, are often the result of a gene with nucleotide-binding site and leucine-rich repeat domains. Here these genes will be designated as NLR genes and proteins. These NLR genes form a large gene family that is of quite variable size in most crop species, and this is especially the case for legumes where *Medicago* contains 571 such genes while adzuki has 87. In general, the size of the NLR family is significantly higher in legumes than the Brassicaceae (*t*-test = 0.016) but of a similar size to Poaceae (*t*-test = 0.962) and Rosaceae (*t*-test = 0.986) species. The expansion of the NLR family appears to proceed at a different rate among species. This is evidenced by the fact that common bean contains 337 NLR genes, while soybean, another member of the Phaseoleae clad, contains 465, a number that is below the 674 expected if the duplication event in this species is considered. The common bean rate appears to be higher than other legumes because it proportionally contains more NLR genes when genome size is considered as a basis for the number of these genes. Whether this higher expansion rate is because of the seemingly greater number of serious pathogens that attack common bean remains to be determined. Recent evidence from a pan-genome study in *Medicago* shows that the NLR gene family is expanding at a higher rate than other gene families (Zhou et al., 2017). There is also evidence for a variable expansion rate of the family in common bean where the Middle American genotype BAT93 contains less NLR genes than G19833, the Andean reference (Vlasova et al., 2016).

An interesting feature of plant genomes is the clustering of the NLR genes. In the common bean genome for example, there are three major clusters of NLR genes of a magnitude not seen in other crop legumes. The clusters located on the distal ends of Pv04 and Pv10 and the proximal end of Pv11 each contain >30 NLR genes, with greater than 50 NLR genes located in the Pv11 cluster alone (Meziadi et al., 2016). Other moderate-sized clusters located on the proximal ends of Pv01, Pv02 and Pv08 contain between 14 and 18 genes. While these are moderate-sized common bean clusters, for other legumes these are considered large clusters. The extensive research on disease resistance in common bean located resistance specificities to certain chromosome arms. The mapping of the molecular markers to the reference genome, however, has now placed many of these genes into specific clusters (Meziadi et al., 2016). The Pv01 and Pv04 clusters are enriched for anthracnose resistance genes, suggesting that expansion within a local region occurred as new avirulence specificities appeared in the pathogen. This expansion also appears to be the case on Pv02 where multiple virus resistance loci are located, and Pv11 a cluster enriched for bean rust resistance genes. The reference-genome sequence along with the gene modelling work that defined the R gene clusters continues to be a critical resource, as new common bean resistance genes is being mapped into R gene clusters (Zuiderveen et al., 2016; Campa and Ferreira 2017; Hurtado-Gonzalez et al., 2017; Valentini et al., 2017).

6 Whole-genome and reduced representation resequencing

With a reference genome available, the next step is to characterize the structure of the polymorphism within a species. From a basic research perspective, these structure analyses can describe the major germplasm subdivisions within a species. For crop species, these subdivisions are often represented in germplasm currently used for human consumption. From an applied perspective, polymorphism studies within and between breeding programmes can identify variable germplasm groups that may be exploited to increase diversity in a programme. SNPs are the principal genetic data used today to study variation within a species. The most extensive data set can be captured by whole-genome resequencing of multiple individuals from a species. With improvements in sequencing technology, these data are currently captured at a depth of 5X–15X. This has been successful for the overall characterization of the organization of variation in species such as soybean (Lam et al., 2010), *Medicago* (Branca et al., 2011) and pigeon pea (Varshney et al., 2017). While this research approach is valuable when studying variation at the species level, it is too expensive at the breeding population level. Soybean resequencing (Lam et al., 2010; Zhou et al., 2015) illustrates the types of discoveries made in multiple species. That wild species are much more diverse than the cultivated materials supports earlier research (Hyten et al., 2006) that has shown that only a small portion of the soybean diversity is captured in modern varieties. Sequence diversity was lost during the domestication process, and the diversity was further eroded during the improvements of the crop for production purposes by plant breeding techniques (Zhou et al., 2015). This successive loss of diversity during the wild to landrace to cultivar transition is typical of other legumes such as common bean (Schmutz et al., 2014), pigeon pea (Varshney et al., 2017) and chickpea (Varshney et al., 2013).

Common bean is unique among the legumes because of its complex evolutionary, domestication and breeding history. The species appears to have evolved in Central Mexico (Bitocchi et al., 2012). Two gene pools of wild common bean ancestors, Middle American and Andean, arose ~110–150 Kya (thousands of years) (Mamidi et al., 2013; Schmutz et al., 2014) and through successive waves of migration were distributed from northern Mexico to southern Bolivia and northern Argentina (Bitocchi et al., 2017). Unlike many other species, common bean underwent two independent domestications (Mamidi et al., 2011b; Schmutz et al., 2014). Two locations in Mexico, the Río Lerma/Río Grande de Santiago basin (Kwak et al., 2009) and the Oaxaca Valley (Bitocchi et al., 2013), are suggested as domestication sites for the Middle American gene pool, while Bolivia has been suggested as the domestication site for the Andean gene pool (Beebe et al., 2001). The regions of the genome associated with domestication are quite distinct between the gene pools, with only 59 domestication candidate genes shared between the two gene pools. Each of the gene pools are further classified into races (Singh et al., 1991), and the races also appear to have evolved independently with only a few shared genes associated with the appearance of the distinct races (Schmutz et al., 2014).

Whole-genome resequencing has been applied to the mapping of loci important for the improvement of legumes. This technique generally referred to as QTL-seq utilizes the bulking of lines in a segregating population that express extreme phenotypes of a quantitative trait followed by the sequencing of those DNA bulks. White mould is a major disease affecting the crop, and tolerance is controlled by many loci (Vasconcellos et al., 2017). White mould-tolerant and susceptible DNA pools of lines from an advanced generation mapping population segregating for two tolerance QTL were sequenced. The bulk sequence SNP data were compared, and high levels of polymorphic SNPs were clustered in two regions that mapped at the genetic peaks of the SNPs (Mamidi et al., 2016). From these results, molecular markers that could track the introgression of the white mould-tolerant QTL during a breeding programme were identified. This technique has also been applied in chickpea to map seed traits (Singh et al., 2016), flowering time

(Srivastava et al., 2017) and growth habit (Kujur et al., 2016) traits as well as disease resistance loci in peanut (Pandey et al., 2017).

The declining cost of whole-genome resequencing has inspired researchers to develop other sequencing techniques that provide lower genome coverage but at a depth that could still address the same questions related to the distribution of variation within the species. These approaches all rely on a sequence reduction step that captures only a subset of the genome followed by the construction of a sequencing library that typically includes from 96 to 384 individuals (summarized in Scheben et al., 2016). Because of the reliance on sequencing to generate the sequence data, these approaches are generally referred to as genotype-by-sequencing or GBS. For example, ~150 000 SNPs were generated by a two-enzyme GBS protocol (Schröder et al., 2016) for a population of 280 improved Middle American varieties of common bean (Moghaddam et al., 2016), and these varieties could be clustered at the race and market class level. Additionally, across the full population distinct patterns of LD were noted for each chromosome, and the LD patterns were unique for each race across all chromosomes. A collection of ~3300 peanut SNPs allowed to distinguish wild and cultivated peanut as well as define US and Chinese varieties as distinct populations.

7 SNP chips

GBS is a useful method of assessing SNP variation within a population, and as such, it is a useful method of developing a SNP-based genetic map. To date, this has been the primary use of the GBS technology in legumes. Multiple legume SNP maps utilizing recombinant inbred populations and the GBS technology have been developed including those for chickpea (Jaganathan et al., 2015; Verma et al., 2015), and lentil (Wong et al., 2015). More recently, GBS SNP data have been collected from diversity panels and used for the association mapping of important production phenotypes in legumes. Examples include the examination of the genetic architecture of phenology and growth habit traits in common bean (Moghaddam et al., 2016), cyst nematode (Vuong et al., 2015) and sudden death syndrome resistance (Zhang et al., 2015a) in soybean, and yield under terminal drought in pea (Annicchiarico et al., 2017). For a more applied programme, GBS can be a challenging technology because it requires downstream processing by a trained bioinformaticist to mine the large SNP dataset generated by the GBS approach.

An important basic and applied genetic tool that has arisen from all sequencing projects is an SNP chip. SNP-based chip technologies (Rasheed et al., 2017) are an alternative to GBS-generated SNP data, which does not have a bioinformatics requirement for the user. This technology is based on the binding of short oligonucleotides to a fixed matrix in an ordered array. Sample DNA is added to the array, and a specific signal is generated that associates with a specific allele for each SNP position. The preferred approach early on was the Golden Gate array assay and more recently the Infinium array assay has replaced the Golden Gate assay. Both of these SNP platforms are marketed by Illumina (www.illumina.com). Affymetrix (www.affymetrix.com) also offers such products, and a ~250 k SNP Arabidopsis chip (Kim et al., 2007) was widely used by the early adopters of this technology on this model species. The chip provided is very useful for pioneering plant genetic studies of LD (Kim et al., 2007), population structure (Horton et al., 2012), local adaptation (Fournier-Level et al., 2011) and mapping genes controlling multiple phenotypic traits (Atwell et al., 2010). Each system is coupled with software that can easily analyse the output data and generate the SNP allele calls for each individual. Those data are output in a format that is amendable to standard genetic analyses. A properly designed chip is created to contain a fixed set of SNPs with known genomic positions that have broad utility across the population structure of the crop species. The SNPs are generally distributed in both genic and non-genic regions of the genome. The ease of use of the array-based technologies is highly valued for plant breeders who do not have a trained bioinformatics staff to support their improvement efforts.

The first community-wide legume SNP chip was for soybean and contained 1536 SNPs (Hyten et al., 2010). This chip was used to develop the genetic map that was critical to the assembly of the pseudomolecules for the reference soybean genome (Schmutz et al., 2010). One application of the chip was to use a genome-wide association study (GWAS) to identify loci association with iron deficiency tolerance (Mamidi et al., 2011a). Other examples include mapping cyst nematode (Vuong et al., 2010) and soybean rust resistance (Kendrick et al., 2011). Recently, a chip containing ~50 k soybean SNPs was developed (Song et al., 2013) that has proven quite useful to fully characterize the USDA soybean core collection (Bandillo et al., 2015), and study the genetic architecture of phenology and growth habit traits (Zhang et al., 2015b) and seed protein and oil content (Hwang et al., 2014). For common bean, a modest-sized (by modern standards) SNP chip of ~5 k SNPs was designed to capture variation across all major market classes of dry and snap common bean of both Middle American and Andean origin (Song et al., 2015). That chip has been applied to mapping bean rust (Hurtado-Gonzales et al., 2017) and white mould (Vasconcellos et al., 2017) resistance loci, GWAS analysis for drought tolerance (Hoyos-Villegas et al., 2017), genetic mapping of loci affecting nitrogen fixation (Heilig et al., 2017) and snap bean pod traits (Hagerty et al., 2016) and characterizing the population structure of an Andean diversity panel (Cichy et al., 2015), among others. Finally, a 50 k SNP chip for cowpea was recently developed, and the chip has proven useful in describing the structure of the within West Africa cowpea breeding programmes (Muñoz-Amatriaín et al., 2017).

The lessons learned from these first-generation SNP chips will guide the development of new chips with wider utility. The original common bean chip (Song et al., 2015) was designed to have an equal distribution of SNPs across the physical length of the genome. Yet genetic mapping and GWAS experiments are dependent on recombination. Newer chips would be more useful for genetic experiments if they better mimic the distribution of recombination in the species. Recombination is higher at the ends of the chromosomes in the euchromatic region of the genome that is gene-rich (Schmutz et al., 2010, 2014). Therefore, SNP distribution of newer chips should be enriched with SNPs located at the end of the chromosome with lower representation in the low recombination, gene-poor heterochromatic regions of the genome. This type of distribution would increase the likelihood of discovering SNPs closer to candidate genes when genetic populations or association panels are evaluated. These SNP distribution considerations are only relevant if the new chips contain a similar number of SNPs. If the number of SNPs was to be increased to levels used in humans, a different approach would be needed for chip design. Nearly all phenotypic variation is the result of changes in the exon rather than in the regulatory region of genes. This principle supported the decision to develop human chips that consists of multiple SNPs located in the exons of genes as well a small subset (~10%) that are located in the regulatory region or within 10 kb of the gene. The power of this approach is widely observed in human GWAS experiments where peak SNPs are often located in genes (Manolio, 2013). Another useful genomic polymorphism that are the copy number variants (CNVs), which are regions of the genome that are duplicated multiple times. CNVs are quite abundant in humans, 5–10% of the genome (Zarrei et al., 2015), and are associated with human disease loci (Wellcome Trust Case Control Consortium, 2010). More recent human array designs have included CNVs as markers (Conrad et al., 2010). The first high-density legume SNP chip was recently described (Wang et al., 2016). This chip had several of the design considerations including an unequal distribution of SNPs in the high and low recombination regions of the soybean genome.

High-density chips are very powerful for genetic studies, but the cost of genotyping multiple individual lines in a breeding programme is prohibitive, especially for public breeding programmes with limited resources. The alternative is to design genotyping tools that can select individual lines with the preferred set of phenotypes required of varieties that will be released several years into the future. For most breeding programmes, the primary focus would be markers linked to disease resistance alleles. For most species, this is probably in the range of 10–15 markers. Genomic selection, the practice of selecting on genotype rather than phenotype to advance lines (Heffner et al., 2009), also requires a limited number of markers. Currently, there is no consensus on the number of markers required for genomic selection, but combining a limited set of markers that tag single gene traits with a set of genomic selection markers would be a practical solution for a project with limited funds.

8 Conclusion

The suite of genomic tools available for legume breeding will continue to grow, maybe not at the same pace of other major crops such as cereals, but with enough development to keep facilitating the breeding efforts. Several models currently being used on major crops can be adapted to legume breeding given a combination of resources available and programme needs and size. The goal of this review was to provide the reader with a high viewpoint of the current status of these genomic tools in grain legumes and how they can continue to be improved.

9 Where to look for further information

Additional resources and research efforts are needed in areas such as optimization of genomic selection for grain legumes, genetic transformation that goes beyond herbicide resistance and development of breeder-friendly markers tightly linked to genes/traits of economic importance. A better understanding of the macro- and micro-synteny among legumes species will be gained in the near future given the new genomic tools and resource available within each grain legume species. Further information can be found in the book by Osorno and McClean (2014) and De Ron (2015). The idea of developing pan-genomes for each legume species will allow a better understanding of the organization of the genetic diversity contained within and among species. Finally, gene editing tools show great potential, but it may be too early to assess its impact in agriculture and much less its impact in grain legumes. Readers are encouraged to follow the newer developments made by CGIAR legume breeding programmes at centres such as CIAT, ICRISAT, ICARDA, IITA, among others. In the same way, universities and research institutes in North America and Europe as well as some private companies are also doing modern breeding using genomic tools and, therefore, are worth following. Finally, the role of national programmes in both developing and developed countries is of crucial importance in order to make a real impact in world's agriculture and more specifically, in grain legume productivity.

10 References

Agrios, G. N. (2004). Losses caused by plant diseases. In: *Plant Pathology*. Elsevier, Oxford, UK, 29–45.

Akibode, S. and Maredia, M. (2011). *Global and Regional Trends in Production, Trade and Consumption of Food Legume Crops*. http://impact.cgiar.org/sites/default/files/images/Legumetrendsv2.pdf.

Albala, K. (2007). *Beans: A History*. Berg, New York.

Amselem, J., Cuomo, C. A., van Kan, J. A., Viaud, M., Benito, E. P., Couloux, A., Coutinho, P. M., de Vries, R. P., Dyer, P. S., Fillinger, S., Fournier, E., Gout, L., Hahn, M., Kohn, L., Lapalu, N., Plummer, K. M., Pradier, J. M., Quevillon, E., Sharon, A., Simon, A., ten Have, A., Tudzynski, B., Tudzynski, P., Wincker, P., Andrew, M., Anthouard, V., Beever, R. E., Beffa, R., Benoit, I., Bouzid, O., Brault, B., Chen, Z., Choquer, M., Collemare, J., Cotton, P., Danchin, E. G., da Silva, C., Gautier, A., Giraud, C., Giraud, T., Gonzalez, C., Grossetete, S., Guldener, U., Henrissat, B., Howlett, B. J., Kodira, C., Kretschmer, M., Lappartient, A., Leroch, M., Levis, C., Mauceli, E., Neuveglise, C., Oeser, B., Pearson, M., Poulain, J., Poussereau, N., Quesneville, H., Rascle, C., Schumacher, J., Segurens, B., Sexton, A., Silva, E., Sirven, C., Soanes, D. M., Talbot, N. J., Templeton, M., Yandava, C., Yarden, O., Zeng, Q., Rollins, J. A., Lebrun, M. H. and Dickman, M. (2011 Aug). Genomic analysis of the necrotrophic fungal pathogens *Sclerotinia sclerotiorum* and *Botrytis cinerea*. *PLoS Genetics*, 7(8), e1002230. doi: 10.1371/journal.pgen.1002230. Epub 18 August 2011.

Annicchiarico, P., Nazzicari, N., Pecetti, L., Romani, M., Ferrari, B., Wei, Y. and Brummer, E. C. (2017). GBS-based genomic selection for pea grain yield under severe terminal drought. *The Plant Genome*, 10, 1–13.

Aragão, F. J. and Faria, J. C. (2009). First transgenic geminivirus-resistant plant in the field. *Nature Biotechnology*, 27(12), 1086–8.

Aragão, F. J., Nogueira, E. O., Tinoco, M. L. and Faria, J. C. (2013). Molecular characterization of the first commercial transgenic common bean immune to the Bean golden mosaic virus. *Journal of Biotechnology*, 166(1–2), 42–50.

Araújo, S. S., Beebe, S., Crespi, M., Delbreil, B., González, E. M., Gruber, V., Lejeune-Henaut, I., Link, W., Monteros, M. J., Prats, E. and Rao, I. (2015). Abiotic stress responses in legumes: Strategies used to cope with environmental challenges. *Critical Reviews in Plant Sciences*, 34(1–3), 237–80.

Araus, J. L. and Cairns, J. E. (2014). Field high-throughput phenotyping: The new crop breeding frontier. *Trends in Plant Science*, 19(1), 52–61.

Aritua, V., Harrison, J., Sapp, M., Buruchara, R., Smith, J. and Studholme, D. J. (2015). Genome sequencing reveals a new lineage associated with lablab bean and genetic exchange between *Xanthomonas axonopodis* pv. *phaseoli* and *Xanthomonas fuscans* subsp. *fuscans*. *Frontiers in microbiology*, 6, 1080.

Atwell, S., Huang, Y. S., Vilhjálmsson, B. J., Willems, G., Horton, M., Li, Y., Meng, D., Platt, A., Tarone, A. M., Hu, T. T., Jiang. R., Muliyati, N. W., Zhang, X., Amer, M. A., Baxter, I., Brachi, B., Chory, J., Dean, C., Debieu, M., de Meaux, J., Ecker, J. R., Faure, N., Kniskern, J. M., Jones, J. D., Michael, T., Nemri, A., Roux, F., Salt, D. E., Tang, C., Todesco, M., Traw, M. B., Weigel, D., Marjoram, P., Borevitz, J. O., Bergelson, J. and Nordborg, M. (2010). Genome-wide association study of 107 phenotypes in a common set of *Arabidopsis thaliana* inbred lines. *Nature*, 465, 627–31.

Bandillo, N., Jarquin, D., Song, Q., Nelson, R., Cregan, P., Specht, J. and Lorenz, A. (2015). A population structure and genome-wide association analysis on the USDA soybean germplasm collection. *The Plant Genome*, 8, 1–13.

Beaver, J. S. and Osorno, J. M.. (2009). Achievements and limitations of contemporary common bean breeding using conventional and molecular approaches. *Euphytica*, 168, 145–76.

Beebe, S. E. (2012). Common bean breeding in the tropics. *Plant Breeding Review*, 36, 357–426.

Beebe, S., Rengifo, J., Gaitan, E., Duque, M. C. and Tohme, J.. 2001. Diversity and origin of Andean landraces of common bean. *Crop Science*, 41, 854–62.

Belamkar, V., Wenger, A., Kalberer, S. R., Bhattacharya, V. G., Blackmon, W. J. and Cannon, S. B. (2015). Evaluation of phenotypic variation in a collection of an edible tuberous legume. *Crop Science*, 55(2), 712–26.

Bennett, S., Francis, C. and Reid, B. (2001). Minor and under-utilised legumes. In: N. Maxted and S. J. Bennett (Eds), *Plant Genetic Resources of Legumes in the Mediterranean* (pp. 207–30). Springer, Netherlands.

Bertioli, D. J., Cannon, S. B., Froenicke, L., Huang, G., Farmer, A. D., Cannon, E. K., Liu, X., Gao, D., Clevenger, J., Dash, S., Ren, L., Moretzsohn, M. C., Shirasawa, K., Huang, W., Vidigal, B., Abernathy, B., Chu, Y., Niederhuth, C. E., Umale, P,, Araújo, A. C., Kozik, A., Kim, K. D., Burow, M. D., Varshney, R. K., Wang, X., Zhang, X., Barkley N, Guimarães, P. M., Isobe, S., Guo, B., Liao, B., Stalker, H. T., Schmitz, R. J., Scheffler, B. E., Leal-Bertioli, S. C., Xun, X., Jackson, S. A., Michelmore, R. and Ozias-Akins, P. (2016). The genome sequences of *Arachis duranensis* and *Arachis ipaensis*, the diploid ancestors of cultivated peanut. *Nature Genetics*, 48, 438–46.

Bitocchi, E., Bellucci, E., Giardini, A., Rau, D., Rodriguez, M., Biagetti, E., Santilocchi, R., Spagnoletti Zeuli, P., Gioia, T., Logozzo, G., Attene, G., Nanni, L. and Papa, R. (2013). Molecular analysis of the parallel domestication of the common bean (*Phaseolus vulgaris*) in Mesoamerica and the Andes. *New Phytology*, 197, 300–13.

Bitocchi, E., Nanni, L., Bellucci, E., Rossi, M., Giardini, A., Zeuli, P. S., Logozzo, G., Stougaard, J., McClean, P., Attene, G. and Papa, R. (2012) Mesoamerican origin of the common bean (*Phaseolus vulgaris* L.) is revealed by sequence data. *PNAS*, 109, E788–96.

Bitocchi, E., Rau, D., Bellucci, E., Rodriguez, M., Murgia, M. L., Gioia, T., Santo, D., Nanni, L., Attene, G. and Papa, R. (2017). Beans (Phaseolus ssp.) as a model for understanding crop evolution. *Frontiers in Plant Science*, 8, 722.

Bohra, A., Jha, U. C., Kishor, P. K., Pandey, S. and Singh, N. P. (2014a). Genomics and molecular breeding in lesser explored pulse crops: Current trends and future opportunities. *Biotechnology Advances*, 32(8), 1410–28.

Bohra, A., Pandey, M. K., Jha, U. C., Singh, B., Singh, I. P., Datta, D., Chaturvedi, S. K., Nadarajan, N. and Varshney, R. K. (2014b). Genomics-assisted breeding in four major pulse crops of developing countries: Present status and prospects. *Theoretical and Applied Genetics*, 127(6), 1263–91.

Bomgardner, M. M. (2017). CRISPR: A new toolbox for better crops. *Chemical and Engineering News*, 95, 30–4.

Bonfim, K., Faria, J. C., Nogueira, E., Érica, A., Mendes, E. A. and Aragão, F. J. L. (2007). RNAi-Mediated Resistance to Bean golden mosaic virus in Genetically engineered common bean (*Phaseolus vulgaris* L.). *Molecular Plant-Microbe Interactions*, 20(6), 717–26.

Bowers, J. E., Chapman, B. A., Rong, J. and Paterson, A. H. (2003) Unravelling angiosperm genome evolution by phylogenetic analysis of chromosomal duplication events. *Nature*, 422, 433–8.

Branca, A., Paape, T. D., Zhou, P., Briskine, R., Farmer, A. D., Mudge, J., Bharti, A. K., Woodward, J. E., May, G. D., Gentzbittel, L., Ben, C., Denny, R., Sadowsky, M. J., Ronfort, J., Bataillon, T., Young, N. D. and Tiffin, P. (2011). Whole-genome nucleotide diversity, recombination, and linkage disequilibrium in the model legume *Medicago truncatula*. *Proceedings of the National Academy of Sciences*, 108(42), E864–70.

Campa, A. and Ferreira, J. J. (2017). Gene coding for an elongation factor is involved in resistance against powdery mildew in common bean. *Theoretical and Applied Genetics*, 130, 849–60.

Cavanagh, C., Morell, M., Mackay. I. and Powell, W. (2008). From mutations to MAGIC: Resources for gene discovery, validation and delivery in crop plants. *Current Opinion on Plant Biology*, 11, 215–21.

Cichy, K. A., Porch, T. G., Beaver, J. S., Cregan, P., Fourie, D., Glahn, R. P., Grusak, M. A., Kkamfwa, K., Katuuramu, D. N., McClean, P., Mndolwa, E., Nchimbi-Msolla, S., Pastor-Corrales, M. A. and Miklas, P. N. (2015). A Phaseolus diversity panel for Andean bean improvement. *Crop Science*, 55, 2149–60.

CGIAR. (2012). *Grain Legumes – Leveraging to Combat Poverty, Hunger, Malnutrition, and Environmental Degradation*. CRP 3.5 Grain Legumes. Consultative Group on International Agricultural Research. ICRISAT, Patancheru, India. https://library.cgiar.org/handle/10947/5316.

Conrad, D. F., Pinto, D., Redon, R., Feuk, L., Gokcumen, O., Zhang, Y., Aerts, J., Andrews, T. D., Barnes, C., Campbell, P., Fitzgerald, T., Hu, M., Ihm, C. H., Kristiansson, K., Macarthur, D. G., Macdonald, J. R., Onyiah, I., Pang, A. W., Robson, S., Stirrups, K., Valsesia, A., Walter, K. and Wei, J., Wellcome Trust Case Control Consortium, Tyler-Smith, C., Carter, N. P., Lee, C., Scherer, S. W. and Hurles, M. E. (2010). Origins and functional impact of copy number variation in the human genome. *Nature*, 464, 704–12.

Curran, J., McLachlan, M., Black, R., Widders, I. and Manary, M. (2017). Collaboration among sectors to increase pulse consumption. *Annals of the New York Academy of Sciences*, 1392, 3–5.

Darrasse, A., Carrère, S., Barbe, V., Boureau, T., Arrieta-Ortiz, M. L., Bonneau, S., Briand, M., Brin, C., Cociancich, S., Durand, K. and Fouteau, S. (2013). Genome sequence of *Xanthomonas fuscans* subsp. *fuscans* strain 4834-R reveals that flagellar motility is not a general feature of xanthomonads. *BMC Genomics*, 14(1), 761.

Demorest, Z. L., Coffman, A., Baltes, N. J., Stoddard, T. J., Clasen, B. M., Luo, S., Retterath, A., Yabandith, A., Gamo, M. E., Bissen, J. and Mathis, L. (2016). Direct stacking of sequence-specific nuclease-induced mutations to produce high oleic and low linolenic soybean oil. *BMC plant biology*, 16(1), 225.

de Ron, A. M. (ed.) (2015). *Grain Legumes* (Volume 10). Springer, New York, USA.

Eapen, S. (2008) Advances in development of transgenic pulse crops. *Biotechnology Advances*, 26, 162–8.

FAO. (1994). Chapter 4. Pulses and derived products. In: *Definition and Classification of Commodities*. Food and Agriculture Organization, Rome, Italy. http://www.fao.org/ES/faodef/fdef04e.htm.

FAO. (2016). *Pulses – Nutrition Seeds for a Sustainable Future*. Food and Agriculture Organization, Rome, Italy. ISBN: 978-92-109463-1. http://www.fao.org/3/a-i5528e.pdf.

Faria, J. C., Valdisser, P. A. M. R., Nogueira, E. O. P. L. and Aragão, F. J. L. (2014). RNAi-based *Bean golden mosaic virus*-resistant common bean (EMBRAPA 5.1) shows simple inheritance for both transgene and disease resistance. *Plant Breeding*, 133, 649–53.

Faria, J., Aragão, F., Souza, T., Quintela, E., Kitajima, E. and Ribeiro, S. (2016). Golden mosaic of common beans in Brazil: Management with a transgenic approach. *APS Features*. doi:10.1094/APSFeature-2016-10.

Fehr, W. R. (1987). *Principles of Cultivar Development: Volume 1 Theory and Technique*. Macmillan Publishing Company, New York.

Feil, H., Feil, W. S., Chain, P., Larimer, F., DiBartolo, G., Copeland, A., Lykidis, A., Trong, S., Nolan, M., Goltsman, E. and Thiel, J. (2005). Comparison of the complete genome sequences of *Pseudomonas syringae* pv. *syringae* B728a and pv. *tomato* DC3000. *Proceedings of the National Academy of Sciences of the United States of America*, 102(31), 11064–9.

Fondevilla, S., Krezdorn, N., Rotter, B., Kahl, G. and Winter, P. (2015). *In planta* identification of putative pathogenicity factors from the chickpea pathogen *Ascochyta rabiei* by *De novo* transcriptome sequencing using RNA-Seq and massive analysis of cDNA ends. *Frontiers in Microbiology*, 6, 1329.

Fournier-Level, A., Korte, A., Cooper, M. D., Nordborg, M., Schmitt, J. and Wilczek, A. M. (2011). A map of local adaptation in *Arabidopsis thaliana*. *Science*, 334, 86–9.

Foyer, C. H., Lam, H. M., Nguyen, H. T., Siddique, K. H., Varshney, R. K., Colmer, T. D., Cowling, W., Bramley, H., Mori, T. A., Hodgson, J. M. and Cooper, J. W. (2016). Neglecting legumes has compromised human health and sustainable food production. *Nature Plants*, 2, 16112.

Freytag G. F. and Debouck D. G. (2002). *Taxonomy, Distribution and Ecology of the Genus Phaseolus (Leguminosae-Papilionoideae) in North America, Mexico and Central America*. Sida, Bot Misc 23, Bot Res Inst Texas, Fort Worth, TX, USA.

Goel, S., Raina, S. N. and Ogihara, Y. (2002). Molecular evolution and phylogenetic implications of internal transcribed spacer sequences of nuclear ribosomal DNA in the Phaseolus–Vigna complex. *Molecular Phylogenetics and Evolution*, 22, 1–19.

Hagerty, C. H., Cuesta-Marcos, A., Cregan, P., Song, Q., McClean, P. and Myers, J. R. (2016). Mapping snap bean pod and color traits, in a dry bean × snap bean recombinant inbred population. *Journal of the American Society for Horticultural Science*, 141(2), 131–8.

Hane, J. K., Anderson, J. P., Williams, A. H., Sperschneider, J. and Singh, K. B. (2014). Genome sequencing and comparative genomics of the broad host-range pathogen *Rhizoctonia solani* AG8. *PLoS Genetics*, 10(5), e1004281.

Heffner, E. L., Sorrells, M. E. and Jannink, J. (2009). Genomic selection for crop improvement. *Crop Science*, 49, 1–12.

Heilig, J. A., Beaver, J. S., Wright, E. M., Song, Q. and Kelly, J. D. (2017). QTL analysis of symbiotic nitrogen fixation in a black bean population. *Crop Science*, 57(1), 118–29.

Horton, M. W., Hancock, A. M., Huang, Y. S., Toomajian, C., Atwell, S., Auton, A., Muliyati, N. W., Platt, A., Sperone, F. G., Vilhjálmsson, B. J., Nordborg, M., Borevitz, J. O. and Bergelson, J. (2012). Genome-wide patterns of genetic variation in worldwide *Arabidopsis thaliana* accessions from the RegMap panel. *Nature Genetics*, 44, 212–16.

Hoyos-Villegas, V., Song, Q. and Kelly, J. D. (2017). Genome-wide association analysis for drought tolerance and associated traits in common bean. *The Plant Genome*, 10, 1–17.

Huang, B. E., Clifford, D. and Cavanagh, C. (2013). Selecting subsets of genotyped experimental populations for phenotyping to maximize genetic diversity. *Theoretical and Applied Genetics*, 126, 379–88

Huang, B. E., Verbyla, K. L., Verbyla, A. P., Raghavan, C., Singh, V. K., Gaur, P., Leung, H., Varshney, R. K. and Cavanagh, C. R. (2015). MAGIC populations in crops: Current status and future prospects. *Theoretical and Applied Genetics*, 128(6), 999–1017.

Hurtado-Gonzales, O. P., Valentini, G., Gilio, T. A., Martins, A. M., Song, Q. and Pastor-Corrales, M. A. (2017). Fine mapping of Ur-3, a historically important rust resistance locus in common bean. *G3: Genes, Genomes, Genetics*, 7, 557–69.

Huynh, B. L., Ehlers, J. D., Munoz-Amatriain, M., Lonardi, S., Santos, J. R., Ndeve, A., Batieno, B. J., Boukar, O., Cisse, N., Drabo, I. and Fatokun, C. (2017). A multi-parent advanced generation inter-cross population for genetic analysis of multiple traits in cowpea (*Vigna unguiculata* L. Walp.). *bioRxiv*, p. 149476.

Hwang, E. Y., Song, Q., Jia, G., Specht, J. E., Hyten, D. L., Costa, J. and Cregan, P. B. (2014). A genome-wide association study of seed protein and oil content in soybean. *BMC Genomics*, 15, 1–12.

Hyten, D. L., Song, Q., Zhu, Y., Choi, I. Y., Nelson, R. L., Costa, J. M., Specht, J. E., Shoemaker, R. C. and Cregan, P. B. (2006). Impacts of genetic bottlenecks on soybean genome diversity. *Proceedings of the National Academy of Sciences*, 103, 16666–71. Hyten, D. L., Choi, I. Y., Song, Q., Specht, J. E., Carter, T. E., Shoemaker, R. C., Ewang, E. Y., Matukumalli, L. K. and Cregan, P. B. (2010). A high density integrated genetic linkage map of soybean and the development of a 1536 universal soy linkage panel for quantitative trait locus mapping. *Crop Science*, 50, 960–8.

Jacob, C., Carrasco, B. and Schwember, A. R. (2016). Advances in breeding and biotechnology of legume crops. *Plant Cell, Tissue and Organ Culture (PCTOC)*, 127(3), 561–84.

Jaganathan, D., Thudi, M., Kale, S., Azam, S., Roorkiwal, M., Gaur, P. M., Kishor, P. B., Nguyen, H., Sutton, T. and Varshney, R. K. (2015). Genotyping-by-sequencing based intra-specific genetic map refines a 'QTL-hotspot' region for drought tolerance in chickpea. *Molecular Genetics and Genomics*, 290(2), 559–71.

Jaillon, O., Aury, J. M., Noel, B., Policriti, A., Clepet, C., Casagrande, A., Choisne, N., Aubourg, S., Vitulo, N., Jubin, C., Vezzi, A., Legeai, F., Hugueney, P., Dasilva, C., Horner, D., Mica, E., Jublot, D., Poulain, J., Bruyère, C., Billault, A., Segurens, B., Gouyvenoux, M., Ugarte, E., Cattonaro, F., Anthouard, V., Vico, V., del Fabbro, C., Alaux, M., di Gaspero, G., Dumas, V., Felice, N., Paillard, S., Juman, I., Moroldo, M., Scalabrin, S., Canaguier, A., Le Clainche, I., Malacrida, G., Durand, E., Pesole, G., Laucou, V., Chatelet, P., Merdinoglu, D., Delledonne, M., Pezzotti, M., Lecharny, A., Scarpelli, C., Artiguenave, F., Pè, M. E., Valle, G., Morgante, M., Caboche, M., Adam-Blondon, A. F., Weissenbach, J., Quétier, F. and Wincker, P. (2007). The grapevine genome sequence suggests ancestral hexaploidization in major angiosperm phyla. *Nature*, 449, 463–7.

Joshi, P. K. and Rao, P. P. (2017). Global pulses scenario: Status and outlook. *Annals Of New York Academy of Sciences*, 1392, 6–17.

Kang, Y. J., Kim, S. K., Kim, M. Y., Lestari, P., Kim, K. H., Ha, B. K., Jun, T. H., Hwang, W. J., Lee, T., Lee, J., Shim, S., Yoon, M. Y., Jang, Y. E., Han, K. S., Taeprayoon, P., Yoon, N., Somta, P., Tanya, P., Kim, K. S., Gwag, J. G., Moon, J. K., Lee, Y. H., Park, B. S., Bombarely, A., Doyle, J. J., Jackson, S. A., Schafleitner, R., Srinives, P., Varshney, R. K. and Lee, S. H. (2014). Genome sequence of mungbean and insights into evolution within Vigna species. *Nature Communications*, 5, 5443.

Kang, Y. J., Satyawan, D., Shim, S., Lee, T., Lee, J., Hwang, W. J., Kim, S. K., Lestari, P., Laosatit, K., Kim, K. H., Ha, T. J., Chitikineni, A., Kim, M. Y., Ko, J. M., Gwag, J. G., Moon, J. K., Lee, Y. H., Park, B. S., Varshney, R. K. and Lee, S. H. (2015). Draft genome sequence of adzuki bean, *Vigna angularis*. *Scientific Reports*, 5, 8069.

Kelly, J. D., Kolkman, J. M., Schneider, K. (1998). Breeding for yield in dry bean (*Phaseolus vulgaris* L.). *Euphytica*, 102, 343–56.

Kendrick, M. D., Harris, D. K., Ha, B. K., Hyten, D. L., Cregan, P. B., Frederick, R. D., Boerma, H. R. and Pedley, K. F. (2011). Identification of a second Asian soybean rust resistance gene in Hyuuga soybean. *Phytopathology*, 101, 535–43.

Kim, S., Plagnol, V., Hu, T. T., Toomajian, C., Clark, R. M., Ossowski, S., Ecker, J. R., Weigel, D. and Nordborg, M. (2007). Recombination and linkage disequilibrium in *Arabidopsis thaliana*. *Nature Genetics*, 39, 1151.

Kover, P. X., Valdar, W., Trakalo, J., Scarcelli, N, Ehrenreich, I. M., Purugganan, M. D. Durrant, C. and Mott, R. (2009) A multiparent advanced generation inter-cross to fine-map quantitative traits in *Arabidopsis thaliana*. *PLoS Genetics*, 5, e1000551.

Kujur, A., Upadhyaya, H. D., Bajaj, D., Gowda, C. L. L., Sharma, S., Tyagi, A. K. and Parida, S. K. (2016). Identification of candidate genes and natural allelic variants for QTLs governing plant height in chickpea. *Scientific Reports*, 6, 27968.

Kumar, J., Choudhary, A. K., Solanki, R. K. and Pratap, A., 2011. Towards marker-assisted selection in pulses: A review. *Plant Breeding*, 130(3), 297–313.

Kwak, M., Kami, J., and Gepts, P. (2009). The putative Mesoamerican domestication center of *Phaseolus vulgaris* is located in the Lerma-Santiago Basin of Mexico. *Crop Science*, 49, 554–63.

Kwak, M., Toro, O., Debouck, D. G. and Gepts, P. (2012). Multiple origins of the determinate growth habit in domesticated common bean (*Phaseolus vulgaris*). *Annals of Botany*, 110, 1573–80.

Lam, H. M., Xu, X., Liu, X., Chen, W., Yang, G., Wong, F. L., Li, M. W., He, W, Qin, N., Wang, B., Li, J., Jian, M., Wang, J., Shao, G., Wang, J., Sun, S. S. and Zhang, G. (2010). Resequencing of 31 wild and cultivated soybean genomes identifies patterns of genetic diversity and selection. *Nature Genetics*, 42(12), 1053–9.

Lévesque, C. A., Brouwer, H., Cano, L., Hamilton, J. P., Holt, C., Huitema, E., Raffaele, S., Robideau, G. P., Thines, M., Win, J. and Zerillo, M. M. (2010). Genome sequence of the necrotrophic plant pathogen *Pythium ultimum* reveals original pathogenicity mechanisms and effector repertoire. *Genome Biology*, 11(7), R73.

Link, T. I., Lang, P., Scheffler, B. E., Duke, M. V., Graham, M. A., Cooper, B., Tucker, M. L., Mortel, M., Voegele, R. T., Mendgen, K. and Baum, T. J. (2014). The haustorial transcriptomes of *Uromyces*

appendiculatus and *Phakopsora pachyrhizi* and their candidate effector families. *Molecular plant pathology*, 15(4), 379–93.
Lonardi, S., Zhu, T., Muñoz-Amatriaìn, M., Liang, Q., Wanamaker, S. I., Ounit, R., Alhakami, H., Luo, M. C. and Close, T. J. (2017). Assembly of eleven pseudomolecules representing the cowpea genome sequence. In: *Plant and Animal Genome XXIV, San Diego, California, 14–18 January 2017*. https://pag.confex.com/pag/xxv/webprogram/Paper24482.html.
Lucas, M. R., Diop, N. N., Wanamaker, S., Ehlers, J. D., Roberts, P. A. and Close, T. J. (2011). Cowpea–soybean synteny clarified through an improved genetic map. *Plant Genome*, 4, 218–25.
Ma, L. J., van der Does, H. C., Borkovich, K. A., Coleman, J. J., Daboussi, M. J., Di Pietro, A., Dufresne, M., Freitag, M., Grabherr, M., Henrissat, B. and Houterman, P. M. (2010). Comparative genomics reveals mobile pathogenicity chromosomes in Fusarium. *Nature*, 464(7287), 367.
Mamidi, S., Chikara, S., Goos, R. J., Hyten, D. L., Annam, D., Moghaddam, S. M., Lee, R. K., Cregan, P. B. and McClean, P. E. (2011a). Genome-wide association analysis identifies candidate genes associated with iron deficiency chlorosis in soybean. *The Plant Genome*, 4(3), 154–64.
Mamidi, S., Rossi, M., Annam, D., Moghaddam, S., Lee, R., Papa, R. and McClean, P. (2011b). Investigation of the domestication of common bean (*Phaseolus vulgaris*) using multilocus sequence data. *Functional Plant Biology*, 38, 953–67.
Mamidi, S., Rossi, M., Moghaddam, S. M., Annam, D., Lee, R., Papa, R. and McClean, P. E. (2013). Demographic factors shaped diversity in the two gene pools of wild common bean Phaseolus vulgaris L. *Heredity*, 110, 267–76.
Mamidi, S., Miklas, P. N., Trapp, J., Felicetti, E., Grimwood, J., Schmutz, J., Lee, R. and McClean, P. E. (2016). Sequence-based introgression mapping identifies candidate white mold tolerance genes in common bean. *The Plant Genome*, 9, 1–11.
Manolio, T. A. (2013). Bringing genome-wide association findings into clinical use. *Nature Reviews Genetics*, 14, 549–58.
McDermott, J. and Wyatt, A. J. (2017). The role of pulses in sustainable and healthy food systems. *Annals of New York Academy of Sciences*, 1392, 30–42.
Meziadi, C., Richard, M. M., Derquennes, A., Thareau, V., Blanchet, S., Gratias, A., Pflieger, S. and Geffroy, V. (2016). Development of molecular markers linked to disease resistance genes in common bean based on whole genome sequence. *Plant Science*, 242, 351–7.
Miklas, P. N., Kelly, J. D., Beebe, S. E. and Blair, M. W. (2006). Common bean breeding for resistance against biotic and abiotic stresses: From classical to MAS breeding. *Euphytica*, 147, 105–31.
Mitter, N., Worrall, E. A., Robinson, K. E., Li, P., Jain, R. G., Taochy, C., Fletcher, S. J., Carroll, B. J., Lu, G. Q. and Xu, Z. P., 2017. Clay nanosheets for topical delivery of RNAi for sustained protection against plant viruses. *Nature Plants*, 3, 16207.
Moghaddam, S. M., Mamidi, S., Osorno, J. M., Lee, R., Brick, M., Kelly, J., Miklas. P,, Urrea, C., Song, Q., Cregan, P., Grimwood, J., Schmutz, J. and McClean, P. E. (2016). Genome-wide association study identifies candidate loci underlying agronomic traits in a Middle American diversity panel of common bean. *The Plant Genome*, 9, 1–21.
Muehlbauer, F. J. (1993). Food and grain legumes, pp. 256–65. In: J. Janick and J. E. Simon (Eds), *New Crops*. Wiley, New York.
Muehlbauer, F. J. and Kaiser, W. J. (1992). Expanding the production and use of cool season food legumes. In: *Proceedings of Second International Food Legume Research Conference*. El Cairo, Egypt.
Muñoz, N., Liu, A., Kan, L., Li, M.-W. and Lam, H.-M. (2017). Potential uses of wild germplasms of grain legumes for crop improvement. *International Journal of Molecular Sciences*, 18(2), 328.
Muñoz-Amatriaín, M., Mirebrahim, H., Xu, P., Wanamaker, S. I., Luo, M., Alhakami, H., Alpert, M., Atokple, I., Batieno, B. J., Boukar, O., Bozdag, S., Cisse, N., Drabo, I., Ehlers, J. D., Farmer, A., Fatokun, C., Gu, Y. Q., Guo, Y. N., Huynh, B. L., Jackson, S. A., Kusi, F., Lawley, C. T., Lucas, M. R., Ma, Y., Timko, M. P., Wu J., You, F., Barkley, N. A., Roberts, P. A., Lonardi, S. and Close, T. J. (2017). Genome resources for climate-resilient cowpea, an essential crop for food security. *The Plant Journal*, 89(5), 1042–54.

Osorno, J. M. and McClean, P. E. (2014). Common bean genomics and its applications in breeding programs, pp. 185–206. In: S. Gupta, N. Nadarajan and D. S. Gupta (Eds), *Legumes in the Omic Era*. Springer, New York, NY.

Palmer, R. G., Gai, J., Sun, H. and Burton, J. W. (2001). Production and evaluation of hybrid soybean. *Plant Breeding Reviews*, 21, 263–307.

Pandey, M. K., Roorkiwal, M., Singh, V. K., Ramalingam, A., Kudapa, H., Thudi, M., Chitikineni, A., Rathore, A. and Varshney, R. K. (2016) Emerging genomic tools for legume breeding: Current status and future prospects. *Frontiers Plant Science*, 7, 455.

Pandey, M. K., Wang, H., Khera, P., Vishwakarma, M. K., Kale, S. M., Culbreath, A. K., Holbrook, C. C., Wang, X., Varshney, R. K. and Guo, B. (2017). Genetic dissection of novel QTLs for resistance to leaf spots and tomato spotted wilt virus in peanut (*Arachis hypogaea* L.). *Frontiers in Plant Science*, 8, 25.

Perez de la Vega, M., Torres, A. M., Cubero, J. I. and Kole, C. (2012). *Genetics, Genomics and Breeding of Cool Season Grain Legumes*. CRC Press, Taylor and Francis Group, Boca Raton, FL.

Pratap, K. and Kumar, J. (2011). *Biology and Breeding of Food Legumes*. CABI, Oxford, U. K.

Rasheed, A., Hao, Y., Xia, X., Khan, A., Xu, Y., Varshney, R. K. and He, Z. (2017). Crop breeding chips and genotyping platforms: Progress, challenges and perspectives. *Molecular Plant*, 10(8), 1047–64.

Repinski, S. L., Kwak, M., and Gepts, P. (2012). The common bean growth habit gene PvTFL1y is a functional homolog of Arabidopsis TFL1. *Theoretical and Applied Genetics*, 124, 1539–47.

Rubiales, D., Fondevilla, S., Chen, W., Gentzbittel, L., Higgins, T. J. V., Castillejo, M. A., Singh, K. B. and Rispail, N. (2015). Achievements and challenges in legume breeding for pest and disease resistance. *Critical Reviews in Plant Sciences*, 34, 1–3, 195–236.

Sankaran, S., Khot, L. R., Espinoza, C. Z., Jarolmasjed, S., Sathuvalli, V. R., Vandemark, G. J., Miklas, P. N., Carter, A. H., Pumphrey, M. O., Knowles, N. R. and Pavek, M. J. (2015). Low-altitude, high-resolution aerial imaging systems for row and field crop phenotyping: A review. *European Journal of Agronomy*, 70, 112–23.

Sawargaonkar, S. L., Saxena, K. B. and Saxena, R. K. (2016). Overview of hybrid pigeonpea seed production technology and its on-farm validation. *Plant Knowledge Journal (PKJ)*, 5, 13–17.

Saxena, M. K., Saxena, U., Saxena, K. B., Khandalkar, V. S. and Sultana, R. (2011). Profitability and production cost of hybrid pigeonpea seed. *Electronic Journal of Plant Breeding*, 2, 409–12.

Saxena, K. B., Kumar, R. V., Tikle, A. N., Saxena, M. K., Gautam, V. S., Rao, S. K., Khare, D. K., Chauhan, Y. S., Saxena, R. K., Reddy, B. V.S. and Sharma, D. (2013). ICPH 2671–the world's first commercial food legume hybrid. *Plant Breeding*, 132(5), 479–85.

Scheben, A., Batley, J. and Edwards, D. (2016). Genotyping-by-sequencing approaches to characterize crop genomes: Choosing the right tool for the right application. *Plant Biotechnology Journal*, 15, 149–61.

Scheben, A., Wolter, F., Batley, J., Puchta, H. and Edwards, D. (2017). Towards CRISPR/Cas crops – bringing together genomics and genome editing. *New Phytologist*, 216, 682–98.

Schlueter, J. A., Dixon, P., Granger, C., Grant, D., Clark, L., Doyle, J. J. and Shoemaker, R. C. (2004). Mining EST databases to resolve evolutionary events in major crop species. *Genome*, 47(5), 868–76.

Schmutz, J., Cannon, S. B., Schlueter, J., Ma, J., Mitros, T., Nelson, W., Hyten, D. L., Song, Q, Thelen, J. J., Cheng, J., Xu, D., Hellsten, U., May, G. D., Yu, Y., Sakurai, T, Umezawa, T., Bhattacharyya, M. K., Sandhu, D., Valliyodan, B., Lindquist, E., Peto, M., Grant, D., Shu, S., Goodstein, D., Barry, K., Futrell-Griggs, M., Abernathy, B., Du, J., Tian, Z., Zhu, L., Gill, N., Joshi, T., Libault, M., Sethuraman, A., Zhang, X. C., Shinozaki, K., Nguyen, H. T., Wing, R. A., Cregan, P., Specht, J., Grimwood, J., Rokhsar, D., Stacey, G., Shoemaker, R. C. and Jackson, S. A. (2010). Genome sequence of the palaeopolyploid soybean. *Nature*, 463, 178–83.

Schmutz, J., McClean, P. E., Mamidi, S., Wu, G. A., Cannon, S. B., Grimwood, J., Jenkins, J., Shu, S., Song, Q., Chavarro, C., Torres-Torres, M., Geffroy, V., Moghaddam, S. M., Gao, D., Abernathy, B, Barry, K., Blair, M., Brick, M. A., Chovatia, M., Gepts, P., Goodstein, D. M., Gonzales, M., Hellsten, U., Hyten, D. L., Jia, G., Kelly, J. D., Kudrna, D., Lee, R., Richard, M. M., Miklas, P. N., Osorno, J. M., Rodrigues, J., Thareau, V., Urrea, C. A., Wang, M., Yu,

Y., Zhang, M., Wing, R. A., Cregan, P. B., Rokhsar, D. S., Jackson, S. A. (2014). A reference genome for common bean and genome-wide analysis of dual domestications. *Nature Genetics*, 46, 707–13.

Schröder, S., Mamidi, S., Lee, R., McKain, M. R., McClean, P. E. and Osorno, J. M. (2016). Optimization of genotyping by sequencing (GBS) data in common bean (*Phaseolus vulgaris* L.). *Molecular Breeding*, 36, 6.

Schwartz, H. F., Steadman, J. R., Hall, R. and Forster, R. L. (2005). *Compendium of Bean Diseases* (2nd Edition). American Phytopathological Society (APS Press), St. Paul, MN, USA.

Singh, S. P., Gepts, P. and Debouck, D. G. (1991). Races of common bean (*Phaseolus vulgaris*, Fabaceae). *Economic Botany*, 45, 379–96.

Singh, V. K., Khan, A. W., Jaganathan, D., Thudi, M., Roorkiwal, M., Takagi, H., Garg, V., Kumar, V., Chitikineni, A., Gaur, P. M., Sutton, T., Terauchi, R. and Varshney, R. K. (2016). QTL-seq for rapid identification of candidate genes for 100-seed weight and root/total plant dry weight ratio under rainfed conditions in chickpea. *Plant Biotechnology Journal*, 14, 2110–19.

Singh, M., Bisht, I. S. and Dutta, M. (2014). *Broadening the Genetic Base of Grain Legumes*. Springer, New Delhi, India.

Sivasankar, S., Ellis, N., Buruchara, R., Henry, C., Rubiales, D., Shandu, J. S. and Negra, C. (2016). *10-Year Research Strategy for Pulse Crops*. http://iyp2016.org/resources/technical-reports/183-10-year-research-strategy-pulse-crops-final.

Smartt, J. (1990). *Grain Legumes: Evolution Andgenetic Resources*. Cambridge Univ. Press, Cambridge, UK.

Song, Q., Hyten, D. L., Jia, G., Quigley, C. V., Fickus, E. W., Nelson, R. L. and Cregan, P. B. (2013). Development and evaluation of SoySNP50K, a high-density genotyping array for soybean. *PLoS One*, 8(1), e54985.

Song, Q., Jia, G., Hyten. D. L., Jenkins, J., Hwang, E. Y., Schroeder, S. G., Osorno, J. M., Schmutz, J., Jackson, S. A., McClean, P. E. and Cregan, P. B. (2015). SNP assay development for linkage map construction, anchoring whole-genome sequence, and other genetic and genomic applications in common bean. *G3: Genes, Genomes, Genetics*, 5(11), 2285–90.

Srivastava, R., Upadhyaya, H. D., Kumar, R. K., Daware, A. V., Basu, U., Shimray, P. W., Tripathi, S., Bharadwaj, Tyagi, A. K. and Parida, S. (2017). A multiple QTL-seq strategy delineates potential genomic loci governing flowering time in chickpea. *Frontiers in Plant Science*, 8, 1105.

Suso, M. J., Bebeli, P. J. and Palmer, R. G., 2015. Reproductive biology of grain legumes. In: A. De Ron (Ed.), *Grain Legumes* (pp. 365–99). Springer, New York.

Tang, H., Krishnakumar, V., Bidwell, S., Rosen, B., Chan, A., Zhou, S., Gentzbittel, L., Childs, K. L., Yandell, M., Gundlach, H., Mayer, K. F., Schwartz, D. C. and Town, C. D. (2014). An improved genome release (version Mt4. 0) for the model legume *Medicago truncatula*. *BMC Genomics*, 15, 312.

Valentini, G., Gonçalves-Vidigal, M. C., Hurtado-Gonzales, O. P., de Lima Castro, S. A., Cregan, P. B., Song, Q. and Pastor-Corrales, M. A. (2017). High-resolution mapping reveals linkage between genes in common bean cultivar Ouro Negro conferring resistance to the rust, anthracnose, and angular leaf spot diseases. *Theoretical and Applied Genetics*, 130, 1705–22.

Vandemark, G. J., Brick, M. A., Osorno, J. M., Kelly, J. D. and Urrea, C. A. (2014). Yield gains in edible grain legumes, pp. 87–124. In: J. Specht, B. Diers, B. Carver and S. Smith (Eds), *Genetic Gains of Major US Field Crops*. CSSA Press, Madison, WI.

van de Wiel, C. C. M., Schaart, J. G., Lotz, L. A. P. and Smulders, M. J. M. (2017). New traits in crops produced by genome editing techniques based on deletions. *Plant Biotechnology Reports*, 11(1), 1–8.

Varshney, R. K., 2016. Exciting journey of 10 years from genomes to fields and markets: Some success stories of genomics-assisted breeding in chickpea, pigeonpea and groundnut. *Plant Science*, 242, 98–107.

Varshney R. K., Chen, W., Li, Y., Bharti, A. K., Saxena, R. K., Schlueter, J. A., Donoghue, M. T., Azam, S., Fan, G., Whaley, A. M., Farmer, A. D., Sheridan, J., Iwata, A., Tuteja, R., Penmetsa, R. V., Wu,

W., Upadhyaya, H. D., Yang, S. P., Shah, T., Saxena, K. B., Michael T, McCombie, W. R., Yang, B., Zhang, G., Yang, H., Wang, J., Spillane, C., Cook, D. R., May, G. D., Xu, X. and Jackson, S. A. (2012). Draft genome sequence of pigeonpea (*Cajanus cajan*), an orphan legume crop of resource-poor farmers. *Nature Biotechnology*, 30, 83–9.

Varshney, R. K., Song, C., Saxena, R. K., Azam, S., Yu, S., Sharpe, A. G., Cannon, S., Baek, J., Rosen, B. D., Tar'an, B., Millan, T., Zhang, X., Ramsay, L. D., Iwata, A., Wang, Y., Nelson, W., Farmer, A. D., Gaur, P. M., Soderlund, C., Penmetsa, R. V., Xu, C., Bharti, A. K., He, W., Winter, P., Zhao, S., Hane, J. K., Carrasquilla-Garcia, N., Condie, J. A., Upadhyaya, H. D., Luo, M. C., Thudi, M., Gowda, C. L., Singh, N. P., Lichtenzveig, J., Gali, K. K., Rubio, J., Nadarajan, N., Dolezel, J., Bansal, K. C., Xu, X., Edwards, D., Zhang, G., Kahl, G., Gil, J., Singh, K. B., Datta, S. K., Jackson, S. A., Wang, J. and Cook, D. R. (2013). Draft genome sequence of chickpea (*Cicer arietinum*) provides a resource for trait improvement. *Nature Biotechnology*, 31(3), 240–6.

Varshney, R. K., Saxena, R. K., Upadhyaya, H. D., Khan, A. W., Yu, Y., Kim, C., Rathore, A., Kim, D., Kim, J., An, S., Kumar, V., Anuradha, G., Yamini, N. Y., Zhang, W., Muniswamy, S., Kim, J.-S., Penmetsa, V., von Wettberg, E. and Datta, S., (2017). Whole-genome resequencing of 292 pigeonpea accessions identifies genomic regions associated with domestication and agronomic traits. *Nature Genetics*, 49, 1082–8.

Vasconcellos, R. C., Oraguzie, O. B., Soler, A., Arkwazee, H., Myers, J. R., Ferreira, J. J., Song, Q., McClean, P. and Miklas, P. N. (2017). Meta-QTL for resistance to white mold in common bean. *PLoS One*, 12(2), e0171685.

Verdcourt, B. (1970). Studies in the Leguminosae–Papilionoideae for the 'Flora of Tropical East Africa' IV. *Kew Bulletin*, 24, 507–69.

Verdcourt, B. (1980). The classification of Dolichos L. emend. Verdc., Lablab Adans., Phaseolus L., Vigna Savi and their allies. In: R. J. Summerfield and A. H. Bunting (Eds), *Advances in Legume Science* (pp. 45–8). Royal Botanic Gardens, Kew.

Verma, S., Gupta, S., Bandhiwal, N., Kumar, T., Bharadwaj, C. and Bhatia, S. (2015). High-density linkage map construction and mapping of seed trait QTLs in chickpea (*Cicer arietinum* L.) using Genotyping-by-Sequencing (GBS). *Scientific Reports*, 5, 17512.

Vlasova, A., Capella-Gutiérrez, S., Rendón-Anaya, M., Hernández-Oñate, M., Minoche, A. E., Erb, I., Câmara, F., Prieto-Barja, P., Corvelo, A., Sanseverino, W., Westergaard, G., Dohm, J. C., Pappas Jr., G. J., Saburido-Alvarez, S., Kedra, D., Gonzalez, I., Cozzuto, L., Gómez-Garrido, J., Aguilar-Morón, M. A., Andreu, N., Aguilar, O. M., Garcia-Mas, J., Zehnsdorf, M., Vázquez, M. P., Delgado-Salinas, A., Delaye, L., Lowy, E., Mentaberry, A., Vianello-Brondani, R. P., García, J. L., Alioto, T., Sánchez, F., Himmelbauer, H., Santalla, M., Notredame, C., Gabaldón, T., Herrera-Estrella, A. and Guigó, R. (2016). Genome and transcriptome analysis of the Mesoamerican common bean and the role of gene duplications in establishing tissue and temporal specialization of genes. *Genome Biology*, 17, 32.

Voysest, O. (2000). *Mejoramiento genético del frijol (Phaseolus vulgaris L.): Legado de variedades de América Latina 1930–1999.* CIAT, Cali, Colombia. [In Spanish].

Vuong, T. D., Sleper, D. A., Shannon, J. G. and Nguyen, H. T. (2010). Novel quantitative trait loci for broad-based resistance to soybean cyst nematode (*Heterodera glycines* Ichinohe) in soybean PI 567516C. *Theoretical and Applied Genetics*, 121, 1253–66.

Vuong, T. D., Sonah, H., Meinhardt, C. G., Deshmukh, R., Kadam, S., Nelson, R. L., Shannon, J. G. and Nguyen, H. T. (2015). Genetic architecture of cyst nematode resistance revealed by genome-wide association study in soybean. *BMC Genomics*, 16, 593.

Waltz, E. (2016). Gene-edited CRISPR mushroom escapes US regulation. *Nature*, 532, 293.

Wang, M. and Jin, H. (2017). Spray-induced gene silencing: A powerful innovative strategy for crop protection. *Trends in Microbiology*, 25(1), 4–6. Wang, J., Chu, S., Zhang, H., Zhu, Y., Cheng, H. and Yu, D. (2016). Development and application of a novel genome-wide SNP array reveals domestication history in soybean. *Scientific Reports*, 6, 20728.

Wang, J., Sun, P., Li, Y., Liu, Y., Yu, J., Ma, X., Sun, S., Yang, N., Xia, R., Lei, T., Liu, X., Jiao, B., Xing, Y., Ge, W., Wang, L., Wang, Z., Song, X., Yuan, M., Guo, D., Zhang, L., Zhang, J., Jin, D., Chen, W., Pan, Y., Liu, T., Jin, L., Sun, J., Yu, J., Cheng, R., Duan, X., Shen, S., Qin, J., Zhang, M. C.,

Paterson, A. H. and Wang, X. (2017). Hierarchically aligning 10 legume genomes establishes a family-level genomics platform. *Plant Physiology*, 174, 284–300.

Weeks, D. P. (2017). Gene editing in polyploid crops: Wheat, camelina, canola, potato, cotton, peanut, sugar cane, and citrus. *Progress in Molecular Biology and Translational Science*, 149, 65–80.

Wellcome Trust Case Control Consortium. (2010). Genome-wide association study of copy number variation in 16,000 cases of eight common diseases and 3,000 shared controls. *Nature*, 464, 713–20.

Williams, A. H., Sharma, M., Thatcher, L. F., Azam, S., Hane, J. K., Sperschneider, J., Kidd, B. N., Anderson, J. P., Ghosh, R., Garg, G. and Lichtenzveig, J. (2016). Comparative genomics and prediction of conditionally dispensable sequences in legume–infecting *Fusarium oxysporum formae speciales* facilitates identification of candidate effectors. *BMC Genomics*, 17(1), 191.

Wong, M. M., Gujaria-Verma, N., Ramsay, L., Yuan, H. Y., Caron, C., Diapari, M., Vandenberg, A. and Bett, K. E. (2015). Classification and characterization of species within the genus Lens using genotyping-by-sequencing (GBS). *PLoS One*, 10, e0122025.

Young, N. D., Debellé, F., Oldroyd, G. E., Geurts, R., Cannon, S. B., Udvardi, M. K., Benedito, V. A., Mayer, K. F., Gouzy, J., Schoof, H., van de Peer, Y., Proost, S., Cook, D. R., Meyers, B. C., Spannagl, M., Cheung, F., de Mita, S., Krishnakumar, V., Gundlach, H., Zhou, S., Mudge, J., Bharti, A. K., Murray, J. D., Naoumkina, M. A., Rosen, B., Silverstein K. A., Tang, H., Rombauts, S., Zhao, P. X., Zhou, P., Barbe, V., Bardou, P., Bechner, M., Bellec, A., Berger, A., Bergès, H., Bidwell, S., Bisseling, T., Choisne, N., Couloux, A., Denny, R., Deshpande, S., Dai, X., Doyle, J. J., Dudez, A. M., Farmer, A. D., Fouteau, S., Franken, C., Gibelin, C., Gish, J., Goldstein, S., González, A. J., Green, P. J., Hallab, A., Hartog, M., Hua, A., Humphray, S. J., Jeong, D. H., Jing, Y., Jöcker, A., Kenton, S. M., Kim, D. J., Klee, K., Lai, H., Lang, C., Lin, S., Macmil, S. L., Magdelenat, G., Matthews, L., McCorrison, J., Monaghan, E. L., Mun, J. H., Najar, F. Z., Nicholson, C., Noirot, C., O'Bleness, M., Paule, C. R., Poulain, J., Prion, F., Qin, B., Qu, C., Retzel, E. F., Riddle, C., Sallet, E., Samain, S., Samson, N., Sanders, I., Saurat, O., Scarpelli, C., Schiex, T., Segurens, B., Severin, A. J., Sherrier, D. J., Shi, R., Sims, S., Singer, S. R., Sinharoy, S., Sterck, L., Viollet, A., Wang, B. B., Wang, K., Wang, M., Wang, X., Warfsmann, J., Weissenbach, J., White, D. D., White, J. D., Wiley, G. B., Wincker, P., Xing, Y., Yang, L., Yao, Z., Ying, F., Zhai, J., Zhou, L., Zuber, A., Dénarié, J., Dixon, R. A., May, G. D., Schwartz, D. C., Rogers, J, Quétier, F, Town, C. D. and Roe, B. A. (2011). The *Medicago* genome provides insight into the evolution of rhizobial symbioses. *Nature*, 480, 520–4.

Yang, K., Tian, Z., Chen, C., Luo, L., Zhao, B., Wang, Z., Yu, L., Li, Y., Sun, Y., Li, W. and Chen, Y. (2015). Genome sequencing of adzuki bean (*Vigna angularis*) provides insight into high starch and low fat accumulation and domestication. *Proceedings of the National Academy of Sciences*, 112(43), 13213–18.

Zargar, S. M., Raatz, B., Sonah, H., Bhat, J. A., Dar, Z. A., Agrawal, G. K. and Rakwal, R. (2015). Recent advances in molecular marker techniques: Insight into QTL mapping, GWAS and genomic selection in plants. *Journal of Crop Science and Biotechnology*, 18(5), pp.293–308.

Zarrei, M., Macdonald, J. R., Merico, D. and Scherer, S. W. (2015). A copy number variation map of the human genome. *Nature Reviews Genetics* 16, 172–83.

Zhang, J., Singh, A., Mueller, D. S. and Singh, A. K. (2015a). Genome-wide association and epistasis studies unravel the genetic architecture of sudden death syndrome resistance in soybean. *The Plant Journal*, 84(6), 1124–36.

Zhang, J., Song, Q., Cregan, P. B., Nelson, R. L., Wang, X., Wu, J. and Jiang, G. L. (2015b). Genome-wide association study for flowering time, maturity dates and plant height in early maturing soybean (*Glycine max*) germplasm. *BMC Genomics*, 16(1), 217.

Zhang, Y. M., Shao, Z. Q., Wang, Q., Hang, Y. Y., Xue, J. Y., Wang, B. and Chen, J. Q. (2016). Uncovering the dynamic evolution of nucleotide-binding site-leucine-rich repeat (NBS-LRR) genes in Brassicaceae. *Journal of Integrative Plant Biology*, 58(2), 165–77.

Zhou, Z., Jiang, Y., Wang, Z., Gou, Z., Lyu, J., Li, W., Yu, Y., Shu, L., Zhao, Y., Ma, Y., Fang, C., Shen, Y., Liu, T., Li, C., Li, Q., Wu, M., Wang, M., Wu, Y., Dong, Y., Wan, W., Wang, X., Ding, Z., Gao, Y., Xiang, H., Zhu, B., Lee, S. H., Wang, W. and Tian, Z. (2015). Resequencing 302 wild and cultivated accessions identifies genes related to domestication and improvement in soybean. *Nature Biotechnology*, 33(4), 408–14.

Zhou, P., Silverstein, K. A., Ramaraj, T., Guhlin, J., Denny, R., Liu, J., Farmer, A. D., Steele, K. P., Stupar, R. M., Miller, J. R., Tiffin, P., Mudge, J. and Young, N. D. (2017). Exploring structural variation and gene family architecture with *De Novo* assemblies of 15 *Medicago* genomes. *BMC Genomics*, 18(1), 261.

Zuiderveen, G. H., Padder, B. A., Kamfwa, K., Song, Q. and Kelly, J. D. (2016). Genome-wide association study of anthracnose resistance in Andean beans (*Phaseolus vulgaris*). *PLoS One*, 11, e0156391.

Chapter 5

Genetic modification of grain legumes

Pooja Bhatnagar-Mathur and Kiran Kumar Sharma, International Crops Research Institute for the Semi-Arid Tropics (ICRISAT), India

1 Introduction

Legumes have a major impact on agriculture and smallholder agriculture in the arid, semi-arid and subhumid countries of the world. Ranked third after cereals and oil seeds in production, edible legume crops include soybean, pigeonpea, chickpea, common pea, lentils and peanuts among others. Grain legumes such as peanut and soybean are also good sources of edible oil. In addition to being rich in proteins, grain legumes are also rich sources of soluble fibre, having twice the protein content of cereals. They are also rich in energy, containing carbohydrates with a low glycaemic index. Since they are good sources of vitamin B, zinc, calcium and magnesium, as well as phytonutrients such as isoflavones, lignans and phytoestrogens, legumes are also known to reduce risks of such diseases as cancer, heart disease and osteoporosis. Being gluten-free, legumes are helpful for people with gluten

http://dx.doi.org/10.19103/AS.2017.0023.36

sensitivity (Ha et al., 2014). Legumes are good for not only people but also the planet as they help in fixing atmospheric nitrogen in the soil through symbiotic association with rhizobia.

Traditional plant breeding methodologies alone have not been successful enough for breeding newer crop varieties that are resistant to various biotic and abiotic stresses. Conventional breeding techniques are limited by sterility barriers and problems of incompatibility. Over the last few decades, genetic transformation technologies have proven to be faster and more precise in controlling desired traits and their introduction into crop plants. The application of recombinant DNA technologies for crop improvement has been found to have great potential in the semi-arid tropics (Sharma and Ortiz, 2000). With the advent of biotechnological tools and technologies, gene transfer across grain legumes has opened up novel opportunities for introducing traits such as disease, pest and virus resistance; abiotic stress tolerance; and quality and nutritional improvements that otherwise are limited by sexual incompatibility.

2 Genetic engineering of grain legumes

Plant genetic transformation involves delivery, integration and expression of foreign genes into plant cells which then regenerate in whole plants, thus allowing the opportunity to express and/or inactivate genes in plant genomes. This technology is not just useful in developing desirable traits, but is also useful in improving our understanding of the genetic basis of plant physiological mechanisms and processes.

The gene transfer process involves choosing a trait, identifying and isolating gene(s) responsible for the trait, followed by introduction of the gene(s) into the host genome. To ensure their functionality, such gene(s) must include regulatory regions that ensure their expression in the plant. The existence of a reliable regeneration system then allows for successful genetic transformation, ensuring that amended DNA sequences are expressed and maintained in the genome through subsequent cell division and development of progeny. Appropriate tissue culture and genetic transformation systems are available for many grain legumes to allow for improvement of this important family of crops (Sharma et al., 2005).

In the past few decades scientists have been successful in developing transgenic legumes with higher yields, improved nutritional properties and resistance to various stresses. Legumes have traditionally been seen as recalcitrant to transformation with the need to develop efficient transformation protocols. With genomic, transcriptomic and proteomic data now available for model legumes such as *Medicago truncatula* and *Lotus japonicus*, genetic engineering tools have now been developed to generate improved, high-yielding legume varieties. Introduction of new traits into commercially important plants through genetic engineering has resulted in plants with features that it would not have been possible to achieve by traditional plant breeding (Tabe et al., 1993; Somers et al., 2003; Szabala et al., 2014). Some recent advances in the genetic modification of grain legumes are discussed in the following sections.

3 Genetic engineering of chickpea

Chickpea (*Cicer arietinum*), an important legume crop and one of the earliest cultivated crops known for its protein content, dates back 7500 years in the Middle East. It has evolved from its wild progenitor *Cicer reticulatum* (Labadi, 1996); cultivation originated in Turkey and then spread into Southwest Asia, the Mediterranean and Ethiopia. Chickpeas are grown in the Indian subcontinent, Australia, the Mediterranean, Western Asia and the northern Great Plains of the United States (van der Maesen, 1987). Chickpeas are highly nutritious and provide over 20% of the daily requirement of essential nutrients. This crop is rich in protein, fibre, folate, minerals like iron and phosphorus and essential amino acids like lysine, isoleucine, tryptophan and total aromatic amino acids.

Chickpeas suffer heavy losses due to susceptibility to insect pests such as the legume pod borer as well as their low tolerance to drought and low temperatures. Chickpea production is affected by several pathogens, with the most important ones being *Ascochyta* blight (AB), *Botrytis* grey mould (BGM), *Fusarium* wilt, *Phytophthora* root rot, dry root rot, *Sclerotinia* stem rot and rust (Haware, 1998).

Over the past four decades, while the area under cultivation of chickpea has remained stagnant, its production has marginally increased due to improvements in yields (ICRISAT, 2007). The average yield of chickpea is below its potential level, although conventional breeding efforts have resulted in the development of resistant and high-yielding varieties. However, further breakthroughs in chickpea production have been constrained by the limited genetic variability present in the available pool of germplasm for the crop. Hence, transferring genes from sources that are otherwise difficult to introduce through conventional breeding provides an effective and alternative crop improvement strategy in chickpea. To accomplish this, it is imperative to identify and isolate the genes responsible for novel traits, and subsequently make these suitable for transfer and expression into chickpea using genetic engineering.

There have been various attempts to develop reliable and reproducible transformation systems for chickpea, although with varying degrees of success and efficiency. Earlier reports on *in vitro* plant regeneration in chickpea used shoot meristem explants (Bajaj and Dhanju, 1979; Sharma et al., 1979) and immature cotyledons (Shri and Davis, 1992; Sagare et al., 1993). Several groups have reported direct or indirect shoot regeneration in chickpea (Khan and Ghosh, 1984; Shri and Davis, 1992; Barna and Wakhlu, 1994; Kar et al., 1996; Altinkut et al., 1997; Subhadra et al., 1998; Chakrabarty et al., 2000; Shikha et al., 2001; Jayanand et al., 2003; Sarmah et al., 2004; Uncuoglu et al., 2008).

Different growth conditions and parameters have been studied and optimized for obtaining maximum transformation efficiency in chickpea. Seeds derived from embryo axes used as explants for chickpea transformation under the *npt*II and *GUS* reporter gene selection initially yielded a transformation efficiency of 4% (Fontana et al., 1993). Subsequently, while transformed embryo axes selected with higher doses of kanamycin made the screening of putative transformants easier, the resistant shoots did not survive following their grafting on wild-type (WT) seedling stocks (Krishnamurthy et al., 2000). Although transformation frequencies and reproducibility were low in these early studies, significant improvements in both transformation frequency and reproducibility were achieved subsequently (Uncuoglu et al., 2008).

An efficient transformation method in chickpea was reported by using axillary meristems from the cotyledonary nodes of *in vitro*-germinated seedlings (Jayanand et al., 2003). The culture conditions for different stages of tissue culture were optimized while placing the elongated shoot on a filter paper bridge immersed in liquid rooting medium increased the rooting efficiency to 90%. Four subsequent studies used embryonic axes as explants for genetic transformation studies in chickpea (Polowick et al., 2004; Sarmah et al., 2004; Senthil et al., 2004; Sanyal et al., 2005). While half embryonic axes were used for transformation by Sarmah et al. (2004), longitudinally sliced embryonic axes were also used (Polowick et al., 2004; Senthil et al., 2004). Sanyal et al. (2005) pre-cultured the explants (L2 layer) for 24 h on a solidified MS-basal medium supplemented with 1 mg L^{-1} BAP prior to co-cultivation. Although these studies reported successful transformation, the overall frequency of transformation was still very low (0.1–1.0%) and required a large number of explants to be co-cultivated to generate sufficient numbers. Another efficient and reproducible transformation protocol was developed for chickpea using germinated seedlings as explants (Senthil et al., 2004). The meristematic explants material was reported to be more susceptible to *Agrobacterium* infection and the marker genes consistently expressed for three generations.

Several reports have suggested the use of various phytohormones to enhance transformational efficiency in chickpea. Direct culture of mature seeds on a MS medium (Murashige and Skoog, 1962) supplemented with thidiazuron (TDZ) and L-proline supported the formation of somatic embryos which later germinated and formed plantlets. It was found that proline shifted the pathway from shoot formation to embryo development in plant cells, with the result that 70% of the resulting plantlets were stable (Murthy, 1996). Culturing immature cotyledons and embryo axes explants on media supplemented with zeatin and indole-3-acetic acid (IAA) resulted in shoot and root formation and stable plant regeneration (Sagare et al., 1993). Similarly immature leaflet explants induced callus formation in media supplemented with 2,4-D or NAA and BAP. The phytohormones supported the development of callus and shoot induction, and addition of potassium phosphate increased multiple shooting efficiency (Barna and Wakhalu, 1994). Low auxin concentration has been reported to induce efficient shoot elongation in chickpea (Chakraborti et al., 2006). Further, the role of TDZ treatment in increasing the rate of multiple shoot formation and subsequently transformation efficiency was reported by several research groups (Jayanand et al., 2003; Senthil et al., 2004; Bhatnagar-Mathur et al., 2009b; Pawar et al., 2012). In order to develop a non-antibiotic marker selection method for chickpea transformation, Tewari-Singh et al. (2004) reported the use of a reliable non-antibiotic selection strategy using the phosphinothricin *N*-acetyltransferase (*pat*) and aspartate kinase (*AK*) genes. Only transformed plants containing the *pat* and sensitized *AK* genes were able to grow on the selection media (Tewari-Singh et al., 2004). The combination of a herbicide or lysine with threonine has also been used for non-antibiotic selection of transformants in chickpea (Singh et al., 2004).

3.1 Biotic stress resistance

Genetic transformation strategies are being exploited for incorporating immunity or high resistance to several viral and fungal diseases. Developing transgenic disease resistance for major pests and diseases in this legume crop has the potential to reduce or eliminate the indiscriminate use of insecticides, fungicides and pesticides that are not only economically burdensome but also a potential threat to health and the environment. Chickpea is affected by many pests and insects. Of these, pod borer (*Helicoverpa armigera*) is an intractable pest estimated to cause losses of over US$400 million in chickpea production in South Asia alone, making it crucial to improve resistance to this insect pest (Sharma et al., 2004). Since no single pest management strategy can be effective, given low levels of resistance in available chickpea germplasm, a potential option is to improve host plant resistance through genetic transformation.

Insecticidal gene encoding for endotoxins such as *Bt* crystal protein genes have been extensively used in generating chickpea transgenic plants for resistance to the legume pod borer, *H. armigera*. Transgenic chickpeas using *cry* genes have been developed at ICRISAT and are being subjected to insect bioassays and evaluated under contained field trials (unpublished data). Constitutive as well as tissue-specific expression of the *cry1Ac* gene in chickpea transgenic events has been found to confer resistance against the legume pod borer. Several lines selected using insect bioassays are currently in the pipeline for product development (Sarmah et al., 2004).

Chickpea productivity is also adversely affected by *Aphis craccivora*, a phloem sap-sucking pest for which genetic transformation strategies have been developed. A powerful anti-feedent *Allium sativum* leaf agglutinin (ASAL) constitutively expressed under CaMV35S promoter was found to be effective against this chickpea pest (Mondal et al., 2009). Moreover, phloem-specific expression of this insecticidal gene also showed a significant reduction in the incidence of this aphid in transgenic chickpea.

Chickpeas are susceptible to a number of diseases that affect its growth and productivity. Several diseases such as *Fusarium* wilt (*Fusarium oxysporum* f. sp. *ciceris*), AB (*Ascochyta rabiei*), BGM (*Botrytis cinerea*), collar rot (*Sclerotium rolfsii*) and dry root rot (*Rhizoctonia bataticola*) have been found to be prevalent in various chickpea growing areas and significantly affect crop yields. Several genetic transformation approaches have been designed to enhance disease resistance in chickpea. Class III fungal chitinases have been expressed in chickpea transgenic plants to improve resistance against fungal infections like *F. oxysporum* (Zarandi et al., 2011).

3.2 Abiotic stress tolerance

Improving the adaptation of chickpea to drought and salinity is critical since it is a rainfed crop at risk to osmotic pressures due to reductions in soil moisture (Bhatnagar-Mathur et al., 2009b). An ethylene-responsive element present in chickpea known as CAP2 protein has been reported to play a major role in growth and stress response. Constitutive expression in tobacco has been found to confer drought and stress tolerance, suggesting the role it plays in cell growth and stress-responsive signalling pathways (Shukla et al., 2006). Constitutive expression of *P5CSF129A* gene encoding for D1-pyrroline-5-carboxylate synthetase in chickpea was carried out to provide 'osmotic protection' to the cellular machinery under terminal drought stress. While the transgenic events accumulated higher proline, this osmotic effect had very little bearing on transpiration efficiency, showing a lack of correlation between the biochemistry and the yield architecture in these transgenic plants (Bhatnagar-Mathur et al., 2009a). Similarly, introduction of stress-inducible *DREB1A* cDNA from *Arabidopsis thaliana,* capable of trans-activating DRE-dependent transcription in chickpea, revealed enhanced drought tolerance component traits such as reduced transpiration rate and extensive rooting under water stress conditions (Anbazhagan et al., 2014, 2015).

3.3 Nutritional biofortification

Like other legumes, chickpea is also deficient in sulphur. To increase sulphur levels, a sunflower seed albumin (*SSA*) gene, under the control a seed-specific promoter from the pea vicilin gene, was introduced into chickpea, alongside a herbicide marker gene under the control of a 35S promoter (Sarmah et al., 2004). These transgenic events resulted in increased levels of methionine and less oxidized sulphur in seeds than the WT plants. This work has paved the way to the production of nutritionally-fortified chickpea in the near future.

4 Genetic engineering of pigeonpea

Pigeonpea (*Cajanus cajan*), one of the important grain legume crops in the semi-arid tropics (SAT), is grown as a stand-alone crop or as an intercrop between cereals. Pigeonpea is cultivated in over 25 tropical and subtropical regions (van der Maesen, 1983). It is thought that cultivation originated around 3500 years ago in India, before spreading to East and West Africa and then Europe (where it was called Congo pea). In terms of area, pigeonpea is ranked fifth after soybean, common bean, peanut and chickpea. About 77% of the world's production is contributed by India. While the wild genepool in pigeonpea shows varied resistance to biotic and abiotic stresses, its transfer to cultivars has been challenging due to incompatibility barriers.

Pigeonpea is reasonably drought-tolerant and can grow in areas with less than 650 mm of annual rainfall. The most common pigeonpea pests include *Fusarium udum*, sterility mosaic disease caused by the Pigeonpea sterility mosaic virus (*PPSMV*) (Kannaiyan et al., 1984), and over 30–80% losses in field occur due to *Phytophthora* blight and the pod borer complex, in addition to the post-harvest losses due to storage pests such as bruchids. Although conventional breeding approaches have helped to produce higher-yielding varieties, they have been less successful in dealing with biotic and abiotic stresses which have contributed to low overall annual yield improvements (of 0.8%) over the last six decades (Ahlawat et al., 2005). Considering these practical production issues, genetic engineering approaches provide opportunities for breeding high-yielding pigeonpea with improved resistance to various biotic and abiotic constraints. There have been significant efforts to improve yield, quality charactersistics and resistance to meet regional needs.

Recent developments in genetic transformation technologies have assisted pigeonpea researchers to develop transgenic varieties with desirable agronomic traits (Satyavathi et al., 2003; Kumar et al., 2004; Surekha et al., 2005; Sharma et al., 2006). Several reports on *Agrobacterium tumefaciens*-mediated pigeonpea transformation are available (Arundhati, 1999; Geetha et al., 1999; Satyavathi et al., 2003; Kumar et al., 2004). Different *Agrobacterium tumefaciens* strains such as LBA4404, C58, EHA105, GV3101 and GV2260 have been used for transformation of this crop. Particle bombardment has also been used in pigeoenpea transformation (Thu et al., 2003; Dayal et al., 2003).

Pigeonpea regeneration has been possible via shoot regeneration from the cotyledonary nodal meristems (Mehta and Mohan Ram, 1980; Eapen and George, 1994; Geetha et al., 1998) and somatic embryogenesis (Patel et al., 1994; Sreenivasu et al., 1998). However, only a few protocols have been successful for efficient genetic transformation (Lawrence and Koundal, 2001; Satyavathi et al., 2003; Kumar et al., 2004; Surekha et al., 2005; Sharma et al., 2006). Lawrence and Koundal (2001) used callus obtained from the embryonal axes for *Agrobacterium* infection with the vector carrying *CPTI* gene. Although the system was amenable for transformation, the transformation efficiency achieved was less than 1%. Subsequently, direct organogenesis was achieved by culturing cotyledonary nodes and embryonic axes on MS medium containing BAP (8.87 μM) for shoot bud initiation and shoot elongation (2.22 μM). The shoots rooted on MS media supplemented with IBA (1.48 μM) exhibited transformation frequency of about 67% when cotyledonary node explants were used and 51% with embryonal axes explants (Satyavathi et al., 2003). Similarly, Kumar et al. (2004) achieved direct regeneration from cotyledonary nodal explants on MS medium supplemented with 2 mg L^{-1} IBA. The shoots elongated on MS medium with 0.5 mg L^{-1} GA_3 and root induction medium consisted of MS and 1 mg L^{-1} IBA, obtaining a transformation frequency of 42.5%.

A genotype-independent direct organogenesis was achieved using meristematic tissue derived from axillary bud region of the *in vitro*-grown seedlings of pigeonpea (Sharma et al., 2006). The meristematic tissue was cultured on MS medium supplemented with BAP (22.0 μM) for shoot induction. After two weeks of infection with *Agrobacterium,* the transformed cells were selected by applying a selection pressure of 125 mg L^{-1} kanamycin. Developed shoots elongated on a shoot elongation medium having MS + 0.5 μM GA_3, while rooting was achieved by giving a pulse treatment with 25 μM IAA followed by culturing on MS medium with reduced sucrose (1%). This method yielded a transformation efficiency of 60% based on the number of positive transformants. A non-tissue-culture-based method has also been developed for pigeonpea transformation using plumular and inter-cotyledonary regions of young seedlings for infection with *Agrobacterium*. Following 1 h co-cultivation, seedlings were transferred to soilrite for establishment and growth. This method has been reported to be efficient and genotype-independent (Rao et al., 2008).

4.1 Biotic stress

The legume pod borer *H. armigera* is the major pest affecting pigeonpea across Africa and Asia, causing particular damage to early-formed flowers and pods. The pod fly (*Maruca obtuse)* takes over as the dominant pest later in the season in many parts of India. While *Helicoverpa* causes losses estimated at over US$317 million in the SAT (Sharma et al., 2001), pod fly damage alone accounts for over US$256 million losses in Asia, Africa and the Caribbean (ICRISAT, 1992).

Transgenic pigeonpeas expressing the *cry* genes of *Bacillus thuringiensis* and the *cowpea trypsin inhibitor* (*cpti*) genes have been developed to combat pod borer attack. Several researchers have reported transgenic pigeonpea with insect resistance traits. Surekha et al. (2005) reported the expression and inheritance of *cry1Ec* in T_1 generation. Similarly, *cry1Ac* expression in pigeonpea has also been reported by Krishna et al. (2011), although this report limited the data to T_0 generation only. Varying expression of the *cry1Ab* gene in different pigeonpea plant parts with maximum protein abundance in flowers has been reported (Sharma et al., 2006). Similarly, expression of the chimeric *cry1AcF* (encoding *cry1Ac* and *cry1F* domains) in transgenic pigeonpea revealed pod borer resistance in T_2 and T_3 progenies, showing high larval mortality of 80–100% (Ramu et al., 2012). More recently, Ghosh et al. (2017) reported transgenic pigeonpea constitutively expressing two insecticidal genes, *cry1Ac* and *cry2A*, demonstrating significant weight loss of the second–fourth instar larvae of *H. armigera*, ultimately achieving 80–100% larval mortality with early generation transgenic events.

4.2 Abiotic stress

Since pigeonpea already demonstrates good drought resistance, there have not been many attempts to develop transgenic techniques to improve abiotic stress resistance. However, the likelihood of more extreme weather associated with climate change has made abiotic stress resistance a higher priority. Transgenic pigeonpea with enhanced osmolyte accumulation has been developed to improve salinity resistance. The transformation efficiency was found to be greater than 15% in all the three genotypes studied (Surekha et al., 2014). The constitutive expression of *P5CSF129A* gene showed higher proline accumulation and a lower lipid peroxidation compared to their WT counterparts upon treatment with 200 mM NaCl. Proline overproduction was shown to play an important role in cellular integrity and resistance to salt shock, as well enhancing growth, chlorophyll and relative water content in saline conditions.

4.3 Nutritional fortification

In a quest to develop nutritionally fortified pigeonpeas, a lysine-rich transgenic line of pigeonpea has been developed using a mutated dihydrodipicolinate synthase (*DHDPS*) gene which encodes the key enzyme in lysine biosynthesis driven by phaseolin and *Arabidopsis* 2S2 promoters. The mutant gene did not show any feedback inhibition, and the transgenic seeds showed a 2 to 6-fold increase in lysine content. Analysis of the amino acid composition of several *DHDPS* overexpressing plants revealed that modification of one gene in the pathway affected both the content of the target amino acid other amino acids derived from the same precursor, aspartate and threonine. Several seed-specific promoters have been used to increase lysine levels (Thu et al., 2007).

Research in pigeonpea biofortification is also underway at ICRISAT, focussing on enhancing β-carotene by manipulating genes in the β-carotene pathway, in order to address the problem of vitamin A malnourishment in developing countries (Bhatnagar et al., unpublished results).

4.4 Healthcare

Pigeonpea has also been used for the production of edible vaccines. Using cotyledonary explants, pigeonpea was subjected to *Agrobacterium*-mediated transformation. The Hemagglutinin–neuraminidase (*HN*) gene of *Peste des petits ruminants virus* (PPRV) was expressed in pigeonpea to develop an edible vaccine for PPRV. The transgenics expressed HN protein and were tested for its neuraminidase activity (Prasad et al., 2004). Another animal vaccine for rinderpest virus was produced using transgenic pigeonpea where the H surface glycoprotein of rinderpest virus was expressed in pigeonpea leaves through *Agrobacterium*-mediated transformation using cotyledonary nodes as explants. These studies provided the proof of concept for using pigeonpea for bio-farming of edible vaccines that potentially can help overcome the problems associated with storage, cost and heat lability in existing fermentation-based production systems (Satyavathi, 2003).

5 Genetic engineering of cowpea

Cowpea (*Vigna unguiculata*) is another important legume crop grown in the semi-arid tropics of Asia, Africa, Southern Europe and Central and South America. This warm-weather crop grows well in poor soils and is drought tolerant; it is usually grown as an intercrop with cereals (Steele, 1976). Cowpeas are grown mostly for their edible beans, although the leaves, green peas and green pea pods are important sources of food for humans and livestock. Cowpea thrives as a sole crop and is effective for intercropping with sorghum, millet, maize, cassava or cotton. Although there is no archaeological evidence for early cowpea cultivation, the centre of diversity of the cultivated cowpea is West Africa, making this the likely centre of origin and early domestication. While there is an uncertainty on when its cultivation began, cowpea is believed to have made its way into South East Asia where secondary domestication events may have occurred (Faris, 1965; Ohler and Mitchell, 1992).

Cowpeas are a rich source of calories, proteins, fibre, minerals and vitamins. They contain up to 25% protein and 60% carbohydrates, as well as amino acids such as lysine, and are low in anti-nutritional factors. The crop also has a high content of vitamins such as thiamine, riboflavin, ascorbic acid, folic acid and niacin. Although considered reasonably drought tolerant, cowpea is affected by various biotic and abiotic stresses that lead to major crop losses. Insects like aphids, *Maruca* and thrips cause damage to crop, while post-harvest produce is affected by pests like beetles. Common diseases affecting cowpea include blight, bacterial pustule, leaf spots, false rusts and stem rots. Viruses such as cowpea mosaic virus, mottle viruses and nematodes also affect growth and yield in cowpea. Weeds like *Striga gesnerioides* and *Alectra vogelii* can cause nearly total crop loss (Langyintuo and DeBoer, 2003). Unfortunately, there has not been sufficient genetic variability to breed resistance genes for insect pests and diseases in cowpea. While resistance to the pod borer *M. vitrata* has been identified in *V. vexillata*, a distant relative of cowpea, interspecific genetic barriers prevent transfer of such alleles into the cultivated germplasm.

Although conventional plant breeders have worked hard to develop cowpea varieties with traits such as improved yield and insect resistance, success has been limited. In order to develop more efficient and resistant varieties for this important African legume crop, biotechnological approaches are critical. Nevertheless, the recalcitrance of cowpea for efficient transformation and regeneration has been a major bottleneck for genetic engineering of this crop.

The initial work on genetic transformation studies in cowpea was able to obtain kanamycin-resistant calli, but these failed to regenerate shoots (Garcia et al., 1986, 1987). Subsequently mature embryo slices and mature de-embryonated cotyledons were used for transformation experiments (Penza et al., 1991; Muthukumar et al., 1996). Although Muthukumar et al. (1996) were successful in obtaining transgenic plants, the transgenes could not be transmitted to the next generation. However, although cowpea transformation using particle bombardment resulted in transgene transmission to a small proportion of the progeny, stable integration of the transgenes was not observed (Ikea et al., 2003). Kononowicz et al. (1997) developed an efficient regeneration and transformation system in cowpea that eventually formed the basis of a refined reproducible system that revealed Mendelian inheritance (Popelka et al., 2006). This cowpea transformation system used explants from cotyledonary nodes or embryonic axes on media devoid of any auxins at early stages; however, including the cytokinin BAP during shoot initiation was capable of producing transgenic events. Another major bottleneck in successful transformation of cowpea is rooting: while regenerated shoots are viable *in vivo,* rooting has been less successful (Brar et al., 1999; Machuka, 2000). Some success was achieved by grafting the shoots onto 10-day-old seedling stocks with the aid of a silicon ring, followed by their transfer and establishment in high humidity conditions before being transferred to the greenhouse. However, this protocol is genotype-dependent and needs further enhancements (Chaudhury et al., 2007). More recently, an improvised transformation protocol was optimized for enhanced gene expression, resulting in a significant increase in transformation efficiencies of up to 80% (Mellor et al., 2012). In this case preconditioning of the explant donor seedlings with TDZ or BA had a positive effect on shoot regeneration efficiency and shoot elongation, thereby yielding considerable better results when compared to the untreated seedlings. An efficient selection process using the *phosphomannose isomerase* gene from *Escherichia coli* as a selectable marker for mannose-based selection has also been reported to be compatible with cowpea transformation (Bakshi, 2012).

5.1 Biotic stress

The legume pod borer, *Maruca (testulalis) vitrata*, is a major cause of low yields in the African cowpea, affecting every developmental stage of the plant. Severe infestations can result in yield losses of over 90%. The borer causes damage to the flower buds, flowers and pods of the plant. Other important pests include pod-sucking bugs, thrips and the post-harvest weevil *Callosobruchus maculatus*. Cowpea germplasm contains absolutely no resistance to these major insect pests and adding insect-resistant traits has therefore been an attractive strategy (Ehlers and Hall, 1997). Transgenic cowpea developed by constitutively expressing *cry1Ab* gene through *in planta* nodal bud electroporation-mediated gene transfer has conferred resistance against *Maruca*. Transformed plants successfully inhibited larval growth and reduced infestation to as low as 11% (Adesoye et al., 2008). By employing extra copies of *vir* genes and geneticin-based selection regime, fertile cowpea transgenic plants expressing the alpha amylase gene (*αAl-1*) targetting *Callosobruchus maculatus and Callosobruchus chinensis* have been recovered (Chaudhary et al., 2007). The same group reported a much higher high seed-specific expression by using additional copies of *vir* genes and by co-cultivating explants in the presence of mixtures of thiol compounds (DTT and L-cysteine). Four selected transgenic lines showed 32–37% inhibition and reduced fecundity of *Callosobruchus maculatus* by over 97%. Insect feeding assays revealed a significant reduction in the number, weight and longevity of the newly emerged larvae upon rearing on the transgenic cowpea seeds (Solleti et al., 2008). Synthetic *cry1Ab* expression in the cowpea provided excellent protection against *Maruca vitrata*. Field trials with selected transgenic lines of cowpea in several African locations have shown improved insect resistance when compared to non-transgenic counterparts (Anwar et al., 2016). It is clear that this has been a breakthrough in cowpea crop improvement, opening up avenues for the development of novel strategies for pest and disease resistance. Similarly, cowpea plants resistant to infection by bruchids *Acanthoscelides obtectus* (Say), *Callosobruchus chinensis* L. and *Callosobruchus maculates* have been developed by inserting the *alpha Al-1* gene by *Agrobacterium*-mediated transformation (Lüthi et al., 2013). Efforts are underway to develop transgenic cowpea expressing chicken cystatin for insecticidal activity against the bruchids (Citaden et al., 2011).

5.2 Herbicide tolerance

Cowpea cultivation is severely affected by weeds, causing damage during growth and development of plants, competing for light, nutrients and water. Imazapyr-resistant cowpea lines have been developed by expressing the mutated *acetohydroxyacid synthase* coding gene (*Atahas*) that conferred tolerance to Imazapyr. The transgenic progeny showed a herbicide tolerance four times greater than the commercially-recommended dose for weed control (Citadin et al., 2013).

5.3 Nutrition

As with all other legumes, cowpea is deficient in sulphur-containing amino acids. While not much effort has been devoted to the nutritional enhancement of cowpea, some work was initiated for introducing methionine-rich protein in cowpea using the maize *delta-zein* gene (Citadin et al., 2011).

6 Genetic engineering of lentil

An annual herbaceous crop, lentil has been grown since Neolithic times. They are an important source of protein, vitamin B and lysine, though they are poor in methionine and cysteine and exhibit some anti-nutritional factors. Lentils can grow in a range of soil types and are widely grown in the Mediterranean, Europe and Asia. Turkey, Canada, the United States and Australia are the main producers. World lentil production has tripled over the last three decades from 1.05 million metric t in 1971 to 3.8 million metric t in 2004, through a 124% increase in sown area and a 58% increase in average national yield from 611 to 966 kg ha^{-1} (FAO, 2004). ICARDA maintains the largest collection of lentil germplasm (Redden et al., 2007).

Due to hypogeal germination, lentil seedlings are less prone to freezing, wind, insect damage and grazing. Although lentils tolerate drought better, the crop is vulnerable to waterlogging and saline soils. Lentils are particularly affected by weeds which can significantly reduce crop yields (Yenish et al., 2009). Lentil is also affected by various bacterial, fungal and viral pathogens like *Alternaria* blight, AB, black streak root rot, black rot, *Cercospora* leaf spots, dry root rots and *Fusarium* wilt. Lentils are also infected by stem and parasitic cyst nematodes and viruses such as *bean yellow mosaic virus* and *cucumber mosaic virus*. Other major biotic stresses of lentil include anthracnose caused by *Colletotrichum truncatum*, BGM caused by *Botrytis fabae* and *Botrytis cinerea*, *Stemphylium* blight caused by *Stemphylium botryosum*, lentil rust caused by *Uromyces fabae* and *Sclerotinia* white mould caused by *Sclerotinia sclerotiorum* (Muehlbauer et al., 2006).

Lentil breeding programmes throughout the world have targeted developing larger seed and stable seed yield varieties (Muehlbauer et al., 1995). Conventional lentil breeding programmes are limited by a relatively small gene pool and problems of cross-compatibility of cultivated varieties with their wild relatives. These constraints highlight the potential for transgenic approaches in lentil breeding programmes. There have been a number of initiatives to use transgenic approaches to improve particular traits (Ufuk et al., 2009).

Although genetic engineering in lentils has been attempted by several research groups, the crop still lacks an efficient and reproducible transformation method. The first known report of lentil transformation by Hinchee et al. (1988) used longitudinally cut embryonic axes that were co-cultivated with *Agrobacterium tumefaciens* carrying the *GUS* reporter gene. Liposome-mediated transfection of lentil protoplasts with *CAT* and *GUS* reporter genes using cationic liposomes made up of dipalmitoylphosphatidylcholine (DPPC) and sterylamine (SA) was carried out by Maccarrone et al. (1992). The same group carried out protoplast transformation using electroporation and successfully inhibited the *LOX* gene expression to elucidate its function and role in lentils (Maccarrone et al., 1995). In the following year, Chowrira et al. (1996) reported electroporation-mediated gene transfer of *uidA* gene driven by a CaMV 35S promoter into intact lentil tissues. Cotyledonary nodes from lentil seedlings were also used for particle bombardment of *GUS* gene (Oktem et al., 1999), where the integration and expression of *GUS* were observed in 4% of the regenerated shoots. Subsequently, lentil transformation systems were developed using stem and leaf explants for optimization of different *Agrobacterium* strains, inoculum densities and wounding methods (Khawar and Ozcan, 2002; Akcay et al., 2009). A reproducible genetic transformation system for lentil using particle bombardment following an explant pre-culture period of up to six days was reported by Gulati et al. (2002). In addition, a vacuum infiltration–based *Agrobacterium*-mediated gene transfer method was used to introduce the coding sequences of *npt*II gene into the lentil cotyledonary nodes (Mahmoudian et al., 2002a,b). Using a particle bombardment–mediated direct delivery method, transformation of cotyledonary nodes of lentil has been achieved with acetolactate synthase gene (*ALS*) from tobacco, conferring resistance to sulfonylurea herbicides (Gulati et al., 2002).

6.1 Abiotic stress

Herbicide tolerance trait has been genetically engineered into lentil using *Agrobacterium*-mediated genetic transformation (Khatib et al., 2007) using the *bar* gene which confers resistance to glufosinate; screened shoots were grafted onto non-transgenic rootstock and transferred to greenhouse for acclimatization. The integration and expression of the transgene was confirmed using PCR and herbicide application studies. More recently, Khatib et al. (2011) also reported transformation of transcription factor *DREB1A* into lentil for enhancing drought and salinity tolerance. The transgenic events in early generations were confirmed using molecular characterization that confirmed the integration and expression of *DREB1A* under salt stress and showed increased tolerance to salinity stress (Khatib et al., 2011).

7 Genetic engineering of pea (*Pisum sativum*)

Pea, *Pisum sativum*, is an annual cool-season legume grown in many parts of the world. Pea is a rich source of nutrition as it contains protein (18–33%), fibre (4–7%), starch (37–49%), vitamins A, B6 and K, phosphorus, magnesium, iron, zinc and leutin. While not much is known about the origin of *Pisum sativum*, it is believed to have originated in the Near East and Mediterranean where it was grown alongside cereal crops (Hedrick, 1928; Zohary and Hopf, 1973). Pea is now widely cultivated in Europe, America and China. Leading producers include Canada, Russia, China, India and France, with Ethiopia being the largest producer in Africa (FAOSTAT, 2006). In the mid-19th century, Austrian monk Gregor Mendel's observations of pea pods led to the principles of Mendelian genetics, the foundation of modern genetics (Ze'ev and Zohary, 1973).

Conventional pea breeding programmes have been remarkably successful due to large genetic variation available in pea germplasm. Significant advances have been made over the last 4–5 decades in improving plant architecture and yield. Following the discovery of '*af*' single gene mutation, the modern-day semi-leafless pea cultivars with improved standing ability have been developed. This combined with recessive '*le*' allele for shortened vine length (Lester et al., 1997) led to further development of a sturdier plant type that remained upright throughout the growing season (Villani and Demason, 2000). Moreover, the upright canopy resulting due to improved stem strength and the intertwining of tendrils from neighbouring plants allowed greater ventilation, reducing the incidence of disease whilst facilitating harvesting.

Pea plants are mainly affected by fungal and viral diseases, as well as by the peel weevil which can reduce yields by over 70%. Among the abiotic constraints, pea is very sensitive to salinity. While a reasonable genetic resistance for many of the key pathogens is available, very little genetic variability exists for resistance to insect pests. Pea breeders and geneticists have been able to introduce single gene-controlled resistance to several pathogens. Similarly, genetic resistance to economically important viruses such as *pea enation mosaic virus, bean leaf roll virus* and *pea seed borne mosaic virus* have also been incorporated into many modern pea cultivars. Although significant improvements have been made in pea breeding programmes, quantitatively-inherited resistance, able to go beyond natural hybridization barriers, has been difficult to achieve.

Early reports from the 1980s and 1990s are available on interactions between *Agrobacterium tumefaciens* and *Agrobacterium rhizogenes* with pea (Hobbs et al., 1989; Hussey et al., 1989; Schaerer and Pilet, 1991; Puonti-Kaerlas, 1993). Genetic transformation of pea has been achieved using *Agrobacterium tumefaciens*-mediated and direct gene delivery methods. While the direct delivery of genes using biolistics has resulted in transient gene expression only (Warkentin and McHughen, 1992), *Agrobacterium*-mediated genetic transformation has been preferred. Other direct delivery methods such as microinjection and electroporation have not been widely adopted (Chowrira et al., 1998).

Most of the early work on pea transformation focused on identification of explants and tissue culture methods, owing to challenges in organogenesis and embryogenesis. A number of explants ranging from epicotyl nodes, sliced embryonic axes, cotyledonary nodes and axillary meristems have been used for genetic transformation studies in pea, albeit with varying efficiencies (<1% to 5%). The selection of transformed pea shoots has been carried out using both antibiotics such as *npt*II for kanamycin resistance and *hpt*II for hygromycin resistance, and herbicide tolerance genes such as *ahas3r* for chlorsulfuron and *bar* for resistance to Basta (De Kathen and Jacobsen, 1990; Lulsdorf et al., 1991; Schroeder et al., 1993; Grant et al., 1995, 1998a; Švábová et al., 2005; Bean et al., 1997; Jones et al., 1998; Polowick et al., 2000).

An increase in the *Agrobacterium*-mediated transformation efficiency in pea by using axenic shoot cultures and seedling epicotyls as explants has been reported (Puonti-Kaerlas et al., 1990). Another simple pea transformation system using *Agrobacterium tumefaciens* that used embryonic axis of immature seeds as explants to transform the herbicide resistance and antibiotic resistance genes in two pea cultivars has been developed (Schroeder et al., 1993). The gene integration that followed Mendelian pattern of inheritance in the first generation was found to be stable. Grant et al. (1995) developed a reliable transformation system in pea using immature cotyledons as explants that yielded stable and fertile transformed plants upon infection with different *Agrobacterium* strains such as *AGL1* and *KYRT1*. While *KYRT1* was more efficient, transformation was temperature dependent (Grant and Cooper, 2003). Another study on *Agrobacterium*-mediated pea transformation used embryonic segments of mature seeds as explants yielded optimum transformation efficiency (Krejčí et al., 2007). The effect of different co-cultivation conditions has also been optimized for pea using a highly virulent bacterial strain *EHA105* bearing two plasmids carrying the *uid*A reporter gene and *bar* or *npt*II marker genes, respectively. Studies on the efficacy of various additives like acetosyringone, L-cysteine, glutathione, dithiothreitol, cellulases and pectinases on pea co-cultivation have also revealed that acetosyringone and cysteine significantly improved the transgene expression (Švábová and Griga, 2008).

7.1 Biotic stress

Transgenic pea expressing the *alpha amylase inhibitor* gene of common bean driven by a seed-specific promoter has been shown to confer resistance to bruchids (Schroeder et al., 1995). Feeding assays revealed minimal seed damage in the transgenic lines, and the gene expression was found to be stable for over five generations. Grant et al. (1998b) expressed the *alfalfa mosaic virus* coat protein to develop transgenic pea for resistance to *alfalfa mosaic virus*. In another instance, using *in planta* transformation through injection or electroporation of the axillary meristems, transgenic pea resistant to *Pea Mosaic Virus* was developed again by overexpressing the coat protein gene. The progenies of T_4 transgenic pea plants were monitored for effective resistance to the virus infection (Chowrira et al., 1998). Researchers have also explored the use of other antifungal genes from raspberry and grape (Richter et al, 2006). More recently, in an effort to develop multiple disease resistance in pea, researchers have used the antifungal *chitinase* gene (Hassan et al., 2009). Amian et al. (2011) reported expression of multiple transgenes in a single line following crossing of two parental transgenic parents expressing *chitinase* gene from *Streptomyces olivaceoviridis* and *1,3-β-glucanase* from *Hordeum vulgare*. The transgenic pea lines expressing four different antifungal proteins when grown in association with AM fungus *Glomus intraradices* did not show any negative effect on AM colonization and the introduced gene was not lethal to the symbiotic association between legumes and AM fungi (Hassan et al., 2012). Nevertheless, the progenies of these lines in subsequent generations showed sensitivity to the herbicide, possibly due to inactivation or silencing of the *bar* gene reportedly located near the left border of T-DNA. More recently, insect-resistant transgenic peas were developed using a codon optimized *cry1Ac* gene from *Bacillus thuringiensis*. The introduced transgene was inherited up to the T_4 generation. Insect bioassays revealed total larval mortality and significantly reduced feeding damage compared the WT plants (Negawo et al., 2013, 2016).

7.2 Abiotic stress

Very little work has been done on developing abiotic stress tolerance in pea. Expressing the Na/H^+ antiporter gene of *Arabidopsis thaliana* conferred higher levels of salinity tolerance in transgenic pea plants without any negative effect on growth-promoting *Rhizobacteria* (Ali et al., 2015).

7.3 Nutritional fortification

As with all legumes, raffinose poses challenges for digestion by humans and ruminants. To deal with this, overexpression of the *alpha galactosidase* from coffee in transgenic pea accumulated 40% less complex carbohydrates (Polowick et al., 2008). The reduction of raffinose content did not affect seed germination in the transgenics, and the accumulation of oligosaccharides was highest in embryo axes, followed by cotyledons and the least in seed coat.

8 Genetic engineering of peanut

Peanut (*Arachis hypogaea* L.) ranking sixth in edible oil production and 13th among food crops in the world is taxonomically classified as legume crop mainly for its edible seeds (Gregory et al., 1980). Peanut seeds have a high nutritional value, with an oil content of 45–65%, protein content of 25–30% and carbohydrate content of 10–25%. Peanuts are rich in vitamins E, K and the B complex, and are a very good source of calcium, phosphorous, magnesium, zinc and iron. The cultivated *Arachis hypogaea* comes from a cross between the two wild species of peanut, *Arachis duranensis* and *Arachis ipaensis*. The genus *Arachis* has 80 different species located across regions and the initial domestication is assumed to have taken place somewhere in northwestern Argentina, or in southeastern Bolivia, where even today one can find peanut landraces with wild-like features. The archaeological evidence suggests that peanut has been cultivated for 7500 years, and it is now mainly grown in North America, Africa, Southeast Asia and Europe (Krapovickas, 1969).

Peanut is grown under rainfed conditions by resource-poor farmers in developing countries of Asia and Africa, and its production is affected by various biotic and abiotic stresses that cause annual yield losses of over US$3.2 billion (Dwivedi et al., 2003). These stresses increase the gap between potential and actual yields in both subsistence and commercial production systems in these regions. Major abiotic factors that affect peanut production include drought, high temperatures and poor soils. Among the biotic factors, fungal diseases, virus diseases, bacterial wilt disease, aflatoxin contamination, nematodes, foliar insect pests and soil insect pests such as *Spodoptera litura* and others play a significant role in yield reduction (Sharma and Oritz, 2000; Dwivedi et al., 2003). Efforts to improve peanut productivity using conventional breeding have not been very effective due to lack of sufficient genetic variability within the cultivated germplasm. While several wild annual *Arachis* species possess a wealth of agronomically desirable traits, these are cross-incompatible with the cultivated genotypes. This suggests the potential of modern biotechnological approaches in overcoming these constraints. These newer tools and technologies will allow breeders to access new gene pools, allowing introgression of the valuable traits into modern cultivated peanut that otherwise cannot be addressed by conventional means.

Several DNA transfer methods have been reported for production of transgenic peanut over the last several decades. Of these the most commonly used approaches include *Agrobacterium tumefaciens* or direct gene transfer through biolistics into organogenic or embryogenic cultures (Krishna et al., 2015). For direct introduction of genes in peanut, as with many other crops, particle bombardment developed by Sanford and co-workers (Klein et al., 1988) has been the most commonly used technique. Callus derived from immature peanut leaflet tissues have been used for bombardment (Clemente et al., 1992; Schnall and Weissinger, 1995) and transient expression observed. However, Ozias-Akins et al. (1993) and Ozias-Akins and Gill (2001) reported an increased transformation frequency of 1% upon bombarding 1- to 2-year-old embryogenic callus derived from immature embryos, while a relatively low transformation frequency of 0.9–1.0% was achieved using the shoot meristems of mature embryonic axis (Brar et al., 1994). Singsit et al. (1997) first reported transformation of somatic embryos developed from immature cotyledons with *cry1Ac* gene for resistance to cornstalk borer (*Elasmopalpus lignosellus*). Co-bombarding embryogenic callus led to the development of transgenic peanut lines exhibiting high levels of resistance to *Peanut stripe virus* (Higgins et al., 2004).

There have been several reports of *Agrobacterium tumefaciens*-mediated genetic transformation of peanut. A number of explants have been used for such studies including hypocotyls (Lacorte et al., 1991; Mansur et al., 1993), leaf (Eapen and George, 1994) and embryonic axes (McKently et al., 1995). A high transformation frequency was demonstrated by using pre-cultured cotyledons as explants (Venkatachalam et al., 1998, 1999, 2000), whereas leaf segments yielded only 0.3% frequency of fertile transgenic plants (Cheng et al., 1997). Using a non-tissue-culture-based *Agrobacterium* transformation involving direct co-cultivation of mature seeds, a 3% transformation frequency was reported (Rohini and Rao 2000). A much higher transformation frequency of >55% was reported by Sharma and Anjaiah (2000) using cotyledon explants from mature seeds following presoaking. These shoots readily rooted and transplanted with over 95% success rate. Similarly, promoter tagged peanut transgenics using the cotyledonary nodes as explants and a promoter-less fusion gene *npt*II:*gus* have been produced (Anuradha et al., 2006). More recently, de-embryonated cotyledons were used as explants and co-cultivated with different *Agrobacterium* strains using myo-ionositol and L-cysteine. The rooting was carried out by micro-grafting, with all plants surviving (Tiwari et al., 2012).

To allow selection of transformed cells, most of the peanut transformation protocols included selectable marker genes encoding resistance to an antibiotic such as kanamycin or hygromycin or to herbicides such as phosphinothricin, glyphosate or bialaphos. (Wilmink and Dons, 1993). While a number of studies chose hygromycin for the selection, kanamycin has also been reported to be an effective selection agent (Clemente et al., 1992; Sharma and Anjaiah, 2000). The herbicide Basta® (active ingredient phosphinothricin) has also been used to select transgenic peanut tissues (Brar et al., 1994). To overcome the challenges associated with identification of a small proportion of transformed cells in a much larger cell population, screenable markers have been coupled with selectable markers in peanut transformation systems. β-glucuronidase (*uid*A) was used as screenable marker in peanut transformation experiments for easy detection in sensitive, specific, quantitative selections (McKently et al., 1995). Green fluorescent protein (GFP) (Reichel et al., 1996), an auto fluorescent jellyfish protein, has also been used as a reporter gene in peanut (Joshi et al., 2005; Tiwari et al., 2012).

8.1 Biotic stress

Diseases caused by various fungi, bacteria, viruses, mycoplasma, nematodes and insect pests are major constraints to peanut yield and productivity. Major biotic stresses include leaf spot diseases (early and late) and rust caused by foliar fungal pathogens. Seed- and soil-borne pathogens cause diseases such as collar rot, stem rot and dry root rot in peanut. Several economically important viruses include *peanut bud necrosis virus* (PBNV), *peanut mottle virus*, *groundnut rosettes virus* (GRV) and *peanut clump virus* (PCV) (Nigam et al., 2012). With regard to insect pests, a wide range of pests such as leaf miner, tobacco caterpillar, white grub, jassids, thrips, aphids, red hairy caterpillar and termites cause severe damage to peanut crops (Ghewande et al., 1987; Basu, 1995). Genetic enhancement through conventional breeding and chemical control methods has yielded very limited success (Pandey et al., 2012), hampering the development of improved varieties, leaving transgenic development as the only option for most of these constraints.

Several bacterial and rice chitinase genes have been used for peanut transformation to impart resistance to early leaf spot (Iqbal et al., 2012). Two lines transformed with bacterial chitinase gene showed 56–62% disease suppression compared to their WT counterparts. Similarly, Sundaresha et al. (2010) reported use of tobacco chitinase gene for developing transgenic peanut for resistance to *Cercospora arachidicola. In vitro* leaf bioassays indicated a much lower per cent disease index (i.e. 0–25% of leaf area covered by spots) and delay in the onset of disease. Several attempts have been made to develop transgenic resistance towards late leaf spot disease caused by *Phaeoisariopsis personata*. Expressing chitinases from tobacco (Rohini and Rao, 2001) and rice (Chenault et al., 2005) have shown considerable resistance to late leaf spot disease. More recently, transgenic peanut plants expressing mustard defensin gene (*Rs-AFP2*) showed an increased disease resistance to *Cercospora arachidicola* and *Phaeoisariopsis personata* in detached leaf assays and greenhouse evaluations (Anuradha et al., 2008). Similarly, overexpressing a multiple gene stack carrying *SniOLP* (osmotin-like protein from *Solanum nigrum*) and *Rs-AFP2* [defensin gene from radish (*Raphanus sativus*)] resulted in enhanced resistance against *Cercospora arachidicola* and *Phaeoisariopsis personata* (Vasavirama and Kirti, 2010). Peanut transgenic events overexpressing *rice chitinase* gene also demonstrated >50% reduction in the incidence of late leaf spot and rust diseases (Prasad et al., 2013).

Transgenic resistance to blight disease caused by *Sclerotinia minor* and *Sclerotinia sclerotiorum* has been reported (Rohini and Rao, 2001; Chenault et al., 2005). Over fourteen transgenic lines in the background of Okrun showed up to 43–100% reduction in disease incidence over a 3-year period (2000–02) in field trials (Chenault et al., 2003, 2005). Similarly, transgenic lines developed using Virginia-type cultivars overexpressing barley *oxalate oxidase* gene showed an enhanced resistance to *Sclerotinia minor* (Livingstone et al., 2005).

Peanut being a subterranean legume is susceptible to aflatoxin contamination by *Aspergillus flavus* and *Aspergillus parasiticus*. Since conventional methods for aflatoxin control have not been effective, genetic engineering options for enhancing host plant resistance to *Aspergillus flavus* invasion and aflatoxin contamination have been explored (Bhatnagar-Mathur et al., 2015). Broad-spectrum hydrolytic enzymes such as chitinases and glucanases have been used extensively for the development of *Aspergillus flavus*-resistant transgenic peanuts (Eapen, 2003; Sundaresha et al., 2010). Similarly, transgenic peanut events expressing synthetic version of maize ribosome-inhibiting protein gene, *mod1*, demonstrated enhanced resistance to *Aspergillus flavus* and resulting aflatoxin accumulation (Weissinger et al., 2002). A reduction in aflatoxin content in transgenic peanut expressing soybean *lox1* gene under the control of carrot embryo-specific promoter (*DC3*) was reported by Ozias-Akins et al. (2002). Antifungal genes have been added to peanut to increase resistance to *Aspergillus flavus*. These include: tomato anionic peroxidase (*tap1*) and synthetic peptide *D4E* (Ozias-Akins et al., 2002); and *D5C* (Weissinger et al., 1999). Over 60–70% reduction in *Aspergillus flavus* colony growth was observed in transgenic peanut lines expressing the bacterial *chloroperoxidase* gene (Jacks et al., 2000; Yang et al., 2003). However, work is still needed to eliminate fungal infection and reduce the levels of aflatoxin contamination to below 4 ppb, which is the safe limit for human consumption.

Peanut production is constrained by insect pests such as *Spodoptera litura*, *Aproaerema modicella*, *Amsacta* spp., *Heliothis* spp., aphids, jassids, thrips and termites. Interspecific reproduction barriers and linkage drag impede use of wild relatives for development of resistant cultivars (Stalker and Moss, 1987). The development of transgenic peanut for resistance to major pests has gained importance in the last two decades. Peanut transgenics expressing *cry1Ac* gene showed varied resistance against *Lesser Cornstalk Borer* in terms of larval mortality and larval weight (Singsit et al., 1997). Transgenic insect resistance trait developed for *S. litura by* expressing *cry1EC* gene (Tiwari et al., 2008) demonstrated almost 100% mortality of second instars of larvae. Transgenic peanuts have been developed and tested for *H. armigera* (Hubner) that occasionally infests this crop by expressing a chimeric *Bt* gene, *cry1X* (Entoori et al., 2008).

In addition to fungal and bacterial pathogens, there are a number of viruses that cause considerable damage to peanut production and yields around the world. Peanut transgenics for resistance to some of these viral diseases have been developed using genetic transformation approaches. Transgenic peanut plants have been developed to combat peanut bud necrosis disease (PBND) by expressing PBNV nucleocapsid gene, showing a modest tolerance to PBND (Chander Rao et al., 2006). Glasshouse virus challenging experiments revealed a 40–67% decrease in disease incidence, whereas under confined field conditions, only one event showed <25% disease incidence. Similarly, transgenic peanut plants were developed for protection against TSWV, which is carried over by both RNA and protein-mediated control nucleocapsid protein gene (*N* gene) introduced into both runner and Valencia-type varieties (Chenault and Payton, 2003). A significantly lower disease incidence was observed in transgenics carrying an antisense nucleocapsid gene (Magbanua et al., 2000) and coat protein gene of TSWV (Yang et al., 2004), indicating its potential for stable pathogen-derived resistance. More recently, an attempt has been made to develop a transgenic peanut resistant to stem necrosis disease by expressing the coat protein gene (Bag et al., 2007; Mehta et al., 2013) as well as the hairpin RNA coding gene. The events did not show any *Tobacco streak virus* infection until harvest, while the wild plants were infected within 15–20 days; the transgenic plants showed stable resistance until the T_3 generation, indicating successful inheritance of the transgene (Reddy et al., 2016). A high level of RNA-mediated resistance was achieved for *Peanut stripe potyvirus* (*PStV*), where the transgene in the transgenic plants with resistance to *PStV* were stably inherited over five generations (Higgins et al., 2004).

8.2 Abiotic stress

Drought is a major cause of low and erratic pod yields in peanut, contributing to annual losses of 6.7 million t (Subbarao et al., 1995; Sharma and Lavanya, 2002). There have therefore been various initiatives to develop transgenic drought tolerance (Bhatnagar-Mathur et al., 2008; Ramu et al., 2015). Efforts have been made to enhance drought tolerance in peanut through *Agrobacterium*-mediated genetic transformation using a transcription factor AtDREB1A from *Arabidopsis thaliana* (Bhatnagar-Mathur et al., 2007). Several transgenic events showing high levels of stress tolerance have been identified, where substantial yield improvement was observed in two transgenic events under confined field trials (Bhatnagar-Mathur et al., 2009a, 2014). Similarly, transgenic peanut plants expressing *AtNHX1* gene were found to be resistant to high concentration of salt and water stress (Asif et al., 2011). Another report on peanut transgenics expressing isopentenyl transferase gene (*IPT*) yielded significant improvement in drought tolerance under both laboratory and field conditions (Qin et al., 2011). The *NAC2* from *Arabidopsis* upon expression in peanut developed resistance against water loss by maintaining high cell membrane integrity, resulting in reduced water loss, high leaf water status and high cell membrane integrity when compared to the WT (Patil et al., 2014). Salinity tolerance in peanut has been achieved by overexpressing bacterial *mannitol dehydrogenase* gene (*mtlD*). The transgenic events accumulated higher mannitol dehydrogenase activity, high levels of oxidases, catalases and glutathione reductases and a corresponding decrease in the levels of peroxides and free radicals when compared to the wild varieties (Patel et al., 2016).

9 Genetic engineering of other grain legumes: tepary bean, *Vigna* species and faba bean

Over the last few decades there have been attempts to achieve efficient and reproducible genetic transformation methods for most grain legumes. While there have been successes in some, most grain legumes lack efficient tissue culture systems to regenerate plants from transformed cells. Major improvements in transformation efficiency are still needed for crops such as tepary bean, faba beans, Asiatic beans and others.

9.1 Tepary bean

Phaseolus acutifolius, also known as Tepary bean, is one of the most drought-tolerant grain legumes grown around the world. Mexico is assumed to be the origin of the domesticated tepary bean (Muñoz-Perea et al., 2006). The main areas of cultivation are the Southwest United States and Mexico. Genetic variability is low, possibly due to a low initial genetic base at the time of domestication. Tepary bean produces good yields even in the arid areas that otherwise do not support cultivation of most beans. Owing to these characteristics and its low rainfall requirements, cultivation of tepary bean is being explored in several African regions (Shisanya, 2002), Puerto Rico and Honduras (Miklas et al., 1994).

Genetic transformation of tepary bean is still in its infancy, and only a few groups have been involved. Nevertheless, a very stable and reproducible transformation system based on callus derived from floral buds and cotyledonary node region is available (Dillen et al., 1997; De Clercq et al., 2002; Zambre et al., 2005). Several transgenic events targetting traits such as insect resistance (Zambre et al., 2005) and methionine content enhancement (De Jaeger et al., 2002) have been developed.

9.2 *Vigna* species

The genus *Vigna*, subgenus *Ceratotropis*, constitute an economically important group of cultivated and wild species distributed across Asia (Egawa et al, 1996). Also known as Asiatic *Vigna* species, this group of grain legumes contributes significantly to the diet of resource-poor populations in developing countries. Mungbean is widely cultivated in the Indian subcontinent and has spread to other Asian countries, tropical and subtropical regions of Africa, as well as to the West Indies, North America and Australia. Urdbean is an important pulse crop in India, Burma, Bangladesh, Pakistan and Thailand and is grown on a limited scale in Australia and Fiji. Another legume in this group is Adzuki bean, an East Asian grain legume important in Japan, China and Korea (Vaughan et al., 2005). While yield and productivity of these *Vigna* species are limited by several biotic and abiotic stresses, the genes conferring resistance to these stresses have been found in wild species (Sahoo et al., 2003). However, cross-incompatibility and linkage drag are major impediments in the transfer of useful alleles to cultivated germplasm (Watanasit and Pichitporn, 1996). Although genetic engineering has tremendous potential to modify crops to overcome these constraints (Estruch et al., 1997), the availability of efficient and routine genetic transformation systems in these *Vigna* species poses a continuing challenge in using this technology (Chandra and Pental, 2003).

Agrobacterium-mediated genetic transformation of *Vigna radiata* and *Vigna mungo* has achieved considerable success using cotyledonary node explants, demonstrating low to moderate transformation frequencies. Nevertheless, the transformation protocols for two Asiatic *Vigna* species (*Vigna radiata* and *Vigna aconitifolia*) need to be developed further. Preconditioning of the explants of *Vigna* species has a positive effect on regeneration of transgenic shoots (Saini et al., 2007). Particle bombardment of shoot apical meristems derived from mature seeds has had limited success in *Vigna radiata* (Bhargava and Smigocki, 1994). A number of studies have reported transformation amenability of *Vigna mungo*. The transformation of calli from primary leaves of *Vigna mungo* was achieved using *Agrobacterium* strains LBA4404 and EHA105 (Karthikeyan et al., 1996), demonstrating 23% and 10% transformation frequencies, respectively. *Agrobacterium*-mediated transformation of epicotyl explants using direct shoot regeneration system has been achieved (Saini and Jaiwal, 2002). Fertile transgenic plants were generated following mechanical wounding of cotyledonary node explants prior to *Agrobacterium*-inoculation (Saini et al., 2003). However, since the efficiency of transformation was very low (1%), which was attributed to the presence of limited number of regenerable cells, another regeneration method using shoot apex explants has been developed (Saini and Jaiwal, 2005).

9.3 *Vicia faba*

Vicia faba (faba bean) was introduced into agriculture in the late Neolithic period (Link et al., 2008). The centre of origin of this legume is known to be in Iraq, Iran with secondary centres in Afghanistan and Ethiopia (Cubero, 1974). Faba bean accounts for 4% of total global pulse production. 41% of cultivation is in Asia, 33% in Africa, 12% in Europe and 7% in Oceania and America. Since its nutritional value is quite (30% protein), it is grown for both food and feed, mainly as a winter crop.

Faba bean is probably one of the most recalcitrant legume crops, exhibiting very little amenability to tissue cultures, primarily due to poor regeneration capacity and the high content of phenolic compounds (Bieri et al., 1984; Selva et al., 1989). Owing to these difficulties, its genetic transformation is challenging even today. Schiemann and Eisenreich (1989) first attempted genetic transformation in faba beans using hairy root transformations and *Agrobacterium rhizogenes* that resulted in GUS-positive roots, which subsequently yielded transgenic calli lines. Following this, Ramsay and Kumar (1990) successfully transferred *npt*II marker gene using cotyledons and stem tissues as explants and an infection strain of *Agrobacterium rhizogenes*. Nevertheless, transgenic plants could not be regenerated. Another attempt at faba bean transformation using mutant bacterial strains carrying shooty type Ti plasmids was not successful (Jelenic et al., 2000). Böttinger et al. (2001) reported the first successful transformation in faba bean plants resulting from *de novo* regeneration using TDZ. Subsequently, direct shoot organogenesis was achieved using mature and immature embryo axes (Hanafy et al., 2005). More recently, using the same method, Hanafy et al. (2013) successfully demonstrated abiotic stress resistance in *Vicia faba* transgenics overexpressing potato PR10a. The transgenic faba bean lines expressing PR10a were fertile and showed stable inheritance of the introduced genes to subsequent generations. Recent success has also been achieved using a tissue culture-based embryo rescue method (Mobini et al., 2015).

10 Challenges in the commercialization of genetically engineered grain legumes

In order to meet the increasing demand for grain legume and pulses production, there is an urgent need to integrate the existing knowledge of genetics, biochemistry and molecular biology to realize the full potential of agricultural biotechnology. In addition, it is necessary to pay due attention to stringent biosafety guidelines that continuously evolve with time and need. Considering that a large number of transgenic grain legume technologies with desirable traits are in the pipeline, translating these to farm products will bring another revolution in breeding.

The establishment of appropriate regulatory frameworks for transgenic biotechnology is critical for commercialization and trade in transgenic products. This requires due diligence by individual countries in terms of issues such as safety, food security and market competitiveness. Transgenic product commercialization requires appropriate risk assessment studies pertaining to biosafety/nutrition and environmental aspects. These risk assessment frameworks will help to establish standard operating procedures for testing transgenic crops before they start field cultivation. A robust framework both provides a route to commercialization whilst ensuring comprehensive evaluation of safety and environmental issues.

While the development of transgenic legumes to overcome important abiotic and biotic constraints is still a challenge mainly due to the lack of ability to transform and efficiently produce transgenic plants, a greater challenge is in the translation of proof of concept into commercial products. With the exception of cowpea for insect resistance, where advanced field trials and biosafety studies are ongoing, other grain legumes such as chickpea, pigeonpea and peanut are still in the research and development pipeline. Following the overcoming of initial technical hurdles for some of these grain legumes, major bottlenecks include the following: lack of a clear regulatory framework in target countries, the need to develop validation processes under controlled conditions, intellectual property rights and inexperience of public sector scientists in product commercialization, lack of funding and adequate infrastructure. With the exception of pigeonpea where ICRISAT and its partners have successfully developed cytoplasmic male sterility–based hybrid production technology (Saxena et al., 2010), the private seed sector which is good at commercial development is not keen on the business of grain legumes.

Hence, it is important that the public sector enhances its ability to exploit modern biotechnologies for crop improvement. However, some of the major bottlenecks in the successful transfer of this technology from laboratory to land revolve around the delays in transferring the products of genetic engineering to farmer's fields. One of the major problems in successful application of genetic engineering technology for product development and delivery in the public sector has been the lack of proper knowledge, expertise and infrastructure to translate transgenic technologies into commercial products. In practice transgenic product development involves an elaborate innovation value chain that may take many years to work through. Key stages include: basic research, translational research, development, verification, validation, prototype development, field trials, production/manufacturing and marketing. No scientist or agency in the public sector has all the expertise, technical and financial resources and management systems to accomplish tasks in this translational research.

This calls for developing transnational capabilities to help carry out this translational research (Rothstein et al., 2014). This would strengthen the following capabilities: ability to produce the transformation vectors and to transform elite lines, ability to undertake the regulatory work necessary to test these lines, ability to conduct multi-location biosafety field trials and ability to interact with public and private sector researchers. This would also ensure sharing of the high cost of product development in a network model. With these objectives, ICRISAT has developed a Platform for Translational Research on Transgenic Crops (PTTC) in partnership with the Department of Biotechnology, Government of India, in 2010. The PTTC can be viewed as a 'clearing' house for innovative ideas and technology that could positively impact agriculture in the semi-arid tropics, with an objective of providing expertise and facilities to the production and assessment of transgenic plants.

A partnership project for the development of *Maruca*-resistant cowpea, the 'Cowpea Productivity Improvement Project', in Nigeria is an international biotechnology public–private partnership aimed at providing pest-resistant cowpea varieties to Nigerian farmers. This involves multiple stakeholders from the government to technology developers to donors and farmers, an example that can be adopted for the translation of other genetic engineering technologies for other grain legumes as well (Ezezika and Daar, 2012).

11 Conclusion

The tools of modern biotechnology offer a tremendous opportunity for crop improvement, particularly for grain legumes where progress in gene modification has still to reach the stage of commercial application. One of the major bottlenecks has been a lack of robust and reproducible genetic transformation protocols with the exception of peanut, common pea, chickpea, pigeonpea and cowpea. Nevertheless, with the advances made so far, and some examples of proof of concept, the stage is set for more rapid progress. Examples include insect resistance in cowpea where advanced field trials are underway in Nigeria and insect resistance in pigeonpea and chickpea where the proof of concepts are being advanced to confined field trials in India. Similar advances are being made in grain legume transgenics for other agronomically and nutritionally important traits such as drought tolerance in peanut, β-carotene biofortification in peanut and pigeonpea, and aflatoxin resistance in peanut.

Despite being rich in proteins, grain legumes do not seem to contribute to the protein needs of humans and livestock as expected, given problems such as low concentrations of sulphur-amino acids, poor digestibility and presence of anti-nutritional factors, some of which themselves are proteins (Bliss, 1990). Moreover, even if grain legumes are capable of fixing enough nitrogen, they do not fix enough nitrogen to meet the demands of plant growth and maturation of protein-rich seeds and they show great variability in productivity with low yields and poor response to changes in cultivation practice. Bliss (1990) enumerated several objectives for the nutritional improvement of legumes that include 1) enhancing protein nutritional quality by increasing the levels of limiting essential amino acids, improving digestibility and lowering anti-nutritional factors; 2) improving the culinary properties of seeds, especially after prolonged storage; 3) altering physical and chemical properties to enhance use in new innovative ways and 4) improving yield potential concurrently with protein quantity and quality enhancement.

The availability of genomic sequence information and bioinformatics tools in most legumes (O'Rourke et al., 2014) has aided better understanding of the underlying molecular biology, physiology and biochemistry of important agronomic traits. The progress in addressing important constraints to crop productivity would benefit from marker-assisted backcrossing for traits that are already in the germplasm, or through functional genomics and candidate gene validation for use in genetic modification through genetic engineering. In conjunction with the availability of genomic resources, advances in molecular biology tools such as host-induced gene silencing, miRNA- and siRNA-based gene silencing, CRISPR/Cas9, TALEN and ZFN-based gene/genome editing, systems biology and synthetic biology should provide significant opportunities for addressing some fundamental constraints to crop biology, including nutritional enhancement and resistance to biotic and abiotic stresses (Songstad et al., 2017; Zhang et al., 2017).

With the development highlighted in the preceding sections and Table 1, rapid progress can be made in the near future on developing a pipeline of product concepts where product development and its commercialization should be relatively easy due to their self-pollinating nature that can contain any potential gene flow. However, some of the major bottlenecks in successful translation of proof of concepts to commercial products still remain and must be addressed to take full advantage of gene modification technology for sustainable agriculture. The lack of active participation of the seed industry in the seed business of grain legumes needs to be addressed by making the business more lucrative through promotion of seed systems in the developing countries and through joint product development. These include sustained funding by donors, development of partnerships through collaborative research and capacity building of researchers from the National Agricultural Research Systems (NARS) of countries where these technologies will be transferred, and product stewardship that form important components of translation research. While International Agricultural Research Institutes under the CGIAR System Organization along with its partners in the Advanced Research Institutes and NARS can play a major role in technology development and transfer, the model of the PTTC that has been established at ICRISAT can be strengthened and scaled up to address needs of the stakeholders from NARS in technology development, validation, transfer and translation.

12 Future trends

Though conventional breeding programmes have produced more resistant and higher-yielding varieties, yields remain below their potential. Most of these efforts are constrained by various insect pests causing pre- and post-harvest losses. It is apparent that in order to meet the world food requirements and solve the nutritional challenges of vulnerable populations, the world's agriculture needs more resistant and high-yielding crop varieties. Although not a panacea, emerging genetic engineering technologies have tremendous potential to develop agri-biotech traits for producing more efficient and resilient varieties. While genetic transformation technology has paved way for insect resistance trait in several grain legumes such as chickpea, pigeonpea and cowpea, a lot more work is required for building solutions to existing pathogens as well as emerging diseases. Moreover, parasitic weeds such as *Striga gesnerioides* and *Alectra vogelii* can cause almost complete crop loss and need to be addressed urgently. In addition to various biotic and abiotic constraints, bio-pharming is an emerging area for grain legume improvement to produce

Table 1 Gene transfer in legumes

Crop	Explant/mode of regeneration		Trait	Genes used	Reference
Chickpea					
	Axillary meristem/Direct organogenesis		Drought stress	*P5CSF129A*, *npt*II, *uid*A	Bhatnagar-Mathur et al. (2009b)
	Axillary meristem	Direct organogenesis	Drought stress	*DREB1A*, *npt*II	Bhatnagar-Mathur et al. (2014)
		Direct organogenesis	–	–	Yousefiara et al. (2008)
	Embryo axis	Direct organogenesis	–	–	Pawar et al. (2012)
	Cotyledon + Embryo axis		Aphid resistance	*ASAL*, *npt*II	Chakraborti et al. (2006)
	Cotyledonary node	Direct organogenesis	Insect resistance	*cry1Ac*, *npt*II, *uid*A	Sanyal et al. (2005)
	Cotyledon + Embryo axis	Direct organogenesis	Insect resistance	*cry1Ac*, *npt*II	Sarmah et al. (2004)
	Embryo axis, Epicotyl, Stem segment	Direct organogenesis	Insect resistance	*npt*II, *uid*A and *crylAc*	Indurker et al. (2010)
	Cotyledon + Embryo axis	Direct organogenesis	Nutrition fortification	(*SSA*), *bar*, *uid*A	Chiaiese et al. (2004)
	Embryo axis	Direct organogenesis		*uid*A	Chakrabarty et al. (2000)
	Cotyledonary node	Direct organogenesis	–	–	Jayanand et al. (2003)
	Embryo axis	Direct organogenesis	–	*npt*II, *uid*A	Kar et al. (1996)
	Embryo axis	Direct organogenesis	–	*uid*A, *npt*II,	Fontana et al. (1993)

	Embryo axis	Direct organogenesis	–	*uidA, nptII*	Krishnamurthy et al. (2000)
	Embryo axis	Direct organogenesis	–	*uidA, nptII*	Polowick et al. (2004)
	Gs	Direct organogenesis	Herbicide	*uidA, bar*	Senthil et al. (2004)
	Cotyledonary node	Direct organogenesis	Insect	*cry1Ac, nptII*	Sanyal et al. (2005)
	Cotyledon + Embryo axis	Direct organogenesis	Insect	*Alpha AI1, nptII*	Sarmah et al. (2004)
	Cotyledon + Embryo axis	Direct organogenesis		*nptII, uidA*	Singh et al. (2004)
	Embryo axis, Epicotyl, Ss	Direct organogenesis	Insect	*nptII, uidA, cryIAc*	Indurker et al. (2010)
	Embryo axis	Direct organogenesis	Insect	*cryIAc, nptII*	Kar et al. (1997)
	Embryo axis	Direct organogenesis	–	*uidA, nptII*	Kar et al. (1996)
Cowpea					
	Nodal buds	–	Insect resistance	*Cry1Ab, nptII*	Adesoye et al. (2008)
	Cotyledonary node	Direct organogenesis	Storage pests, bruchids and beetles	*aAI-1, nptII, uidA*	Solleti et al. (2008)
	Cotyledonary node	Direct organogenesis	–	*nptII, uidA*	Bakshi et al. (2012)
	Embryo axis	Direct organogenesis	Herbicide resistance	*Ahas, uidA*	Ivo et al. (2008)

(Continued)

Table 1 (*Continued*)

Crop	Explant/mode of regeneration		Trait	Genes used	Reference
	Cotyledonary node	Direct organogenesis	–	*nptII, uidA*	Chaudhury et al. (2007)
	Cotyledonary node	Direct organogenesis	Insect	*Cry1AC, nptII, uidA*	Bakshi et al. (2011)
	Cotyledonary node	Direct organogenesis	–	*Pmi*	Bakshi et al. (2012)
	Embryo axis	Direct organogenesis	Virus	*CPSMV* proteinase cofactor gene, *CABMV* coat protein gene	Cruzab and Aragão (2013)
	Embryo axis	Direct organogenesis	Herbicide	*Atahas*	Citadin et al. (2013)
	–	–	Drought	*VuDREB2A*	Sadhukhan et al. (2014)
	Embryo axis	Direct organogenesis	–	*nptII, uidA*	Raji et al. (2008)
	Sr	–	Parasitic root weed tolerance	*GFP*	Mellor et al. (2012)
		–	Bruchid pest resistance	*αAI-1*	Lüthi et al. (2013)
	Leaf disc	Through callus		*nptII*	Garcia et al. (1986)
	Leaf disc	Through callus	Virus	*Full-length DNA copy of M-RNA of cowpea mosaic virus*	Garcia et al. (1987)
	Leaf disc	Through callus	Insect	*CpTI*	Hilder et al. (1989)
	Embryo axis	Through callus	–	*Gus, neo*	Penza et al. (1991)
Pigeonpea					
	Leaf	Organogenesis	–	*nptII, uidA*	Dayal et al. (2003)

Cotyledonary node, Shoot apex	Direct organogenesis	–	*nptII*, *uidA*	Geetha et al. (1999)
Embryo axis	Through callus	–	*CPT1*, *nptII*	Lawrence and Koundal (2001)
Cotyledonary node	Direct organogenesis	Virus	*nptII*, *HN*	Prasad et al. (2004)
Cotyledonary node, Embryo axis	Direct organogenesis	virus	*H* gene, *nptII*	Satyavathi et al. (2003)
Axillary meristem	Direct organogenesis	Insect	*cry1Ab*, *nptII*	Sharma et al. (2006)
Embryo axis	Direct organogenesis	Insect	*cry1E-C*, *nptII*	Surekha et al. (2005)
Cotyledonary node	Through callus	–	*uidA*, *nptII*	Thu et al. (2003)
Embryo axis	Direct organogenesis	Salinity	*P5CSF129A*, *hptII*, *uidA*	Surekha et al. (2014)
–	–	Insect	*cry1AcF*, *nptII*	Ramu et al. (2012)
Embryo axis	Direct organogenesis	Insect	*cry1Ac*, *nptII*	Krishna et al. (2011)
Cotyledonary node	Through callus	Nutritional fortification	*dhdps-r1* gene, *nptII*	Thu et al. (2007)
Embryo axis	–	Insect	*cry1Ac*, *nptII*	Kaur et al. (2016)
Cotyledonary node, Ep, Leaf	Callus and direct organogenesis	Insect	*cry1Ac*, *nptII*, *uidA*, *hptI*	Thangella and Fakrudin (2016)
Cotyledon	Direct organogenesis	–	*nptII*, *uidA*	Raghavendra and Sudhakar (2014)

(Continued)

Table 1 (*Continued*)

Crop	Explant/mode of regeneration		Trait	Genes used	Reference
Common pea					
	Axillary meristem	–	Virus	*PEMV-CP*	Chowrira et al. (1998)
	Immature seed, Cotyledon	Callus	Herbicide	*nptII*, *bar*	Grant et al. (1995)
	Immature seed, Cotyledon	Callus	–	*nptII*, *uidA*	Grant et al. (2003)
	Cotyledonary node	–	–	*uidA-int*, *bar* or *nptII*	Svabova and Griga (2008)
		–	–	*nptII*, *uidA*, *hpt*	Kathen and Jacobsen (1990)
	Cotyledonary node	Direct organogenesis	–	*nptII*, *uidA*	Jordan and Hobbs (1993)
	Embryo axis	–	–	*uidA*	Lurquin et al. (1998)
	Nodal thin cell layer segments	Organogenesis	–	*uidA*	Nauerby et al. (1991)
	Epicotyl, Leaf, Stem segment	Callus	–	*nptII*	Puonti-Kaerlas et al. (1989)
	Cotyledon + Ea	Direct organogenesis	Fungal	*Bar*, *PGIP*, *Vst1*	Richter et al. (2006)
	Embryo axis, Immature seed	Callus	Herbicide	*bar*, *nptII*	Schroeder et al. (1993)
	Embryo axis, Immature seed	Callus	Pea weevil	*α-amylase inhibitor (α-AI)*, *bar*	Schroeder et al. (1995)
	Embryo axis, Root segment	–	–	*uidA*	Warkentin et al. (1992)
	Axillary bud, Embryo axis, Leaf, Stem segment	Callus and direct organogenesis	Herbicide	uidA, *bar*	Krejčí et al. (2007)
	Embryo axis	Organogenesis	Fungal	*Chit30*	Hassan et al. (2009)
	Embryo axis	Organogenesis	Insect	*Cry1Ac*, *bar*	Negawo et al. (2013)

Groundnut					
	Cotyledon	Direct organogenesis	Drought and salinity	*SbASR-1, hptII, uidA*	Tiwari et al. (2015)
	Cotyledonary node	Direct organogenesis	–	*uidA, nptII*	Venkatachalam et al. (1998)
	Seedling	–	Fungus	*Glucanase, nptII, uidA*	Sundaresha et al. (2010)
	Cotyledon	Direct organogenesis	Salinity and drought	*AtNHX1, nptII*	Asif et al. (2011)
	Seedling	–	Insect	*cry1AcF, nptII*	Keshavareddy et al. (2013)
	Cotyledon	–	–	*RolC, Opines*	Akasaka et al. (1998)
	Cotyledon	Direct organogenesis	Insect	*cry1EC, hptII*	Tiwari et al. (2008)
	Cotyledon, EA, Leaf	–	Salinity	*Cox, nptII, uidA*	Vadawale et al. (2012)
	Cotyledon	Direct organogenesis	–	*Zmpsy1, Rchit*	Bhatnagar et al. (2010)
	Cotyledonary node	Direct organogenesis	Late leaf spot disease	*SniOLP, Rs-AFP2, hpt*	Vasavirama and Kirti (2012)
	Cotyledonary node	Direct organogenesis	Virus	*TSV-Rep, nptII*	Reddy et al. (2016)
	Cotyledonary node	Direct organogenesis	Fungus	*Rchit, hpt, gfp*	Prasad et al. (2013)
	Embryo axis	Direct organogenesis	Fungus	*BjD, nptII*	Swathi Anuradha et al. (2008)
	Embryo axis	Direct organogenesis	Salinity	*SbpAPX, uidA, hptII*	Singh et al. (2014)
			Drought and salinity	*AtNAC2, nptII*	Patil et al. (2014)

(Continued)

Table 1 (*Continued*)

Crop	Explant/mode of regeneration		Trait	Genes used	Reference
	Cotyledonary node	Direct organogenesis	Drought	*DREB1A, nptII*	Bhatnagar-Mathur et al. (2014)
	Cotyledonary node	Direct organogenesis	Fungus	*Chit, hpt*	Iqbal et al. (2012)
			Salinity	*mtlD*	Patel et al. (2016)
	Cotyledon, Leaf	Direct organogenesis	Virus	*TSV-CP, hptII*	Mehta et al. (2013)
	Embryo axis, Epicotyl, Leaf	Direct organogenesis	–	*nptII, uidA*	McKently et al. (1995)
			–	*nptII, hpt, uidA*	Egnin et al. (1998)
	Cotyledonary node	Direct organogenesis	Virus	*HN, nptII, uidA*	Khandelwal et al. (2011)
	Cotyledon	Direct organogenesis	–	*nptII, hpt, uidA*	Tiwari and Tuli (2012)
Lentil					
	Cotyledonary node	Direct organogenesis	–	*nptII, uidA*	Akcay et al. (2009)
	Axillary bud	–	–	*uidA*	Chowrira et al. (1996)
	Cotyledonary node	Direct organogenesis	Herbicide	*ALS*	Gulati et al. (2002)
	Embryo axis	Direct organogenesis	–	*uidA*	Lurquin et al. (1998)
	Cotyledonary node	Direct organogenesis	–	*uidA*	Mahmoudian et al. (2002a,b)
	Epicotyl, Root segment, Shoot apex	–	–	*uidA*	Warkentin and McHughen (1992b)

high-value commercial products in a cost-effective manner with high efficiency. Another area that needs biotechnological attention is to reduce the levels of anti-nutritional factors in grain legumes to make them more attractive for food and feed use. Traits like seed size and quality are important for commercialization of grain legumes such as peas, lentils, chickpea and pigeonpea, and so initiatives to enhance protein content and quality using advanced genome engineering techniques are also key.

13 Where to look for further information

13.1 Further reading

Dar, W. D., Bhatnagar-Mathur, P. and Sharma, K. K. (2008). Application of modern biotechnology for improving livelihoods in the semi-arid tropics: ICRISAT's initiatives. In: *Agriculture for Food, Nutritional Security and Rural Growth*, V. Dhawan (Ed.), TERI, New Delhi, pp. 171–94.

Grumet, R., Hancock, J. F., Meridia, K. M. and Weebadde, C. (2011). *Environmental Safety of Genetically Engineered Crops*, Michigan State University, East Lansing, pp. 234.

Jez, J. M., Lee, S. G. and Sherp, A. M. (2016), The next green movement: Plant biology for the environment and sustainability, *Science*, 353(6305), 124101244. http://www.sciencemag.org/cgi/doi/10.1126/science.aag1698.

National Academies of Sciences, Engineering, and Medicine (2016). *Genetically Engineered Crops: Experiences and Prospects*, The National Academies Press, Washington DC. doi: 10.17226/23395.

Prado, J. R., Segers, G., Voelker, T., Carson, D., Dobert, R., Phillips, J., Cook, K., Cornejo, C., Monken, J., Grapes, L., Reynolds, T. and Martino-Catt, S. (2014). Genetically engineered crops: From idea to product, *Annu. Rev. Plant Biol.*, 65, 7690790. doi: 10.1146/annurev-arplant-050213-040039.

Ricroch, A., Chopra, S. and Fleischer, S. (2014). *Plant Biotechnology: Experiences and Future Prospects*, Springer International Publishing, Switzerland, pp. 291. doi: 10.1007/978-3-319-06892-3.

13.2 International agricultural research centres

1. The International Crops Research Institute for the Semi-Arid Tropics (ICRISAT), Patancheru, Hyderabad 502 324, Telangana, India; www.icrisat.org
2. The International Institute for Tropical Agriculture (IITA), PMB 5320, Oyo Road, Ibadan 200 001, Nigeria; www.iita.org
3. International Center for Tropical Agriculture (CIAT), KM 17 Recta Cali Palmira CP 763537, Apartado Aéreo 6713, Columbia; www.ciat.cgiar.org
4. International Center for Agricultural Research in the Dry Areas (ICARDA, Dalia Building 2nd Floor, Bashir El Kassar Street, Verdun, Beirut, Lebanon 1108–2010; www.icarda.org
5. CGIAR Research Program on Grain Legumes, www. http://grainlegumes.cgiar.org
6. International Life Science Institute (ILSI) Research Foundation | 1156 Fifteenth Street NW, Suite 200, Washington DC 20005. www.ilsirf.org

14 References

Adesoye, A., Machuka, J. and Togun, A. (2008), 'CRY 1AB transgenic cowpea obtained by nodal electroporation', *Afr. J. Biotechnol.*, 7(18), 3200–10.

Ahlawat, I. P. S., Gangaiah, B. and Singh, I. P. (2005), 'Pigeonpea (*Cajanus cajan*) research in India: An overview', *Ind. J. Agric. Sci.*, 75(6), 309–20.

Akasaka, Y., Mii, M. and Daimon, H. 1998, 'Morphological alterations and root nodule formation in Agrobacterium rhizogenes-mediated transgenic hairy roots of peanut (*Arachis hypogaea* L.)', *Ann. Bot.*, 81(2), 355–62.

Akcay, U. C., Mahmoudian, M., Kamci, H., Yucel, M. and Oktem, H. A. (2009), '*Agrobacterium tumefaciens*-mediated genetic transformation of a recalcitrant grain legume, lentil (*Lens culinaris* Medik)', *Plant Cell Rep.*, 28(3), 407–17.

Ali, Z., Ullah, N., Naseem, S., Inam-Ul-Haq, M. and Jacobsen, H. J. (2015), 'Soil bacteria conferred a positive and improved salt tolerance in transgenic pea (*Pisum sativum* L.) harbouring Na^+/K^+ antiporter', *Turk. J. Bot.*, 39, 962–72.

Altinkut, A., Bajrovic, K. and Gozukirmizi, N. (1997), 'Regeneration and hairy root formation of chickpea using callus-derived plantlets and seedlings', *Int. Chickpea Pigeonpea Newslett.*, 4, 30–1.

Amian, A. A., Papenbrock, J., Jacobsen, H. J. and Hassan, F. (2011), 'Enhancing transgenic pea (*Pisum sativum* L.) resistance against fungal diseases through stacking of two antifungal genes (chitinase and glucanase)', *GM Crops*, 2(2), 104–9.

Anbazhagan, K., Vadez, V., Bhatnagar Mathur, P., Sharma, K. K., Baddam, R., Kavikishor, P. and Vadez, V. (2014). 'Changes in timing of water uptake and phenology favors yield gain in terminal water stressed chickpea AtDREB1A transgenics', *Funct. Plant Biol.*, 42, 84–94.

Anbazhagan, K., Bhatnagar-Mathur, P., Vadez, V., Dumbala, S. R., Kishor, P. K. and Sharma, K. K. (2015), 'DREB1A overexpression in transgenic chickpea alters key traits influencing plant water budget across water regimes', *Plant Cell Rep.*, 34(2), 199–210.

Anuradha, T. S., Jami, S. K., Datla, R. S. and Kirti, P. B. (2006), 'Genetic transformation of peanut (*Arachis hypogaea* L.) using cotyledonary node as explant and a promoter less *gus::nptII* fusion gene based vector', *J. Biosci.*, 31 (2), 235–46.

Anuradha, T. S., Divya, K., Jami, S. K. and Kirti, P. B. (2008), 'Transgenic tobacco and peanut plants expressing a mustard defensin show resistance to fungal pathogens', *Plant Cell Rep.*, 27(11), 1777–86.

Anwar, S., Ali, B. and Sajid, I. (2016), 'Screening of rhizospheric actinomycetes for various *in-vitro* and *in-vivo* plant growth promoting (PGP) traits and for agroactive compounds', *Front. Microbiol.*, 7, 1334. doi:10.3389/fmicb.2016.01334.

Arundhati, A. (1999), '*Agrobacterium*-mediated transformation of pigeonpea (*Cajanus cajan* L. Millsp.) by using leaf disks', *Int. Chickpea Pigeonpea Newslett.*, 6, 62–4.

Asif, M. A., Zafar, Y., Iqbal, J., Iqbal, M. M., Rashid, U., klAli, G. M., Arif, A. and Nazir, F. (2011), 'Enhanced expression of AtNHX1, in transgenic groundnut (*Arachis hypogaea* L.) improves salt and drought tolerance', *Mol. Biotechnol.*, 49(3), 250–6.

Bag, S., Singh, R. S. and Jain, R. K. (2007), '*Agrobacterium*-mediated transformation of groundnut with coat protein gene of *Tobacco streak virus*', *Ind. J. Virol.*, 18(2), 65–9.

Bajaj, Y. P. S. and Dhanju, M. S. (1979), 'Regeneration of plants from apical meristem tips of some legumes', *Curr. Sci.*, 48(20), 906–7.

Bakshi, S., Sadhukhan, A., Mishra, S. and Sahoo, L. 2011, 'Improved Agrobacterium-mediated transformation of cowpea via sonication and vacuum infiltration', *Plant Cell Rep.*, 30(12), 2281–92.

Bakshi, S., Saha, B., Roy, N. K., Mishra, S., Panda, S. K. and Sahoo, L. (2012), 'Successful recovery of transgenic cowpea (*Vigna unguiculata*) using the 6-phosphomannose isomerase gene as the selectable marker', *Plant Cell Rep.*, 31(6), 1093–103.

Barna, K. S. and Wakhlu, A. K. (1994), 'Whole plant regeneration of *Cicer arietinum* from callus cultures via organogenesis', *Plant Cell Rep.*, 13(9), 510–13.

Basu, M. S. (1995), 'Peanut bud necrosis disease: Activities in the Indian national program', In: *Recent Studies on Peanut Bud Necrosis Disease. Proceedings of a Meeting of ICRISAT Asia Center*, India, 61–3.

Bean, S. J., Gooding, P. S., Mullincaux, P. M. and Davies, D. R. (1997), 'A simple system for pea transformation', *Plant Cell Rep.*, 16(8), 513–19.

Bhargava, S. C. and Smigocki, A. C. (1994), 'Transformation of tropical grain legumes using particle bombardment', *Curr. Sci.*, 66(6), 439–42.

Bhatnagar, M., Prasad, K., Bhatnagar-Mathur, P., Narasu, M. L., Waliyar, F. and Sharma, K. K., (2010), 'An efficient method for the production of marker-free transgenic plants of peanut (*Arachis hypogaea* L.)', *Plant Cell Rep.*, 29(5), 495–502.

Bhatnagar-Mathur, P., Reddy, D. S., Lavanya, M., Yamaguchi-Shinozaki, K., Sharma, K. K. (2007), 'Stress-inducible expression of *Arabidopsis thaliana DREB1A* in transgenic peanut (*Arachis hypogaea* L.) increases transpiration efficiency under water limiting conditions', *Plant Cell Rep.*, 26, 2071–82.

Bhatnagar-Mathur, P., Vadez, V. and Sharma, K. K. (2008), 'Transgenic approaches for abiotic stress tolerance in plants: Retrospect and prospects', *Plant Cell Rep.*, 27, 411–24.

Bhatnagar-Mathur, P., Devi, M. J, Vadez, V. and Sharma, K. K. (2009a), 'Differential antioxidative responses in transgenic peanut bear no relationship to their superior transpiration efficiency under drought stress', *J. Plant Physiol.*, 166, 1207–17.

Bhatnagar-Mathur, P., Vadez, V., Devi, M. J., Lavanya, M., Vani, G. and Sharma, K. K. (2009b), 'Genetic engineering of chickpea (*Cicer arietinum* L.) with the P5CSF129A gene for osmoregulation with implications on drought tolerance', *Mol. Breed.*, 23(4), 591–606.

Bhatnagar-Mathur, P., Rao, J. S., Vadez, V., Reddy, D. S., Rathore, A., Yamaguchi, S. K. and Sharma, K. K. (2014), 'Transgenic peanut overexpressing the DREB1A transcription factor has higher yields under drought stress', *Mol. Breed.*, 33, 327–40.

Bhatnagar-Mathur, P., Sunkara, S., Bhatnagar-Panwar, M., Waliyar, F. and Sharma, K. K. (2015), 'Biotechnological advances for combating *Aspergillus flavus* and aflatoxin contamination in crops', *Plant Sci.*, 234, 119–32.

Bieri, V., Schmid, J. and Keller, E. R. (1984), 'Shoot tip culture in *Vicia faba* L', In: *Efficiency in Plant Breeding Proceedings of the 10th Congress of the European Association for Research on Plant Breeding, EUCARPIA*, Wageningen, 295.

Bliss, F. A. (1990), 'Genetic alteration of legume seed proteins', *Hort. Sci.*, 25(12), 1517–20.

Böttinger, P., Steinmetz, A., Schieder, O. and Pickardt, T. (2001), '*Agrobacterium*-mediated transformation of *Vicia faba*', *Mol. Breed.*, 8, 243–54.

Brar, G. S., Cohen, B. A., Vick, C. L. and Johnson, G. W. (1994), 'Recovery of transgenic peanut (*Arachis hypogaea* L.) plants from elite cultivars utilizing ACCELL® technology', *Plant J.*, 5(5), 745–53.

Brar, M. S., Al-Khayri, J. M., Morelock, T. E. and Anderson, E. J. (1999), 'Genotypic response of cowpea *Vigna unguiculata* (L.) to in vitro regeneration from cotyledon explants', *Vitro Cell. Dev. Biol. Plant*, 35(1), 8–12.

Chakrabarty, R., Singh, D. T. and Garg, G. K. (2000), 'Transformation studies on chickpea embryo axis', *J. Plant Biochem. Biotechnol.*, 9(2), 107–10.

Chakraborti, D., Sarkar, A. and Das, S. (2006), 'Efficient and rapid *in vitro* plant regeneration system for Indian cultivars of chickpea (*Cicer arietinum* L.)', *Plant Cell Tissue Organ Cult.*, 86(1), 117–23.

Chandra, A. and Pental, D. (2003), 'Regeneration and genetic transformation of grain legumes: An overview', *Curr. Sci.*, 84(3), 381–7.

Chander Rao, S., Kumar, P. L., Reddy, A. S., Swamy Krishna, T., Waliyar, F., Nigam, S. N., Laxminarasu, M. and Sharma, K. K. (2006), 'Evaluation of transgenic peanut plants against peanut bud necrosis disease (PBND) under greenhouse and field conditions', *Ind. J. Virol.*, 17, 135–6.

Chaudhury, D., Madanpotra, S., Jaiwal, R., Saini, R., Kumar, P. A. and Jaiwal, P. K. (2007), '*Agrobacterium tumefaciens*-mediated high frequency genetic transformation of an Indian cowpea (*Vigna unguiculata* L. Walp.) cultivar and transmission of transgenes into progeny', *Plant Sci.*, 172(4), 692–700.

Chenault, K. D. and Payton, M. E. (2003), 'Genetic transformation of a runner-type peanut with the nucleocapsid gene of tomato spotted wilt virus', *Peanut Sci.*, 30(2), 112–15.
Chenault, K. D., Melouk, H. A. and Payton, M. E. (2005), 'Field reaction to *Sclerotinia* blight among transgenic peanut lines containing antifungal genes', *Crop Sci.*, 45(2), 511–15.
Cheng, M., Li, Z., Demski, J. W. and Jarret, R. L. (1997), 'Expression and inheritance of foreign genes in transgenic peanut plants generated by *Agrobacterium*-mediated transformation', *Plant Cell Rep.*, 16(8), 541–4.
Chiaiese, P., Ohkama-Ohtsu, N., Molvig, L., Godfree, R., Dove, H., Hocart, C., Fujiwara, T., Higgins, T. J. V. and Tabe, L. M. (2004), 'Sulphur and nitrogen nutrition influence the response of chickpea seeds to an added, transgenic sink for organic sulphur', *J. Exp. Bot.*, 55(404), 1889–901.
Chowrira, G. M., Akella, V., Fuerst, P. E. and Lurquin, P. F. (1996), 'Transgenic grain legumes obtained by *in planta* electroporation-mediated gene transfer', *Mol. Biotechnol.*, 5(2), 85–96.
Chowrira, G. M., Cavileer, T. D., Gupta, S. K., Lurquin, P. F. and Berger, P. H. (1998), 'Coat protein-mediated resistance to pea enation mosaic virus in transgenic *Pisum sativum* L', *Transgenic Res.*, 7(4), 265–71.
Citadin, C. T., Ibrahim, A. B. and Aragao, F. J. (2011), Genetic engineering in Cowpea (*Vigna unguiculata*): History, status and prospects, *GM Crops*, 2(3), 144–9.
Citadin, C. T., Cruz, A. R. R. and Aragao, F. J. L. (2013), 'Development of transgenic imazapyr-tolerant cowpea (*Vigna unguiculata*)', *Plant Cell Rep.*, 32(4), 537–43.
Clemente, T. E., Robertson, D., Isleib, T. G., Beute, M. K. and Weissinger, A. K. (1992), 'Evaluation of Peanut (*Arachis hypogaea* L.) Leaflets from mature zygotic embryos as recipient tissue for biolistic gene transfer', *Transgenic Res.*, 1(6), 275–84.
Cruz, A. R. R. and Aragão, F. J. L. 2014, 'RNAi-based enhanced resistance to Cowpea severe mosaic virus and Cowpea aphid-borne mosaic virus in transgenic cowpea', *Plant Pathol.*, 63(4), 831–7.
Cubero, J. I. (1974), 'On the evolution of *Vicia faba* L', *Theor. Appl. Genet.*, 45(2), 47–51.
Dayal, S., Lavanya, M., Devi, P., Sharma, K. K. (2003), 'An efficient protocol for shoot regeneration and genetic transformation of pigeonpea [*Cajanus cajan* (L.) Millsp.] using leaf explants', *Plant Cell Rep.*, 21(11), 1072–9.
De Clercq, J., Zambre, M., Montagu, M. V., Dillen, W. and Angenon, G. (2002), 'An optimized *Agrobacterium*-mediated transformation procedure for *Phaseolus acutifolius* A. Gray', *Plant Cell Rep.*, 21(4), 333–40.
De Jaeger, G., Scheffer, S., Jacobs, A., Zambre, M., Zobell, O., Goossens, A., Depicker, A. and Angenon, G. (2002), 'Boosting heterologous protein production in transgenic dicotyledonous seeds using *Phaseolus vulgaris* regulatory sequences', *Nat. Biotechnol.*, 20(12), 1265–8.
De Kathen, A. and Jacobsen, H. J. (1990), '*Agrobacterium tumefaciens*-mediated transformation of *Pisum sativum* L. using binary and cointegrate vectors', *Plant Cell Rep.*, 9(5), 276–9.
Dillen, W., Clercq, J. D., Goossens, A., Montagu, M. V. and Angenon, G. (1997), '*Agrobacterium*-mediated transformation of *Phaseolus acutifolius* A. Gray', *Theor. Appl. Genet.*, 94(2), 151–8.
Dwivedi, S. L., Crouch, J. H., Nigam, S. N., Ferguson, M. E. and Paterson, A. H. (2003), Molecular breeding of groundnut for enhanced productivity and food security in the semi-arid tropics: Opportunities and challenges, *Adv. Agron.*, 80, 153–221.
Eapen, S. (2003), 'Regeneration and genetic transformation in peanut: Current status and future prospects', In: *Applied Genetics of Leguminosae Biotechnology*, P. K. Jaiwal and R. P. Singh (Eds), Kluwer Academic Publishers, Netherlands, pp. 165–86.
Eapen, S. and George, L. (1994), *Agrobacterium tumefaciens* mediated gene transfer in peanut (*Arachis hypogaea* L.), *Plant Cell Rep.*, 13(10), 582–6.
Egawa, Y., Bujang, I. B., Chotechuen, S., Tomooka, N. and Tateishi, Y. (1996), 'Phylogenetic differentiation of tetraploid *Vigna* species, *V. glabrescens* and *V. reflexo-pilosa*', *JIRCAS J.*, 3, 49–58.
Egnin, M., Mora, A. and Prakash, C. S., 1998, 'Factors enhancing *Agrobacterium tumefaciens*-mediated gene transfer in peanut (*Arachis hypogaea* L.)', *Vitro Cell. Dev. Biol. Plant*, 34(4), 310–18.

Ehlers, J. D. and Hall, A. E. (1997), 'Cowpea (*Vigna unguiculata* L. Walp)', *Field Crops Res.*, 53(1), 187–204.

Entoori, K., Sreevathsa, R., Arthikala, M. K., Kumar, P. A., Kumar, A. R. V., Madhusudhan, B. and Makarla, U. (2008), 'A chimeric cry1X gene imparts resistance to *Spodoptera litura* and *Helicoverpa armigera* in the transgenic groundnut' *Eur. Asian J. Bio. Sci.*, 2, 53–65.

Estruch, J. J., Carozzi, N. B., Desai, N., Duck, N. B., Warren, G. W. and Koziel, M. G. (1997), 'Transgenic plants: An emerging approach to pest control', *Nat. Biotechnol.*, 15(2), 137–41.

Ezezika, O. C. and Daar, A. S. (2012), 'Overcoming barriers to trust in agricultural biotechnology projects: A case study of Bt cowpea in Nigeria', *Agric. Food Security*, 1(1), S5.

FAO (2004), *Production Yearbook*. FAO, Rome.

FAOSTAT (2006), http://apps.fao.org/cgi-bin/nphdb.

Faris, D. G. (1965), 'The origin and evolution of cultivated forms of *Vigna sinensis*', *Can J. Genet. Cytol.*, 7(3), 433–52.

Fontana, G. S., Santini, L., Caretto, S., Frugis, G. and Mariotti, D. (1993), 'Genetic transformation in the grain legume *Cicer arietinum* L. (Chickpea)', *Plant Cell Rep.*, 12(4), 194–8.

Garcia, J. A., Hille, J. and Goldbach, R. (1986), 'Transformation of cowpea *Vigna unguiculata* cells with an antibiotic resistance gene using a Ti-plasmid-derived vector', *Plant Sci.*, 44(1), 37–46.

Garcia, J. A., Hille, J., Vos, P. and Goldbach, R. (1987), 'Transformation of cowpea *Vigna unguiculata* with a full-length DNA copy of cowpea mosaic virus M-RNA', *Plant Sci.*, 48(2), 89–98.

Geetha, N., Venkatachalam, P., Prakash, V. and Sita, L. G. (1998), 'High frequency induction of multiple shoots and plant regeneration from seedling explants of pigeonpea (*Cajanus cajan* L.)', *Curr. Sci.*, 75(10), 1036–41.

Geetha. N., Venkatachalam, P. and Sita, L. G. (1999), '*Agrobacterium*-mediated genetic transformation of Pigeonpea (*Cajanus cajan* L.) and development of transgenic plants via direct organogenesis', *Plant Biotechnol.*, 16 (3), 213–18.

Ghewande, M. P., Nandagopal, V. and Reddy, P. S. (1987), Plant protection in groundnut. Technical Bulletin, National Research Centre for Groundnut, Indian Council of Agricultural Research, Junagadh, Gujarat, India.

Ghosh, G., Ganguly, S., Purohit, A., Chaudhuri, R. K., Das, S. and Chakraborti, D. (2017), 'Transgenic pigeonpea events expressing Cry1Ac and Cry2Aa exhibit resistance to *Helicoverpa armigera*', *Plant Cell Rep.*, 36, 1037–51.

Grant, J. E. and Cooper, P. A. (2003), 'Genetic transformation of pea', In: *Applied Genetics of Leguminosae Biotechnology*, Jaiwal, P. K. and Singh, R. P. (Eds), Kluwer, London, pp. 23–34.

Grant, J. E., Cooper, P. A., McAra, A. E. and Frew, T. J. (1995), 'Transformation of pea (*Pisum sativum*) using immature cotyledons', *Plant Cell Rep.*, 15(3),254–8.

Grant, J. E., Cooper, P. A., Gilpin, B. J., Hoglund, S. J., Reader, J. K., Pither-Joyce, M. D. and Timmerman-Vaughan, G. M. (1998a), 'Kanamycin is effective for selecting transformed peas', *Plant Sci.*, 139(2), 159–64.

Grant, J. E., Pither-Joyce, M., Fifield, W., Cooper, P. and Timmerman-Vaughan, G. (1998b), 'Partial resistance to alfalfa mosaic virus in transgenic pea (*Pisum sativum* L.)', In *Opportunities for High Quality, Healthy and Added Value Crops to Meet European Demands. Proceedings of the 3rd European Conference on Grain Legumes*, Valladolid, Spain, November 14–19, pp. 372–3.

Gregory, N. C., Krapovickas, A., and Gregory, M. P. (1980), 'Structural variation, evolution and classification in *Arachis*', *Advances in Legume Science*, Royal Botanical Garden, Kew, UK, pp. 469–81.

Gulati, A., Schryer, P. and McHughen, A. (2002), 'Production of fertile transgenic lentil (*Lens culinaris* Medik) plants using particle bombardment', *Vitro Cell. Dev. Biol. Plant*, 38(4), 316–24.

Ha, V., Sievenpiper, J. L., de Souza, R. J., Jayalath, V. H., Mirrahimi, A., Agarwal, A., Chiavaroli, L., Mejia, S. B., Sacks, F. M., Di Buono, M. and Bernstein, A. M. (2014), 'Effect of dietary pulse intake on established therapeutic lipid targets for cardiovascular risk reduction: A systematic review and meta-analysis of randomized controlled trial', *Can. Med. Assoc. J.*, 186(8), E252–62.

Hanafy, M., Pickardt, T., Kiesecker, H. and Jacobsen, H. J. (2005), '*Agrobacterium*-mediated transformation of faba bean (*Vicia faba* L) using embryo axes', *Euphytica*, 142(3), 227–326.

Hanafy, M. S., El-Banna, A., Schumacher, H. M., Jacobsen, H. J. and Hassan, F. S. (2013), 'Enhanced tolerance to drought and salt stresses in transgenic faba bean (*Vicia faba* L.) plants by heterologous expression of the PR10a gene from potato', *Plant Cell Rep.*, 32, 663–74.

Hassan, F., Meems, J., Jacobsen, H. and Kiesecker, H. (2009). 'A family of 19 chitinase (Chit30) from *Streptomyces olivaceoviridis* ATCC 11238 expressed in transgenic pea affects the development of *T. harzianum in vitro*', *J Biotechnol*, 143(4): 302–8.

Hassan, F., Noorian, M. S. and Jacobsen, H. J. (2012), 'Effect of antifungal genes expressed in transgenic pea (*Pisum sativum* L.) on root colonization with *Glomus intraradices*', *GM Crops Food*, 3(4), 301–9.

Haware, M. P. (1998), 'Disease of chickpea', In: *The Pathology of Food and Pasture Legumes*, D. J. Allen and J. M. Lenné (Eds), CAB International, Wallingford, UK, pp. 473–516.

Hedrick, U. P. (1928) *Vegetables of New York, Peas of New York*, 1, J. B. Lyon Co., Albany.

Higgins, C. M., Hall, R. M., Mitter, N., Cruickshank, A. and Dietzgen, R. G. (2004), 'Peanut stripe potyvirus resistance in peanut (*Arachis hypogaea* L.) plants carrying viral coat protein gene sequences', *Transgenic Res.*, 13(1), 59–67.

Hilder, V. A., Barker, R. F., Samour, R. A., Gatehouse, A. M., Gatehouse, J. A. and Boulter, D. (1989), 'Protein and cDNA sequences of Bowman-Birk protease inhibitors from the cowpea (*Vigna unguiculata* Walp.)', *Plant Mol. Biol.*, 13(6), 701–10.

Hinchee, M. A. W., Connor-Ward, D. V., Newell, C. A., McDonnell, R. E., Sato, S. J., Gasser, C. S., Fischoff, D. A., Re, D. B., Fraley, R. T. and Horsch, R. B. (1988), 'Production of transgenic soybean plants using *Agrobacterium*-mediated DNA transfer', *Bio/Technol.*, 6(8),915–22.

Hobbs, S. L. A., Jackson, J. A. and Mahon, J. D. (1989), 'Specificity of strain and genotype in the susceptibility of pea to *Agrobacterium tumefaciens*', *Plant Cell Rep.*, 8(5), 274–7.

Hussey, G., Johnson, R. D. and Warren, S. (1989), 'Transformation of meristematic cells in the shoot apex of cultured pea shoots by *Agrobacterium tumefaciens* and *A.* rhizogenes', *Protoplasma*, 148(2), 101–5.

ICRISAT (1992), The Medium Term Plan, ICRISAT (International Crops Research Institute for the Semi-Arid Tropics), Patancheru, 1, 80.

ICRISAT (2007), Chickpea, International Crop Research Institute for the Semi-Arid Tropics. Available from www.icrisat.org.

Indurker, S., Misra, H. S. and Eapen, S. (2010), 'Agrobacterium-mediated transformation in chickpea (Cicer arietinum L.) with an insecticidal protein gene: Optimisation of different factors', *Physiol. Mol. Bio. Plants*, 16(3), 273–84.

Ikea, J., Ingelbrecht, I., Uwaifo, A. and Thottappilly, G. (2003), 'Stable gene transformation in cowpea (*Vigna unguiculata* L. Walp.) using particle gun method', *Afr. J. Biotechnol.*, 2(8), 211–18.

Iqbal, M. M., Nazir, F., Ali, S., Asif, M. A., Zafar, Y., Iqbal, J., Ali, G. M. (2012), 'Over expression of rice chitinase gene in transgenic peanut (*Arachis hypogaea* L.) improves resistance against leaf spot', *Mol. Biotechnol.*, 50(2), 129–36.

Ivo, N., Nascimento, C., Vieira, L., Campos, F. and Aragao, F. (2008), 'Biolistic-mediated genetic transformation of cowpea and stable Mendelian inheritance of transgenes', *Plant Cell Rep.*, 27, 1475–83.

Jacks, T. J., De Lucca, A. J., Rajasekaran, K., Stromberg, K. and van Pée, K. H. (2000), 'Antifungal and peroxidative activities of nonheme chloroperoxidase in relation to transgenic plant protection', *J. Agric. Food Chem.*, 48(10), 4561–4.

Jayanand, B., Sudarsanam, G. and Sharma, K. K. (2003), 'An efficient protocol for the regeneration of whole plants of chickpea (*Cicer arietinum* L.) by using axillary meristem explants derived from *in vitro*-germinated seedlings', *Vitro Cell. Dev. Biol. Plant*, 39(2), 171–9.

Jelenic, S., Mitrikeski, P. T., Papes, D. and Jelaska, S. (2000), '*Agrobacterium*-mediated transformation of broad bean *Vicia faba* L', *Food Technol. Biotechnol.*, 38 (3), 167–72.

Jones, A. L., Johansen, I. E., Bean, S. J., Bach, I. and Maule, A. J. (1998), 'Specificity of resistance to pea seed-borne mosaic potyvirus in transgenic peas expressing the viral replicase (NIb) gene', *J. Gen. Virol.*, 79(12), 3129–37.

Jordan, M. C. and Hobbs, S. L. A. (1993), 'Evaluation of a cotyledonary node regeneration system for Agrobacterium-mediated transformation of pea (Pisum sativum L.)', *Vitro Cell. Dev. Biol. Plant*, 29(2), 77–82.

Joshi, M., Niu, C., Fleming, G., Hazra, S., Chu, Y., Nairn, C. J., Yang, H. and Akins, P. O. (2005), 'Use of green fluorescent protein as a non-destructive marker for peanut genetic Transformation', *Vitro Cell. Dev. Biol. Plant*, 41(4), 437–45.

Kannaiyan, J., Nene, Y. L., Reddy, M. V., Ryan, J. G. and Raju, T. N. (1984), 'Prevalence of pigeonpea diseases and associated crop losses in Asia, Africa and the Americas', *Int. J. Pest Manage.*, 30(1), 62–72.

Kar, S., Johnson, T. M., Nayak, P. and Sen, S. K. (1996), 'Efficient transgenic plant regeneration through *Agrobacterium* mediated transformation in chickpea (*Cicer arietinum* L.)', *Plant Cell Rep.*, 16(1), 32–7.

Kar, S., Basu, D., Das, S., Ramkrishnan, N. A., Mukherjee, P., Nayak, P. and Sen, S. K. (1997), 'Expression of cryIA (c) gene of *Bacillus thuringiensis* in transgenic chickpea plants inhibits development of pod-borer (*Heliothis armigera)* larvae', *Transgenic Res.*, 6(2), 177–85.

Karthikeyan, A. S., Sarma, K. S. and Veluthambi, K. (1996), '*Agrobacterium tumefaciens*-mediated transformation of *Vigna mungo* (L.) Hepper', *Plant Cell Rep.*, 15(5), 328–31.

Kaur, A., Sharma, M., Sharma, C., Kaur, H., Kaur, N., Sharma, S., Arora, R., Singh, I. and Sandhu, J. S. (2016), 'Pod borer resistant transgenic pigeon pea (*Cajanus cajan* L.) expressing cry1Ac transgene generated through simplified *Agrobacterium* transformation of pricked embryo axes', *Plant Cell Tiss Org Cult (PCTOC).*, 127(3), 717–27.

Keshavareddy, G., Rohini, S., Ramu, S. V., Sundaresha, S., Kumar, A. R. V., Kumar, P. A. and Udayakumar, M. (2013),' Transgenics in groundnut (*Arachis hypogaea* L.) expressing cry1AcF gene for resistance to *Spodoptera litura* (F.)', *Physiol. Mol. Biol. Plants*, 19(3), 343–52.

Khan, S. K. and Ghosh, P. D. (1984), 'Plantlet regeneration from cotyledonary nodes of chickpea', *Int. Chickpea Newslett.*, 11, 22–4.

Khandelwal, A., Renukaradhya, G. J., Rajasekhar, M., Sita, L. G. and Shaila, M. S. (2011), 'Immune responses to hemagglutinin-neuraminidase protein of peste des petits ruminants virus expressed in transgenic peanut plants in sheep', *Vet. Immunol. Immunopathol.*, 140(3), 291–6.

Khatib, F., Koudsieh, S., Ghazal, B., Barton, J., Tsujimoto, H. and Baum, M. (2007), 'Developing herbicide resistant lentil (*Lens culinaris* Medikus subsp. *culinaris*) through *Agrobacterium*-mediated transformation', *Arab J. Plant Prot.*, 25(2), 185–92.

Khatib, F., Makris, A., Yamaguchi-Shinozaki, K., Kumar, S., Sarker, A. and Erskine, W. (2011), 'Expression of the *DREB1A* gene in lentil (*Lens culinaris* Medik. subsp *culinaris*) transformed with the *Agrobacterium* system, *Crop Past. Sci.*, 62(6), 488–95.

Khawar, K. M. and Özcan, S. (2002), '*In vitro* induction of crown galls by *Agrobacterium tumefaciens* super virulent strain A281 (pTiBo 542) in lentil (*Lens culinaris* Medik.)', *Turk. J. Bot.*, 26(3), 165–70.

Klein, T. M., Harper, E. C., Svab, Z., Sanford, J. C., Fromm, M. E. and Maliga, P. (1988), 'Stable genetic transformation of intact *Nicotiana* cells by the particle bombardment process', *Proc. Nat. Acad. Sci. USA*, 85(22), 8502–5.

Kononowicz, A. K., Cheah, K. T., Narasimhan, M. L., Murdock, L. L., Shade, R. E., Chrispeels, M. J., Filippone, E., Monti, L., Bressan, R. A. and Hasegawa, P. M. (1997), 'Developing a transformation system for cowpea (*Vigna unguiculata* [L.] Walp.)', In: *Advances in Cowpea Research*, B. B. Singh, D. R. Mohan Raj, K. E. Dashiell and L. E. N. Jackai (Eds), IITA, Nigeria, pp. 361–71.

Krapovickas, A. (1969), 'The origin, variability and spread of the peanut (*Arachis hypogaea)*, In: *The Domestication and Exploitation of Plants and Animals*, J. Ucko and Dimbleby (Eds), Duckworth, London, pp. 427–40.

Krejčí, P., Matušková, P., Hanáček, P., Reinöhl, V. and Procházka, S. (2007), 'The transformation of pea (*Pisum sativum* L.): Applicable methods of *Agrobacterium tumefaciens*-mediated gene transfer', *Acta Physiol. Plant*, 29(2), 157–63.

Krishna, G., Reddy, P. S., Ramteke, P. W., Rambabu, P., Tawar, K. B. and Bhattacharya, P. (2011), '*Agrobacterium*-mediated genetic transformation of pigeon pea [*Cajanus cajan* (L.) Millsp.] for resistance to legume pod borer *Helicoverpa armigera*', *J. Crop Sci. Biotechnol.*, 14 (3), 197–204.

Krishna, G., Singh, B. K., Kim, E-K., Morya, V. K. and Ramteke, P. W. (2015), 'Progress in genetic engineering of peanut (*Arachis hypogaea* L.) - a review', *Plant Biotechnol. J.*, 13, 147–62.

Krishnamurthy, K. V., Suhasini, K., Sagare, A. P., Meixner, M., Kathen, A. D., Pickardt, T. and Schieder, O. (2000), '*Agrobacterium* mediated transformation of chickpea (*Cicer arietinum* L.) embryo axes', *Plant Cell Rep.*, 19(3), 235–40.

Kumar, S. M., Kumar, B. K., Sharma, K. K. and Devi, P. (2004), 'Genetic transformation of pigeonpea with rice chitinase gene', *Plant Breed.*, 123(5), 485–9.

Labadi, M., Robertson, L. D., Singh, K. B. and Charrier, A. (1996), 'Genetic diversity and phylogenetic relationships among the annual *Cicer* species as revealed by isozyme polymorphisms', *Euphytica*, 88, 181–8.

Lacorte, C., Mansur, E., Timmermann, B. (1991), 'Gene transfer into peanut (*Arachis hypogaea* L.) by *Agrobacterium tumefaciens*', *Plant Cell Rep.*, 10, 354–7.

Langyintuo, A. and DeBoer, L. J. (2003), 'Assessing the potential impacts of biotechnologies on cowpea trade in west and Central Africa'. Department of Agricultural Economics, Purdue University, available at IPIA, Purdue University.

Lawrence, P. K. and Koundal, K. R. (2001), '*Agrobacterium tumefaciens*-mediated transformation of pigeon pea (*Cajanus cajan* L. Millsp.) and molecular analysis of regenerated plants', *Curr. Sci.*, 80, 11.

Lester, D. R., Ross, J. J., Davies, P. J. and Reid, J. B. (1997), Mendel's stem length gene (Le) encodes a gibberellin 3ß-hydroxylase', *Plant Cell*, 9, 1435–43.

Link, W., Hanafy, M., Malenica, N., Jacobsen, H.-J. and Jelenić, S. (2008), 'Faba bean, In: *Compendium of Transgenic Crop Plants: Transgenic Grain Legumes and Forages*, C. Kole and T. C. Hall (Eds), Blackwell Publishing Ltd., Oxford, UK, pp. 71–88.

Livingstone, D. M., Hampton, J. L., Phipps, P. M. and Grabau, E. A. (2005), 'Enhancing Resistance to *Sclerotinia* minor in peanut by expressing a barley oxalate oxidase gene', *Plant Physiol.*, 137, 1354–62.

Lulsdorf, M. M., Rempel, H., Jackson, J. A., Baliski, D. S. and Hobbs, S. L. A. (1991), 'Optimizing the production of transformed pea (*Pisum sativum* L.) callus using disarmed *Agrobacterium tumefaciens* strains', *Plant Cell Rep.*, 9, 479–83.

Lurquin, P. F., Cai, Z., Stiff, C. M. and Fuerst, E. P. (1998), 'Half-embryo cocultivation technique for estimating the susceptibility of pea (*Pisum sativum* L.) and lentil (*Lens culinaris* Medik.) cultivars to *Agrobacterium tumefaciens*', *Mol. Biotechnol.*, 9, 175–9. doi:10.1007/BF02760819.

Luthi, C., Alfageme, F. A., Ehlers, J. D., Higgins, T. J. V. and Romeis, J. (2013), 'Resistance of αAI-1 transgenic chickpea (*Cicer arietinum*) and cowpea (*Vigna unguiculata*) dry grains to bruchid beetles (*Coleoptera: Chrysomelidae*)', *Bull. Entomol. Res.*, 103, 373–81.

Maccarrone, M., Dini, L., Dimarzio, L., Digiulio, A., Rossi, A., Mossa, G. and Finazziagro, A. (1992), 'Interaction of DNA with cationic liposomes: Ability of transfecting lentil protoplasts', *Biochem. Biophys. Res. Commun.*, 186, 1417–22.

Maccarrone, M., Hilbers, M. P., Veldink, G. A., Vliegenthart, J. F. G. and Agro, A. F. (1995), 'Inhibition of lipoxygenase in lentil protoplasts by expression of antisense RNA', *Biochim. Biophys. Acta*, 1259, 1–3.

Machuka, J. (2000), Potential role of transgenic approaches in the control of cowpea insect pests. In: *Challenges and Opportunities for Enhancing Sustainable Cowpea Production*, C. Fatokun, S. A. Tarawali, B. B. Singh, P. M. Kormawa and M. Tamo (Eds), *Proceedings of World Cowpea Conference III*. September 4–7, 2000, IITA, Nigeria,

Magbanua, Z. V., Wilde, H. D., Roberts, J. K., Chowdhury, K., Abad, J., Moyer, J. W., Wetzstein, H. Y. and Parrott, W. A. (2000), 'Field resistance to tomato spotted wilt virus in transgenic peanut (*Arachis hypogaea* L.) expressing an antisense nucleocapsid gene sequence', *Mol. Breed.*, 6, 227–36.

Mahmoudian, M., Celikkol, U., Yucel, M. and Oktem, H. A. (2002a), 'Vacuum infiltration based *Agrobacterium* mediated gene transfer to lentil tissues', *Biotechnol. Equip.*, 16, 24–9.

Mahmoudian, M., Yucel, M. and Oktem, H. A. (2002b), 'Transformation of lentil (*Lens culinaris* M.) cotyledonary nodes by vacuum infiltration of *Agrobacterium tumefaciens*', *Plant Mol. Biol. Rep.*, 20, 251–7.

Mansur, E. A., Lacorte, C. and De Freitas, V. G. (1993), 'Regulation of transformation efficiency of peanut (*Arachis hypogaea* L.) explants by *Agrobacterium tumefaciens*', *Plant Sci.*, 89, 93–9.

McKently, A. H., Moore, G. A., Doostdar, H. and Niedz, R. P. (1995), '*Agrobacterium*-mediated transformation of peanut (*Arachis hypogaea* L.) embryo axes and the development of transgenic plants', *Plant Cell Rep.*, 14, 699–703.

Mehta, U. and Mohan Ram, H. Y. (1980), 'Regeneration of plantlets from cotyledons of *Cajanus cajan*'. *Ind. J. Exp. Biol.*, 18, 800–2.

Mehta, R., Radhakrishnan, T., Kumar, A., Yadav, R., Dobaria, J. R., Thirumalaisamy, P. P., Jain, R. K. and Chigurupati, P. (2013), 'Coat protein-mediated transgenic resistance of peanut (*Arachis hypogaea* L.) to peanut stem necrosis disease through *Agrobacterium*-mediated genetic transformation', *Ind. J. Virol.*, 24(2), 205–13.

Mellor, K. E., Hoffman, A. M. and Timko, M. P. (2012), 'Use of *ex vitro* composite plants to study the interaction of cowpea (*Vigna unguiculata L.*) with the root parasitic angiosperm Striga gesnerioides'. *Plant Meth.* 8, 22.

Miklas, P. N., Rosas, J. C., Beaver, J. S., Telek, L. and Freytag, G. F. (1994), 'Field performance of select tepary bean germplasm in the tropics', *Crop Sci.*, 34(6), 1639–44.

Mobini, S. H., Lulsdorf, M., Warkentin, T. D., Vandenberg, A. (2015), 'Plant growth regulators improve *in vitro* flowering and rapid generation advancement in lentil and faba bean', *Vitro Cell. Dev. Biol. Plant*, 51, 71–9.

Mondal, H. A., Das, S. and Sarkar, D. C. A. (2009), 'Tissue specific expression of potent insecticidal, *Allium sativum* leaf agglutinin (ASAL) in important pulse crop, chickpea (*Cicer arietinum L.*) to resist the phloem feeding *Aphis craccivora*', *Transgenic Res.*, 18, 529–44.

Muehlbauer, F. J., Cho, S., Sarker, A., McPhee Coyne, K. E. C. J., Rajesh, P. N. and Ford, R. (2006), 'Application of biotechnology in breeding lentil for resistance to biotic and abiotic stresses', *Euphytica*, 147, 149–65.

Muehlbauer, F. J., Kaiser, W. J., Clement, S. L. and Summerfield, R. J. (1995), 'Production and breeding of lentil', *Adv. Agron.*, 54, 283–332.

Muñoz-Perea, C. G., Terán, H., Allen, R. G., Wright, J. L., Westermann, D. T., and Singh, S. P. (2006). 'Selection for drought resistance in dry bean landraces and cultivars', *Crop Sci*, 46, 2111–20.

Murashige, T. and Skoog, F. (1962), 'A revised medium for rapid growth and bio assays with tobacco tissue cultures', *Physiol. Plant.*, 15(3), 473–97.

Murthy, B. S. N., Jerrin Victo Rana, I., Singh, R. A. and Saxena, P. K. (1996), '*In vitro* regeneration of chickpea (*Cicer arietinum* L.): Stimulation of direct organogenesis and somatic embryogenesis by thidiazuron', *Plant Growth Regul.*, 19, 233–40.

Muthukumar, B., Mariamma, M., Veluthambi, K. and Gnanam, A. (1996), 'Genetic transformation of cotyledon explants of cowpea (*Vigna unguiculata* Walp.) using *A. tumefaciens*', *Plant Cell Rep.*, 15, 980–5.

Nauerby, B., Madsen, M., Christiansen, J. and Wyndaele, R. (1991),' A rapid and efficient regeneration system for pea (*Pisum sativum*), suitable for transformation', *Plant Cell Rep.*, 9(12), 676–9.

Negawo, A. T., Aftabi, M., Jacobsen, H.-J., Altosaar, I., Hassan, F. S. (2013), 'Insect resistant transgenic pea expressing cry1Ac gene product from *Bacillus thuringiensis*', *Biol. Control.*, 67, 293–300 http://dx.doi.org/10.1016/j.biocontrol.2013.09.016.

Negawo, A. T., Baranek, L., Jocobsen, H-J. and Hassan, F. (2016). 'Molecular and functional characterization of *cry1Ac* transgenic pea lines', *GM Crops Food*, 7(3–4): 159–74.
Nigam, S. N., Prasada Rao, R. D. V. J., Bhatnagar-Mathur, P. and Sharma, K. K. (2012), 'Genetic management of virus diseases in peanut', *Plant Breed. Rev.*, 36, 293–356.
Ohler, T. A. and Mitchell, C. A. (1992), 'Evaluation of cowpea (*Vigna unguiculata* L. Walp) as a candidate species for inclusion in bioregenerative life-support systems', *Amer. Soc. Space Gravit. Biol. Bull.* 6, 37.
Oktem, H. A., Mahmoudian, M., Eyidogan, F. and Yucel, M. (1999), 'GUS gene delivery and expression in lentil cotyledonary nodes using particle bombardment', *LENS Newslett.* 26, 3–6.
O'Rourke, J. A., Bolon, Y. T., Bucciarelli, B. and Vance, C. P. (2014), 'Legume genomics: Understanding biology through DNA and RNA sequencing', *Ann. Bot.*, 113(7), 1107–20.
Ozias-Akins, P. and Gill, R. (2001), 'Progress in the development of tissue culture and transformation methods applicable to the production of transgenic peanut', *Peanut Sci.*, 28(2), 123–31.
Ozias-Akins, P., Schnall, J. A., Anderson, W. F., Singsit, C., Clemente, T. E., Adang, M. J. and Weissinger, A. K. (1993), 'Regeneration of transgenic peanut plants from stably transformed embryogenic callus,' *Plant Sci.*, 93(1–2), 185–94.
Ozias-Akins, P., Yang, H., Gill, R., Fan, H. and Lynch, R. E. (2002), 'Reduction of aflatoxin contamination in peanut: A genetic engineering approach', *Crop Biotechnol.* 12, 151–60.
Pandey, M. K., Monyo, E., Akins, P. O., Liang, X., Guimaraes, G., Nigam, S. N., Upadhyaya, H. D., Janila, P., Zhang, X., Guo, G., Cook, D. R., Bertioli, D. J., Michelmore, R. and Varshney, R. K. (2012), Advances in *Arachis* genomics for peanut improvement, *Biotechnol. Adv.*, 30, 639–51.
Patel, D. B., Brave, D. M., Nagar, N. and Mehta. A. R. (1994), 'Regeneration of pigeonpea, *Cajanus cajan*, through somatic embryogenesis', *Ind. J. Exp. Biol.*, 32, 740–4.
Patel, K. G., Mandaliya, V. B., Mishra, G. P., Dobaria, J. R and Thankappan, R. (2016), 'Transgenic peanut overexpressing *mtlD* gene confers enhanced salinity stress tolerance via mannitol accumulation and differential antioxidative responses', *Acta Physiol. Plant.*, 38, 181. doi:10.1007/s11738–016-2200–0.
Patil, M., Ramu, S. V., Jathish, P., Sreevathsa, R., Reddy, P. C. and Udayakumar, T. G. P. M. (2014), 'Overexpression of AtNAC2 (ANAC092) in groundnut (*Arachis hypogaea* L.) improves abiotic stress tolerance', *Plant Biotechnol. Rep.*, 8, 161–9.
Pawar, B. D., Jadhav, A. S, Kale, A. A, Chimote, V. P. and Pawar, S. V. (2012), 'Thidiazuron improves *in vitro* multiple shoot induction in chickpea (*Cicer arietinum* L.) cotyledon with embryonic axis', *J. Food Leg.*, 25(1), 9–13.
Penza, R., Lurquin, P. F. and Filippone, E. (1991), 'Gene transfer by cocultivation of mature embryos with *Agrobacterium tumefaciens:* Application to cowpea (*Vigna-unguiculata* walp.)', *J. Plant Physiol.*, 138, 39–43.
Polowick, P. L., Baliski, D. S., Bock, C., Ray, H., and Georges, F. (2008), 'Over-expression of a-galactosidase in pea seeds to reduce raffinose oligosaccharide', *Botany*, 87, 526–32.
Polowick, P. L., Quandt, J. and Mahon, J. D. (2000), 'The ability of pea transformation technology to transfer genes into peas adapted to western Canadian growing conditions', *Plant Sci.*, 153, 161–70.
Polowick, P. L., Baliski, D. S. and Mahon, J. D. (2004), '*Agrobacterium tumefaciens*-mediated transformation of chickpea (*Cicer arietinum* L.): Gene integration, expression and inheritance', *Plant Cell Rep.*, 23, 485–91.
Popelka, J. C., Gollasch, S., Moore, A., Molvig, L. and Higgins, T. J. V. (2006), 'Genetic transformation of cowpea (*Vigna unguiculata* L.) and stable transmission of the transgenes to progeny', *Plant Cell Rep.*, 25, 304–12.
Prasad. V., Satyavathi, V. V., Sanjaya Valli, K. M., Khandelwal, A., Shaila, M. S. and Sita, L. G. (2004), 'Expression of biologically active Hemagglutinin-neuraminidase protein of *Peste des Petits ruminants* virus in transgenic pigeonpea [*Cajanus cajan* (L) Millsp.]', *Plant Sci.*, 166, 199–205.

Prasad, P., Mathur, P. B., Waliyar, F. and Sharma, K. K. (2013), 'Overexpression of a chitinase gene in transgenic peanut confers enhanced resistance to major soil borne and foliar fungal pathogens', *J. Plant Biochem. Biotechnol.*, 22(2), 222–33.

Puonti-Kaerlas, J. (1993), 'Genetic engineering in pea crop improvement', *Acta Agric. Scand., Sec. B Soil Plant Sci.*, 43(2), 65–73.

Puonti-Kaerlas, J., Stabel, P. and Eriksson, T. (1989), 'Transformation of pea (*Pisum sativum* L.) by *Agrobacterium tumefaciens*', *Plant Cell Rep.*, 8(6), 321–4.

Puonti-Kaerlas, J., Eriksson, T. and Engstri, J. P. (1990), 'Production of transgenic pea (*Pisum sativum* L.) plants by *Agrobacterium tumefaciens*- mediated gene transfer', *Theor. Appl. Genet.*, 80, 246–52.

Qin, H., Gu, Q., Zhang, J., Sun, L., Kuppu, S., Zhang, Y., Burow, M., Payton, P., Blumwald, E. and Zhang, H. (2011), 'Regulated expression of an isopentenyltransferase gene (IPT) in peanut significantly improves drought tolerance and increases yield under field conditions', *Plant Cell Physiol.*, 52(11), 1904–14.

Raghavendra, T. and Sudhakar, P., (2014), 'Pattern of plant regeneration from shoot tip explants of pigeonpea (*Cajanus cajan* L. Millsp.) var LRG-41', *Int. J. Appl. Biol. Pharm.*, 5(1): 261–70.

Raji, A. A., Oriero, E., Odeseye, B., Odunlami, T. and Ingelbrecht, I. L. (2008), 'Plant regeneration and Agrobacterium-mediated transformation of African cowpea [Vigna unguiculata (L.) Walp] genotypes using embryonic axis explants', *J. Food Agric. Env.*, 6(3&4), 350–6.

Ramsay, G. and Kumar, A. (1990), 'Transformation of *Vicia faba* cotyledon and stem tissues by *Agrobacterium* rhizogenes: Infectivity and cytological studies', *J. Exp. Bot.*,41, 841–7.

Ramu, S. V., Rohini, S., Keshavareddy, G., Neelima, M. G., Shanmugam, N. B., Kumar, A. R. V., Sarangi, S. K., Ananda Kumar, P. and Udayakumar, M. (2012), 'Expression of a synthetic *cry1AcF* gene in transgenic Pigeon pea confers resistance to *Helicoverpa armigera*', *J. Appl. Entomol.*, 136(9), 675–87.

Ramu, V. S., Swetha, T. N., Sheela, S. H., Babitha, C. K., Rohini, S., Reddy, M. K., Tuteja, N., Reddy, C. P., Prasad, T. G. and Udayakumar, M. (2015), 'Simultaneous expression of regulatory genes associated with specific drought-adaptive traits improves drought adaptation in peanut', *Plant Biotechnol. J.*, 14(3), 1008–20. doi:10.1111/pbi.12461.

Rao, K. S., Sreevathsa, R., Sharma, P. D., Keshamma, E. and Udaya Kumar, M. (2008), In planta transformation of pigeon pea: A method to overcome recalcitrancy of the crop to regeneration *in vitro*, *Physiol. Mol. Biol. Plants*, 14(4), 321–8.

Redden, B., Maxted, N., Furman, B. and Coyne, C. (2007), *Lens* biodiversity, In: *Lentil, an Ancient Crop for Modern Times*, S. S. Yadav, McNeil, D. and Stevenson, P. C. (Eds), Springer, Dordrecht, pp. 11–22.

Reddy, D., Bhatnagar-Mathur, P., Reddy, P. and Sri Cindhuri, K. (2016), 'Identification and validation of reference genes and their impact on normalized gene expression studies across cultivated and wild *Cicer* species', *PLoS ONE*, 11(2), e0148451. (https://doi.org/10.1371/journal.pone.0148451)

Reichel, C., Mathur, J., Eckes, P., Langenkemper, K., Koncz, C., Schell, J., Reiss, B. and Maas, C. (1996), 'Enhanced green fluorescence by the expression of an *Aequorea victoria* green fluorescent protein mutant in mono- and dicotyledonous plant cells', *Proc. Nat. Acad. Sci. USA*, 93(12), 5888–93.

Richter, A., Jacobsen, H. J., De Kathen, A., De Lorenzo, G., Briviba, K., Hain, R., Ramsay, G. and Kiesecker, H., 2006, 'Transgenic peas (*Pisum sativum*) expressing polygalacturonase inhibiting protein from raspberry (*Rubus idaeus*) and stilbene synthase from grape (*Vitis vinifera*)', *Plant Cell Rep.*, 25(11), 1166–73.

Rohini, V. K. and Rao, K. S. (2000), 'Transformation of peanut (*Arachis hypogaea* L.): A non-tissue culture based approach for generating transgenic plants', *Plant Sci.*, 150, 41–9.

Rohini, V. K. and Rao, K. S. (2001), 'Transformation of peanut (*Arachish hpogaea* L.) with tobacco chitinase gene: Variable response of transformants to leaf spot disease', *Plant Sci.*, 160, 889–98.

Rothstein, S. J., Bi, Y. M., Coneva, V., Han, M. and Good, A. (2014), 'The challenges of commercializing second-generation transgenic crop traits necessitate the development of international public sector research infrastructure', *J. Exp. Bot.*, 65 (19), 5673–82.

Sadhukhan, A., Kobayashi, Y., Kobayashi, Y., Tokizawa, M., Yamamoto, Y. Y., Iuchi, S., Koyama, H., Panda, S. K. and Sahoo, L. (2014), 'VuDREB2A, a novel DREB2-type transcription factor in the drought-tolerant legume cowpea, mediates DRE-dependent expression of stress-responsive genes and confers enhanced drought resistance in transgenic *Arabidopsis*', *Planta*, 240(3), 645–64.

Sagare, A. P, Suhasini, K. and Krishnamurthy, K. V. (1993), 'Plant regeneration via somatic embryogenesis in chick pea', *Plant Cell Rep.*, 12, 652–5.

Sahoo, L., Sugla, T. and Jaiwal, P. K. (2003), '*In vitro* regeneration and genetic transformation of cowpea, mungbean, urdbean and azuki bean', In: *Applied Genetics of Leguminosae Biotechnology*, P. K. Jaiwal and R. P. Singh (Eds), Focus on Biotechnology, 10B, Kluwer Academic Publishers, Dordrecht, the Netherlands, pp. 89–120.

Saini, R. and Jaiwal, P. K. (2002), 'Age, position in mother seedling, orientation, and polarity of the epicotyl segments of blackgram (*Vigna mungo* L. Hepper) determines its morphogenic response', *Plant Sci.*, 163, 101–9.

Saini, R and Jaiwal, P. K. (2005), 'Transformation of a recalcitrant grain legume, *Vigna mungo* L. Hepper, using *Agrobacterium tumefaciens*-mediated gene transfer to shoot apical meristem cultures', *Plant Cell Rep.*, 24, 164–71.

Saini, R., Jaiwal, S. and Jaiwal, P. K. (2003), 'Stable genetic transformation of *Vigna mungo* L. Hepper via *Agrobacterium tumefaciens*', *Plant Cell Rep.*, 21, 851–9.

Saini, R., Singh, R. P. and Jaiwal, P. K. (2007), '*Agrobacterium tumefaciens* mediated transfer of *Phaseolus vulgaris* α-amylase inhibitor-1 gene into mungbean (*Vigna radiata* (L.) Wilczek) using bar as selectable marker', *Plant Cell Rep.*, 26(2), 187–98.

Sanyal. I., Singh. A. K., Kaushi. M and Amla. D. V. (2005), '*Agrobacterium*-mediated transformation of chickpea (*Cicer arietinum* L.) with *Bacillus thuringiensis cry1Ac* gene for resistance against pod borer insect *Helicoverpa armigera*', *Plant Sci.*, 168, 1135–46.

Sarmah, B. K., Moore, A., Tate, W., Molvig, L., Morton, R. L., Rees, D. P., Chiaiese, P., Chrispeels, M. J., Tabe, L. M. and Higgins, T. J. V. (2004), 'Transgenic chickpea seeds expressing high levels of a bean α-amylase inhibitor', *Mol. Breed.*, 14, 73–82.

Satyavathi, V. V., Prasad, V., Khandelwal, A., Shaila, M. S. and Sita, L. G. (2003), 'Expression of hemagglutinin protein of Rinderpest virus in transgenic pigeon pea [*Cajanus cajan* (L.) Millsp.] plants', *Plant Cell Rep.*, 21(7), 651–8.

Saxena, K. B., Sultana, R., Mallikarjuna, N., Saxena, R. K., Kumar, R. V., Sawargaonkar, S. L. and Varshney, R. K. (2010), 'Male-sterility systems in pigeonpea and their role in enhancing yield', *Plant Breed.*, 129(2), 125–34.

Schaerer, S. and Pilet, P. E. (1991), 'Roots, explants and protoplasts from pea transformed with strains of *Agrobacterium tumefaciens* and *rhizogenes*', *Plant Sci.*, 78, 247–58.

Schiemann, J. and Eisenreich, G. (1989), 'Transformation of field bean (*Vicia faba* L.) cells: Expression of a chimaeric gene in cultured hairy roots and root-derived callus', *Biochem. Physiol. Pflanzen*, 185(1–2), 135–40.

Schnall, J. A. and Weissinger, A. K. (1995), 'Genetic transformation in *Arachis hypogaea* L. (peanut)', In: *Plant Protoplasts and Genetic Engineering VI*, Y. P. S. Bajaj (Ed.), Biotechnology in Agriculture and Forestry, 34, Springer-Verlag, Germany, pp. 135–44.

Schroeder, H. E., Schotz, A. H., Richardson, T. W., Spencer, D. and Higgins, T. J. V. (1993), 'Transformation and regeneration of two cultivars of pea (*Pisum sativum* L.)', *Plant Physiol.*, 101, 751–7.

Schroeder, H. E., Gollasch, S., Moore, A., Tabe, L., Craig, S., Hardie, D., Chrispeels, M., Spencer, D. and Higgins, T. J. V. (1995), 'Bean α-amylase inhibitor confers resistance to the pea weevil (*Bruchus pisorum*) in transgenic peas (*Pisum sativum* L.)', *Plant Physiol.*, 107, 1233–9.

Selva, E., Stouffs, B. and Briquet, M. (1989), 'In vitro propagation of *Vicia faba* L. by micro-cutting and multiple shoot induction', *Plant Cell Tissue Organ Cult.*, 18(2), 167–79.

Senthil, G., Williamson, B., Dinkins, R. D. and Ramsay, G. (2004), 'An efficient transformation system for chickpea (*Cicer arietinum* L.)', *Plant Cell Rep.*, 23,297–303.
Sharma, K. K. and Anjaiah, V. (2000), 'An efficient method for the production of transgenic plants of peanut (*Arachis hypogaea* L.) through *Agrobacterium tumefaciens*-mediated genetic transformation', *Plant Sci.*, 159, 7–19.
Sharma, K. K. and Ortiz, R. (2000), 'Program for the application of genetic transformation for crop improvement in the semi-arid tropics', *Vitro Cell. Dev. Biol. Plant*, 36, 83–92.
Sharma K. K. and Lavanya M. (2002), 'Recent developments in transgenics for abiotic stress in legumes of the semi-arid tropics, In: *Genetic Engineering of Crop Plants for Abiotic Stress*, M. Ivanaga (Ed.). JIRCAS Working Report No. 23, JIRCAS, Tsukuba, Japan, pp. 61–73.
Sharma, D. R., Kumari, R. and Chowdhury, J. B. (1979), 'Plant regeneration in *Cicer* species through tissue culture [chickpeas, India], *Ind. J. Exp. Biol.*, 17, 607–9.
Sharma, H. C., Sharma, K. K., Seetharama, N. and Oritz, R. (2001), 'Genetic transformation of crop plants: Risks and opportunities for the rural poor', *Curr. Sci.*, 80, 1495–508.
Sharma, H. C., Sharma, K. K. and Crouch, J. H. (2004). 'Genetic transformation of crops for insect resistance: Potential and limitations', *Crit. Rev. Plant Sci.*, 23: 1–26.
Sharma, K. K., Bhatnagar-Mathur, P. and Thorpe, T. A. (2005), 'Genetic transformation technology: Status and problems,' *Vitro Cell. Dev. Biol. Plant*, 41(2), 102–12.
Sharma, K. K., Lavanya, M. and Anjaiah, V. (2006), '*Agrobacterium*-mediated production of transgenic pigeonpea (*Cajanus cajan* L. Millsp.) expressing the synthetic Bt *cry1Ab* gene', *Vitro Cell. Dev. Biol. Plant*, 42, 165–73.
Shika, S., Batra, P., Sindhu, A. and Chowdhury, V. K. (2001), 'Multiple shoot induction and complete plant regeneration of chickpea (*Cicer arietinum* L.)', *Crop Res.*, 21(3), 308–11.
Shisanya, C. A. (2002), 'Improvement of drought adapted tepary bean (*Phaseolus acutifolius* A. Gray var. *latifolius*) yield through biological nitrogen fixation in semi-arid SE-Kenya', *Eur. J. Agron.*,16, 13–24.
Shri, P. V. and Davis, T. M. (1992), 'Zeatin-induced shoot regeneration from immature chickpea (*Cicer arietinum* L.) cotyledons', *Plant Cell Tissue Organ Cult.*, 28(1), 45–51.
Shukla, R. K., Raha, S., Tripathi, V. and Chattopadhyay, D. (2006), 'Expression of CAP2, an APETALA2-family transcription Factor from chickpea enhances growth and tolerance to dehydration and salt stress in transgenic tobacco, *Plant Physiol.*, 142, 113–23.
Singh, T. N., Sen, J., Kiesecker, H., Reddy, V. S., Jacobsen, H. J. and Mukherjee, G. S. (2004), 'Use of a herbicide or lysine plus threonine for non-antibiotic selection of transgenic chickpea', *Plant Cell Rep.*, 22, 576–83.
Singh, N., Mishra, A. and Jha, B., 2014, 'Ectopic over-expression of peroxisomal ascorbate peroxidase (SbpAPX) gene confers salt stress tolerance in transgenic peanut (Arachis hypogaea)', *Gene.*, 547(1), 119–25.
Singsit, C., Adang, M. J., Lynch, R. E., Anderson, W. F., Wang, A., Cardineau, G and Akins, O. P. (1997), 'Expression of a *Bacillus thuringiensis cryIA(c)* gene in transgenic peanut plants and its efficacy against lesser cornstalk borer', *Transgenic Res.*, 6(2), 169–76.
Solleti, S. K., Bakshi, S., Purkayastha, J., Panda, S. K. and Sahoo, L. (2008), 'Transgenic cowpea (*Vigna unguiculata*) seeds expressing a bean *a*-amylase inhibitor 1 confer resistance to storage pests, bruchid beetles', *Plant Cell Rep.*, 27, 1841–50.
Somers, D. A., Samac, D. A. and Olhoft, P. M. (2003), 'Recent advances in legume transformation', *Plant Physiol.*, 131(3), 892–9.
Songstad, D. D., Petolino, J. F., Voytas, D. F. and Reichert, N. A. (2017), 'Genome editing of plants', *Crit. Rev. Plant Sci.*, 36, 1–23. doi:10.1080/07352689.2017.1281663
Sreenivasu, K., Malik, S. K., Kumar, A. P. and Sharma, R. P. (1998), 'Plant regeneration via somatic embryogenesis in pigeonpea (*Cajanus cajan* L. Millsp.)', *Plant Cell Rep.*, 17, 294–7.
Stalker, H. T. and Moss, J. P. (1987), 'Speciation, cytogenetics, and utilization of *Arachis* species', *Adv. Agron.*, 41, 1–40.
Steele, W. M. (1976), 'Cowpeas' In: *Evolution of Crop Plants*, N. W. Simmonds (Ed.), Longman, London, pp. 301–24.

Subbarao, G. V., Johansen, C., Slinkard, A. E., Nageswara Rao, R. C., Saxena, N. P., Chauhan, Y. S. and Lawn, R. J. (1995), 'Strategies for improving drought resistance in grain legumes', *Crit. Rev. Plant Sci.*, 14(6), 469–523.

Subhadra, V. R. K., Chowdhury, J. B., Singh M. and Sareen, P. K. (1998), Multiple shoots from cotyledonary node explants of non-nodulating genotype (ICC435M) of chickpea, *Cicer arietinum* L., *Ind. J. Exp. Biol.*, 36(12), 1276–9.

Sundaresha, S., Kumar, A. M., Rohini, S., Math, S. A., Keshamma, E., Chandrashekar, S. C. and Udaya Kumar, M. (2010), 'Enhanced protection against two major fungal pathogens of groundnut, *Cercospora arachidicola* and *Aspergillus flavus* in transgenic groundnut over-expressing a tobacco β 1–3 glucanase,' *Eur. J. Plant Pathol.*, 126(4), 497–508.

Surekha, C. H., Beena, M. R., Arundhati, A., Singh, P. K., Tuli, R., Gupta, D. A. and Kirti, P. B. (2005), '*Agrobacterium* mediated genetic transformation of pigeonpea (*Cajanus cajan* (L.) Millsp.) using embryonal segments and development of transgenic plants for resistance against Spodoptera', *Plant Sci.*, 169(6), 1074–80.

Surekha, C. H., Kumari, K. N., Aruna, L. V., Suneetha, G., Arundhati, A. and Kishor, P. K. (2014), 'Expression of the *Vigna aconitifolia P5CSF129A* gene in transgenic pigeonpea enhances proline accumulation and salt tolerance', *Plant Cell Tissue Organ Cult.*, 116(1), 27–36.

Švábová, L. and Griga, M. (2008), The effect of cocultivation treatments on transformation efficiency in pea (*Pisum sativum* L.)', *Plant Cell Tissue Organ Cult.*, 95(3), 293–304.

Švábová, L., Smykal, P., Griga, M. and Ondrej, V (2005), '*Agrobacterium*-mediated transformation of *Pisum sativum in vitro* and *in vivo*', *Biol. Plant.*, 49, 361–70.

Swathi Anuradha, T., Jami, S., Beena R. and Kirti, P. (2008), 'Cotyledonary node and embryo axes as explants in legume transformation', in, *Handbook of New Technologies for Genetic Improvement of Grain Legumes*, pp. 253–71. P. Kirti (ed.), CRC Press, New York, NY.

Szabala, B. M., Osipowski, P. and Malepszy, S. (2014), 'Transgenic crops: The present state and new ways of genetic modification', *J. Appl. Genet.*, 55, 287–94. doi: 10.1007/s13353–014-0210–4

Tabe, L. M., Higgins, C. M., McNabb, W. C. and Higgins, T. J. V. (1993), 'Genetic engineering of grain and pasture legumes for improved nutritive value', *Genetica*, 90(2), 181–200.

Tewari-Singh, N., Sen, J., Kiesecker, H., Reddy, V. S., Jacobsen, H. J. and Guha-Mukherjee, S. (2004), 'Use of a herbicide or lysine plus threonine for non-antibiotic selection of transgenic chickpea', *Plant Cell Rep.*, 22(8), 576–83.

Thangella, P. A. V. and Fakrudin, B. (2016), 'Agrobacterium mediated genetic transformation of Pigeonpea (*Cajanus cajan* Millsp.) using cry1Ac gene for resistance against pod borer', *Adv. Life Sci.*, 5(19), 8181–90.

Thu, T. T., Mai, T. T. X., Dewaele, E., Farsi, S., Tadesse, Y., Angenon, G. and Jacobs, M. (2003), 'In vitro regeneration and transformation of pigeonpea [*Cajanus cajan* (L.) Millsp.]', *Mol. Breed.*, 11(2), 159–68.

Thu, T. T., Dewaele, E., Claeys, M., Jacobs, M. and Angenon, G. (2007), 'Increasing lysine levels in pigeonpea (*Cajanus cajan* (L.) Millsp.) seeds through genetic engineering', *Plant Cell Tissue Organ Cult.*, 91(2), 135–43.

Tiwari, S. and Tuli, R. (2008), 'Factors promoting efficient in vitro regeneration from de-embryonated cotyledon explants of *Arachis hypogaea* L', *Plant Cell Tissue Organ Cult.*, 92(1), 15–24.

Tiwari, S. and Tuli, R. (2012), 'Optimization of factors for efficient recovery of transgenic peanut (*Arachis hypogaea* L.)', *Plant Cell Tissue Organ Cult.*, 109(1), 111–21.

Tiwari, V., Chaturvedi, A. K., Mishra, A. and Jha, B., 2015, 'Introgression of the SbASR-1 gene cloned from a halophyte *Salicornia brachiata* enhances salinity and drought endurance in transgenic groundnut (*Arachis hypogaea*) and acts as a transcription factor', *PLoS One.*, 10(7), e0131567.

Ufuk C. A., Mahmoudian, M., Kamci, H., Yucel, M. and Oktem, H. A. (2009), '*Agrobacterium tumefaciens*-mediated genetic transformation of a recalcitrant grain legume, lentil (*Lens culinaris* Medik)', *Plant Cell Rep.*, 28(3), 407–17.

Uncuoglu, A. A., Sarmah, B. K., Sharma, K. K., Bhatnagar-Mathur, P., Ratnaparkhe, M. B., Pawar, P. and Ranjekar, P. K. (2008), 'Chickpea', In: *Compendium of Transgenic Crop Plants: Transgenic*

Legume Grains and Forages, C. Kole and T. C. Hall (Eds), Blackwell Publishing Ltd., Hoboken, NJ, pp. 171–88.

Vadawale, A., Mihani, R., Mathew, A. and Robin, P., (2012), 'Transformation of groundnut - *Arachis hypogaea* L. Var. GG20 with the *cox* gene - An attempt to develop salinity tolerance', *Int. J. Pharm. Bio. Sci.*, 3(1), 591–8.

van der Maesen. L. J. G. (1983), World distribution of pigeonpea. ICRISAT Information Bulletin No. 14, ICRISAT, Patancheru, AP, India, p. 40.

van der Maesen, L. J. G. (1987), 'Origin, history and taxonomy of chickpea', In: *The Chickpea*, M. C. Saxena, and K. B. Singh (Eds), CAB International Publishing, Wallingford, UK, pp.11–34.

Vasavirama, K. and Kirti, P. B. (2010), 'Generation of "tikka" resistant transgenic groundnut plants by introducing double gene construct, In: DNA 2010: 2nd International conference on the science of DNA fingerprinting in crime investigation, 29–30th October 2010, Bioaxis DNA Research Centre (BDRC) private limited, Hyderabad, India.

Vasavirama, K. and Kirti, P. B., 2012, 'Increased resistance to late leaf spot disease in transgenic peanut using a combination of PR genes', *Funct. Integrative Genomics*, 12(4), 625–34.

Vaughan, D. A., Tomooka, N. and Kaga, A. (2005), 'Azuki Bean [*Vigna angularis* (Willd.) Ohwi & Ohashi', In: *Genetic Resources, Chromosome Engineering and Crop Improvement: Grain Legumes*, R. J. Singh and P. P. Jauhar (Eds), Taylor and Francis, Boca Raton, FL, pp. 347.

Venkatachalam, P., Geetha, N., Jayabalan, N. and Sita, L. G. (1998), '*Agrobacterium*-mediated genetic transformation of groundnut (*Arachis hypogaea* L.): An assessment of factors affecting regeneration of transgenic plants', *J. Plant Res.*, 111(4), 565–72.

Venkatachalam, P., Geetha, N., Khandelwal, A., Shaila, M. S. and Sita, L. G. (1999), 'Induction of direct somatic embryogenesis and plant regeneration from mature cotyledon explants of *Arachis hypogaea* L', *Curr. Sci.*, 77, 269–73.

Venkatachalam, P., Geetha, N., Khandelwal, A., Shaila, M. S. and Sita, L. G. (2000), '*Agrobacterium*-mediated genetic transformation and regeneration of transgenic plants from cotyledon explants of groundnut (*Arachis hypogaea* L.) via somatic embryogenesis', *Curr. Sci.*, 78(9), 1130–6.

Villani, P. J. and Demason, D. A. (2000), 'Role of *Af* and *Tl* genes in pea leaf morphogenesis: Shoot ontogeny and leaf development in the heterozygotes', *Ann. Bot.*, 85, 123–35.

Warkentin, T. D. and McHughen, A. (1992), '*Agrobacterium tumefaciens*-mediated beta-glucuronidase (GUS) gene expression in lentil (*Lens culinaris* Medik.) tissues', *Plant Cell Rep.*, 11(5), 274–8.

Warkentin, T. D., Jordan, M. C. and Hobbs, S. L. A. (1992), 'Effect of promoter-leader sequences on transient reporter gene expression in particle bombarded pea (*Pisum sativum* L.) tissues', *Plant Sci.*, 87(2), 171–7.

Watanasit, A. and Pichitporn, S. (1996), 'Improvement of mungbean for resistance to bruchids', In: *Mungbean Germplasm: Collection, Evaluation and Utilization for Breeding Program*, P. Srinives, C. Kitbamroong and S. Miyazaki (Eds), Japan International Research Center for Agricultural Sciences, Tsukuba, Japan, pp. 67–71.

Weissinger, A., Liu, Y. S., Scanlon, S., Murray, J., Cleveland, T. E., Jaynes, J., Mirkov, E. and Moonan, F. (1999), 'Transformation of peanut with the defensive peptidyl MIM D5C', In: *Proceedings of the USDA-ARS Aflatoxin Elimination Workshop*, pp. 66–8.

Weissinger, A., Wu, M., Liu, Y. S., Ingram, K., Rajasekaran, K. and Cleveland, T. E. (2002), 'Development of transgenic peanut with enhanced resistance against pre-harvest aflatoxin contamination', *Mycopathol.*, 155(97), 93.

Wilmink, A. and Dons, J. J. M. (1993), 'Selective agents and marker genes for use in transformation of monocotyledonous plants', *Plant Mol. Biol. Rep.*, 11(2), 165–85.

Yang, H., Nairn, J. and Ozias-Akins, P. (2003), 'Transformation of peanut using a modified bacterial mercuric ion reductase gene driven by an actin promoter from *Arabidopsis thaliana*', *J. Plant Physiol.*, 160(8), 945–52.

Yang, H., Ozias-Akins, P., Culbreath, A. K., Gorbet, D. W., Weeks, J. R., Mandal, B. and Pappu, H. R. (2004), 'Field evaluation of Tomato spotted wilt virus resistance in transgenic peanut (*Arachis hypogaea*)', *Plant Dis.*, 88(3), 259–64.

Yenish, J. P., Brand, J., Pala, M. and Haddad, A. (2009), 'Weed management', In: *The Lentil: Botany, Production and Uses, CABI*, W. Erskine, F. J. Muehlbauer, A. Sarker and B. Sharma (Eds), p. 326. doi:10.1079/9781845934873.0000.

Yousefiara, M., Bagheri, A. and Moshtaghi, N. (2008), 'Optimizing regeneration conditions in chickpea (*Cicer arietinum* L.)', *Pakistan J. Biol. Sci.*, 11(7), 1009–14.

Zambre, M., Goossens, A., Cardona, C., Van Montagu, M., Terryn, N. and Angenon, G. (2005), 'A reproducible genetic transformation system for cultivated *Phaseolus acutifolius* (tepary bean) and its use to assess the role of arcelins in resistance to the Mexican bean weevil', *Theor. Appl. Genet.*, 110(5), 914–24.

Zarandi, H. S., Bagheri, A., Baghizadeh, A. and Moshtaghi, N. (2011), 'Quantitative analysis of chitinase gene expression in chickpea', *Russ. J. Plant Physiol.*, 58(4), 681.

Ze'ev, N. B. and Zohary, D. (1973). 'Species relationships in the genus *Pisum* L.,' *Isr. J. Bot.*, 22(2), 73–91.

Zhang, K., Raboanatahiry, N, Zhu, B. and Li, M. (2017), 'Progress in genome editing technology and its application in plants', *Front. Plant Sci.*, 8, 177. doi:10.3389/fpls.2017.00177

Zohary, D. and Hopf, M. (1973), 'Domestication of pulses in the old world', *Science*, 182(4115), 887–94.

Chapter 6

Developing drought- and heat-tolerant varieties of grain legumes

Shoba Sivasankar, Former Director, CGIAR Research Program on Grain Legumes

1 Introduction

The stability of crop yields is challenged by a variety of abiotic and biotic stresses during each growing season. Adverse environmental conditions such as drought, extreme temperatures, submergence and soil nutrient deficiencies or mineral toxicities constitute important abiotic stress concerns. These impact yield stability directly through their effects on crop establishment, growth and yield, and indirectly through their influence on the nature and intensity of incidence of biotic stresses such as diseases, insects and weeds. Drought and high temperatures are expected to adversely impact crop yields, and consequently food security, even more intensely in the future based on predicted changes in climate. Over the last 1400 years, the period from 1982 to 2012 is identified to be likely the warmest 30-year period in the Northern hemisphere, and the warming of global surface temperatures is likely to exceed 2°C by the end of the twenty-first century relative to the nineteenth-century baseline (IPCC, 2014). While changes in precipitation are predicted to be not uniform across the globe, many mid-latitude and subtropical dry regions are likely to experience decreased precipitation. Without proper adaptation measures for managing crop production, climate change will have significant impacts on crop yields, leading to food insecurity and the breakdown of food systems, especially in developing nations. It can also affect the geographical distribution of crop species.

http://dx.doi.org/10.19103/AS.2017.0023.04

Grain legumes are agriculturally important food crops that are recognized for their vital health benefits due, primarily, to their high protein, minerals, folate and dietary fibre contents, and low glycemic index. Legumes have an advantage over other plants in terms of leaf nitrogen, photosynthesis and water use efficiency, thus enhancing their fitness and survival especially in arid and semi-arid conditions (Adams et al., 2016). There exists significant diversity in the range of cultivated leguminous species, allowing their distribution across almost all geographical regions. Grain legumes such as cowpea (*Vigna unguiculata*), chickpea (*Cicer arietinum*), faba bean (*Vicia faba*), groundnut (*Arachis hypogaea*), lentil (*Lens culinaris*) and pigeon pea (*Cajanus cajan*) are cultivated mostly in the dry environments of Africa and Asia, and have a critical role in subsistence agriculture. Legumes such as common bean (*Phaseolus vulgaris*) and soybean (*Glycine max*) grow in less dry environments, but breeding efforts for adaptation to drought or to heat stress are emerging to be successful. Thus, the ability of leguminous crops to tolerate drought or heat stress varies between and within species. A recent analysis of more than 30 years of research literature from 1980 to 2014 pertinent to legume crop productivity under drought in monoculture conditions in various regions of the globe identified that lentil and groundnut suffered the least yield reduction from drought among the 13 legume crop species included in the analysis (Daryanto et al., 2015). Yield reduction was directly correlated with observed soil-moisture reduction, while geographic regions and associated factors were not found to have a significant effect.

Tolerance to drought and heat is complex and polygenic, involving multiple genetic loci. Therefore, development of tolerance by breeding or genetic engineering is more complex than it is for simple traits with single-gene solutions. Important strides have been made in understanding the physiological and molecular responses of plants to drought and to heat, and at least two reviews have addressed these from the perspective of drought and grain legumes (Turner et al., 2001; Farooq et al., 2017). This understanding is crucial to developing tolerance to these stresses. The initial section of this chapter provides an overview of the current understanding of the physiological, biochemical and molecular responses and adaptation mechanisms to drought, and to heat, with specific examples from grain legumes. It will also touch upon the effects of drought on nitrogen fixation. The chapter further discusses the status of current research in developing tolerance to drought and to heat, examining available information on (a) traditional breeding and the utilization of natural genetic diversity; (b) efforts towards molecular breeding, specifically, the identification of genetic associations and quantitative trait loci (QTLs); and (c) functional genomics studies on stress-relevant candidate genes or gene families. A systematic recent research effort that led to the identification of heat-tolerant common bean lines derived from interspecific crosses with tepary bean is discussed in some detail in this section. This is followed by a discussion of critical contemporary research considerations in the development of drought or heat tolerance, namely, (a) systematic breeding-programme design and implementation for efficient and effective selections, (b) the determination and use of predictive traits for stress tolerance, precision phenotyping and crop modelling, and (c) deeper exploration of crop wild relatives for natural genetic diversity. This section also provides a brief overview of the recent establishment of a precision phenotyping platform that enables high-throughput measurement of the transpiration response to vapour pressure deficit in screening for drought tolerance. Finally, the chapter discusses, in brief, major crop management considerations that impart tolerance to drought or to heat.

Scientific studies pertinent to abiotic stress are fewer for most cultivated legumes relative to the major cereals, and relative to soybean, the leguminous species with the highest global acreage and production. Also, studies on heat tolerance lag behind those on drought tolerance in legumes, as in other crops. This chapter uses both legume-specific and general abiotic stress literature contextually. While the definition for grain legumes (or pulses) by the Food and Agricultural Organization of the United Nations excludes soybean and groundnut, and identifies them as oilseed crops, the current chapter includes discussion of research on these crops where relevant. Further, although crops are subjected to a combination of stresses in the growing environment, and the response to each combination is unique (see Suzuki et al., 2014 for a recent review), this chapter focuses on crop responses and the development of tolerance specific to drought and/or to heat.

2 Physiological responses in adaptation to stress

Drought and heat constitute the most prevalent environmental stresses during plant growth, frequently occurring together especially in tropical and subtropical regions. The specific responses of plants to drought or to heat share some similarities across species. While plant responses and adaptations to simultaneous stresses that occur in the growing environment during each season are unique to each combination, and complex, individual stresses such as drought or heat also trigger complex responses that include multiple molecular mechanisms and secondary physiological traits.

2.1 Drought stress

Plant response to drought is generally classified into three types, namely, 'drought escape', 'drought avoidance' and 'drought tolerance', and adaptive mechanisms contributing to these response behaviours have been reviewed most recently by Bargaz et al. (2015) and Basu et al. (2016). 'Drought escape' involves the completion of growth cycles before the incidence of drought. This is achieved by the adjustment of vegetative and reproductive growth to the availability of water through rapid phenological development and through developmental plasticity. 'Drought avoidance' involves mechanisms that allow the plant to maintain high tissue water levels despite reduced soil moisture, through adaptive measures that enable them to be water savers or spenders. Water savers use the available water more efficiently by reducing losses through reduced transpiration, while water spenders maintain plant water content effectively by enhanced water absorption with deeper rooting and better hydraulic conductance. 'Drought tolerance' refers to the ability to maintain cellular metabolism even under moisture stress through adaptive mechanisms including osmotic adjustment, cellular elasticity and increased protoplasmic resistance.

In this chapter, we use the term 'drought tolerance' to denote the ultimate protection of yield under drought. Genetic makeup predisposes species and varieties to use one or a combination of physiological adaptations to combat drought and to produce grain yield under limiting soil moisture conditions. Yield under drought is generally understood to be a function of transpiration efficiency (TE), water use and harvest index, based on the Passioura equation (Passioura, 1977). Stomatal sensitivity under high vapour pressure deficit and the consequent control of transpiration rate play an important role in combating drought stress (Sinclair et al., 2017). The effect of drought on the grain yield of crops is most devastating when moisture stress occurs at or around the reproductive stage of plants, a phenomenon termed terminal drought. Studies show that some drought-tolerant common bean cultivars protect grain yield under terminal drought by maximizing carbon uptake and minimizing canopy water loss by increasing stomatal closure during the day and attaining higher relative water contents during the night, relative to drought-sensitive cultivars (Rosales et al., 2013). Improved carbohydrate partitioning towards grain filling has also been identified to be a mechanism by which some drought-tolerant common bean cultivars protect yield under terminal drought, and increased effectiveness of sucrose transport is implicated (Cuellar-Ortiz et al., 2008).

In chickpea, a crop mostly grown in semi-arid regions, flower and pod numbers, pollen viability and pistil function are reduced under terminal drought, leading to significant flower and pod abortion (Fang et al., 2010). Grain yield under terminal drought in this crop has been shown to be highly correlated with partitioning rate and crop growth rate, with canopy temperature depression, growth duration and other related traits also having an impact (Ramamoorthy et al., 2016). Shoot-based physiological traits such as partitioning coefficient and crop growth rate are critical components of the yield response and act as integrators of the drought response across the entire growing season. As such, these are good predictors of the drought effect on crop yield and are measurable parameters in breeding for drought resistance. Root traits that are important determinants of crop performance under drought include root morphology and plasticity. An increased root-to-shoot ratio under drought does not always lead to better crop performance under drought. A conservative water use pattern, where less water uptake by roots occurs during the vegetative stage, and more water uptake occurs during the reproductive stage, is known to be a critical mechanism to tolerate terminal drought in several species, and has been described for chickpea (Zaman-Allah et al., 2011).

Drought and heat also trigger early senescence in plants, with early chlorophyll degradation and nitrogen remobilization, ultimately resulting in reduced yields. Delayed senescence occurs in 'stay-green' mutants of plant species, and such delayed senescence can be either cosmetic or functional (see Thomas and Ougham, 2014, for a recent review). Cosmetic 'stay-green' is the result of genetic lesions early in the chlorophyll biosynthesis pathway. Functional 'stay-green' enables prolonged carbon capture and delayed initiation of nitrogen remobilization, ultimately leading to the maintenance of yields under adverse environments in several cases. Genes in the hormonal pathways of cytokinin and ethylene, and transcription factors of the *WRKY* and *NAC* gene families, have been implicated in functional 'stay-green'. Conventional breeding has identified several functional 'stay-green' mutants, especially in cereals like maize and sorghum. QTLs associated with the trait are known to be present in many plant species, and cowpea is an example (Muchero et al., 2013).

2.1.1 Drought and nitrogen fixation

Symbiotic nitrogen fixation of legumes is especially vulnerable to drought, and there exists a close relationship between leaf and nodule water potential. Both nodule development and nitrogen fixation are impacted. Several contributory factors have been identified, including oxygen availability and carbon flux to nodules, activities of nitrogenase and nodule sucrose synthase, and accumulation of ureides and free amino acids, mainly asparagine. Feedback inhibition of nitrogen fixation has been proposed to occur under water deficit, though functional confirmation of the nature of the metabolite(s) involved in such feedback inhibition is still awaited. Nodule ureides and aspartate, along with several leaf amino acids have been proposed to exert feedback inhibition of nitrogen fixation in soybeans, based on the coincidence of their accumulation with declining nitrogenase activity (King and Purcell, 2005). Accumulation of nodule asparagine as well as organic acids, especially malate, has been reported also in chickpea under drought stress (Esfahani et al., 2014). In common bean, on the contrary, no accumulation of ureides was observed in nodules or roots under drought stress, and feedback inhibition of nitrogen fixation by ureides from nodules was ruled out (Coleto et al., 2014).

An efficient method for the determination of nitrogen fixing ability of plants under drought stress has recently been described and constitutes an important tool for breeding common bean, and likely other legume species (Polania et al., 2016). It involves the measurement of ^{15}N natural abundance in shoot tissue during mid pod-filling stage or in grain at harvest, both of which are positively correlated and can discriminate between common bean varieties with superior or inferior nitrogen fixing ability under moisture stress.

2.2 Heat stress

Similar to drought, heat stress also impacts seed germination, plant growth, leaf expansion, dry matter accumulation, photosynthesis, reproduction and grain yield. The impact on crop yield is the greatest when the stress occurs at or around flowering time, primarily due to adverse impacts on pollen maturation and viability. Hormonal changes and oxidative stress associated with high temperatures have an important effect on reproductive development. Pea varieties of the semi-arid growing environments of Western Canada had reduced pollen germination, pollen tube growth, pod length, seeds per pod and seed-to-ovule ratio when grown under higher than optimal growth temperatures ranging from 24°C to 36°C (Jiang et al., 2015). Pollen-wall thickness was also found to be increased under heat stress. Changes in pollination mechanisms under heat stress have been reported in faba bean, where increasing temperatures were shown to increase insect-mediated cross pollination, concomitantly decreasing self-pollination, under both controlled environment conditions and in the field (Bishop et al., 2017). Modulation of stomatal sensitivity to vapour pressure deficit also has a key role in combating heat stress. Wild genotypes of soybean have been shown to have more transpiration control under lower vapour pressure deficits than cultivated genotypes and shown to maintain this control when temperatures increased (Seversike et al., 2013).

3 Biochemical and molecular responses in adaptation to stress

Several biochemical and molecular responses have been reported to occur in the plant upon exposure to drought and to heat. Several plant hormones, including abscisic acid (ABA), ethylene and cytokinins, are known to be involved in plant response to drought. Tissue levels of ABA increase under a variety of environmental stresses, and regulate plant responses by triggering signalling cascades through the expression of stress-related genes by activating the ABA-responsive element in gene promoters. ABA is involved in seed dormancy, seed germination and stomatal closure, and in legumes additionally, it plays a role in the symbiosis with rhizobia exerting inhibitory effects on nodulation and possibly nitrogen fixation. A *Lotus japonicus* mutant with low sensitivity to the hormone was found to have increased nodulation and nitrogen fixation (Tominaga et al., 2009). The hormone ethylene is also a negative regulator of nodulation, acting at the very early stages of nodule formation, while cytokinin is a positive regulator.

Another cellular response to both drought and heat stress is the accumulation of reactive oxygen species and the onset of oxidative stress. This can lead to damage of cellular components including lipids and proteins, causing impairment to photosynthesis and respiration, and consequently to growth and yield. Enhanced activities of anti-oxidant enzymes such as superoxide dismutase, glutathione reductase, catalase, etc. can mitigate oxidative stress. Osmotic adjustment mediated by low-molecular-weight organic metabolites, generally called compatible solutes, is also a cellular response to dehydration stress and helps maintain turgor, thus protecting cell membranes and constituents (see Blum, 2017 for a review). Compatible solutes include sugars such as trehalose, amino acids such as proline and quarternary ammonium compounds such as glycine betaine.

A group of hydrophilic proteins called Late Embryogenesis Abundant (LEA) proteins or dehydrins accumulate to high levels early upon the onset of moisture stress in plant tissues, and these are widely distributed across plant species. LEA proteins are proposed to have roles in preventing inactivation of proteins and destabilization of cell membranes under dehydration stress. An analysis of the seven different classes of LEA proteins across different legumes, in comparison with other plant species, identified a conserved, legume-specific, 15-amino acid, repetitive, glycine-rich motif named the K segment in the LEA2 class of proteins in some legumes, the functional significance of which is yet to be described (Battaglia and Covarrubias, 2013). Another large and diverse family of proteins of significance under drought stress are aquaporins, which are membrane-integral, water-channel proteins. Limited transpiration under high vapour pressure deficits confers tolerance to drought, and it involves low hydraulic conductivity, with implications for aquaporins (Sinclair et al., 2017). Genetic associations have been established between this physiological trait and aquaporins, leading to the identification of breeding lines and cultivars that tolerate drought.

Several classes of transcription factors are also involved in responses to drought and heat. Specifically, the DREB, NAC and WRKY transcription factors have been demonstrated to be involved in drought responses or tolerance in a variety of plant species through genetic associations and/or transgenic modifications. Many heat shock factors are also known to be associated with both drought and heat tolerance, as are several microRNAs.

4 Development of tolerance

As indicated by the above concise summary, there are multiple underlying mechanisms, and consequently multiple genetic loci, that modulate plant response to drought and heat, making adaptation and performance under these stresses quite complex. Such complexity is integrated into crop yield traits, which is the ultimate selection criterion in plant breeding for development of tolerance to these stresses, at whatever intensity, frequency and combinations they occur during a growing season. Through the years of domestication and crop improvement, plant breeders have naturally selected for yield stability under a variety of pressures, and positive genetic gain has likely been achieved to some level for all crops, and possibly the highest being for the cereal crop, maize, in the United States as described by Duvick (1977). The rate of this genetic gain would vary significantly between crops, the difference resulting primarily from the size, focus and longevity of established breeding programmes for the various crops, which in turn have determined the scope and scale of development, accessibility and application of advanced, sophisticated technologies in genetics, genomics, precision phenotyping and crop modelling. In this context, while crops such as grain legumes lag significantly behind major crops like maize, rice, wheat or soybean, they are rapidly being positioned to take advantage of the lessons from the latter, and from the knowledge and increasing cost-effectiveness in the application of advanced technologies.

Traditionally, breeding for yield stability under abiotic stresses in grain legumes has been supported by the significant natural diversity available within species and across wild relatives. More recently, rapid advances have been made in the determination and utilization of molecular information in the form of QTLs, molecular markers and genome-wide association data. The following section discusses instances of development of tolerance through traditional breeding, and follows up with a discussion of specific examples of identifying molecular tools that can be deployed for accelerated breeding.

4.1 Natural genetic diversity and traditional breeding

Significant genetic diversity within legume species continues to be utilized through traditional breeding for stress tolerance, and yet more is being explored in crop wild relatives. Crop wild relatives serve as a rich source of novel alleles that can be introgressed into agronomically important cultivars.

While it is more than likely that traditional breeding over the last several decades have generally led to varying levels of yield stability under stress in legumes, specific examples for the development of drought or heat tolerance in this group of crops are few. One of these sparse examples involve common bean. In this grain legume, focused breeding for drought tolerance goes back more than two decades, and has led to the identification of several tolerant lines. In the bean breeding programme at the International Center for Tropical Agriculture (CIAT), head-quartered in Cali, Colombia, two races, namely Durango from the dry highlands of Mexico, and Mesoamerica from Central America, have been used in the development of drought tolerance. Combinations of the two have led to improved yield under drought in the case of the small-seeded Mesoamerican types, and Durango is now being used for introgression in the sensitive Andean lines (Beebe et al., 2009). In a recent example, field screening of breeding populations developed from double or triple crosses involving small red, small black, cream or cream-striped grain classes of common bean resulted in identifying lines with yield superiority of 15–20% under drought, over commercial check cultivars (Beebe et al., 2008). The focus in breeding for drought tolerance in the CIAT programme has been on remobilization of photosynthates from vegetative structures to pods, and significant genetic variability has been observed for the trait within existing germplasm collections. An analysis of two diversity pools of 81 Andean, and 121 Mesoamerican gene pool accessions under well-watered and terminal drought conditions indicated that drought tolerance results from different traits including pods per plant, seed weight, pod partitioning index and harvest index, each of which is differentially expressed in the two different gene pools (Asfaw et al., 2017). Evaluation of exotic dry bean germplasm derived from the Mesoamerican gene pool has led to the identification of superior lines for the mid-western United States as well, and these show yield advantage under both drought-stressed and non-stressed conditions, relative to the local check cultivars (Urrea et al., 2009).

Despite the considerably higher intensity of research on soybean relative to other legumes, genetic improvement specifically for drought or heat tolerance has progressed slowly in this crop. The reason is primarily attributed to the narrow genetic base of elite soybean breeding programmes in the Americas (Manavalan et al., 2009). Most recently, new generations of drought-adapted breeding lines have been developed using parental stocks exhibiting drought tolerance and multiple rounds of field screening for grain yield under drought, and some of these lines were characterized for their physiological performance under water-limited conditions, as reported by Devi et al., 2014. The authors showed that unlike most elite soybean cultivars, the new lines showed a slow-wilting phenotype with reduced transpiration under conditions of high vapour pressure deficits, thus indicating the ability to conserve soil moisture. Furthermore, these lines exhibited drought tolerance for nitrogen fixation.

4.1.1 Development of heat-tolerant common beans

The adaptation range of present-day cultivated bean has been influenced by its wild progenitor which grew in sub-humid conditions and in mid-altitudes under moderate temperatures (Beebe et al., 2013). Cultivated bean is currently grown predominantly in Central and South America and the Andes, and in East and Central Africa, either as a sole crop or in mixtures. It grows under mean temperatures ranging from 14°C to 28°C, though day temperatures above 30°C and night temperatures above 20°C significantly reduce grain yield. High night temperatures lead to flower and pod abortion, loss of pollen viability and restricted pollen-tube growth. Researchers at CIAT have undertaken a systematic series of studies most recently that led to the identification of bean lines with improved tolerance to high temperatures (CGIAR 2015).

The first step involved an environmental analysis, focusing on drought and heat (Beebe et al., 2011, 2013). Specifically, this analysis evaluated the occurrence of heat stress, drought stress or a combination of the two in the geographical areas growing bean in Central America, Brazil, East Africa and Southern Africa. Heat stress was determined to be the predominant deterrent to bean production globally. These bean-producing geographies were further evaluated at the level of administrative units, and at a resolution of 10 km pixels, for estimated rise in temperatures during the 16-year window from 2013 to 2030. A good proportion of current bean production areas were predicted to face temperature increases by 2030. Based on field experience that some breeding lines could tolerate up to 5°C higher night temperatures than the optimal 18°C, researchers next undertook a modelling exercise with a simple model of climatic suitability, namely EcoCrop, that helped determine production areas in East Africa and Central America against a simulated scenario of 3°C increasing future temperature (2050), with and without the heat-tolerance trait in cultivated varieties. The simulations showed that without heat tolerance, much of the current production areas would become unsuitable for bean cultivation, but with tolerance to a 3°C increase in temperatures, current production areas would be able to continue bean cultivation, while new production areas would also become accessible.

With this impetus as immediate background, elite breeding lines from the existing breeding programme at CIAT were subjected to field evaluation for heat tolerance in Colombia in 2013. Different germplasm sets of more than 1000 lines, including both interspecific crosses and purely common bean parentage, were evaluated under a seasonal mean maximum temperature of 35°C and minimum of 23°C. A group of 40 genotypes were identified to be heat tolerant, of which several had been derived from interspecific crosses involving tepary bean, *P. acutifolius*. A selection of 36 elite SEF lines and 25 INB lines, both from interspecific crosses with tepary bean, were further evaluated under heat stress in field screens over two seasons for grain yield and yield components, biomass, pollen viability and pod partitioning index. These screens identified one germplasm accession of *P. acutifolius* (G 40001), and the interspecific lines INB 841, SEF 15, SEF 16, SEF 14, SEF 60 and SEF 43 as having superior heat tolerance. Pollen viability in some of the interspecific lines was above 64% under severe heat stress compared to less than 20% in the check lines. Many of these interspecific lines are now being incorporated into crosses in breeding programmes in Central America, Colombia and Africa. Finally, the focus on screening for heat tolerance allowed the release of a line previously coded in 2005 for drought tolerance, namely, SEN 52, through rapid advancement in national evaluation trials in Nicaragua and release in 2013 as INTA Negro Precoz.

4.2 Genetic associations, QTLs and molecular breeding

The term molecular breeding is used here to denote breeding efforts that utilize molecular data to make selections, as in marker-assisted selection (MAS). While MAS is not yet in routine use in grain legumes, genetic studies have reported the identification of putative markers for breeding for drought tolerance, and these are rapidly being supplemented by genomic information from next-generation sequencing technologies. Genome-wide association studies (GWAS), as well as more focused determination of genetic associations of a specific allele/gene family/gene pathway with phenotypic performance, are leading to the identification of critical markers, the validation of which for MAS is in progress for several grain legumes.

Chickpea, an important component crop of subsistence farming, smallholder systems and predominantly grown in these instances under marginal conditions of soil, water and other inputs, was recently evaluated for genetic associations for drought tolerance, and several important QTLs for the trait were identified (Varshney et al., 2014). The study involved a multi-seasonal, multi-locational phenotyping of 20 different shoot and root traits under both rain-out shelter and field conditions, in two mapping populations of chickpea segregating for drought tolerance, namely, ICCRIl03 (ICC 4958 × ICC1882) and ICCRIl04 (ICC 283 × ICC 8261). Genotypic analysis revealed nine QTL clusters of which one that harboured 48% of the main-effect QTLs identified in the study, and that accounted for 58% phenotypic variation, can potentially be used for trait introgression into elite cultivars. This same region, or QTL hot spot, was also shown to hold seven simple sequence repeat (SSR) markers for potential use in MAS for drought tolerance. Another important toolkit for genetic studies and MAS in chickpea that has been recently developed includes genome-wide single nucleotide polymorphisms (SNPs) and indels identified through the re-sequencing of 35 parental lines of 16 mapping populations that segregated for tolerance to abiotic stresses, namely, drought, heat and salinity, as well as for other traits (Thudi et al., 2016).

In cowpea, another important crop under subsistence agriculture and predominantly grown in West Africa, the stay-green trait and its association with grain and biomass yield under drought stress was evaluated in a panel of 383 diverse accessions and one recombinant inbred line (RIL) population (Muchero et al., 2013). This diversity panel was phenotyped under seedling-stage senescence, early vegetative senescence and post-flowering senescence, imposed by drought, across thirteen experiments conducted in four countries, namely, Burkina Faso, Nigeria and Senegal in West Africa and the United States of America. In parallel, the panel was subjected to SNP genotyping. The study identified seven loci, five of which indicated positive associations between delayed senescence and grain yield in the three African environments. Of these five, three putative stay-green QTLs were present in both the RILs and the diverse germplasm, suggesting their potential use in MAS. The identification of positive correlations in this study between delayed senescence at the vegetative stage and grain yield points to the potential use of screening under vegetative senescence to make selections for grain yield under drought stress, although this would require further confirmation.

A greenhouse study in lentil under progressive drought conditions across two seasons and involving 132 RILs identified 18 QTLs, three of which showed major phenotypic effects for root-shoot ratio, lateral root number or root length, and were found to be very close to specific SNP markers evaluated in the study (Idrissi et al., 2016). In groundnut, genome-wide association studies on a reference set of 300 genotypes from 48 semi-arid tropical countries have identified marker-trait associations for drought tolerance, along with other important traits, and these remain to be validated (Pandey et al., 2014). In soybean, GWAS applied to a collection of 373 diverse genotypes for association with carbon isotope ratio, as surrogate measure for water use efficiency, in a two-season, two-location field experiment identified 39 SNPs, which were likely distributed across 21 loci (Dhanapal et al., 2015). Confirmation of these would enable selection of soybean genotypes with better water use efficiency.

Beyond whole-genome analysis, specific genes, gene families or gene pathways with established involvement in the drought or heat response can also be evaluated for allelic diversity and genetic associations with the tolerant phenotype, and associated sequence polymorphisms can be used as molecular markers. Thus, nucleotide diversity was examined in two genes encoding the Dehydration Response Element Binding factor (DREB) and belonging to the DREB transcription factor family that has been heavily implicated in the drought stress response, in a common bean diversity panel including 104 wild and 297 cultivated genotypes with a wide range of tolerance to drought (Cortes et al., 2012). The study identified genetic associations with the drought tolerance trait in both DREB2 genes, but at a significantly higher intensity in the DREB2A gene. The same group has also evaluated genetic associations for drought tolerance in the ERECTA gene, which has been implicated in the drought response of several plant species to date, in common bean (Blair et al., 2016). Cloning and sequencing of the gene followed by analysis of polymorphisms in its 5′ end in a panel of 145 wild and cultivated common bean accessions with different drought responses and geographical adaptation revealed genetic associations with the potential for use as molecular markers.

4.3 Functional genomics approaches

Functional genomics studies endeavour to understand the relationship between the genome, or parts of it, and the phenotype. It can involve high-throughput analyses at varying levels, including the transcriptome, proteome or metabolome. It can also address the evaluation of the role of a single gene in drought or heat tolerance through forward or reverse genetics approaches by either mutation or genetic engineering. Determination of the existence of functional associations between specific genes or gene families and a specific phenotype can then lead to focused efforts in allele mining. Establishment of gene functionality in a phenotype can also lead to the use of a specific gene in genetic engineering for trait improvement. However, such efforts are mostly exploratory in academia, and practical development of drought or heat tolerance by genetic engineering is only being pursued by the biotechnology industry, and that too mostly for major crops like maize (see Sivasankar et al., 2012 for a review).

Functional genomic studies in legumes of a few specific gene families or genes with known involvement in plant responses to drought or heat are presented here. In recent years, there have been several reports involving microRNAs (miRNAs) in the responses of many plant species to drought and heat. MiRNAs are known to control several biological functions, exerting this control by regulating the expression of their target genes. In common bean, miRNAs have been identified that respond to heat (Jyothi et al., 2015) and to drought (Wu et al., 2017). The evaluation of miRNAs in common bean with elevated expression under heat stress resulted in the identification of eight novel miRNAs and their target genes, mostly transcription factors. Comparison of the expression of small RNAs from drought-sensitive and drought-tolerant cultivars of common bean under drought-stressed or well-watered conditions led to the identification of 16 miRNAs with potential involvement in the drought response, and their target genes. Similar analysis in cowpea using deep sequencing of small RNAs from both drought-tolerant and drought-susceptible cultivars led to the identification of 44 drought-associated miRNAs, of which 11 were predominantly or exclusively expressed in only one of the two genotypes tested, and nine were drought responsive only in one genotype (Barrera-Figueroa et al., 2011).

Aquaporins constitute another class of genes with known association to the drought response, as mentioned earlier. The aquaporin gene family of common bean was evaluated in a genome-wide study and 44 members were identified and characterized, of which one, namely, PvTIP1;1 was found to be specifically associated with drought stress, its expression increasing significantly under drought relative to well-watered conditions (Ariani and Gepts, 2015). Members of two further gene families with known involvement in the drought and heat responses, namely, the AP2/ERF transcription factor and the HSP90 chaperonin, have been identified from comprehensive analyses of five legume genomes, and expression of a few specific identified sequences was characterized in chickpea and pigeonpea (Agarwal et al., 2016).

The ability of specific genes or members of specific gene families with proven association to tolerance to drought or heat to actually confer the trait can be evaluated through genetic engineering or mutation approaches. Thus, downregulation of two farnesyltransferase genes, known to be implicated in the response to the hormone ABA, in soybean was shown to result in enhanced stomatal response to ABA and reduced moisture loss under limited water (Ogata et al., 2017). Another gene, encoding isopentenyltransferase, critical to the biosynthesis of the hormone cytokinin, when expressed in groundnut from a maturation- and stress-induced promoter, led to improved drought tolerance in transgenic plants under both controlled-environment and field conditions (Qin et al., 2011). In other examples, ectopic expression of a DREB transcription factor from Arabidopsis, AtDREB1A, in groundnut conferred tolerance to drought in the transgenic groundnut plants, while expression of a cowpea DREB gene, VuDREB2A, in Arabidopsis provided drought tolerance in the transgenic Arabidopsis plants (Sarkar et al., 2014; Sadhukhan et al., 2014). Several of these genes are pursued in patents by the plant biotechnology industry, their utility for drought tolerance validated in major crops like maize, and two examples include specific members of the DREB family of transcription factors, and the ERECTA gene, covered by the patents US 7317141 and US 7847158, respectively (Sivasankar et al., 2008; Guo et al., 2015).

5 Critical contemporary considerations for strengthened research

Genetic improvement of grain legumes for tolerance to drought and/or heat has been pursued for most of the duration of the various legume-breeding programmes, though these efforts have been very sporadic during the early histories of these programmes. More focused efforts in breeding for drought tolerance are rather recent, and these vary in their duration and intensity between the various legumes. However, starkly absent in most breeding programmes are systematic design and implementation. A systematic breeding programme aiming for accelerated genetic gain under environmental stress should include planning and implementation based on a sound knowledge of target environments, together with discovery research that combines genotyping and precision phenotyping. Good contemporary examples are available from major crops within breeding programmes of the private sector (see Cooper et al., 2014 for an overview in the context of maize and drought stress). The development of heat-tolerant beans described earlier in this chapter also constitutes a good example. With increasing availability of information on phenotypic performance associated with different environmental parameters, and knowledge of the underlying genetics, sophisticated breeding tools including crop modelling and genomic selection will become increasingly useful in breeding for yield stability under drought and heat stress.

5.1 Target population of environments and systematic testing

Crop yield is understood to be the consequence of the interaction between genetics, environment and management (G × E × M). Yet, often overlooked in the case of breeding for tolerance to drought and heat in minor crops is a systematic programme design that includes: (1) understanding of the target population of environments (TPE), (2) use of testing locations representative of the TPE for specific adaptation, (3) ensuring uniformity within testing locations to the extent possible, and the utilization of appropriate statistical tools to address existing spatial variability, (4) multi-environment testing for broad adaptation, (5) multi-year testing to ensure confidence in selections with adaptation to the TPE, and (6) utilization of managed-stress environments (ME), specifically in the case of drought, to enable precise management of the stress where necessary.

The characterization and classification of the TPE is a fundamental step in a systematic breeding effort for yield stability. Data on climate parameters from past records and predictions based on climate modelling allow such classification. Environmental classification of the TPE for a breeding programme, and modelling of future climate change in the TPE, allows the identification of representative testing locations that would cover the whole TPE. This helps to manage the size of breeding programmes for both specific and broad adaptation, thus imparting efficiency without compromising efficacy. The studies leading to the development of the heat-tolerant common bean described above involved a basic environmental characterization of the target regions in Central and South America, and in Africa. A similar, but more detailed, study was undertaken recently for chickpea breeding in Australia (Chauhan et al., 2017). The authors characterized the Northern Grain Regions of Australia, a major TPE of chickpea, for drought and temperature regimes using the Agricultural Production Systems sIMulator (APSIM) model. The characterization of the TPE for four types of drought stress, and five types each of maximum, minimum and mean temperature patterns, led to the identification of eight unique agro-ecological zones, each of which was fairly uniform, supporting the idea of using one representative testing location per zone for local adaptation trials. Use of one representative testing location each for all eight agro-ecological zones would allow breeding for broad adaptation. Land and soil heterogeneity within the representative testing locations can then be addressed using statistical experimental designs and analyses that account for spatial variability.

Another important consideration in breeding for either drought or heat tolerance is multi-season evaluation of the selected material for high confidence in anticipated performance under stress. An ideal selection criteria is that breeding lines selected for better performance under the abiotic stress maintain their performance under normal environmental conditions, on par with the best check varieties. In other words, there should be no yield penalty under no-stress conditions.

Finally, the use of ME is important both for field testing for yield under drought and for characterization of specific breeding material. Such testing environments are normally established in areas with very sparse rainfall during the cropping season, and water availability to the crop is managed for the timing of availability and the amount made available by controlled irrigation using automated drip tapes or other delivery systems. Thus, MEs allow for precise management of the stress in screening by providing predictability for the occurrence of the stress, its timing and its intensity under field conditions. They are also useful in characterizing selected breeding material for tolerance to varying types and intensities of managed drought stress, such as that during the vegetative stage, flowering or grain filling. The MEs also allow field-based validations of high-throughput precision phenotyping technologies.

5.2 Predictive secondary traits, precision phenotyping and crop modelling

The expression of multiple molecular, biochemical and physiological responses, determined by genetics, conditioned by the prevailing environmental and management conditions, and integrated over the duration of the growing season, determines final crop biomass and grain yield under stress. Such responses can serve as predictive measures for drought tolerance, provided there is a high correlation between individual measures (or a combination of measures), and yield under drought. Molecular responses, or the molecular bases of biochemical responses, can become markers for the trait based on the existence of sequence polymorphism. Biochemical responses can facilitate measurement of drought tolerance if there exists a correlation between specific metabolites and the trait. Plant physiological responses and their components can also serve as predictive measures in the measurement of drought tolerance, and traditionally these have included stomatal conductance (Bates and Hall, 1982), carbon isotope discrimination (Farquhar et al., 1982), photosynthetic capacity (Sharkey, 1984), root morphology (Champoux et al., 1995), leaf expansion rate (Salah and Tardieu, 1997) and delayed senescence (Young et al., 2004).

TE has remained a major focus in physiological measurements for drought tolerance, and is determined as the amount of biomass accumulated per unit of moisture transpired (see Vadez et al., 2014 for a recent review). It is part of the three-component Passioura equation mentioned earlier, where Yield = TE × WU × HI, with WU being the amount of water used in transpiration, and HI being the harvest index. Measurement of TE has traditionally been quite cumbersome and involved the use of controlled environments and destructive sampling, or the use of lysimeters to facilitate the gravimetric measurement of evapo-transpiration and biomass accumulation. To overcome the difficulties involved in the determination of TE, surrogate measures such as carbon isotope discrimination and specific leaf area have been used. More recently, the measurement of TE has been facilitated in precise high-throughput mode in an automated platform named LeasyScan at the International Crop Research Institute for the Semi-Arid Tropics (Vadez et al., 2015). This platform allows simultaneous measurement of leaf area and canopy development through 3-D scanners, along with automated gravimetric measurement of transpiration and plant growth in tall pots under outdoor conditions during the entire growth duration. Establishment of the LeasyScan platform opens the prospect for high-throughput (or at least moderate throughput) screening for drought tolerance based on TE. The validity of such physiological screening of germplasm for drought tolerance in legumes is strengthened by recent results that have established the existence of genetic variability for transpiration response to vapour pressure deficit in crops such as chickpea and cowpea (Zaman-Allah et al., 2011; Belko et al., 2013). Rapid response of transpiration to vapour pressure deficit implies causal relationships between hydraulic conductance and the expression of genes such as the water channel proteins, aquaporins (Ehlert et al., 2009). Confirmation of such correlations enables the integration of genetic information and precision phenotyping data in breeding for drought tolerance under natural stress or in ME, as recently described for wheat (Lopes et al., 2014).

Similar to yield stability under stress, the component traits that contribute to it are also conditioned by interactions between genetics, environment and management. Evaluating their contribution to yield stability across multiple geographies, across multiple time periods, under different weather parameters and under different management conditions is best done by crop modelling. A basic example of such a study is the determination of the extent of yield reduction resulting from drought, and as affected by planting density, in groundnut in West Africa using crop simulation modelling together with available, and generated, weather data (Vadez et al., 2017). More sophisticated modelling exercises that include genetic variables known to condition the predisposition of plants to drought (or heat) tolerance, together with associated physiological responses and environmental variables, are becoming increasingly crucial in predicting crop performance under these stresses (Parent and Tardieu, 2014). Conversely, a combination of crop modelling and field experimentation across several maize genotypes under varying intensities of drought and heat stress in Europe resulted in the identification of several QTLs, which can be used in evaluating genotype performance in current and future stress scenarios (Millet et al., 2016).

5.3 Exploring genetic diversity in wild relatives

Crop wild relatives offer a largely untapped source of genetic diversity, especially for traits such as adaptation to adverse environmental conditions. Large-scale collections of germplasm resources are maintained by the crop research centres of the CGIAR system, under the purview of the Global Crop Diversity Trust. Characterizing such collections and continuing their expansion based on analyses of existing gaps are important for long-range breeding efforts, and examples from common bean and pigeon pea have been described (Ramirez-Villegas et al., 2010; Khoury et al., 2015). An elegant method for the utilization of plant agro-biodiversity in breeding is the Focused Identification of Germplasm Strategy or FIGS (Mackay and Street, 2004). The strategy essentially uses information on environmental parameters from collection sites in the selection of collected material for breeding research. The strategy has been successfully used in identifying drought adaptation traits in germplasm collections of faba bean (Khazaei et al., 2013).

Cost-effective next-generation sequencing and high-throughput phenotyping technologies together can allow the assessment of large collections of wild accessions for secondary physiological traits that predict their capacity to adapt to drought or heat, and the allelic diversity that is contributory to such adaptation. Genotypic and phenotypic information from such assessment can be used in the introgression of superior alleles from wild relatives into agronomically viable elite crop germplasm. Interspecific crosses are not always straightforward, as in the case of crosses between cultivated cowpea and its wild relatives, *Vigna unguiculata* subspecies *spontanea*, *stenophylla* or *tenuis* (Boukar et al., 2016). Methods such as embryo rescue followed by backcrossing to lose any level of infertility introduced by embryo rescue have been successful in several cases of interspecific crossing. Aside from genome-wide sequence information, crop wild relatives are also critical for allele mining, focusing on specific genes with known involvement in adaptation mechanisms. However, tolerance to drought and heat being complex polygenic traits, modification of single alleles through mutation breeding or gene editing will only account for a fraction of the potential response for yield stability. Often such modifications are also associated with pleiotropy, affecting agronomic performance of the plant. Again, this calls for breeding selections, especially for drought tolerance, to be made under both drought-stressed and well-watered environments, such that a selected genetic loci or genetic modification that confers tolerance to drought does not hinder yield under well-watered conditions.

6 Crop management for abiotic stress tolerance

The importance of an integrated approach that combines both crop improvement and crop management to tackle the challenge of food security under climate change has been discussed in detail in a recent review by McKersie (2015). Crop management of grain legumes under environmental stress has also been specifically discussed (Siddique et al., 2012; Farooq et al., 2017). Crop management practices that are important for yield stability under drought and heat stress include conservation agriculture, timing of sowing, use of supplemental irrigation, weed control, nutrient management and inoculation with mycorrhizal fungi and rhizobacteria. Not all these practices are optimally used in developing countries owing to various limitations.

Conservation agriculture includes a package of three practices, namely, reduced or no-till that provides minimal soil disturbance, permanent soil cover by crop residues or cover crops and crop rotation. Debates on the benefits of conservation agriculture has been put to rest recently through a global meta-analysis of 5463 paired yield observations from 610 studies addressing no-till versus conventional tillage in 48 crops and 63 countries (Pittelkow et al., 2015). The study revealed that no-till by itself reduced yields, but combined with the other two conservation agriculture practices, it led to increased productivity especially under water-limited conditions in drier regions of the globe, such as sub-Saharan Africa and South Asia. It was also recommended that no-till be introduced only in areas where the other two practices were already in routine use. Grain legumes are ideal crops for two of the three components of conservation agriculture, namely, as cover crops and in crop rotations.

Adjustment of planting time to avoid extreme drought and heat during critical stages such as flowering and grain filling, and the use of water conservation measures such as contour bunding for supplemental irrigation are other important crop management practices that help combat these environmental stresses. Control of weeds through herbicides or manual weed control is important in no-till agriculture. Use of nutrients, especially micronutrients, is also critical, and practices such as micro-dosing, which enable the placement of small quantities of fertilizer close to the root zone at planting, are in low-input agriculture in developing countries. Use of microbial symbionts such as arbuscular mycorrhizal fungi or rhizobacteria through soil or seed inoculation can also have beneficial effects on crop performance under abiotic stress conditions (Celebi et al., 2010; Mengual et al., 2014).

7 Conclusion

Reduced precipitation and rising temperatures adversely affect the growth and yield of grain legumes, especially in the drier regions of the developing world where they are mostly grown under subsistence agriculture. Tolerance of plants to drought and heat is complex and polygenic, involving multiple genetic loci. The underlying molecular, biochemical and physiological bases of plant response and adaptation to these stresses are becoming increasingly understood. Such understanding allows the development of tolerant varieties through traditional and molecular breeding, and through genetic engineering. It also allows the tailoring of precision phenotyping tools that helps predict plant performance under these stresses, thus facilitating reliable screening for tolerance at moderate or high throughput. While there is emerging success in the development of tolerance to drought and heat in grain legumes, most breeding programmes still lack systematic planning and implementation. Knowledge of the TPE, representative testing locations, superior statistical tools that address spatial variability, assembly and use of molecular tools including those from crop wild relatives and integration of genotype and phenotype information together in crop modelling are all essential areas to be strengthened for accelerated breeding. Finally, the adoption of optimal crop management practices that help combat these stresses is critical to realize the yield potential of grain legume varieties that have been bred to incorporate superior genetics for stress tolerance.

8 Where to look for further information

CGIAR Research Program on Grain Legumes (http://grainlegumes.cgiar.org/).

Bean breeder's genome toolbox and database (http://phaseolusgenes.bioinformatics.ucdavis.edu/).

Cowpea iSelect Consortium Array from Illumina (http://www.illumina.com/areas-of-interest/agrigenomics/consortia.html).

Draft genome of Kabuli chickpea (http://www.icrisat.org/gt-bt/ICGGC/GenomeSequencing.htm).

Draft genome of Desi chickpea (http://www.nipgr.res.in/CGWR/home.php).

International Initiative on Pigeonpea Genomics (IIPG, http://www.icrisat.org/gt-bt/iipg/Home.html).

Literature and information on environmental stress in plants (http://plantstress.com/).

9 References

Adams, M. A., Turnbull, T. L., Sprent, J. I. and Buchmann, N. (2016). Legumes are different: Leaf nitrogen, photosynthesis, and water use efficiency. *Proc. Natl. Acad. Sci. USA* 113(15), 4098–103.

Agarwal, G., Garg, V., Kudapa, H., Doddamani, D., Pazhamala, L. T., Khan, A. W., Thudi, M., Lee, S. H. and Varshney, R. K. (2016). Genome-wide dissection of AP2/ERF and HSP90 gene families in five legumes and expression profiles in chickpea and pigeonpea. *Plant Biotechnol. J.* 14(7), 1563–77.

Ariani, A. and Gepts, P. (2015). Genome-wide identification and characterization of aquaporin gene family in common bean (*Phaseolus vulgaris* L.). *Mol. Genet. Genomics* 290(5), 1771–85.

Asfaw, A., Ambachew, D., Shah, T. and Blair, M. W. (2017). Trait associations in diversity panels of the two common bean (*Phaseolus vulgaris* L.) gene pools grown under well-watered and water-stress conditions. *Front Plant Sci.* 8, 733.

Bargaz, A., Zaman-Allah, M., Farissi, M., Lazali, M., Drevon, J.-J., Maougal, R. T. and Carlsson, G. (2015). Physiological and molecular aspects of tolerance to environmental constraints in grain and forage legumes. *Int. J. Mol. Sci.* 16, 18976–9008.

Barrera-Figueroa, B. E., Gao, L., Diop, N. N., Wu, Z., Ehlers, J. D., Roberts, P. A., Close, T. J., Zhu, J. K. and Liu, R. (2011). Identification and comparative analysis of drought-associated microRNAs in two cowpea genotypes. *BMC Plant Biol.* 11, 127.

Basu, S., Ramegowda, V., Kumar, A. and Pereira, A. (2016). Plant adaptation to drought stress [version 1; referees: 3 approved]. *F1000Research* 5(F1000 Faculty Rev):1554.

Bates, L. M. and Hall, A. E. (1982). Diurnal and seasonal responses of stomatal conductance for cowpea plants subjected to different levels of environmental drought. *Oecologia* 54(3), 304–8.

Battaglia, M. and Covarrubias, A. A. (2013). Late Embryogenesis Abundant (LEA) proteins in legumes. *Front Plant Sci.* 4, 190.

Beebe, S. E., Rao, I. M., Cajiao, C. and Grajales, M. (2008). Selection for drought resistance in common bean also improves yield in phosphorus limited and favorable environments. *Crop Sci.* 48(2), 582–92.

Beebe, S., Rao, I. M., Blair, M. W. and Butare, L. (2009). Breeding for abiotic stress tolerance in common bean: Present and future challenges. *SABRAO J. Breed Genet.* 41(special supplement), 1–11.

Beebe, S., Ramírez-Villegas, J., Jarvis, A., Rao, I. M., Mosquera, G., Bueno, J. M. and Blair, M. W. (2011). Chapter 16: Genetic improvement of common beans and the challenges of climate change. In: Yadav, S. S., Redden, R. J., Hatfield, J. L., Lotze-Campen, H. and Hall, A. E. (Eds), *Crop Adaptation to Climate Change*. Wiley-Blackwell, Oxford, UK, pp. 356–69.

Beebe, S., Rao, I., Mukankusi, C. and Buruchara, R. (2013). Improving resource use efficiency and reducing risk of common bean production in Africa, Latin America, and the Caribbean. In: Hershey, C. and Neate, P. (Eds), *Eco-efficiency: From Vision to Reality*. International Center for Tropical Agriculture (CIAT), Cali, Colombia, pp. 117–34.

Belko, N., Zaman-Allah, M., Diop, N. N., Cisse, N., Zombre, G., Ehlers, J. D. and Vadez, V. (2013). Restriction of transpiration rate under high vapour pressure deficit and non-limiting water conditions is important for terminal drought tolerance in cowpea. *Plant Biol. (Stuttg)* 15(2), 304–16.

Bishop, J., Jones, H. E., O'Sullivan, D. M. and Potts, S. G. (2017). Elevated temperature drives a shift from selfing to outcrossing in the insect-pollinated legume, faba bean (Vicia faba). *J. Exp. Bot.* 68(8), 2055–63.

Blair, M. W., Cortés, A. J. and This, D. (2016). Identification of an ERECTA gene and its drought adaptation associations with wild and cultivated common bean. *Plant Sci.* 242, 250–9.

Blum, A. (2017). Osmotic adjustment is a prime drought stress adaptive engine in support of plant production. *Plant Cell Environ.* 40(1), 4–10.

Boukar, O., Fatokun, C. A., Huynh, B. L., Roberts, P. A. and Close, T. J. (2016). Genomic tools in cowpea breeding programs: Status and perspectives. *Front Plant Sci.* 7, 757.

Celebi, S. Z., Demir, S., Celebi, R., Durak, E. D. and Yilmaz, I. H. (2010). The effect of arbuscular mycorrhizal fungi (AMF) applications on the silage maize (*Zea mays* L.) yield in different irrigation regimes. *Eur. J. Soil Biol.* 46, 302–5.

CGIAR (2015). Developing beans that can beat the heat. International Center for Tropical Agriculture (CIAT), Cali, Colombia, 12pp.

Champoux, M. C., Wang, G., Sarkarung, S., Mackill, D. J., O'Toole, J. C., Huang, N. and McCouch, S. R. (1995). Locating genes associated with root morphology and drought avoidance in rice via linkage to molecular markers. *Theor. Appl. Genet.* 90(7–8), 969–81.

Chauhan, Y., Allard, S., Williams, R., Williams, B., Mundree, S., Chenu, K. and Rachaputi, N. C. (2017). Characterization of chickpea production systems in Australia for major abiotic production constraints. *Field Crops Res.* 204, 120–34.

Coleto, I., Pineda, M., Rodiño, A. P., De Ron, A. M. and Alamillo, J. M. (2014). Comparison of inhibition of N2 fixation and ureide accumulation under water deficit in four common bean genotypes of contrasting drought tolerance. *Ann. Bot.* 113(6), 1071–82.

Cooper, M., Gho, C., Leafgren, R., Tang, T. and Messina, C. (2014). Breeding drought-tolerant maize hybrids for the US corn-belt: Discovery to product. *J. Exp. Bot.* 65(21), 6191–204.

Cortés, A. J., This, D., Chavarro, C., Madriñán, S. and Blair, M. W. (2012). Nucleotide diversity patterns at the drought-related DREB2 encoding genes in wild and cultivated common bean (*Phaseolus vulgaris* L.). *Theor. Appl. Genet.* 125(5), 1069–85.

Cuellar-Ortiz, S. M., De La Paz Arrieta-Montiel, M., Acosta-Gallegos, J. and Covarrubias, A. A. (2008) Relationship between carbohydrate partitioning and drought resistance in common bean. *Plant Cell Environ.* 31(10), 1399–409.

Daryanto, S., Wang, L. and Jacinthe, P. A. (2015). Global synthesis of drought effects on food legume production. *PLoS One* 10(6), e0127401.

Devi, M. J., Sinclair, T. R., Chen, P. and Carter, T. E. (2014). Evaluation of elite southern maturity soybean breeding lines for drought-tolerant traits. *Agron. J.* 106(6), 1947–54.

Dhanapal, A. P., Ray, J. D., Singh, S. K., Hoyos-Villegas, V., Smith, J. R., Purcell, L. C., Andy King, C., Cregan, P. B., Song, Q. and Fritschi, F. B. (2015). Genome-wide association study (GWAS) of carbon isotope ratio (δ13C) in diverse soybean [*Glycine max* (L.) Merr.] genotypes. *Theor. Appl. Genet.* 128(1), 73–91.

Duvick, D. N. 1977. Genetic rates of gain in hybrid maize yields during the past 40 years. *Maydica* 22, 187–96.

Ehlert, C., Maurel, C., Tardieu, F. and Simonneau, T. (2009). Aquaporin-mediated reduction in maize root hydraulic conductivity impacts cell turgor and leaf elongation even without changing transpiration. *Plant Physiol.* 150, 1093–104.

Esfahani, N. M., Sulieman, S., Schulze, J., Yamaguchi-Shinozaki, K., Shinozaki, K. and Tran, L. S. (2014). Mechanisms of physiological adjustment of N2 fixation in *Cicer arietinum* L. (chickpea) during early stages of water deficit: single or multi-factor controls. *Plant J.* 79(6), 964–80.

Fang, X., Turner, N. C., Yan, G., Li, F. and Siddique, K. H. (2010). Flower numbers, pod production, pollen viability, and pistil function are reduced and flower and pod abortion increased in chickpea (*Cicer arietinum* L.) under terminal drought. *J. Exp. Bot.* 61(2), 335–45.

Farooq, M., Gogoi, N., Barthakur, S., Baroowa, B., Bharadwaj, N., Alghamdi, S. S. and Siddique, K. H. M. (2017). Drought stress in grain legumes during reproduction and grain filling. *J. Agron. Crop. Sci.*, 203(2), 81–102.

Farquhar, G. D., O'Leary, M. H. and Berry, J. A. (1982). On the relationship between carbon isotope discrimination and the intercellular carbon dioxide concentration in leaves. *Aust. J. Plant Physiol.* 9, 121–37.

Guo, M., Rupe, M., Simmons, C. and Sivasankar, S. (2015). Maize ERECTA genes for improving plant growth, transpiration efficiency and drought tolerance in plants. Patent US7847158.

Idrissi, O., Udupa, S. M., De Keyser, E., McGee, R. J., Coyne, C. J., Saha, G. C., Muehlbauer, F. J., Van Damme, P. and De Riek, J. (2016). Identification of quantitative trait loci controlling root and shoot traits associated with drought tolerance in a lentil (*Lens culinaris* Medik.) recombinant inbred line population. *Front Plant Sci.* 7, 1174.

IPCC (2014). Climate Change 2014: Synthesis Report - Summary for policymakers.

Jiang, Y., Lahlali, R., Karunakaran, C., Kumar, S., Davis, A. R. and Bueckert, R. A. (2015). Seed set, pollen morphology and pollen surface composition response to heat stress in field pea. *Plant Cell Environ.* 38(11), 2387–97.

Jyothi, M. N., Rai, D. V. and Nagesh babu, R. (2015). Identification and characterization of high temperature stress responsive novel miRNAs in French Bean (*Phaseolus vulgaris*). *Appl. Biochem. Biotechnol.* 176(3), 835–49.

Khazaei, H., Street, K., Bari, A., Mackay, M. and Stoddard, F. L. (2013). The FIGS (focused identification of germplasm strategy) approach identifies traits related to drought adaptation in Vicia faba genetic resources. *PLoS One* 8(5), e63107.

Khoury, C. K., Castañeda-Alvarez, N. P., Achicanoy, H. A., Sosa, C. C., Bernau, V., Kassa, M. T., Norton, S. L., van der Maesen, L. J. G., Upadhyaya, H. D., Ramírez-Villegas, J, Jarvis, A and Struik, P. C. (2015). Crop wild relatives of pigeonpea [*Cajanus cajan* (L.) Millsp.]: Distributions, ex situ conservation status, and potential genetic resources for abiotic stress tolerance. *Biol. Conserv.* 184, 259–70.

King, C. A. and Purcell, L. C. (2005). Inhibition of N2 fixation in soybean is associated with elevated ureides and amino acids. *Plant Physiol.* 137(4), 1389–96.

Lopes, M. S., Rebetzke, G. J. and Reynolds, M. (2014). Integration of phenotyping and genetic platforms for a better understanding of wheat performance under drought. *J. Exp. Bot.* 65(21), 6167–77.

Mackay, M. and Street, K. (2004). Focused identification of germplasm strategy – FIGS. In: *Proceedings of the 54th Australian Cereal Chemistry Conference and the 11th Wheat Breeders' Assembly.* Cereal Chemistry Division, Royal Australian Chemical Institute (RACI), Melbourne, Victoria, Australia, pp. 138–41.

Manavalan, L. P., Guttikonda, S. K., Tran, L. S. and Nguyen, H. T. (2009). Physiological and molecular approaches to improve drought resistance in soybean. *Plant Cell Physiol.* 50(7), 1260–76.

McKersie, B. (2015). Planning for food security in a changing climate. *J. Exp. Bot.* 66(12), 3435–50.

Mengual, C., Schoebitz, M., Azco, R. and Roldan, A. (2014). Microbial inoculants and organic amendment improves plant establishment and soil rehabilitation under semiarid conditions. *J. Environ. Manag.* 134, 1–7.

Millet, E. J., Welcker, C., Kruijer, W., Negro, S., Coupel-Ledru, A., Nicolas, S. D., Laborde, J., Bauland, C., Praud, S., Ranc, N., Presterl, T., Tuberosa, R., Bedo, Z., Draye, X., Usadel, B., Charcosset, A., Van Eeuwijk, F. and Tardieu, F. (2016). Genome-wide analysis of yield in Europe: Allelic effects vary with drought and heat scenarios. *Plant Physiol.* 172(2), 749–64.

Muchero, W., Roberts, P. A., Diop, N. N., Drabo, I., Cisse, N., Close, T. J., Muranaka, S., Boukar, O. and Ehlers, J. D. (2013). Genetic architecture of delayed senescence, biomass, and grain yield under drought stress in cowpea. *PLoS One* 8(7), e70041.

Ogata, T., Nagatoshi, Y., Yamagishi, N., Yoshikawa, N. and Fujita, Y. (2017). Virus-induced down-regulation of GmERA1A and GmERA1B genes enhances the stomatal response to abscisic acid and drought resistance in soybean. *PLoS One* 12(4), e0175650.

Pandey, M. K., Upadhyaya, H. D., Rathore, A., Vadez, V., Sheshshayee, M. S., Sriswathi, M., Govil, M., Kumar, A., Gowda, M. V., Sharma, S., Hamidou, F., Kumar, V. A., Khera, P., Bhat, R. S., Khan, A. W., Singh, S., Li, H., Monyo, E., Nadaf, H. L., Mukri, G., Jackson, S. A., Guo, B., Liang, X. and Varshney, R. K. (2014). Genomewide association studies for 50 agronomic traits in peanut using the 'reference set' comprising 300 genotypes from 48 countries of the semi-arid tropics of the world. *PLoS One* 9(8), e105228.

Parent, B. and Tardieu, F. (2014). Can current crop models be used in the phenotyping era for predicting the genetic variability of yield of plants subjected to drought or high temperature? *J. Exp. Bot.* 65(21), 6179–89.

Passioura, J. B. (1977). Grain yield, harvest index and water use of wheat. *J. Aust. Instit. Agric. Sci.* 43, 117–21.
Pittelkow, C. M., Liang, X., Linquist, B. A., van Groenigen, K. J., Lee, J., Lundy, M. E., van Gestel, N., Six, J., Venterea, R. T. and van Kessel, C. (2015). Productivity limits and potentials of the principles of conservation agriculture. *Nature* 517, 365–8.
Polania, J., Poschenrieder, C., Rao, I. and Beebe, S. (2016). Estimation of phenotypic variability in symbiotic nitrogen fixation ability of common bean under drought stress using 15N natural abundance in grain. *Eur. J. Agron.* 79, 66–73.
Qin, H., Gu, Q., Zhang, J., Sun, L., Kuppu, S., Zhang, Y., Burow, M., Payton, P., Blumwald, E. and Zhang, H. (2011). Regulated expression of an isopentenyltransferase gene (IPT) in peanut significantly improves drought tolerance and increases yield under field conditions. *Plant Cell Physiol.* 52(11), 1904–14.
Ramamoorthy, P., Lakshmanan, K., Upadhyaya, H. D., Vadez, V. and Varshney, R. K. (2016). Shoot traits and their relevance in terminal drought tolerance of chickpea (*Cicer arietinum* L.). *Field Crops Res.* 197, 10–27.
Ramírez-Villegas, J., Khoury, C., Jarvis, A., Debouck, D. G. and Guarino, L. (2010). A gap analysis methodology for collecting crop genepools: A case study with phaseolus beans. PLoS One 5(10), e13497.
Rosales, M. A., Cuellar-Ortiz, S. M., de la Paz Arrieta-Montiel, M., Acosta-Gallegos, J. and Covarrubias, A. A. (2013). Physiological traits related to terminal drought resistance in common bean (*Phaseolus vulgaris* L.). *J. Sci. Food Agric.* 93(2), 324–31.
Sadhukhan, A., Kobayashi, Y., Kobayashi, Y., Tokizawa, M., Yamamoto, Y. Y., Iuchi, S., Koyama, H., Panda, S. K. and Sahoo, L. (2014). VuDREB2A, a novel DREB2-type transcription factor in the drought-tolerant legume cowpea, mediates DRE-dependent expression of stress-responsive genes and confers enhanced drought resistance in transgenic Arabidopsis. *Planta* 240(3), 645–64.
Salah, H. and Tardieu, F. (1997). Control of leaf expansion rate of droughted maize plants under fluctuating evaporative demand (A superposition of hydraulic and chemical messages?). *Plant Physiol.* 114(3), 893–900.
Sarkar, T., Thankappan, R., Kumar, A., Mishra, G. P. and Dobaria, J. R. (2014). Heterologous expression of the AtDREB1A gene in transgenic peanut-conferred tolerance to drought and salinity stresses. *PLoS One* 9(12), e110507.
Seversike, T. M., Sermons, S. M., Sinclair, T. R., Carter Jr., T. E. and Rufty T. W. (2013). Temperature interactions with transpiration response to vapor pressure deficit among cultivated and wild soybean genotypes. *Physiol Plant* 148(1), 62–73.
Sharkey, T. D. (1984). Transpiration-induced changes in the photosynthetic capacity of leaves. *Planta* 160(2), 143–50.
Siddique, K. H. M., Johansen, C., Turner, N. C., Jeuffroy, M.-H, Hashem, A., Sakar, D., Gan, Y. and Alghamdi, S. S. (2012). Innovations in agronomy for food legumes. A review. *Agron. Sustain. Dev.* 32, 45–64.
Sinclair, T. R., Devi, J., Shekoofa, A., Choudhary, S., Sadok, W., Vadez, V., Riar, M. and Rufty, T. (2017). Limited-transpiration response to high vapor pressure deficit in crop species. *Plant Sci.* 260, 109–18.
Sivasankar, S., Helentjaris, T. G., Xu, D. and Lyznik, A. (2008). Transcriptional activators involved in abiotic stress tolerance. Patent US7317141.
Sivasankar, S., Williams, R. and Greene, T. (2012). Abiotic-stress tolerance in plants: An industry perspective. In: Tuteja, N. and Gill, S. S. (Eds), *Stress-Resistant Crops: Omics Approaches*, Volume I. Wiley InterSciences, Germany (Invited chapter), pp. 27–48.
Suzuki, N., Rivero, R. M., Shulaev, V., Blumwald, E. and Mittler, R. (2014). Abiotic and biotic stress combinations. *New Phytol.* 203(1), 32–43.
Thomas, H. and Ougham, H. (2014). The stay-green trait. *J. Exp. Bot.* 65(14), 3889–900.

Thudi, M., Khan, A. W., Kumar, V., Gaur, P. M., Katta, K., Garg, V., Roorkiwal, M., Samineni, S. and Varshney, R. K. (2016). Whole genome re-sequencing reveals genome-wide variations among parental lines of 16 mapping populations in chickpea (*Cicer arietinum* L.). *BMC Plant Biol.* 16 Suppl 1, 10.

Tominaga, A., Nagata, M., Futsuki, K., Abe, H., Uchiumi, T., Abe, M., Kucho, K., Hashiguchi, M., Akashi, R., Hirsch, A. M., Arima, S. and Suzuki, A. (2009). Enhanced nodulation and nitrogen fixation in the abscisic acid low-sensitive mutant enhanced nitrogen fixation1 of *Lotus japonicus*. *Plant Physiol.* 151(4), 1965–76.

Turner, N. C., Wright, G. C. and Siddique, K. H. M. (2001). Adaptation of grain legumes (pulses) to water-limited environments. *Adv. Agron.* 71, 193–231.

Urrea, C. A., Yonts, C. D., Lyon, D. J. and Koehler, A. E. (2009). Selection for drought tolerance in dry bean derived from the Mesoamerican gene pool in western Nebraska. *Crop Sci.* 49, 2005–10.

Vadez, V., Kholova, J., Medina, S., Kakkera, A. and Anderberg, H. (2014). Transpiration efficiency: New insights into an old story. *J. Exp. Bot.* 65(21), 6141–53.

Vadez, V., Kholová, J., Hummel, G., Zhokhavets, U., Gupta, S. K. and Hash, C. T. (2015). LeasyScan: A novel concept combining 3D imaging and lysimetry for high-throughput phenotyping of traits controlling plant water budget. *J. Exp. Bot.* 66(18), 5581–93.

Vadez, V., Halilou, O., Hissene, H. M., Sibiry-Traore, P., Sinclair, T. R. and Soltani, A. (2017). Mapping water stress incidence and intensity, optimal plant populations, and cultivar duration for African Groundnut Productivity Enhancement. *Front Plant Sci.* 8, 432.

Varshney, R. K., Thudi, M., Nayak, S. N., Gaur, P. M., Kashiwagi, J., Krishnamurthy, L., Jaganathan, D., Koppolu, J., Bohra, A., Tripathi, S., Rathore, A., Jukanti, A. K., Jayalakshmi, V., Vemula, A., Singh, S. J., Yasin, M., Sheshshayee, M. S. and Viswanatha, K. P. (2014). Genetic dissection of drought tolerance in chickpea (*Cicer arietinum* L.). *Theor. Appl. Genet.* 127(2), 445–62.

Wu, J., Wang, L. and Wang, S. (2017). MicroRNAs associated with drought response in the pulse crop common bean (*Phaseolus vulgaris* L.). *Gene* 628, 78–86. pii: S0378-1119(17)30547-4.

Young, T. E., Meeley, R. B. and Gallie, D. R. (2004). ACC synthase expression regulates leaf performance and drought tolerance in maize. *Plant J.* 40(5), 813–25.

Zaman-Allah, M., Jenkinson, D. M. and Vadez, V. (2011). A conservative pattern of water use, rather than deep or profuse rooting, is critical for the terminal drought tolerance of chickpea. *J. Exp. Bot.* 62(12), 4239–52.

Chapter 7

Developing pest- and disease-resistant cultivars of grain legumes

Diego Rubiales, Institute for Sustainable Agriculture, Spain

1 Introduction

Grain legumes are important crops for consumption by both animals and humans. However, average yields of most of them are still relatively low due to limited adaptability of available cultivars to a broad range of environmental conditions and susceptibility to pests and diseases.

There are a number of diseases that affect legumes. This list includes a number of foliar diseases incited by either biotrophic pathogens, such as rusts, downy mildews and powdery mildews (Sillero et al., 2006) or necrotrophic pathogens such as ascochyta blights, chocolate spot and anthracnose (Tivoli et al., 2006). Grain legumes can be severely affected by a number of soil-borne diseases such as damping-off/root rots, vascular wilts and stem wilts (Infantino et al., 2006). Legume crops can also be damaged by a number of nematodes, bacteria, insect pests (Edwards and Singh, 2006) and parasitic weeds (Rubiales et al., 2006). Prevalence and relative importance of each of these diseases vary with the agroecological conditions and the cropping systems. We will not discuss in detail the aetiology, epidemiology or control practices of each of them, but will focus only on the current knowledge of resistances available.

http://dx.doi.org/10.19103/AS.2017.0023.05

2 Faba bean (*Vicia faba*)

Faba bean can be grown in regions under a range of climatic conditions from temperate to subtropical where it can be susceptible to attack by a wide variety diseases and pests.

Faba bean rust (Fig. 1) is a disease that is prevalent worldwide, especially in the Middle East, North Africa and parts of Australia. It is usually observed late in the season, when pod filling has started with little effects on yield. However, epidemics that start early in the season can be severe, causing yield losses as high as 70%. It is incited by *Uromyces viciae-fabae*, which also causes rusts of other legumes such as lentil and pea, with a clear differentiation for host specificities (Emeran et al., 2005). The most frequently reported resistance mechanism against *U. viciae-fabae* in faba bean is incomplete non-hypersensitive resistance (Sillero et al., 2010). Races have been described and race-specific hypersensitive resistance has been reported, being controlled by genes with major effects (Sillero et al., 2000; Rojas-Molina et al., 2006). Molecular markers have been reported for one of these genes (*Uvf-1*) (Avila et al., 2003), but there might be many others that have not been explored yet. This will help in gene/QTL-pyramiding in a single cultivar.

Downy mildew is incited by *Peronospora viciae* f. sp. *fabae* that is more prevalent in cool, maritime climates. Primary infection arises from soil-borne oospores that germinate to infect the hypocotyls, leading to systemic colonization of plants. Humid conditions promote sporulation and give rise to cycles of secondary infection via conidia. Partial resistance to *Peronospora viciae* has been identified in faba bean (Thomas et al., 1999) although little is known about its genetic control.

Ascochyta blight (Fig. 2), incited by *Didymella fabae*, is particularly prevalent in rainy environments. This disease is most prevalent where faba beans are grown through the winter and spreads most rapidly in the early spring, before the plant starts flowering. Physiological specialization has been suggested; however, there is still debate on the existence of true races that require standardization of methodologies and agreement on a unique differential. Incomplete levels of resistance have been identified in faba bean germplasm (Sillero et al., 2001; Tivoli et al., 2006; Rubiales et al., 2012) that are being used in breeding programmes to develop improved cultivars. Both oligogenic (Kharrat et al., 2006) and polygenic inheritance (Avila et al., 2004; Atienza et al., 2016) have been reported.

Figure 1 Faba bean rust: susceptible (left) *vs* partially resistant (right) accessions.

Figure 2 Details of ascochyta blight infection on a faba bean stem and leaves.

Chocolate spot (Fig. 3), caused by *Botrytis fabae*, is also quite a destructive disease. The first symptoms are discrete but infections spread rapidly when temperatures are mild and humidity is high, defoliating and even killing the plants. Although differences in the virulence of isolates have been suggested, no comprehensive description of races is available so far. Incomplete resistance has been reported in faba bean germplasm (Villegas-Fernández et al., 2009, 2011, 2012). No quantitative trait loci (QTLs) or associated molecular markers have been reported so far, although molecular characterization of the resistance is in progress (Villegas-Fernández et al., 2014).

Figure 3 Faba bean plant severely affected by chocolate spot.

Figure 4 Faba bean accessions resistant (left) *vs* susceptible (right) to *Orobanche crenata.*

Several broomrape species including *Orobanche crenata*, *O. foetida* and *Phelipanche aegyptiaca* (formerly *O. aegyptiaca*) can infect faba bean, with *O. crenata* (Fig. 4) being the most widespread. *O. foetida* is the most recent species reported as infective and seems to be evolving from parasitizing wild hosts to crop plants (Vaz Patto et al., 2008). A number of cultivars with various levels of resistance to *O. crenata* and/or *O. foetida* are available (Maalouf et al., 2011; Rubiales et al., 2014), with resistance being polygenic (Román et al., 2002; Gutiérrez et al., 2013) and resulting from a battery of resistance mechanisms acting from low induction of broomrape seed germination (Fernández-Aparicio et al., 2012b) to hampered attachment to roots and vascular connection (Pérez-de-Luque et al., 2007; Rubiales et al., 2016). Identified resistance based on low induction of seed germination is due to low strigolactone exudation and is similarly operative against *O. crenata*, *O. foetida* and *Phelipanche aegyptiaca* (Fernández-Aparicio et al., 2014), reinforcing the value of this resistance.

Faba bean can be attacked by a number of viruses and nematodes. Resistances to bean yellow mosaic virus and bean leaf roll virus (Gadh and Bernier, 1984; Kumari and Makkouk, 2003) as well as to the nematodes *Ditylenchus dipsaci*, *Pratylenchus thornei* and *Pratylenchus neglectus* (Hanounik et al., 1986; Di Vito et al., 2001) have been identified in faba bean although no information on the genetics of resistance is thus far available.

3 Pea (*Pisum sativum*)

Pea rust (Fig. 5) can be incited by *U. viciae-fabae* and *U. pisi*, with *U. viciae-fabae* being prevalent in tropical areas and *U. pisi* in cooler areas. In both cases, only partial resistance has been described (Chand et al., 2006; Barilli et al., 2009a,b) that might be due to a single gene/major QTL (Vijayalakshmi et al., 2005; Barilli et al., 2010). Unfortunately, markers that are currently available are not yet close enough to allow a reliable MAS approach in any of the two cases.

Powdery mildew (Fig. 6) incited by *Erysiphe pisi* is more prevalent in mild and maritime climates. Resistance to *E. pisi* in pea is conferred by the three genes (*er1*, *er2* and *Er3*) (Fondevilla and Rubiales, 2012). Molecular markers linked to these genes are making

Figure 5 Rust on pea.

gene-pyramiding in a single background feasible. In addition, lines with different levels of incomplete resistance to powdery mildew have been reported in pea and wild relatives (Pal et al., 1980; Fondevilla et al., 2007). However, most pea breeding programmes rely on the use of *er1*. It is considered to confer a durable resistance based on pre-penetration resistance (Fondevilla et al., 2006) associated with protein cross-linking rather than on callose apposition or hypersensitive response (Iglesias-García et al., 2015). It has recently been discerned that *er1* is in fact an MLO gene (Humphry et al., 2011; Rispail and Rubiales, 2016). Other species such as *E. trifolii* can also infect pea, defeating *er1* gene, which deserves urgent attention (Fondevilla et al., 2013).

Ascochyta blight (Fig. 7) in pea is caused by a complex of fungal species that includes *Ascochyta pisi*, *Didymella pinodes*, *Phoma medicaginis* var. *pinodella*, *P. koolunga* and *P. glomerata* (Tivoli et al., 2006; Khan et al., 2013), with *Didymella pinodes* being the most prevalent and destructive. Unfortunately, low levels of resistance, of polygenic nature, have been identified so far (Tar'an et al., 2003; Prioul et al., 2004; Fondevilla et al., 2008, 2011), hampering the release of resistant cultivars (Rubiales and Fondevilla, 2012). Resistance is available in wild *Pisum* relatives (Fondevilla et al., 2005) that are being used in pea breeding but not yet resulting in the release of resistant cultivars.

Resistance to *Fusarium solani* f. sp. *pisi* in pea is quantitatively inherited. A number of QTLs and their associated markers have been identified (Coyne and Pilet-Nayel, 2008).

Figure 6 Pea powdery mildew.

Figure 7 Pea ascochyta blight.

Resistance to *F. oxysporum* f. sp. *pisi* controlled by single dominant genes for some races has also been reported (McPhee et al., 1999). However, resistance seems to be quantitative for race 2 (Bani et al., 2012) (Fig. 8). Genetic resistance to *Aphanomyces euteiches* in pea is known to be quantitative (Pilet-Nayel et al., 2005) and largely influenced by interactions with environmental conditions.

Contrary to faba bean, pea seems to be little affected by *O. foetida* and *Phelipanche aegyptiaca* (Fernández-Aparicio and Rubiales, 2011) but is even more susceptible to *O. crenata* than faba bean, with little resistance identified within pea germplasm (Rubiales et al., 2003a). Some levels of resistance have been identified in *Pisum* spp. (Rubiales et al., 2005). Such resistances, although of complex (Pérez-de-Luque et al., 2005) and quantitative inheritance (Fondevilla et al., 2010), have been successfully transferred to cultivated pea (Rubiales et al., 2009) by crossing and selection, resulting in the recent release of the first resistant cultivars (Fig. 9).

Pea bacterial blight is incited by *Pseudomonas syringae* pv. *pisi* (Fig. 10). The disease can be severe in autumn-sown crops especially when the plants have been damaged by frost or hail. Up to seven races have been reported. Race-specific resistance controlled by single dominant genes has been reported (Bevan et al., 1995; Hunter et al., 2001). *Pisum abyssinicum* accessions are resistant or partially resistant to all races including race 6. This resistance is controlled by a major recessive gene together with a number of modifiers (Elvira-Recuenco and Taylor, 2001).

Resistance to pea seed-borne mosaic virus and pea enation mosaic virus (Hagedorn and Gritton, 1973; Khetarpal et al., 1990) and to cyst nematode (*Heterodera goettingiana*) (Veronico et al., 2011) has also been reported to be found in pea.

Figure 8 Responses to *F. oxysporum* f. sp. *pisi*, from very resistant (left) to very susceptible (right).

Figure 9 Pea accessions resistant (left and back) and susceptible (right) to *Orobanche crenata*.

There is little resistance to pea weevil (*Bruchus pisorum*) in pea germplasm (Teshome et al., 2015; Aznar-Fernández et al., 2017). The highest resistance has been found in *Pisum fulvum* (Byrne et al., 2008; Clement et al., 2009) that was successfully hybridized with pea, allowing identification of QTLs (Aryamanesh et al., 2013) in an attempt to transfer the resistance. Neoplastic pod formation (proliferation on callus on the pods) in response to *Bruchus pisorum* egg pods might reduce larval survival (Berdnikov et al., 1992; Doss et al., 1995). Transgenic peas have been produced with the insecticidal activity of α-amylase

Figure 10 Pea bacterial blight.

inhibitor that blocks the digestion of starch, limiting the development of the weevil larvae (Sousa-Majer et al., 2007).

Pea has limited resistance against pea aphid (*Acyrthosiphon pisum*) (Ali et al., 2005; Carrillo et al., 2013).

4 Lentil (*Lens culinaris*)

Lentil can be affected by a wide range of pathogens, with fungal attacks being the most destructive. Lentil rust (Fig. 11) incited by *U. viciae-fabae* is a major problem in Bangladesh, India, Nepal, Morocco and Ethiopia, with reported losses of up to 70%. Although some hypersensitive resistant sources have also been described, resistance that has been mainly reported is partial resistance (Negussie et al., 2005; Rubiales et al., 2013). Monogenic resistance has been reported (Erskine et al., 1993) with only preliminary information on chromosome location and associated molecular markers available so far.

Ascochyta blight is incited by *Didymella lentis*. It is prevalent throughout the world and causes yield losses of up to 70%. In severe infections the lesions often girdle the stem or petiole, leading to stem breakage. Incomplete resistance has been reported in lentil germplasm. Two mating types for sexual reproduction are known, but the genetic variation and existence of races have not been systematically investigated. Single genes, either dominant (Ford et al., 1999) or recessive (Chowdhury et al., 2001), have been reported together with molecular markers flanking the resistance genes. In addition, a number of QTLs have been identified (Rubeena et al., 2006).

Anthracnose, incited by *Colletotrichum truncatum*, is a destructive disease in certain areas. Symptoms start as irregularly shaped necrotic lesions on lower stems that can coalesce giving the stems a blackish brown appearance. Lesions on leaves are circular causing premature leaf drop by early flowering. Plant wilting can occur in the field after flowering. Two races, Ct0 and

Figure 11 Rust on lentil.

Figure 12 *Orobanche crenata* infecting lentil.

Ct1, have been described and resistance identified in lentil (Buchwaldt et al., 2004; Shaikh et al., 2013) and in *Lens ervoides* (Fiala et al., 2009). Resistance to Ct1 is governed by two recessive genes and three closely linked dominant genes (Buchwaldt et al., 2013). Resistance in *Lens ervoides* is also conferred by two recessive genes (Fiala et al., 2009).

Fusarium wilt incited by *F. oxysporum* f. sp. *lentis* is one of the most destructive diseases of the lentil crop worldwide, causing complete yield loss especially under drought and warm conditions. Pathotypes have only recently been reported and a differential set described (Pouralibaba et al., 2016). Resistance is available and has been characterized histologically (Bayaa et al., 1997; Pouralibaba et al., 2015, 2017).

Broomrape is a major problem in the Mediterranean basin. Resistance against *O. crenata* (Fig. 12) has only recently been reported to be found in lentil and in wild relatives, offering opportunities for breeding (Fernández-Aparicio et al., 2008, 2009a).

Varying levels of resistance to weevil (*Bruchus* sp.) have been reported in lentil and wild relative germplasm (Laserna-Ruiz et al., 2012).

5 Chickpea (*Cicer arietinum*)

Chickpea rust (Fig. 13) is incited by *U. ciceris-arietini* and is of importance only if infection occurs early in the season. Only incomplete resistance has been identified (Sillero et al., 2012) although a major QTL has been reported as responsive to the resistance that might correspond to a single gene (Madrid et al., 2008).

Ascochyta blight is incited by *Didymella rabiei*. A number of races have been reported that are grouped today in four pathotypes. Resistance is available in chickpea germplasm (Singh and Reddy, 1993) and is being used in breeding programmes. Early inheritance

Figure 13 Rust on chickpea.

studies concluded that resistance could be controlled by one, two or three genes. Recent studies suggest the existence of two pathotype-specific QTLs (Udupa and Baum, 2003; Cho et al., 2004; Iruela et al., 2006).

Fusarium wilt incited by *F. oxysporum* f. sp. *ciceris* is a disease prevalent worldwide, with eight races (0, 1A, 1B/C, 2, 3, 4, 5 and 6) identified so far. Resistance is believed to be either monogenic or oligogenic depending upon the race or resistance source. In particular, resistance to races 1A, 2 and 4 is under control of two or three genes, while resistance to races 3 and 5 is controlled by one gene (Sharma et al., 2005). No resistance has been reported for races 1B/C and 6. In addition to race-specificity, slow wilting is characterized by a latent period, disease progress rate and final disease severity and might also be controlled by minor genes (Sharma and Muehlbauer, 2007).

Resistance to *Heterodera ciceri* was not found in chickpea and lentil although resistance was detected in several wild relatives of chickpea (Erskine et al., 1993; Di Vito et al., 1996). Resistance to *O. crenata* has been reported both in the cultivated species (Rubiales et al., 2003b) and in the wild relatives (Rubiales et al., 2004).

Chickpea can be severely constrained by pod borers (*Helicoverpa armigera*) and leaf miner (*Liriomyza cicerina*). Resistance has been identified and cultivars moderately resistant to both pests have been released. Resistance to bruchids (*Callosobruchus* spp.) is more frequent in desi chickpeas (Sharma et al., 2014). Transgenic chickpea with *Cry1A* gene of *Bacillus thuringiensis* inhibits the development of and feeding by *Helicoverpa armigera* (Kar et al., 1997).

6 Grass pea (*Lathyrus sativus*)

Rust and powdery mildew (Fig. 14) are important diseases of grass pea. However, information available on the fungal identity is often insufficient. Rusts in *Lathyrus* sp. are believed to be caused by both *U. pisi* and *U. viciae-fabae*. Powdery mildew is believed to be caused by *E. pisi*, although other species might also infect *Lathyrus* sp. Existence of specialized forms and races is still unclear, but a different ability to infect different plant species has been reported. Resistance to rusts (Vaz Patto and Rubiales, 2009; Vaz

Figure 14 Rust and powdery mildew on grass pea.

Patto et al., 2009) and powdery mildews (Campbell et al., 1994; Vaz Patto et al., 2006) has been reported to be found in grasspea and other *Lathyrus* species.

Ascochyta lathyri is described as causing ascochyta blight on several *Lathyrus* species, but it seems not to infect grass pea which was believed to be infected only by

Figure 15 Orobanche crenata infecting grass pea.

Mycosphaerella pinodes (Gurung et al., 2002). Only recently, ascochyta blight has been reported to be found in grasspea and the causal agent identified as *Ascochyta lentils* var. *lathyri* (Infantino et al., 2016).

Resistance to the parasitic weed *Orobanche crenata* and escape to infection due to precocity (Fig. 15) has also been identified in *L. sativus* and other *Lathyrus* sp. germplasm (Fernández-Aparicio et al., 2009b, 2012a).

7 Common bean (*Phaseolus vulgaris*)

Common bean is the most important food legume consumed worldwide. A majority of the bean production occurs on small-scale farms in developing countries. Therefore, development of resistant cultivars is a primary goal of bean breeding programmes (Miklas et al., 2006).

Common bean rust (Fig. 16) can be very damaging in warm seasons in humid tropical and subtropical areas. Yield losses can be particularly high when infection starts early and can reach 100%. It is incited by *U. appendiculatus* in which high variability has been reported with more than 250 races described. Monogenic resistance (*Ur*-genes) has been identified, with *Ur-3*, *Ur-4*, *Ur-5*, *Ur-6*, *Ur-7* and *Ur-11* widely deployed in cultivar development (Pastor-Corrales and Stavely, 2005). The highly variable nature of the pathogen and the consequent rapid breakdown of major gene resistance present in bean cultivars have challenged breeders to develop durable resistance (Pastor-Corrales, 2003). Pyramiding of different resistance genes and mechanisms, such as partial resistance, should prolong the life of a bean cultivar by creating a more durable resistance complex (Mmbaga et al., 1996).

Although *Erysiphe poligony* has been frequently described to be the causal agent of common bean powdery mildew, recent studies suggest that it is closer to *Erysiphe diffusa* (Almeida et al., 2008). Single inherited resistance has been reported (Ferreira et al., 1999; Trabanco et al., 2012).

Figure 16 Common bean leaf infected by rust.

Bean anthracnose, caused by *Colletotrichum lindemuthianum*, is a seed-borne fungal pathogen. It is prevalent worldwide, although damage is more likely to occur in temperate and subtropical areas. Yield losses can reach 100% when contaminated seeds are planted and conditions favourable for disease development occur throughout the crop cycle. Monogenic resistance is available, with *Co-1* gene of Andean origin and *Co-2* to *Co-10* of Middle American origin. There is co-localization with major resistance genes and QTL that condition partial resistance to anthracnose (Geffroy et al., 2000). Molecular markers linked to most of these genes are available and have been used successfully in marker assisted selection (MAS) (Miklas et al., 2003; Ragagnin et al., 2003).

White mould caused by *Sclerotinia sclerotiorum* is a disease prevalent in cool subtropical and temperate climates. It can be destructive in crops that have a dense canopy when cool, moist conditions occur during and after flowering. Breeding strategies to reduce white mold damage consist in combining the existing quantitatively inherited resistance with avoidance mechanisms that facilitate plant aeration such as an upright growth habit and open architecture (Schwartz et al., 1987). Resistance is available in the secondary gene pool (*Phaseolus coccineus*) offering potential for contributing resistance to common bean (Miklas et al., 2006).

Root rots are caused by a complex of soil-borne pathogens. *Macrophomina phaseolina* is a major problem under conditions of terminal drought. *Rhizoctonia solani* and *F. solani* f. sp. *phaseoli* and *F. oxysporum* f. sp. *phaseoli* cause problems in areas where intermittent drought occurs. Resistance to *Macrophomina phaseolina* has been reported to be controlled by two complementary dominant genes (*Mp-1*, *Mp-2*) (Olaya et al., 1996) and by four QTLs with relatively minor effect (Miklas et al., 1998b). Identified resistance to *F. solani* is complex with over 30 QTLs reported. Race-specific resistance conferred by single to multiple genes has been reported to be found in *F. oxysporum* (Cross et al., 2000).

The bean pod weevil (*Apion godmani*) is a destructive insect pest in Mexico and Central America. Larvae burrow into immature seed, causing yield loss and reduced seed quality. Antibiosis involving a hypersensitive response that encapsulates the oviposition sites, insect eggs or larvae within necrotic tissue has been reported to be caused by epistasis between two independent dominant genes (*Agr* and *Agm*) (Garza et al., 1996). This simple inheritance facilitated the transfer of resistance to elite breeding lines (Beebe et al., 1993).

Figure 17 Common bean resistant (left) and susceptible (right) to powdery mildew.

Bean bruchids include the *Zabrotes subfasciatus*, majorly found in tropical regions, and *Acanthoscelides obtectus*, more common in cooler climates. A major gene discovered in Mexican accessions provides high resistance to *Zabrotes* and slight resistance to *Acanthoscelides* (Acosta-Gallegos et al., 1998). This gene controls seed protein (arcelin). Arcelin and related proteins, including alpha amylase inhibitors and phytohemagglutinins, provide resistance to bruchids by reducing the adult emergence, female fertility and insect growth and lifecycle (Osborn et al., 1988).

8 Future trends and conclusion

Disease resistance is a key component of legume breeding. Phenotypic expression of resistance is usually poorly described. As a consequence of this, and of the poorly developed genomic resources available, genetic basis of resistances is, in most cases, largely unknown. Resistance breeding can be facilitated by the use of molecular markers linked to genes/QTLs controlling resistance, although at present in most instances we are not yet in the position to do this efficiently due to the relatively large distance between the QTLs and their associated markers. Unfortunately, successful application in breeding is far more delayed in most cool-season food legumes compared to common bean where MAS is successfully used. However, lowering costs in high-throughput sequencing and high-throughput phenotyping have encouraged the development of new molecular tools that will indisputably boost the genetic characterization and utilization of germplasm also in other legumes.

However, since resistance breeding is slow due to the relatively low investment on genetic, genomic and biotechnology of the crops, it is also true that a major handicap is the even lower investment on the understanding of the biology of the causal agents and on plant/disease interactions. Comprehensive studies on host status and virulence of the pathogens affecting grain legumes are often missing. Unfortunately, for most of the mentioned diseases, there is even little consensus on the existence of races. Also, available information on levels of resistance and on the responsible mechanisms is often incomplete. This is a major limitation for any breeding programme. Improved knowledge on the biology of the pathogen and of host–pathogen interactions is required to improve the efficiency of the breeding process.

9 Where to look for further information

Readers might look for further information in the special issue on Resistance to Biotic and Abiotic Stresses in Legumes published in *Euphytica* in 2006, volume 147, issue 1–2 wherein a number of articles cover the problems of the various crops, some cover the various groups of diseases and some the various methodologies.

Additional useful information by crop can be found in the following books and reviews:

Biddle, A. J. and Cattlin, N. D. (2007). *Pests, Diseases and Disorders of Peas and Beans*. Manson Publishing Ltd. ISBN-13:978-1-84076-018-7.

Chen, W., Sharma, H. C. and Muehlbauer, F. J. (Eds.) (2011). *Compendium of Chickpea and Lentil Diseases and Pests*. APS Press, American Phytopathological Society, Minnesota, USA, ISBN 978r0-89054-383-2.

Erskine, W., Muehlbauer, F. J., Sarker, A. and Sharma, B. (Eds.) (2009). *The Lentil: Botany, Production and Uses*. CABI, Wallingfore, UK. ISBN-13: 978-1-84593-487-3.
Pérez de la Vega, M., Torres, A. M., Cubero, J. I. and Kole, C. (Eds.) (2011). *Genetics, Genomics and Breeding of Cool Season Grain Legumes*, Science Publishers, Enfield, NH, ISBN 978-1-57808-765-5.
Rubiales, D. and Fernández-Aparicio, M. 2012. *Innovations in Parasitic Weeds Management in Legume Crops*. A review. *Agron. Sustain. Develop.* 32: 433–49. doi: 10.1007/s13593-011-0045-x.
Schwartz, H. F., Steadman, J. R., Hall, R., and Foster R. L. (Eds.) (1991). *Compendium of Bean Diseases, Second Edition*. APS Press, American Phytopathological Society, Minessota, USA, ISBN 0-89054-327-5.
Yadav, S. S., Redden, R., Chen, W. and Sharma, B. (Eds.) (2008). *Chickpea Breeding and Management*. CABI. ISBN13: 9781845932145.

10 References

Acosta-Gallegos, J. A., Quintero, C., Vargas, J., Toro, O., Tohme, J. and Cardona, C. 1998. A new variant of arcelin in wild common bean *Phaseolus vulgaris* L. from southern México. *Genet. Resour. Crop Evol.* 45: 235–42.
Ali, K., Louw, S. V. D. M. and Swart, W. J. 2005. Components and mechanisms of resistance in selected field pea *Pisum sativum* lines to the pea aphid *Acyrthosiphon pisum* (Homoptera: Aphididae). *Int. J. Trop. Insect Sci.* 25: 114–21.
Almeida, A. M. R., Binneck, E., Piuga, F. F., Marin, S. R. R., do RibeiroValle, P. R. Z. and Silveira, C. Z. 2008. Characterization of powdery mildews strains from soybean, bean, sunflower, and weeds in Brazil using rDNA-ITS sequences. *Trop. Plant Pathol.* 33: 20–6.
Aryamanesh, N., Zeng, Y., Byrne, O., Hardie, D. C., Al-Subhi, A. M., Khan, T., Siddique, K. H. M. and Yan, G. 2013. Identification of genome regions controlling cotyledon, pod wall/seed coat and pod wall resistance to pea weevil through QTL mapping. *Theor. Appl. Gen.* 127: 489–97.
Atienza, S., Palomino, C., Gutiérrez, N., Alfaro, C. M., Rubiales, D., Torres, A. M. and Ávila, C. M. 2016. QTLs QTLs for ascochyta blight resistance in faba bean (*Vicia faba* L.): Validation in field and controlled conditions. *Crop Pasture Sci.* 67: 216–24.
Avila, C. M., Sillero, J. C., Rubiales, D., Moreno, M. T. and Torres, A. M. 2003. Identification of RAPD markers linked to the *Uvf-1* gene conferring hypersensitive resistance against rust (*Uromyces viciae-fabae*) in *Vicia faba* L. *Theor. Appl. Genet.* 107: 353–8.
Avila, C. M., Satovic, Z., Sillero, J. C., Rubiales, D., Moreno, M. T. and Torres, A. M. 2004. Isolate and organ-specific QTLs for ascochyta blight resistance in faba bean (*Vicia faba* L). *Theor. Appl. Genet.* 108: 1071–8.
Aznar-Fernández, T., Carrillo-Perdomo, E., Flores, F. and Rubiales, D. 2017. Identification and multi-environment validation of resistance to pea weevil (*Bruchus pisorum*) in (*Pisum*) germplasm. *J. Pest Sci.* on line first. DOI: 10.1007/s10340-017-0925-1.
Bani, M., Rubiales, D. and Rispail, N. 2012. A detailed evaluation method to identify sources of quantitative resistance to *Fusarium oxysporum* f. sp. *pisi* race 2 within a *Pisum* spp. germplasm collection. *Plant Pathol.* 61: 532–42.
Barilli, E., Sillero, J. C., Moral, A. and Rubiales, D. 2009a. Characterization of resistance response of pea (*Pisum* sp.) against rust (*Uromyces pisi*). *Plant Breed.* 128: 665–70.
Barilli, E., Sillero, J. C., Serrano, A. and Rubiales, D., 2009b. Differential response of pea (*Pisum sativum*) to rusts incited by *Uromyces viciae-fabae* and *U. pisi*. *Crop Prot.* 28: 980–6.
Barilli, E., Zatovic, Z., Rubiales, D. and Torres, A. M. 2010. Mapping of quantitative trait loci controlling partial resistance against rust incited by *Uromyces pisi* (Pers.) Wint. in a *Pisum fulvum* L. intraspecific cross. *Euphytica* 175: 151–9.

Bayaa, B., Erskine, W. and Singh, M. 1997. Screening lentil for resistance to fusarium wilt: Methodology and sources of resistance. *Euphytica* 98: 69–74.

Beebe, S., Cardona, C., Díaz, O., Rodríguez, F., Mancia, E. and Ajquejay, S. 1993. Development of common bean (*Phaseolus vulgaris* L.) lines resistant to the bean pod weevil, *Apion godmani* Wagner, in Central America. *Euphytica* 96: 83–8.

Berdnikov, V. A., Trusov, Y. A., Bogdanova, V. S., Kosterin, O. E., Rozov, S. M., Nedel'kina, S. V. and Nikulina, Y. N. 1992. The neoplastic pod gene (*Np*) may be a factor for resistance to the pest *Bruchus pisorum* L. *Pisum Genet.* 24: 37–9.

Bevan, J. R., Taylor, J. D., Crute, I. R., Hunter, P. J. and Vivian, A. 1995. Genetics of specific resistance in pea (*Pisum sativum*) cultivars to seven races of *Pseudomonas syringae* pv. *pisi*. *Plant Pathol.* 44: 98–108.

Buchwaldt, L., Anderson, K. L., Morrall, R. A. A., Gossen, B. D. and Bernier, C. C. 2004. Identification of lentil germ plasm resistant to *Colletotrichum truncatum* and characterization of two pathogen races. *Phytopathology* 94: 236–43.

Buchwaldt, L., Shaikh, R., Adam. J., Tullu, A. and Slinkard, A. E. 2013. Recessive and dominant genes confer resistance to *Colletotrichum truncatum* in cultivated lentil. *Can. J. Plant Pathol.* 35: 222–31.

Byrne, O. M., Hardie, D. C., Khan, T., Speijer, J. and Yan, G. 2008. Genetic analysis of pod and seed resistance to pea weevil in a *Pisum sativum* x *P. fulvum* interspecific cross. *Aust. J. Agri. Res.* 59: 854–62.

Campbell, C. G., Mehra, R. B., Agrawal, S. K., Chen, Y. Z., El Ali, A. M. A., Khawaja, H. I. T., Yadav, C. R., Toy, J. and Araya, W. A. 1994. Current status and future strategy in breeding grass pea (*Lathyrus sativus*). *Euphytica* 73: 167–75.

Carrillo, E., Rubiales, D. and Castillejo, M. A. 2013. Proteomic analysis of pea (*Pisum sativum L.)* response during compatible and incompatible interactions with the pea aphid (*Acyrthosiphon pisum* H.). *Plant Mol. Biol. Rep.* 32: 697–718.

Chand, R., Srivastava, C. P., Singh, B. D. and Sarode, S. B. 2006. Identification and characterization of slow rusting components in pea (*Pisum sativum* L.). *Gen. Res. Crop Evol.* 53: 219–24.

Cho, S., Chen, W. and Muehlbauer, F. J. 2004. Pathotype-specific genetic factors in chickpea (*Cicer arietinum* L.) for quantitative resistance to ascochyta blight. *Theor. Appl. Genet.* 109: 733–9.

Chowdhury, M. A., Andrahennnadi, C. P., Slinkard, A. E. and Vandenberg, A. 2001. RAPD and SCAR markers for resistance to ascochyta blight in lentil. *Euphytica* 118: 331–7.

Clement, S. L., McPhee, K. E., Elberson, L. R. and Evans, M. A. 2009. Pea weevil, *Bruchus pisorum* L. (Coleoptera: Bruchidae), resistance in *Pisum sativum*×*Pisum fulvum* interspecific crosses. *Plant Breed.* 128: 478–85.

Coyne, C. J. and Pilet-Nayel, M. L. 2008. Candidate genes associated with QTL controlling resistance to fusarium root rot in pea. In: *Plant and animal genome XVI, the international conference on the status of plant & animal genome research*. San Diego, CA.

Cross, H., Brick, M. A., Schwartz, H. F., Panella, L. W. and Byrne, P. F. 2000. Inheritance of resistance to Fusarium wilt in two common bean races. *Crop Sci.* 40: 954–8.

Di Vito, M., Abbad Andaloussi, F., Caubel, G. and Greco, N. 2001. Resistance of faba bean and chickpea to root nematodes. In: *Legumed Grain Legumes in Mediterranean Agriculture*, pp. 57–60. European Association for Grain Legume Research. ISBN: 2-9509491-6-9.

Di Vito, M., Singh, K. B., Greco, N. and Saxena, M. C. 1996. Sources of resistance to cyst nematode in cultivated and wild *Cicer* species. *Gen. Res. Crop Evol.* 43: 103–7.

Doss, R. P., Probsting, W. M., Potter, S. W. and Clement S. L. 1995. Response of Np mutant of pea (*Pisum sativum* L.) to pea weevil (*Bruchus pisorum* L.) oviposition and extracts. *J. Chem. Ecol.* 21: 97–106.

Edwards, O. and Singh, K. B. 2006. Resistance to insect pests; what do legumes have to offer. *Euphytica* 147: 273–85.

Elvira-Recuenco, M. and Taylor, J. D. 2001. Resistance to bacterial blight (*Pseudomonas syringae* pv. *pisi*) in Spanish pea (*Pisum sativum*) landraces. *Euphytica* 118: 305–11.

Emeran, A. A., Sillero, J. C., Niks, R. E. and Rubiales, D. 2005. Infection structures of host-specialized isolates of *Uromyces viciae-fabae* and of other species of *Uromyces* infecting leguminous crops. *Plant Dis.* 89: 17–22.
Erskine, W., Tufail, M., Russell, A., Tyagi, M. C., Rahman, M. M. and Saxena, M. C. 1993. Current and future strategies in breeding lentil for resistance to biotic and abiotic stresses. *Euphytica* 73: 127–35.
Fernández-Aparicio, M., Sillero, J. C. and Rubiales, D. 2009a. Resistance to broomrape in wild lentils (*Lens* spp.). *Plant Breed.* 128: 266–70.
Fernández-Aparicio, M., Flores, F. and Rubiales, D. 2009b. Field response of *Lathyrus cicera* germplasm to crenate broomrape (*Orobanche crenata*). *Field Crops Res.* 113: 321–7.
Fernández-Aparicio, M. and Rubiales, D. 2011. Differential response of pea (*Pisum sativum*) to *Orobanche crenata*, *Orobanche foetida* and *Phelipanche aegyptiaca*. *Crop Prot.* 31: 27–30.
Fernández-Aparicio, M., Sillero, J. C., Pérez-de-Luque, A. and Rubiales, D. 2008. Identification of sources of resistance to crenate broomrape (*Orobanche crenata*) in Spanish lentil (*Lens culinaris*) germplasm. *Weed Res.* 48: 85–94.
Fernández-Aparicio, M., Flores, F. and Rubiales D. 2012a. Escape and true resistance to crenate broomrape (*Orobanche crenata* Forsk.) in grass pea (*Lathyrus sativus* L.) germplasm. *Field Crop Res.* 125: 92–7.
Fernández-Aparicio, M., Moral, A., Kharrat, M. and Rubiales, D. 2012b. Resistance against broomrapes (*Orobanche* and *Phelipanche* spp.) in faba bean (*Vicia faba*) based in low induction of broomrape seed germination. *Euphytica* 186: 897–905.
Fernández-Aparicio, M., Kisugi, T., Xie, X., Rubiales, D. and Yoneyama, K. 2014. Low strigolactone root exudation: A novel mechanism of broomrape (*Orobanche* and *Phelipanche* spp.) resistance available for faba bean breeding. *J. Agric. Food Chem.* 62: 7063–71.
Ferreira, R. V., Ramalho, M. A. P. and Corte, H. R. 1999. Genetic control of common bean (*Phaseolus vulgaris* L.) resistance to powdery mildew (*Erysiphe polygoni*). *Genet. Mol. Biol.* 22: 233–6.
Fiala, J. V., Tullu, A., Banniza, S., Seguin-Swartz, G. and Vandenberg, A. 2009. Interspecies transfer of resistance to anthracnose in lentil (*Lens culinaris* Medic.). *Crop Sci.* 49: 825–30.
Fondevilla, S. and Rubiales, D. 2012. Powdery mildew control in Pea. A review. *Agron. Sustain. Develop.* 32: 401–9.
Fondevilla, S., Avila, C. M., Cubero, J. I. and Rubiales, D. 2005. Response to *Mycosphaerella pinodes* in a germplasm collection of *Pisum* spp. *Plant Breed.* 124: 313–15.
Fondevilla, S., Carver, T. L. W., Moreno, M. T. and Rubiales, D. 2006. Macroscopic and histological characterisation of genes *er1* and *er2* for powdery mildew resistance in pea. *Eur. J. Plant Pathol.* 115: 309–21.
Fondevilla, S., Carver, T. L. W., Moreno, M. T. and Rubiales, D. 2007. Identification and characterisation of sources of resistance to *Erysiphe pisi* Syd. in *Pisum* spp. *Plant Breed.* 126: 113–19.
Fondevilla, S., Rubiales, D., Zatovic, S. and Torres, A. M. 2008. Mapping of quantitative trait loci for resistance to *Mycosphaerella pinodes* in *Pisum sativum* subsp. *syriacum*. *Mol. Breed.* 21: 439–54.
Fondevilla, S., Fernández-Aparicio, M., Satovic, Z., Emeran, A. A., Torres, A. M., Moreno, M. T. and Rubiales, D. 2010. Identification of quantitative trait loci for specific mechanisms of resistance to *Orobanche crenata* Forsk. in pea (*Pisum sativum* L.). *Mol. Breed.* 25: 259–72.
Fondevilla, S., Küster, H., Krajinski, F., Cubero, J. I. and Rubiales, D. 2011. Identification of genes differentially expressed in a resistant reaction to *Mycosphaerella pinodes* in pea using microarray technology. *BMC Genom.* 12: 28.
Fondevilla, S., Chattopadhyay, C., Khare, N. and Rubiales, D. 2013. *Erysiphe trifolii* is able to overcome *er1* and *Er3*, but not *er2*, resistance genes in pea. *Eur. J. Plant Pathol.* 136: 557–63.
Ford, R., Pang, E. C. K. and Taylor, P. W. J. 1999. Genetics of resistance to ascochyta blight (*Ascochyta lentis*) of lentil and the identification of closely linked RAPD markers. *Theor. Appl. Genet.* 98: 93–8.
Gadh, I. P. S. and Bernier, C. C. 1984. Resistance in faba bean (*Vicia faba*) to bean yellow mosaic virus. *Plant Dis.* 68: 109–11.

Garza, R., Cardona, C. and Singh, S. P. 1996. Inheritance of resistance to the bean-pod weevil (*Apion godmani* Wagner) in common beans from Mexico. *Theor. Appl. Genet.* 92: 357–62.

Geffroy, V., Sévignac, M., De Oliveira, J., Fouilloux, G., Skroch, P., Thoquet, P., Gepts, P., Langin, T. and Dron, M. 2000. Inheritance of partial resistance against *Colletotrichum lindemuthianum* in *Phaseolus vulgaris* and co-localization of QTL with genes involved in specific resistance. *Mol. Plant Micr. Inter.* 13: 287–96.

Gurung, A. M., Pang, E. C. K. and Taylor, P. W. J. 2002. Examination of *Pisum* and *Lathyrus* species as sources of ascochyta blight resistance for field pea (*Pisum sativum*). *Australas. Plant Pathol.* 31: 41–5.

Gutiérrez, N., Palomino, C., Satovic, Z., Ruiz-Rodríguez, M. D., Vitale, S., Gutiérrez, M. V., Rubiales, D., Kharrat, M., Amri, M., Emeran, A., Cubero, J. I., Atienza, S., Torres, A. M. and Avila, C. M. 2013. QTLs for *Orobanche* spp. resistance in faba bean: Identification and validation across different environments. *Mol. Breed.* 32: 909–22.

Hagedorn, D. J. and Gritton, E. T. 1973. Inheritance of resistance to the pea seed-borne mosaic virus. *Phytopathology* 63: 1130–3.

Hanounik, S. B., Halila, H. and Harrabi, M. 1986. Resistance in *Vicia faba* to stem nematodes (*Ditylenchus dipsaci*). *FABIS Newsl.* 16: 37–9.

Humphry, M., Reinstädler, A., Ivanov, S., Bisseling, T. and Panstruga, R. 2011. Durable broadspectrum powdery mildew resistance in pea *er1* plants is conferred by natural loss-of-function mutations in PsMLO1. *Mol. Plant Pathol.* 12: 866–78.

Hunter, P. J., Ellis, N. and Taylor, J. D. 2001. Association of dominant loci for resistance to *Pseudomonas syringae* pv. *pisi* with linkage groups II, VI and VII of *Pisum sativum. Theor. Appl. Genet.* 103: 129–35.

Iglesias-García, R., Rubiales, D. and Fondevilla, S. 2015. Penetration resistance to *Erysiphe pisi* in pea mediated by *er1* gene is associated with protein cross-linking but not with callose apposition or hypersensitive response. *Euphytica* 201: 381–7.

Infantino, A., Kharrat, M., Riccioni, L., Coyne, C. J., McPhee, K. E. and Grunwald, N. J. 2006. Screening techniques and sources of resistance to root diseases in cool season food legumes. *Euphytica* 147(1–2): 201–21.

Infantino, A., Zaccardelli, M., Costa, C., Ozkilinc, H., Habibi, A. and Peever, T. 2016. A new disease of grasspea (*Lathyrus sativus*) caused by *Ascochyta lentis* var. *lathyri. J. Plant Pathol.* 98 (3): 541–8.

Iruela, M., Rubio, J., Barro, F., Cubero, J. I., Millán, T. and Gil, J. 2006. Detection of two QTL for resistance to ascochyta blight in an intraspecific cross of chickpea (*Cicer arietinum* L.): Development of SCAR markers associated to resistance. *Theor. Appl. Genet.* 112: 278–87.

Kar, S., Basu, D., Das, S., Ramkrishnan, N. A., Mukherjee, P., Nayak, P. and Sen, S. K. 1997. Expression of *CryIA*(c) gene of *Bacillus thuringiensis* in transgenic chickpea plants inhibits development of pod borer (*Heliothis armigera*) larvae. *Transgenic Res.* 6: 177–85.

Khan, T. N., Timmerman-Vaughan, G. M., Rubiales, D., Warkentin, T. D., Siddique, K. H. M., Erskine, W. and Barbetti, M. J. 2013. *Didymella pinodes* and its management in field pea: Challenges and opportunities. *Field Crops Res.* 148: 61–77.

Kharrat, M., Le Guen, J. and Tivoli, B. 2006. Genetics of resistance to 3 isolates of *Ascochyta fabae* on faba bean (*Vicia faba* L.) in controlled conditions. *Euphytica* 151: 49–61.

Khetarpal, R. K., Maury, Y., Cousin, R., Burghofer, A. and Varma, A. 1990. Studies on resistance of pea to pea seed borne mosaic virus pathotypes. *Ann. Appl. Biol.* 116: 297–304.

Kumari, S. G. and Makkouk, K. M. 2003. Differentiation among Bean leaf roll virus susceptible and resistant lentil and faba bean genotypes on the basis of virus movement and multiplication. *J. Phytopathol.* 151: 19–25.

Laserna-Ruiz, I., De-Los-Mozos-Pascual, M., Santana-Méridas, O., Sánchez-Vioque, R. and Rodríguez-Conde, M. F. 2012. Screening and selection of lentil (*Lens* Miller) germplasm resistant to seed bruchids (*Bruchus* spp.). *Euphytica* 188: 153–62.

Maalouf, F., Khalil, S., Ahmed, S., Akintunde, A. N., Kharrat, M., El Shama'a, K., Hajjar, S. and Malhotra, R. S. 2011. Yield stability of faba bean lines under diverse broomrape prone production environments. *Field Crops Res.* 124: 288–94.

Madrid, E., Rubiales, D., Moral, A., Moreno, M. T., Millán, T., Gil, J. and Rubio, J. 2008. Mechanism and molecular markers associated with rust resistance in a chickpea interspecific cross (*Cicer arietinum* L. x *Cicer reticulatum* Lad.). *Eur. J. Plant Pathol.* 121: 43–53.

McPhee, K. E., Tullu, A., Kraft, J. M. and Muehlbauer, F. J. 1999. Resistance to fusarium wilt race 2 in the *Pisum* core collection. *J. Am. Soc. Hort. Sci.* 124: 28–31.

Miklas, P. N., Stone, V., Urrea, C. A., Johnson, E. and Beaver, J. S. 1998. Inheritance and QTL analysis of field resistance to ashy stem blight. *Crop Sci.* 38: 916–21.

Miklas, P. N., Kelly, J. D. and Singh, S. P. 2003. Registration of anthracnose-resistant pinto bean germplasm line USPT-ANT-1. *Crop Sci.* 43: 1889–90.

Miklas, P. N., Kelly, J. D., Beebe, S. E and Blair, M. W. 2006. Common bean breeding for resistance against biotic and abiotic stresses: From classical to MAS breeding. *Euphytica* 147: 105–31.

Mmbaga, M., Steadman, J. R. and Stavely, J. R. 1996. The use of host resistance in disease management of rust in common bean. *Integrated Pest Manag. Rev.* 1: 191–200.

Negussie, T., Pretorius, Z. A. and Bender, C. M. 2005. Components of rust resistance in lentil. *Euphytica* 142: 55–64.

Olaya, G., Abawi, G. S. and Weeden, N. F. 1996. Inheritance of the resistance to *Macrophomina phaseolina* and identification of RAPD markers linked to the resistance genes in beans. *Phytopathology* 86: 674–9.

Osborn, T. C., Alexander, D. C., Sun, S., Cardona, C. and Bliss, F. 1988. Insecticidal activity and lectin homology of arcelin seed protein. *Science* 240: 207–10.

Pal, A. B., Brahmappa, R. D. R. and Ullasa, B. A. 1980. Field resistance of pea germplasm to powdery mildew (*Erysiphe polygoni*) and rust (*Uromyces fabae*). *Plant Dis.* 64: 1085–6.

Pastor-Corrales, M. A. 2003. Sources, genes for resistance, and pedigrees of 52 rust and mosaic resistant dry bean germplasm lines released by the USDA Beltsville Bean Project in collaboration with Michigan, Nebraska, and North Dakota Agricultural Experiment Stations. *Annu. Rpt. Bean Improv. Coop.* 46: 235–41.

Pastor-Corrales, M. A, and Stavely, J. R. 2005. Rust. In: H. F. Schwartz, J. R. Steadman, R. Hall and R. L. Foster (Eds), *Compendium of Bean Diseases*, second edition, pp. 38–9. APS Press, St. Paul, MN.

Pérez-de-Luque, A., Jorrín, J., Cubero, J. I. and Rubiales, D. 2005. *Orobanche crenata* resistance and avoidance in pea (*Pisum* spp.) operate at different developmental stages of the parasite. *Weed Res.* 45: 379–87.

Pérez-de-Luque, A., Lozano, M. D., Moreno, M. T., Testillano, P. S. and Rubiales, D. 2007. Resistance to broomrape (*Orobanche crenata*) in faba bean (*Vicia faba*): Cell wall changes associated with pre-haustorial defensive mechanisms. *Ann. Appl. Biol.* 151: 89–98.

Pilet-Nayel, M. L., Muehlbauer, F. J., McGee, R. J., Kraft, J. M., Baranger, A. and Coyne, C. J. 2005. Consistent QTL in pea for partial resistance to *Aphanomyces euteiches* isolates from United States and France. *Phytopathology* 95: 1287–93.

Pouralibaba, H. R., Rubiales, D. and Fondevilla, S. 2015. Identification of resistance to *Fusarium oxysporum* f. sp. *lentis* in Spanish lentil germplasm. *Eur. J. Plant Pathol.* 143: 399–4050.

Pouralibaba, H. R., Rubiales, D. and Fondevilla, S. 2016. Identification of pathotypes in *Fusarium oxysporum* f. sp. *lentis*. *Eur. J. Plant Pathol.* 144: 539–49

Pouralibaba, H. R., Pérez-de-Luque, A. and Rubiales, D. 2017. Histopathology of the infection on resistant and susceptible lentil accessions by two contrasting pathotypes of *Fusarium oxysporum* f. sp. *lentis*. *Eur. J. Plant Pathol.* 148: 53–63.

Prioul, S., Frankewitz, A., Deniot, G., Morin, G. and Baranger, A. 2004. Mapping of quantitative trait loci for partial resistance to *Mycosphaerella pinodes* in pea (*Pisum sativum* L.) at the seedling and adult plant stages. *Theor. Appl. Genet.* 108: 1322–34.

Ragagnin, V. A., Sanglard, D. A., de Souza, T. L. P. O., Moreira, M. A. and de Barros, E. G. 2003. Simultaneous transfer of resistance genes for rust, anthracnose, and angular leaf spot to cultivar Perola assisted by molecular markers. *Annu. Rep. Bean Improv. Coop.* 46: 159–60.

Rispail, N. and Rubiales D. 2016. Genome-wide identification and comparison of legume *MLO* gene family. *Sci. Rep.* 6: 32673.

Rojas-Molina, M. M., Rubiales, D. and Sillero, J. C. 2006. Pathogenic specialization of *Uromyces viciae-fabae* in Spain and Portugal. In: C. M. Avila, J. I. Cubero, M. T. Moreno, M. J. Suso and A. M. Torres (Eds), *International Workshop on Faba Bean Breeding and Agronomy*, Córdoba, pp. 154–6.

Román, B., Torres, A. M., Rubiales, D., Cubero, J. I. and Satovic, Z. 2002. Mapping of quantitative trait loci controlling broomrape (*Orobanche crenata* Forsk.) resistance in faba bean (*Vicia faba* L.). *Genome* 45: 1057–63.

Rubeena, T., Taylor, P. W. J., Ades, P. K. and Ford, R. 2006. QTL mapping of resistance in lentil (*Lens culinaris* ssp. *culinaris*) to ascochyta blight (*Ascochyta lentis*). *Plant Breed.* 125: 506–12.

Rubiales, D. 2003. Parasitic plants, wild relatives and the nature of resistance. *New Phytol.* 160: 459–61.

Rubiales, D. and Fondevilla, S. 2012. Future prospects for ascochyta blight resistance breeding in cool seasons food legumes. *Front. Plant Sci.* 3: 27.

Rubiales, D., Alcántara, C. and Sillero, J. C. 2004. Variation in resistance to crenate broomrape (*Orobanche crenata*) in species of *Cicer*. *Weed Res.* 44: 27–32.

Rubiales, D., Moreno, M. T. and Sillero, J. C. 2005. Search for resistance to crenate broomrape (*Orobanche crenata*) in pea germplasm. *Gen. Resour. Crop Evol.* 52: 853–61.

Rubiales, D., Pérez-de-Luque, A., Cubero, J. I. and Sillero, J. C. 2003a. Crenate broomrape (*Orobanche crenata*) infection in field pea cultivars. *Crop Prot.* 22: 865–72.

Rubiales, D., Pérez-de-Luque, A., Joel, D. M., Alcántara, C. and Sillero, J. C. 2003b. Characterization of resistance in chickpea to broomrape (*Orobanche crenata*). *Weed Sci.* 51: 702–7.

Rubiales, D., Pérez-de-Luque, A., Sillero, J. C., Román, B., Kharrat, M., Khalil, S., Joel, D. M. and Riches, C. R. 2006. Screening techniques and sources of resistance against parasitic weeds in grain legumes. *Euphytica* 147: 187–99.

Rubiales, D., Fernández-Aparicio, M., Pérez-de-Luque, A., Prats, E., Castillejo, M. A., Sillero, J., Rispail, N. and Fondevilla, S. 2009. Breeding approaches for crenate broomrape (*Orobanche crenata* Forsk.) management in pea (*Pisum sativum* L.). *Pest Manag. Sci.* 65: 553–9.

Rubiales, D., Castillejo, M. A., Madrid, E., Barilli, E. and Rispail, N. 2011. Legume breeding for rust resistance: Lessons to learn from the model *Medicago truncatula*. *Euphytica* 180: 89–98.

Rubiales, D., Avila, C. M., Sillero, J. C., Hybl, M., Narits, L., Sass, O. and Flores, F. 2012. Identification and multi-environment validation of resistance to *Ascochyta fabae* in faba bean (*Vicia faba*). *Field Crops Res.* 126: 165–70.

Rubiales, D., Rojas-Molina, M. M. and Sillero, J. C. 2013. Identification of pre and posthaustorial resistance to rust (*Uromyces viciae-fabae*) in lentil (*Lens culinaris*) germplasm. *Plant Breed.* 132: 676–80.

Rubiales, D., Flores, F., Emeran, A. S., Kharrat, M., Amri, M., Rojas-Molina, M. M. and Sillero, J. C. 2014. Identification and multi-environment validation of resistance against broomrapes (*Orobanche crenata* and *O. foetida*) in faba bean (*Vicia faba*). *Field Crops Res.* 166: 58–65.

Rubiales, D., Sillero, J. C. and Rojas-Molina, M. M. 2016. Characterization resistance mechanisms in faba bean (*Vicia faba*) against broomrape species (*Orobanche* and *Phelipanche* spp.). *Front. Plant Sci.* 7: 1747.

Schwartz, H. F., Casciano, D. H., Asenga, J. A. and Wood, D. R. 1987. Field measurement of white mold effects upon dry beans with genetic resistance or upright plant architecture. *Crop Sci.* 27: 699–702.

Shaikh, R., Diederichsen, A., Harrington, M., Adam, J., Conner, R. L. and Buchwaldt, L. 2013. New sources of resistance to *Colletotrichum truncatum* race Ct0 and Ct1 in *Lens culinaris* Medikus subsp. *culinaris* obtained by single plant selection in germplasm accessions. *Genet. Resour. Crop Evol.* 60: 193.

Sharma, K. D. and Muehlbauer, F. J. 2007. Fusarium wilt of chickpea: Physiological specialization, genetics of resistance and resistance gene tagging. *Euphytica* 157: 1–14.

Sharma, K. D., Chen, W. and Muehlbauer, F. J. 2005. Genetics of chickpea resistance to five races of Fusarium wilt and a concise set of race differentials for *Fusarium oxysporum* f. sp. *ciceris*. *Plant Dis.* 89: 385–90.

Sharma, H. C., Gaur, P. M., Srinivasan S. and Gowda, C. L. L. 2014. Exploiting host plant resistance for pest management in chickpea. *Legume Perspectives* 3: 25–8.

Sillero, J. C., Moreno, M. T. and Rubiales, D. 2000. Characterization of new sources of resistance to *Uromyces viciae-fabae* in a germplasm collection of *Vicia faba*. *Plant Pathol.* 49: 389–95.

Sillero, J. C., Ávila, C. M., Moreno, M. T. and Rubiales, D. 2001. Identification of resistance to *Ascochyta fabae* in *Vicia faba* germplasm. *Plant Breed.* 120: 529–31.

Sillero, J. C., Fondevilla, S., Davidson, J., Vaz Patto, M. C., Warkentin, T. D., Thomas, J. and Rubiales, D. 2006. Screening techniques and sources of resistance to rusts and mildews in grain legumes. *Euphytica* 147(1–2): 255–72.

Sillero, J. C., Villegas-Fernández, A. M., Thomas, J., Rojas-Molina, M. M., Emeran, A. A., Fernández-Aparicio, M. and Rubiales, D. 2010. Faba bean breeding for disease resistance. *Field Crops Res.* 115: 297–307.

Sillero, J. C., Moreno-Alías, I. and Rubiales, D. 2012. Identification and characterization of resistance to rust (*Uromyces ciceris-arietini* (Grognot) Jacz. & Boyd) in a germplasm collection of *Cicer* spp. *Euphytica* 188: 229–38.

Singh, K. B. and Reddy, M. V. 1993. Resistance to six races of *Ascochyta rabiei* in the world germplasm collection of chickpea. *Crop Sci.* 33: 186–9.

Sousa-Majer, M. J. D., Hardie, D. C., Turner, N. C. and Higgins, T. J. V. 2007. Bean α-amylase inhibitors in transgenic peas inhibit development of pea weevil larvae. *J. Econ. Entomol.* 100: 1416–22.

Tar'an, B., Warkentin, T., Somers, D. J., Miranda, D., Vandenburg, A., Blade, S., Woods, S., Bing, D., Xue, A., De Koeyer, D. and Penner, G. 2003. Quantitative trait loci for lodging resistance, plant height and partial resistance to *Mycosphaerella* blight in field pea (*Pisum sativum* L.). *Theor. Appl. Genet.* 107: 1482–91.

Teshome, A., Mendesil, E., Geleta, M., Andargie, D., Anderson, P., Rämert, B., Seyoum, E., Hillbur, Y., Dagne, K. and Bryngelsson, T. 2015. Screening the primary gene pool of field pea (*Pisum sativum* L. subsp. *sativum*) in Ethiopia for resistance against pea weevil (*Bruchus pisorum* L.). *Gen. Resour. Crop Evol.* 62: 525–38.

Thomas, J. E., Kenyon, D. M. and Kightley, S. P. J. 1999. Progress in the exploitation of disease resistance in oilseed rape, field peas and field beans. *Asp. Appl. Biol.* 56: 67–74.

Tivoli, B., Baranger, A., Avila, C. M., Banniza, S., Barbetti, M., Chen, W., Davidson, J., Lindeck, K., Kharrat, M., Rubiales, D., Sadiki, M., Sillero, J. C., Sweetingham, M. and Muehlbauer, F. J. 2006. Screening techniques and sources of resistance to foliar diseases caused by major necrotrophic fungi in grain legumes. *Euphytica* 147: 223–53.

Trabanco, N., Pérez-Vega, E., Campa, A., Rubiales, D. and Ferreira, J. J. 2012. Genetic resistance to powdery mildew in common bean. *Euphytica* 186: 875–82.

Udupa, S. M. and Baum, M. 2003. Genetic dissection of pathotype-specific resistance to ascochyta blight disease in chickpea (*Cicer arietinum* L.) using microsatellite markers. *Theor. Appl. Genet.* 106: 1196–202.

Vaz Patto, M. C. and Rubiales D. 2009. Identification and characterization of partial resistance to rust in a germplasm collection of *Lathyrus sativus*. *Plant Breed.* 128: 495–500.

Vaz Patto, M. C., Fernández-Aparicio, M., Moral, A. and Rubiales D. 2006. Characterization of resistance to powdery mildew (*Erysiphe pisi*) in a germplasm collection of *Lathyrus sativus*. *Plant Breed.* 125: 308–10.

Vaz Patto, M. C., Díaz-Ruiz, R., Satovic, Z., Román, B., Pujadas-Salvà, A. J. and Rubiales, D. 2008. Genetic diversity of Moroccan populations of *Orobanche foetida*: Evolving from parasitising wild hosts to crop plants. *Weed Res.* 28: 179–86.

Vaz Patto, M. C., Fernández-Aparicio, M., Moral, A. and Rubiales D. 2009. Pre and posthaustorial resistance to rusts in *Lathyrus cicera*. *Euphytica* 165: 27–34.

Veronico, P., Melillo, M. T., Saponaro, C., Leonetti, P., Picardi, E. and Jones, J. T. 2011. A polygalacturonase-inhibiting protein with a role in pea defence against the cyst nematode *Heterodera goettingiana*. *Mol. Plant Pathol.* 12: 275–87.

Vijayalakshmi, S., Yadav, K., Kushwaha, C., Sarode, S. B., Srivastava, C. P., Chand, R. and Singh, B. D. 2005. Identification of RAPD markers linked to the rust (*Uromyces fabae*) resistance gene in pea (*Pisum sativum*). *Euphytica* 144: 265–74.

Villegas-Fernández, A. M., Sillero, J. C., Emeran, A. A., Winkler, J., Raffiot, B., Tay, J., Flores, F. and Rubiales, D. 2009. Identification and multi-environment validation of resistance to *Botrytis fabae* in *Vicia faba*. *Field Crops Res.* 114: 84–90.

Villegas-Fernández, A. M., Sillero, J. C., Emeran, A. A., Flores, F. and Rubiales, D. 2011. Multiple-disease resistance in *Vicia faba*: Multi-environment field testing for identification of combined resistance to rust and chocolate spot. *Field Crops Res.* 124: 59–65.

Villegas-Fernández, A. M., Sillero, J. C. and Rubiales, D. 2012. Screening faba bean for chocolate spot resistance: Evaluation methods and effects of age of host tissue and temperature. *Eur. J. Plant Pathol.* 132: 443–53.

Villegas-Fernández, A. M., Krajinski, F., Schlereth, A., Madrid, E. and Rubiales, D. 2014. Characterisation of transcription factor expression profiling of the interaction *Medicago truncatula–Botrytis* spp. interactions. *Plant Molec. Biol. Rep.* 32: 1030–40.

Chapter 8

Biofortification of grain legumes

Bodo Raatz, International Center for Tropical Agriculture (CIAT), Colombia

1 Introduction

1.1 Micronutrient malnutrition: a global health problem

Biofortification is the improvement of the nutritional quality of food crops with essential nutrients, which represents a promising route to combat micronutrient malnutrition (MNM). MNM is caused by deficiencies of micronutrients in human diets that are adequate in calories but short on vitamins and/or mineral elements, also referred to by the term 'hidden hunger'. Mineral deficiencies in human populations are one of the greatest health challenges the world is facing. Currently, more than half of the world population is affected by MNM (Bouis 1999, 2000; WHO 2000; Nestel et al. 2006). It is most prevalent among poor resource-limited families in developing countries where the severity is particularly pronounced in infants, small children and pregnant women (WHO 2000; Graham et al. 2001; Tulchinsky 2010). The 'Green Revolution' was a major success in agricultural development to meet the challenge of feeding the world's poor. The major focus was primarily on three staple crops – rice, wheat and maize, the major staple crops that provide enough calories to prevent widespread famines in many developing nations. However, an unforeseen consequence of the new agricultural systems was a rapid rise in MNM in many areas that adopted the new cropping systems that prevented large-scale starvation (Welch and Graham 1999; Bouis 2003).

http://dx.doi.org/10.19103/AS.2017.0023.06

At least 44 known nutrients are required by humans consistently and in adequate amounts to live healthily (Bouis and Welch 2010). Next to macronutrients carbohydrates, protein, fatty acids and macrominerals, many micronutrients, essential trace minerals and vitamins are indispensable to human health. Deficiencies in Fe (iron), Zn (zinc), vitamin A and I (iodine) are noted as serious health concerns worldwide (Bohra et al. 2015). Fe deficiency is the most prevalent MNM in the world, it is estimated that over 30% of the global population suffers from Fe deficiency (Tulchinsky 2010). Iron-deficiency anaemia (IDA) affects >50% of women and preschool children in developing countries, and is responsible for as much as 20% of deaths among women during childbirth. Furthermore, IDA impairs physical and mental development in childhood and adolescence (Pfeiffer and McClafferty 2007) and renders the affected more vulnerable to a range of infectious diseases including malaria (Tulchinsky 2010). Zinc deficiency (ZD) is also globally widespread, but less information is available due to less testing for this nutrient (Dwivedi et al. 2012). ZD is a major cause of child and infant stunting, affecting small children and adolescents (FAO/WHO 1998; Bouis 2003). ZD also increases morbidity of a number of infectious diseases and possibly malaria (Allen et al. 2006). Vitamin A deficiency (VAD) is associated with blindness and higher morbidity and mortality of diseases (FAO/WHO 1998). It is estimated that VAD affects 140 million preschool children and over 7 million pregnant women every year, representing the most severe vitamin deficiency in human nutrition. Iodine deficiencies can lead to goitre, hypothyroidism and cognitive impairment among other symptoms. A lack of iodine in soil and basic foods is widespread globally, two billion individuals are estimated to have insufficient iodine intake, mainly in South Asia and sub-Saharan Africa (Tulchinsky 2010). Major efforts are directed towards alleviating MNM, to supply a healthy diet for the whole population. Agricultural systems have to be formed that not only provide enough calories to meet the energy needs of the poor but also deliver all the essential nutrients needed for adequate nutritional health.

1.2 Biofortification: a strategy to combat MNM

Biofortification is the enhancement of food crops with significant amounts of bioavailable essential nutrients, sometimes more narrowly defined as advances through breeding. An increase of both nutrient levels and their bioavailability in the edible part of staple crops can be achieved by plant breeding, genetic engineering or agronomic intervention. An obvious solution to MNM is to guide dietary diversification to include sufficient amounts of nutritious food; however, while this is largely the case in developing countries, poor smallholder farmers usually do not have the financial capacity for this option, even if the required information could be communicated. Food supplementation in form of pills has been applied in large projects, but this system is inherently not self-sustainable, and rural populations are often not reached. Fortifying processed food is successfully applied in large scale, for example, it is estimated that two-thirds of the world's households use iodine-fortified table salt, and 85 countries have legislation to mandate fortification of wheat flower with folate or iron (http://www.ffinetwork.org/global_progress/index.php). However, this is difficult to implement in developing countries because it relies on a strong food processing and distribution infrastructure and processed foods are unavailable and/or unaffordable for impoverished people living in remote rural areas. Crop-based biofortification, meaning the development of crop varieties that by themselves form nutrient-rich products, on the other hand, offers a more efficient, cost-effective and sustainable method that can also benefit rural populations. Large investment is only required for developing biofortified varieties and seed systems work for adoption, afterwards recurrent costs are low (Bouis and Welch 2010).

Biofortification is used as a strategy to improve many nutrients in a large number of crops, including essential proteins (Abhary et al. 2011), selenium, calcium, copper, magnesium, iodine (White and Broadley 2009), provitamin A (Paine et al. 2005) and folate (vitamin B9) (Díaz de la Garza et al. 2007; Storozhenko et al. 2007; Nunes et al. 2009). However, the major efforts to combat MNM in developing countries are targeting Fe, Zn and vitamin A, which are associated with the most severe and widespread deficiencies. Legumes are very nutritious crops, mainly grown for their high level of protein; furthermore legume crops contain much higher amounts of essential minerals (Fe, Zn, Ca and Mg) compared to cereals (White and Broadley 2009). As legumes are a major source of these minerals, they are suitable targets for biofortification, providing a higher starting base and thereby more potential for improvement. The HarvestPlus programme is a major project currently funding biofortification in >10 crops. Based on mineral levels, genetic diversity and consumption in MNM affected countries, crops with high impact potential were selected for biofortification (Bouis et al. 2011). Hence, improved biofortified legume germplasm is being developed for higher Fe levels in common bean and cowpea and for high Fe and high Zn lines in lentil (HarvestPlus 2014a). As legume biofortification efforts concentrate on improvement of Fe and Zn contents, this chapter will focus on these essential microminerals.

2 Fe and Zn in grain legumes: assessing natural variation and QTL studies

2.1 Assessing natural variation for Fe and Zn contents

Selecting and/or developing micromineral-rich grain legume varieties through biofortification seems the most appropriate and attractive approach to sustainably reduce MND, particularly in poor smallholder farming systems in developing countries. To identify best micromineral-rich genotypes that may be released or might act as potential donors in the downstream breeding scheme, the first step is to assess the exploitable natural variation available in the crop gene pools for the micronutrient of interest (Dwivedi et al. 2012). Extensive evaluation of genetic variability was carried out in grain legumes to determine the extent of exploitable range of mineral contents.

For example, Beebe et al. (2000) investigated the mineral concentrations (B, Ca, Cu, Fe, K, Mg, Mn, Na, P, S and Zn) in a comprehensive collection of common bean, composed of 1150 accessions (119 wild and 1031 cultivated genotypes), finding mean iron concentrations of 55 ppm (mg/kg), with a range of 34–96 ppm. Zinc concentrations ranged between 21 ppm and 54 ppm, with an average value of 35 ppm. In a similar study in cowpea germplasm 1541 lines of different origins were evaluated, finding a mean Fe value of 53.2 ppm (ranging from 33.6 ppm to 79.5 ppm) and an average Zn concentration of 38.1 ppm (22.1–58.0 ppm) (Boukar et al. 2011). Substantial natural variation was observed in a lentil collection from south-eastern Turkey comprising 39 landraces and 7 cultivars (Karaköy et al. 2012). Similarly, Kumar et al. (2014) observed a wide range of genetic variation for Fe (40–61 ppm) and Zn (38–103 ppm) concentrations among 41 lentil genotypes. Taunk and Yadav (2012) evaluated 16 mung bean lines, reporting Fe concentrations varying from 44.5 to 106.1 ppm, while Zn concentration varied from

23.3 ppm to 40.5 ppm. Similarly, 23 mung bean cultivars were analysed by Singh et al. (2013), resulting in moderate levels of Fe averaging 39 ppm (17–65 ppm) and Zn averaging 25 ppm (15–38 ppm). Investigating 129 landraces and 4 cultivars of faba bean in Turkey, Baloch et al. (2014) identified Fe ranges of 29.7–96.3 ppm, and Zn from 10.4 ppm to 49.3 ppm. Variability of mineral elements in the landraces was significantly higher than those in the commercial cultivars. Other crops investigated include US-grown field peas yielding Fe levels of 46–54 ppm and Zn levels of 39–63 ppm (Amarakoon et al. 2012) furthermore 240 germplasm lines of pigeon pea were phenotyped for grain Zn content and biomass-related traits. Zn concentrations ranged from 15.3 ppm to 69.5 ppm with a mean of 30.6 ppm (Basavarajeshwari et al. 2014). Diapari et al. (2014) evaluated 94 diverse accessions of chickpea in four locations/seasons in Canada, showing significant variability in both Fe ranging from 38.1 ppm to 86.4 ppm and Zn 25.1–62.3 ppm. Multi-location analysis revealed significant location*genotype interactions, suggesting that mineral levels have to be locally confirmed in production areas. Ray et al. (2014) published another multi-locational and multi-seasonal study, analysing samples of 8–18 accessions of each field pea, common bean, chickpea and lentil. Highest Fe and Zn concentrations were found in lentil with 93.4 ppm and 50.6 ppm, respectively.

A large natural variability in seed mineral content was observed in most legume crops, which provides a basis for recombination and selection. The data shows that the employment of conventional breeding strategies should be suitable to increase the accumulation of these nutrients in grain of commercial cultivars.

2.2 QTL studies for Fe and Zn

The use of molecular tools in breeding holds promise to greatly improve the efficiency of breeding programmes. However, adoption of these methods has not been universal as the utility and cost-effectiveness vary greatly with crops and traits, depending on the genetic system, available molecular tools, phenotyping methods and other factors. The major strategy is to first carry out genetic studies to identify marker-trait associations, which are then utilized in marker-assisted selection (MAS) to improve breeding efficiency. Many marker-trait association studies for Fe and Zn in legumes have been carried out, whereas common bean has received most attention in the past (Table 1).

Cichy et al. (2009) presented among the most reproducible and reliable results, with a RIL population of Andean bean genotypes evaluated in two seasons and under two conditions. Constitutive quantitative trait loci (QTLs) were discovered, meaning that they were detected in most conditions/seasons, on chromosomes 1, 6 and 11, the same loci affecting both Fe and Zn concentrations. Blair and Izquierdo (2012) used a backcross population of high-iron wild genotype G10022 backcrossed into the genetic background of the commercial-type variety Cerinza, discovering among others a G10022-derived QTL for both Fe and Zn on chromosome 7, however, with low additive effects. No QTLs were found to have an effect in both investigated trial sites Darien and Palmira; the site specificity indicates significant location*genotype interactions. A further analysis of the same population investigating separately mineral QTLs in seed coat and cotyledon identified QTLs only in seed coat, without clear overlaps with previous analyses (Blair et al. 2013). Guzman-Maldonado et al. (2003) analysed a cross between the wild bean accession G22837 and cultivar Bayo Baranda, identifying two QTLs for Fe, one for Zn and four for tannin concentration. In the Andean x Mesoamerican population DOR 364 × G19833 several QTLs for Zn and Fe were detected, on five chromosomes each (Blair et al. 2009a). A strong season-stable locus on chromosome

Table 1 Overview of marker-trait association studies in several legume crops

Crop	Reference	Population	Population type	Trait	Main results
Common bean	Cichy et al. (2009)	AND 696 × G19833	Andean x Andean RIL	Seed Fe, Zn and P	3 QTLs for both Fe and Zn co-localized on chr 1, 6, 11
Common bean	Blair and Izquierdo (2012)	(Cerinza × (Cerinza × (Cerinza × G10022)))	Wild advanced backcross RIL	Seed Fe, Zn	Co-localized Fe/Zn QTL on chr 7, and 3 further QTL
Common bean	Blair et al. (2013) , 6	(Cerinza × (Cerinza × (Cerinza × G10022)))	Wild advanced backcross RIL	Fe, Zn in seed coat and cotyledons	1 seed coat QTL for Fe for Zn, none in cotyledons. Most Fe found in cotyledons
Common bean	Guzmán-Maldonado et al. (2003)	Baranda × G22837	Wild cross F2:3	Fe, Zn, tannins	2 QTLs for Fe, 1 for Zn and 4 for tannin concentration
Common bean	Blair et al. (2009)	DOR 364 × G19833	Mesoamerican x Andean RIL	Seed Fe, Zn	Major QTL for Fe/Zn on chr 11, a total of 26 QTLs
Common bean	Blair et al. (2010)	DOR 364 × G19833	Mesoamerican x Andean RIL	Fe reductase activity	QTL on chr 2 and QTL hotspot on chr 11
Common bean	Blair et al. (2010)	G14519 x G4825	Mesoamerican x Mesoamerican RIL	Seed Fe, Zn	Co-localization of Fe/Zn QTLs on chr 6 in 2/3 locations, 4 QTLs for Fe, 6 for Zn
Common bean	Blair et al. (2011) , 3	G21242 × G21078	Andean x Andean RIL	Seed Fe, Zn	4 QTLs for Fe for Zn
Common bean	Gelin et al. (2007)	Albion x Voyager	Mesoamerican x Mesoamerican RIL	Seed Fe, Zn	Zn QTL on chr 9
Chickpea	Diapari et al. (2014)	94 diverse accessions	Diversity panel	Seed Fe, Zn	8 SNP loci associated with Fe or Zn concentrations, one SNP on chr 1 associated with both
Lotus japonicus	Klein and Grusak (2009) , 2	Miyakojima MG-20 x Gifo B-129	RIL	Seed Fe, Zn	3 QTLs for Fe for Zn
Lotus japonicus	Klein et al. (2012)	Miyakojima MG-20 x Gifo B-129	RIL	Ferric reductase activity	Major QTL on chr 3
Medicago truncatula	Sankaran et al. (2009)	Jemalong-6 × DZA 315.16	RIL	Seed Fe, Zn	2 QTLs for Fe concentration for Zn
Lentil	Ates et al. (2014)	CDC-Redberry × ILL7502	RIL	Zn	
Lentil	Aldemir et al. (2014)	ILL 8006–BM (Barimasur-4) x CDC Milestone	RIL	Fe	4 QTLs for Fe

11 explaining up to 48% of the variance appears to be a major QTL. Blair et al. (2010a) evaluated iron reductase activity in the same population, resulting in QTL on chromosome 1 and again the major QTL region on chromosome 11. Blair et al. (2010b) investigated an intra-gene pool Mesoamerican x Mesoamerican population evaluated in three locations. Four and six QTLs were found for Fe and Zn, respectively; a hot spot on chromosome 6 had QTLs for both Fe and Zn in two locations. In this study no QTL had significant associations in all three trial locations, again confirming that these phenotypes can be affected by the environment. Furthermore, the Andean G21242 × G21078 population was analysed, without finding QTL overlaps of traits or locations (Blair et al. 2011), and Gelin et al. (2007) reported a Zn concentration QTL on chromosome 9 in a navy bean RIL population.

Several genetic analyses of micromineral content and biofortification-related traits have also been carried out in other legume crops. In chickpea, a GWAS on 94 diverse accessions using array genotyping of 1129 SNPs resulted in the identification of eight SNP loci associated with Fe and/or Zn concentrations in seed. One SNP located on chromosome 1 is associated with both Fe and Zn concentrations (Diapari et al. 2014). Klein and Grusak (2009) performed a study in the model legume *Lotus japonicus*, identifying three QTLs for Fe and two for Zn in a RIL population. In the same RIL population, root ferric reductase activity was investigated, which can be the rate-limiting step in Fe uptake in some conditions. A major QTL on chromosome 3 was identified, which is localized independent from those previously reported (Klein et al. 2012). Furthermore, Sankaran et al. (2009) found two QTLs for Fe and four QTLs for Zn in a *Medicago truncatula* RIL population, utilizing a large phenotypic variability of 50–150 ppm Fe and 40–74 ppm Zn. Ates et al. (2014) and Aldemir et al. (2014) investigated QTL in two lentil RIL populations evaluated in Turkey, reporting significant genetic variability. Zn concentrations ranged from 26 to 51 ppm, and Fe values varied between 37 ppm and 149 ppm, leading to the identification of four QTLs.

Taken together, many QTLs have been detected in several species. Fe and Zn showed large positive correlation, and often QTLs appear at the same locus for both traits. Even though a long list of markers reported to be linked to Fe/Zn traits is available, no publication on their application in breeding has been reported. Successful examples of markers used in application are usually major effect QTLs for disease resistance that are often introgressed from well-defined resistance sources. In case of Fe/Zn QTL the case is different – additive effects are mostly low; they appear in many chromosomes in the various studies, and markers used in RIL populations are unlikely to tag only the mineral source in new breeding crosses. For these reasons the impact of the genetic studies has been low so far. GWASs in panels of breeding lines that may deliver ready-to-use markers for breeding crosses are in progress, and genomic selection models are being investigated that may be applied for breeding for this oligo- or polygenic trait.

3 Biofortification: transgenic approaches and agronomic management

3.1 Transgenic approaches for biofortification

Transgenic technologies offer large advances in genetic improvement; hence, numerous efforts have been made to utilize this technology to improve nutritional content of crop plants (Hirschi 2009). Transgenic methods require detailed understanding of molecular mechanisms and pathways, and high technological infrastructure. The most well-known example of using transgenics for biofortification is the 'Golden Rice' – the first attempted practical application of transgenic to improve nutrition in developing countries. The rice endosperm which is used for human consumption does not contain notable levels of β-carotene, a precursor of vitamin A. Therefore, a strategy was developed to introduce the correct metabolic steps into rice endosperm to facilitate β-carotene biosynthesis (Ye et al. 2000). Initially, a rice line expressing a daffodil (*Narcissus pseudonarcissus*) phytoene synthase was developed leading to the original Golden Rice variety, expressing two daffodil enzymes and one from *Erwinia uredovora*, which accumulates β-carotene, resulting in its eponymous golden colour. Further, Golden Rice varieties were developed, improving the β-carotene accumulation (Paine et al. 2005). At the moment these varieties are still in the release process, facing huge political challenges of GMO technology acceptance.

In spite of this well-known example of transformation for biofortification to relieve MNM, no transgenic legume lines are available at a similar stage. Some soybean lines have been developed in this context, but the mainly industrial soybean crop has no large importance to improve nutritional status of the poor suffering from MNM. Transformation protocols are not well established for all legumes, and expected difficulties and costs in liberating transgenic events have caused non-transgenic strategies to be emphasized. Some transgenic strategies and advances involving legumes will be discussed.

Legumes have been part of several strategies to improve levels of essential amino acids – the most limiting for human or animal consumption are the sulphur (S) containing amino acids methionine and cysteine and furthermore lysine. A simple approach is the expression of recombinant storage proteins with desirable amino acid profiles. Sunflower seed albumin was stably transformed into laboratory model lupin (*Lupinus angustifolius* L.) improving methionine levels by 94 % (Molvig et al. 1997). The expression of sunflower seed albumin was also reported in chickpea inducing higher levels of methionine containing amino acids and lower levels of oxidized sulphur (Chiaiese et al. 2004). In addition to observed improvement of biologically desirable S levels, a reduction of other pre-existing grain S sources was noted, indicating a regulatory mechanism affecting seed storage protein composition. Expression of feedback regulation resistant bacterial aspartokinase and dihydrodipicolinic acid synthase genes in soybean was reported to increase lysine levels (Falco et al. 1995). Following similar strategies, legume originated genes were used in several efforts to improve other crops, for example, the expression of pea legumin for high lysine content in rice (Singh Sindhu et al. 1997).

Improving mineral contents requires different strategies to the examples described above, as minerals are not synthesized in the plants but taken up from the soil, and then transported and stored in the seed. The basic transgenic biofortification strategies regarding Fe and Zn contents are 1) to enhance Fe (III) reduction or release chelating compounds, 2) to modify mineral transporters either in the root or in the grain, or 3) to improve mineral storage in the grain. In addition to these factors, reduction of anti-nutrients represents a biofortification strategy.

A strategy targeting the first steps of mineral uptake is to increase Fe (III) reductase activity to convert the more abundant Fe (III) into plant absorbable Fe (II), which can then be taken up by the plant. Vasconcelos et al. (2006) described that heterologous expression of the *Arabidopsis thaliana* ferric chelate reductase gene, *FRO2*, in transgenic soybean, significantly enhances Fe(III) reduction in roots and leaves. Reductase activity was up to 10-fold higher in transgenic plants and not subjected to post-transcriptional regulation.

However, transgenic events under non-iron stress conditions showed reduced biomass accumulation, indicating that constitutive *FRO2* expression under non-iron stress conditions may reduce plant performance; hence correct regulation of introduced transgenes is of importance. Another strategy to improve Fe uptake is based on the secretion of chelating chemicals. Graminaceous plants secrete phytosiderophores from their roots that form Fe(III)-phytosiderophore complexes, which are absorbed into root cells. Takahashi et al. (2001) introduced barley genomic DNA fragment containing two enzymes involved in the biosynthesis of phytosiderophores in rice. Transgenic lines displayed higher nicotianamine aminotransferase activity, larger secretion of phytosiderophores and enhanced tolerance to low iron availability. A yet different strategy is to improve the levels of ferritin, which is a Fe storage protein in the grain. Drakakaki et al. (2005) expressed soybean ferritin in maize, increasing iron concentration. In combination with the expression of *Aspergillus* phytase, phytate levels were strongly reduced and Fe bioavailability was improved. They concluded that the expression of recombinant ferritin and phytase could help to increase iron availability and enhance the absorption of iron.

Several transgenics projects performed in the past by academic institutes have been advocated to have potential use for smallholder farmers in developing countries; however, they can be mostly considered proof-of-concept projects, as they were not backed up by a strong dissemination plan. As these projects are often based at European or US universities, there is usually no contact with companies developing and selling seed in tropical countries, which would be willing or able to perform deregulation and release; hence usually no impact can be expected. Golden Rice is not only the noticeable exception as it is likely moving towards release, but also an example for the enormous difficulties faced due to legal issues, political opposition and controversy over the adoption of GMO germplasm. With many more private sector-introduced transgenic events in the market, political resistance is likely to be lower in future so that strategies and concepts that have been proven in various crops should now be applied in impact-oriented projects, which requires partnering with industry or larger organizations. CGIAR centres, which are funded for impact, may be able to develop, deregulate and disseminate transgenic products, depending on the political resistance and financial barriers in the respective target countries. Gene editing using CRISPR/Cas9 system may turn out to be free of deregulation processes, and could be used, for example, to reduce anti-nutrient levels (Song et al. 2016; Wolt et al. 2016).

3.2 Biofortification through agronomic management

Biofortification breeding concentrates on enhancing plant's mechanisms to take up minerals from the soil and transport them to the edible part of the plant where they are stored; however, this cannot be effective if the soil does not harbour significant levels of minerals available to the plant. Micromineral-deficient soils are a global problem. Next to the productivity losses associated with growing crops on mineral-deficient soils MND in consumers are aggravated. Adoption of high-yielding varieties and fertilizers of higher purity may increase soil micromineral deficiencies, further affecting plant performance and grain nutritional value. Availability of minerals to plants strongly depends on various soil factors such as texture, pH, organic matter, clay content, soil moisture, nutrient interactions in soil, microbial activity, redox potential and aeration, and so on (Kumar et al. 2016a). Alleviating limiting soil conditions through proper management practices has been shown to significantly increase micronutrient uptake and crop yield, such as liming and applying micronutrients directly to the plant. Additionally, exploitation of soil microbes such as micronutrient solubilizers and arbuscular mycorrhiza (AM) fungi has shown promise to improve micronutrient uptake.

pH is a major factor limiting Fe and Zn availability. Fe uptake is compromised in soils with high soil pH, soluble salts and/or calcium carbonate levels, whereas Zn becomes inaccessible in mucks and mineral soils with pH >6.5. Also the concentration of one micronutrient in soil-plant system may affect the level of other micronutrients (Kumar et al. 2016a). Too much Fe may produce Zn deficiencies, whereas application of Zn has been found to reduce extractable Fe. Generally, sandy soils tend to have lowest mineral contents due to leaching; strongly leached acid soils are low in most micronutrients and require intensive inputs for high productivity.

Agronomic biofortification, fertilizer fortification or just ferti-fortification, which refers to the direct application of the micronutrient to soil or foliar tissue, has proven to improve nutrient uptake. Poonya and Shivay (2013) evaluated Zn concentration with several fertilizer treatments in rice, and recorded the highest values of Zn concentration in grain and straw after foliar spray of 0.2% $ZnSO_4$. Applications of four Zn concentrations were evaluated on three chickpea varieties, recording an increase in grain Zn levels from 36 ppm to 47 ppm, demonstrating the effectiveness of application (Shivay et al. 2014). The same trend was observed in common bean, wheat and rice where foliar Zn application led to an increase in Zn grain concentration from 68 ppm to 78 ppm (Ram et al. 2016). Equally effective increase in grain Fe levels was demonstrated through Fe application (Khalid et al. 2015). Applying correct agronomic practices is the quickest method of biofortification, leading to higher nutrient levels in crop products. However, it requires constant financial input, and it may not be available or sustainable for the poorest of the poor compared to the introduction of genetically biofortified germplasm. Nevertheless, it should be deployed where possible, likely requiring policy changes to be most successful, and find wider application in future.

Utilization of microorganisms is steadily increasing in agriculture, as it offers an attractive way to reduce the use of agrochemicals. Plant growth-promoting rhizobacteria include beneficial bacteria that colonize plant roots and enhance plant growth by a wide variety of mechanisms. Also AM fungi, a type of mycorrhiza in which the fungus penetrates the cortical cells of the roots of a vascular plant, which may help plants to capture nutrients, are utilized. The main mechanism improving micromineral uptake for plants is likely the uptake of bacterial siderophores that are chelating ions and make them available to the plant.

Rana et al. (2012) evaluated inoculations with one bacterial (*Providencia* sp. PW5) and three cyanobacterial strains CW1, CW2 and CW3 (*Anabaena* sp., *Calothrix* sp. and *Anabaena* sp.) in a field experiment on wheat. The largest effects were observed with the PW5 strain, increasing Fe levels in grain by 105%. Interestingly in this case Zn accumulation was not affected. Prasanna et al. (2014) confirmed the utility of biotic inoculations with useful selected bacteria, fungi and cyanobacterial strains in mung bean and soybean, observing improved biomass traits in inoculated materials. In mung bean the application of *Pseudomonas* strain GRP3 led to an improvement in Fe nutrition as compared to control treatment (Sharma et al. 2003). Khalid et al. (2015) evaluated in pot experiments the effect of addition of $FeSO_4$ in combination with five strains of *Rhizobacteria* on Fe uptake in chickpea. Fe concentration of grain was found to be significantly enhanced by both $FeSO_4$ addition and inoculation, the combination of both being most effective. Also AM fungi were shown to be effective in pea, where a higher amount of Fe uptake (10%) and Zn uptake (22%) was registered (Kumar 2012). Next to the diversity of growth-promoting microorganisms also their interaction needs to be considered. A study of the effects on plant growth, nutrient uptake and yield of coinoculation of *Pseudomonas* strains with effective root nodulating symbiotic nitrogen fixing *Rhizobium leguminosarum* showed that Fe uptake more than doubled in co-inoculated plants, both in common bean (Mishra et al. 2014) and in field pea (Mishra et al. 2012). Application of PGRPs and AM fungi will continue to expand to improve plant performance and nutrition; many more species and species-plant-soil interactions need to be evaluated in further research to fully utilize this technology.

4 Micronutrient bioavailability and anti-nutrients

An important component of the nutritional quality of any food product is the bioavailability of its nutrients, that is, the proportion of a consumed nutrient that is absorbed by the human body. Several promoters and anti-nutritional factors are known that affect the bioavailability

Table 2 Examples of promoters of Fe and Zn bioavailability and anti-nutrients that reduce their bioavailability and their major dietary sources. Taken from Bouis and Welch (2010)

Promoter	Trace element	Major dietary sources
Certain organic acids (e.g. ascorbic acid, fumarate, malate and citrate)	Fe and/or Zn	Fresh fruits and vegetables
Haemoglobin	Fe	Animal meats
Certain amino acids (e.g. methionine, cysteine and histidine)	Fe and/or Zn	Animal meats
Long-chain fatty acids (e.g. palmitate)	Zn	Human breast milk
Se	I	Seafoods, tropical nuts
β-carotene	Fe	Green and orange vegetables
Inulin and other non-digestible carbohydrates (prebiotics)	Fe, Zn	Chicory (*Cichorium intybus* L.), garlic (*Allium sativum* L.), onion (*Allium cepa* L.), wheat, Jerusalem artichoke (*Helianthus tuberosus* L.)
Anti-nutrients	**Essential micronutrient metal inhibited**	**Major dietary sources**
Phytic acid or phytin	Fe, Zn, Cu, Ni	Whole legume seeds and cereal grains
Certain fibres (e.g. cellulose, hemicellulose, lignin, cutin and suberin)	Fe, Zn, Cu	Whole cereal grain products (e.g. wheat, rice, maize, oat, barley [*Hordeum vulgare* L.] and rye [*Secale cereale* L.])
Certain tannins and other polyphenolics	Fe	Tea [*Camellia sinensis* (L.) Kuntze], coffee (*Coffea arabica* L.) beans, sorghum [*Sorghum bicolor* (L.) Moench]
Haemagglutinins (e.g. lectins)	Fe	Most legumes and wheat
Goitrogens	I	Brassicas and alliums
Heavy metals (e.g. Cd, Hg and Pb)	Fe, Zn	Contaminated leafy vegetables and roots

of Fe, Zn and other micronutrients (Table 2). Promoters include certain amino acids and organic acids, provitamin A, vitamin C, and certain lipids and carbohydrates (Bouis and Welch 2010). Major anti-nutritional factors for iron and zinc absorption include phytates, tannins and polyphenolics; furthermore certain fibres, proteins and heavy metals reduce bioavailability.

Precise measurement of bioavailability is not an easy task as studies with humans using stable Fe isotopes are complex, time consuming and expensive (Petry et al. 2014). Therefore, the most widely used test for bioavailability uses Caco-2 cells, which are cultured human intestinal cells evaluated for uptake of specific nutrients after passage across a filter membrane (Glahn et al. 1998). Caco-2 measurements were applied in common bean (Ariza-Nieto et al. 2007), showing negative effect on bioavailability of the seed coat and its components, which is a major source of tannin. Also in pea the negative correlation of Fe bioavailability and phytate levels was confirmed (Shunmugam et al. 2015); similarly the method is used in other crops, for example, transgenic maize (Drakakaki et al. 2005). A recent study in 23 lentil varieties with the same technology showed significant potential for biofortification as high natural variation of Fe concentration and relative bioavailability were found (DellaValle et al. 2013). Similarly, bioavailability was evaluated in field pea low phytic acid (*lpa*) mutants (Liu et al. 2015). The Caco-2 test is the most widely used evaluation method of bioavailability; however, not all nutritionists agree that results are completely transferable to humans. Recently a poultry model has been reported for common bean (Tako et al. 2011). Welch et al. (2000) presented a rat model showing the increased absorption of Fe in biofortified high Fe beans.

Fe bioavailability depends on the source. Haem iron, for example, from consumption of meat, poultry and fish has an average absorption of about 25%, ranging from 10–40% (FAO/WHO 1998). Absorption of non-haem iron, which is the main form of dietary iron obtained from cereals, pulses, legumes, fruits and vegetables, may range from 5% to 15% of total iron intake, depending on many factors (Singh 2016; FAO and WHO 2001). The proportion of zinc that is absorbed from typical diets has been quoted to range from about 18% to 34% (Singh 2016) or 26% to 36% (FAO/WHO 1998). High Fe concentrations were shown to correlate with lower relative bioavailability and higher phytate levels in beans (Petry et al. 2014) and lentils (DellaValle et al. 2013), meaning that an increase in Fe concentrations may not be fully utilized. Absorption of nutrients is also affected by the composition of meals, for example, Fe uptake is reduced if a legume is consumed with tannin-rich tea which is customary in some regions. Therefore, to understand the uptake of micronutrients in target populations, the compositions of typical meals have to be taken into account, beyond understanding the bioavailability based on specific crop products.

4.1 Anti-nutrients: genetic studies and breeding

For breeding programmes that aim to reduce MND, several strategies have been advocated to direct attention not only towards enriching the edible plant parts with the bioavailable nutrients, but also to modify the levels of various promoters and anti-nutritional factors (White and Broadley 2009). Several studies have evaluated the phenotypic variability of anti-nutrient levels, and QTL studies have investigated the genetic bases of anti-nutrient concentrations in numerous crops. Phytate has been considered the major damaging anti-nutrient-inhibiting Fe uptake and has received most attention. A range of methods have been discussed to reduce phytate levels, as reviewed by Gupta et al. (2015).

The phytate concentration was evaluated in seeds of 250 accessions of mung bean, and the values range from 2.34 mg/g to 5.75 mg/g (Sompong et al. 2009).

Sompong et al. (2012) identified two QTLs for phytate concentration in mung bean, analysing an F2 population from a low phytate cultivated mung bean (V1725BG) and a high phytate wild mung bean (AusTRCF321925). In 163 RILs of pea, a phytic acid QTL was mapped to LG5 (Shunmugam et al. 2015). In the same work, Fe bioavailability of this population was estimated using the Caco-2 cell culture bioassay, resulting in a QTL on LG5 for both phytate and bioavailability, confirming the expected mechanism. In the common bean inter-gene pool RIL population G2333 × G19839, Blair et al. (2009b) analysed seed phytate concentration, discovering QTLs on chromosomes 6 and 2. In a similar study in the DOR364 × G19833 population, QTLs for phytate concentration were discovered on chromosome 1 and in a similar region on chromosome 6 as in the previous report (Blair et al. 2012). Díaz et al. (2010) evaluated tannins and anthocyanin levels in the same population. Percentage of total tannins in seed coats averaged 20% ranging from 11% to 31%, whereas anthocyanins in seed coats averaged 0.08% (0.01–0.21%), indicating significant genetic variability also for these traits that could be exploited through breeding. Caldas and Blair (2009) evaluated tannins in three RIL populations. A total of 12 QTLs were identified on separate linkage groups in each of the populations, with individual QTLs explaining from 10 to 64% of the phenotypic variation for this trait. Also in the wild cultivar cross, Bayo Baranda x G-22837, four tannin QTLs were reported (Guzman-Maldonado et al. 2003). Genotypes of 23 lentils grown in five different locations in Saskatchewan, Canada, were evaluated for Fe concentration, Fe bioavailability and phytic acid concentration (DellaValle et al. 2013). Positive associations were observed between Fe concentration and phytate concentration ($r = 0.55$, $p < 0.001$), whereas relative Fe bioavailability was negatively correlated with both Fe concentration ($r = -0.41$, $p < 0.001$) and phytate concentration ($r = -0.78$, $p < 0.001$).

A promising lead in common bean is the *low phytic acid* (*lpa*) mutant with 90% reduced phytate levels (Campion et al. 2009), likely caused by a mutation in the *Mrp1* gene (Panzeri et al. 2011). Campion et al. (2013) showed no germination defects in the *lpa* mutants; further unpublished data confirms no agronomic penalty. Petry et al. (2013) demonstrated that the reduction in PA leads to an increase in bioavailable iron. However, Petry et al. (2016) showed adverse digestive side effects in a trial with *lpa* beans, hence their usability in biofortification breeding needs to be further studied. Warkentin et al. (2012) developed two mutant lines in pea, in which phytate phosphorus concentration was reduced by approximately 60%, with a compensating increase in inorganic phosphorus. The investigated low phytate lines were found to be mildly inferior compared to CDC Bronco, in maturity, seed weight and yield. Liu et al. (2015) further evaluated these mutant lines, finding that in one line iron bioavailability showed a 1.4–1.9 times increase over that of the three normal phytate varieties as assessed using a cell culture bioassay, whereas total iron concentration did not differ significantly between normal and low phytate varieties.

Taken together, numerous studies confirm the genetic diversity of anti-nutritional compounds and bioavailability in several crops; also the negative effect of anti-nutrients is quantified and promising germplasm with superior bioavailability is identified. Breeding populations, for example, with the common bean *lpa* mutation have been generated, but standard phenotyping of anti-nutritional compounds or even bioavailability is not yet commonplace in breeding programmes. Methods need to be developed with sufficient ease, high throughput and cost-effectiveness and adopted into biofortification breeding programmes to fully utilize this knowledge to positively impact on MNM.

5 HarvestPlus: breeding, releases, adoption and impact

The HarvestPlus programme has been enabling significant biofortification breeding work to improve Fe content of common bean and cowpea, as well as Fe and Zn contents in lentil. During HarvestPlus Phase I (Discovery, 2003–8), >1000 bean germplasm accessions were screened at CIAT and >1500 cowpea lines were screened at IITA to identify high Fe donor genotypes. High-iron genotypes were then crossed to combine the high-mineral trait with acceptable grain types and agronomic characteristics unless multi-locational evaluation suggested the possibility of direct release.

The first strategy applied for biofortification breeding of Andean beans was backcrossing with gamete selection, where selection for grain type was applied in the BC1F1 and again in the BC1F3 (backcross 1, filial 1 and 3 generations, respectively), and mineral evaluation was applied in BC1F3:4. Following this strategy, Blair et al. (2010c) reported registration of high-mineral common bean lines NUA 35 and NUA 56, with the pedigree CAL 96 × (CAL 96 × G14519) where CAL 96 is a commercial cultivar in Colombia and Uganda, and G14519 is a high seed mineral germplasm accession from the CIAT gene bank. The seed iron and zinc concentrations of NUA 35 and NUA 56, which was determined over 15 sites, were on average 18 and 23 ppm above the check CAL 96. The first biofortified common bean line to be released in 2006 was NUA 35 in Congo. According to the PABRA database (PABRA 2016) 12 more releases were carried out, of 6 further lines in 9 African countries. These NUA (Nutrition Andean) lines were the first generation of biofortified lines that showed Fe levels above baseline but do not meet today's target levels (termed 1st Wave by HarvestPlus). NUA 45 is most popular in several countries due to its beautiful grain. The HarvestPlus programme is the dominating effort in producing, releasing and delivering biofortified germplasm; up to 2016, the programme reports on 26 releases of common bean in Africa (namely Burundi, Rwanda, Democratic Republic of Congo and Uganda) and 18 in Latin America and the Caribbean (Saltzman et al. 2017, in press; HarvestPlus 2014a). National programmes in target countries such as Rwanda (Rwanda Agriculture Board – RAB) and the Democratic Republic of Congo (L'Institut National pour l'Etude et la Recherché Agronomique – INERA) have developed crosses locally and are assuming an increasing proportion of the breeding work. RAB's and INERA's analytical capacities were strengthened by installing X-ray fluorescence (XRF) spectrometers to allow local mineral evaluations. Also, other organizations began disseminating germplasm developed under HarvestPlus, for example, red seeded Mesoamerican bush varieties SMR 101 and SMR 107 were released in Colombia in 2016 as BIO-101 and BIO-107 (CIAT 2016), stating 82 ppm Fe and 43 ppm Zn. Lines SMR 39 and SMR 43 showed Zn >25 ppm and Fe >80 ppm in multiple trials as well as good agronomic properties and are in the release process in Northern Colombia (Tofiño-Rivera et al. 2016).

In Brazil, 87 F6 cowpea breeding lines derived from six crosses were evaluated with the aim to develop cultivars that had improved levels of nutrients (Santos and Boiteux 2013). Fe concentrations ranged from 36.5 ppm to 137 ppm, and Zn concentrations ranged from 36 ppm to 58 ppm. Later, Santos and Boiteux (2015) reported advances in breeding, determining narrow sense heritabilities for Fe (65–86%) and Zn (51–83%) and confirming that minerals in cowpea can be improved via standard breeding methods. Through HarvestPlus efforts, seven cowpea varieties have been released in Brazil and India (Saltzman et al. 2016). Indian varieties named Pant Lobia-1, 2, 3 and 4 harbour Fe levels of 51–100 ppm and Zn ranging from 36 ppm to 40 ppm (HarvestPlus 2014b). For all HarvestPlus supported legumes breeding, work is ongoing and the breeding pipelines contain the next generation of further improved biofortified materials with higher mineral levels and better agronomic performance.

HarvestPlus legume work also enabled releases of biofortified lentil varieties directed at Asia (HarvestPlus 2014a). Initially, more than 1600 germplasm, breeding lines, released cultivars and wild relatives of lentil were analysed for iron and zinc concentrations, discovering a wide variability of 43–132 ppm of iron and 22–95 ppm of zinc. In India, variety L4704 was released, reporting mineral concentrations of 125 ppm Fe and 74 ppm Zn. In Bangladesh the released variety Barimasur-7 contains 81 ppm Fe, and in Nepal variety ILL 7723 is available with 83 ppm Fe and 62 ppm Zn.

In order to validate the utility of biofortification programmes, bioefficacy trials were carried out to quantify the positive effects of improved germplasm. A trial was conducted to determine the efficacy of iron-biofortified common bean to improve iron status in Rwandan women (Haas et al. 2016). A total of 195 women with serum ferritin <20 mg/L were randomly assigned to receive either high Fe beans, with 86 ppm Fe, or standard unfortified beans with 50 ppm Fe. The high Fe beans group had significantly greater increases in haemoglobin (3.8 g/L), log serum ferritin (0.1 log mg/L) and body iron (0.5 mg/kg), demonstrating that consumption of iron-biofortified beans significantly improved iron status in Rwandan women. This is in accordance with data from Petry et al. (2014), who reported that the quantity of iron absorbed from the biofortified bean meals was 19% higher ($P < 0.05$) than that from the control bean meals. Asare-Marfo et al. (2016) carried out an adoption study to evaluate usage of biofortified common bean germplasm in Rwanda. They estimated that almost half a million rural Rwandan households have grown a high Fe bean variety since 2010. In addition, the number of high Fe bean growers increased for almost all varieties through the eight seasons studied, suggesting a significant impact in this country. From 2012 to 2015, delivery of high-iron beans to ~600 000 households in eastern DRC, to 180 000 households in Uganda and to 1.5 million households in Rwanda was reported by Saltzman et al. (2017); however, some may have received seed repeatedly over multiple seasons. A more detailed impact study is under way in Rwanda using DNA fingerprinting to identify released biofortified varieties and landraces.

6 Case study: breeding for biofortification of the common bean at CIAT

CIAT's bean breeding programme has committed to biofortification breeding to increase Fe and Zn contents. Given the solid correlation among them and the lower phenotypic variability of Zn, the main breeding goal is usually high Fe; also HarvestPlus, the main funding project for CIAT biofortification work is emphasizing Fe in common bean. Work started with screening of germplasm collections, the biggest among them being the core collection of the largest common bean germplasm bank at CIAT (Beebe et al. 2000). The identified high Fe and Zn lines were crossed with elite breeding materials to combine the desired biofortification and agronomic traits by continuing cycles of crossing and selection. Also interspecific crosses with *P. coccineus* and *P. dumosus* were evaluated as high Fe sources, giving rise to the MIB (mineral bean) lines that are in the pedigree of many elite biofortified germplasm. During the breeding cycle, evaluation and selection for Fe/Zn concentrations are carried out usually several times between F2 and F6 generations, selecting for genotypes with acceptable agronomic properties and grain types in combination with higher mineral levels. After the first initial stages utilizing more exotic sources as high Fe donors, preferably breeding lines of the biofortification programme are being selected for further crosses, to improve the general quality of the breeding pool. Crosses are carried out to further improve Fe/Zn levels and to combine this trait with disease resistance and stress tolerance traits.

Reports have been published stating that mineral-rich plants tend to be more vigorous and more tolerant of biotic stress (Singh et al. 2016a); furthermore breeding for nutrient efficiency lowers requirements of fertilizer and irrigation and increases disease resistance, seedling vigour and plant yields (Bouis 2003). The resulting impression may be that biofortification breeding is an easy task, given that it positively affects secondary traits. However, while we have not observed significant positive effects, some difficulties are associated with breeding for this trait. A confounding issue is the definition of mineral-rich or nutrient-efficient. In breeding products, Fe concentration is defined as ppm Fe in grain, baseline in most locations is ~50–60 ppm, and current elite biofortified germplasm register 80–100 ppm. There is a consistent negative correlation between Fe concentration and yield. A likely reason is a dilution effect of Fe caused by high carbohydrate translocation to grain (high yield), which was also reported in other crops (Diapari et al. 2014). Adversely, every problem in carbohydrate or protein production (low-yielding genotypes) will lead to an enrichment of Fe and high Fe concentration, which is, for example, often observed under stress conditions such as drought. A similar effect was observed when breeding for high protein germplasm; highest protein concentrations were noted in those with low carbohydrate productivity. So while very high Fe sources with concentrations over 100 ppm have been identified, these are often accompanied by dire agronomic performances. Hence care has to be taken to use 'good' high Fe sources that increase Fe uptake on top of good yield, rather than 'bad high Fe sources' that are defective in carbohydrate accumulation and yields while retaining similar Fe uptake. For this reason the trait iron yield (Fe concentration * yield (g/ha)) is also used in selection, combining good yield and Fe content. Setting a very high Fe concentration as a selection cut-off in the breeding process may not represent the best route to impact, as breeding lines may not have the required agronomic performance to pass national performance trials that are required for release. The breeder's task is to break the negative correlation between Fe concentration and yield, and find a good balance between these to be most efficient in relieving MNM. Taken together, breeding for biofortification – just like other quality traits – has a certain trade-off in terms of productivity advance, which should be taken into consideration in impact strategies.

Recent breeding lines show that germplasm combining high Fe with good agronomic properties can be obtained through persistent recurrent crossing, evaluation and selection. Particularly, climbing beans and bush types of the small seeded Mesoamerican gene pool are available now with good trait combinations, whereas advances in bush types of the Andean gene pool have been slower. A major improvement in the breeding process has been the installation of the XRF spectrometer, replacing the imprecise NIRS method and the time-consuming and costly outsourcing for atomic absorption or ICP methods. XRF combines cost-effectiveness with throughput, acceptable sample preparation effort and precision, allowing faster breeding progress and more genetic gain.

Even though numerous genetic studies have been carried out, no molecular markers are in use for mineral levels in breeding. Parentals of RIL populations are often too distant from elite breeding materials, major effect loci are scarce and markers are too unspecific outside RIL populations. An ongoing marker validation project aims to confirm a QTL effect in five different backgrounds to confirm marker reliability; furthermore, ongoing GWAS analyses in breeding materials could deliver useful loci in elite breeding material with immediately deployable markers that work in breeding populations.

Recent Mesoamerican bush genotypes bred for high Fe and Zn levels also show good yields and mineral levels under drought and low soil fertility conditions, next to good performance for common biotic stresses such as virus. Biofortified climbing beans of Andean grain types have been developed that combine good commercial grain types with yield as well as virus and anthracnose resistance. Lines are best evaluated in several locations as significant GXE interactions have been observed. Next to a baseline that varies slightly depending on location and conditions, particular genotypes show strong GXE interactions like the released variety NUA 35, for example, which is a full target level high Fe line in Darien site, but has near baseline Fe content in Palmira site. Significant location dependence of mineral levels was also reported by Ray et al. (2014) and in other crops (Norton et al. 2014), hence multi-locational testing in target areas is required to assure that the intended impact can be achieved. New breeding lines are shared with partners in the PABRA network (PABRA-Africa 2016) and other NARS institutes for evaluation of local adaptation and releases. The breeding pipeline at CIAT hold promising materials in early and later generations that are expected to create impact in target populations in cooperation with the established extension systems through HarvestPlus and PABRA.

7 Future trends

Improved nutrition is a major theme in the strategy of many funding organizations and research institutes, hence active efforts are currently undertaken in research, germplasm development and dissemination of research products, affecting available germplasm and policies to alleviate MNM. As many reports point out the importance and variability of bioavailability, it can be expected that future strategies will place more emphasis on bioavailability of microminerals. Technologies for anti-nutrient and bioavailability phenotyping have to be developed that allow fast, high-throughput and cost-effective evaluations at breeding stations, to allow optimal advance on this field. Low phytate mutants such as the *lpa* line (Campion et al. 2009) hold large promise to increase bioavailability; *lpa* will be tested to determine culinary quality and decide on future utilization of this mutant. Development of genotypes is planned to aid the 'first 1000 days', describing the time from pregnancy to two years of age, the phase most vulnerable to damages from MND. White low tannin lines with high Fe and high bioavailability (potentially *lpa*) should be made available in target countries.

Several transgenic lines have been generated in many crops, demonstrating the high potential of GM technology to increase micronutrient levels and bioavailability. However, GM strategies have been considered unlikely to be successful, due to intellectual property laws and regulatory/biosafety issues (Mba et al. 2012; Hirschi 2009). In case the Golden Rice will be finally approved and adopted, this could pave the way for further transgenic events to find their way into application. As an alternative, CRISPER/Cas9 technology allows highly specific and efficient gene editing, generating variation that can be indistinguishable from that obtained through conventional methods (Song et al. 2016). It has high potential to revolutionize crop improvement; however, the regulatory status of to what extent this technology is considered GM is still unclear (Wolt et al. 2016). Taken together, mutants such as *lpa*, transgenic or gene edited lines can allow major leaps in micromineral levels and bioavailability. As defined mono- or oligo-genetic traits, they can be integrated with greater ease into elite germplasm, usually without negatively affecting yield. Hence, once available these methods have a high probability to outcompete the oligo/multi-genetic conventional breeding methods.

Breeding pipelines to produce biofortified germplasm are well set up in several legume crops, extension services are in action and growing impact can be expected in future. Active research indicates that the pool of available, disseminated and adopted biofortified varieties will continue to grow. More extensive utilization of interspecific crosses with the secondary gene pool has been promoted in lentil (Kumar et al. 2016b) and also in common bean, mentioning valuable exotic alleles for several traits (Beebe 2012). Interspecific materials have shown high Fe/Zn levels and have been used in breeding crosses. These resources will be further exploited in common bean breeding and should also be evaluated in further crops, to enhance the genetic variability available to the breeder. Agronomic methods need to be further developed in cooperation with breeding to have germplasm that properly responds to the treatments, to generate packages of germplasm and adequate agronomic practices to guarantee high nutritive crop products in the target soil types.

To what extent the added value of biofortification will be perceived and desired by consumers and local markets will depend on advertising and campaigning, requiring continuing investment for sustained impact. It is feasible that biofortified lines will be increasingly established in remote areas using the continuing dissemination efforts. Impact through biofortified varieties has been considered sustainable after a one-time investment of producing the variety. However, a multitude of breeding programmes is continuously producing new variety candidates. Due to the negative correlation with yield traits, those programmes that do not consider biofortification as a breeding trait may advance faster and generate superior yielding varieties. Hence, continuous investments have to be made to produce new biofortified germplasm that is competitive. Other strategies for sustainability suggest that policy makers make biofortification compulsory, encouraging all breeding efforts to take this trait into account. In the HarvestPlus supported breeding programmes, the strategy is to introduce high Fe or Zn levels into all elite breeding materials so that in future this trait is automatically included in all crosses. Further institutions have now also committed to this strategy such as the CGIAR consortium and several donor organizations. However, as not all breeding efforts follow this policy, and genetic gain for yield and minerals has to be balanced, the dominance of biofortified breeding materials in the long term is not unambiguous. Currently, biofortification breeding and seed systems work is well funded, and increasing impact improving the health situation for the poor in developing countries can be expected.

8 Summary and conclusion

MNM is one of the major health threats the world is currently facing, most prevalent in small children and women in developing countries. Actions need to be taken to further alleviate this problem in those poverty-stricken regions, where people suffer from MNM. Legumes are highly nutritious crops and are used to alleviate Fe and Zn deficiencies, which are two of the most critical MNM health threats, affecting more than a third of the world's population. Biofortification is a major strategy, representing the most sustainable solution compared to fortification and food supplementation, to make nutritious diets available to vulnerable populations.

Research to enhance micromineral levels in several legume crops has been reported; the HarvestPlus programme is a major effort in this respect, generating high Fe bean and cowpea lines, and high Fe and Zn lentil germplasm. Diversity analyses and genetic studies have shown exploitable variability in many legume crops, and a polygenic inheritance of the trait. Several QTLs have been identified, but application in MAS is so far hindered mainly by a lack of major effect loci. Bioavailability of microminerals in crop plants is fairly low, which makes it very important to consider anti-nutritional factors that reduce bioavailability in breeding programmes. Significant natural variation in bioavailability has been demonstrated and mutants and transgenic events prospect sizable jumps in bioavailable micromineral levels that should be further explored. Goal setting for biofortification advances could emphasize bioavailability as anti-nutrient levels have a high impact on the nutritional value of a breeding product.

A review of breeding activities and releases demonstrates that ample work has been done and is in progress in all stages of development. Several lines in several crops have been and are being developed, released and disseminated; studies confirm their adoption and positive effects. Biofortification breeding has successfully produced superior lines, however; negative correlations with yield have to be overcome during the breeding process. Transgenic or CRISPR/Cas9 strategies may allow big leaps in increasing bioavailable micromineral levels if regulatory hurdles can be resolved. Overall the research field is very active – biofortified varieties are being produced and fielded. With bioefficacy studies confirming the positive effect of biofortified germplasm on the health status of consumers, positive impact on improving the health status of vulnerable populations is achieved; more can be expected in future.

9 Where to look for further information

Biofortification has found a lot of attention in recent publications. Two books have lately been published specifically on the biofortification topic. The book on biofortification by De Gara et al. (2014) harbours many chapters focusing on reviewing molecular mechanisms and gene functions, whereas a book by Singh et al. (2016b) contains several chapters with more focus on legume crops. A very detailed information source on human nutritional needs is available in publications FAO/WHO (1998); Dwivedi et al. (2012) published a detailed review on nutrition of staple crops. Bouis and Welch (2010) provided a good overview on biofortification and strategies, and a survey of techniques and breeding status in several crops was made available by Saltzman et al. (2016). Singh (2016) reviewed biofortification in legumes, with a focus on bioavailability, and Bohra et al. (2015) specifically addressed genetics and genomics in legume biofortification. More crop-specific reviews are available from Blair (2013) focusing on common bean and Nair et al. (2013) on mung bean. Detailed information on the anti-nutrient phytate and strategies for improvement were presented by Gupta et al. (2015). Moura et al. (2014) gave an overview of efficacy and effectiveness trials, showing the positive impact of biofortified crops; consumer acceptance of several biofortified crops was discussed by Birol et al. (2015).

The CGIAR Research Program on Agriculture for Nutrition and Health (A4NH) emphasizes biofortification as one of the main components, next to value chains for enhanced nutrition, prevention and control of agriculture-associated diseases and integrated agriculture, nutrition, and health programmes and policies (http://www.cgiar.org/about-us/our-programs/cgiar-research-program-on-agriculture-for-nutrition-and-health/). HarvestPlus became a component of A4NH upon its establishment.

10 References

Abhary, M., Siritunga, D., Stevens, G., et al. (2011). Transgenic biofortification of the starchy staple cassava (Manihot esculenta) generates a novel sink for protein. *PLoS ONE*. doi:10.1371/journal.pone.0016256

Aldemir, S. B., Sever, T., Ates, D., et al. (2014). QTL Mapping of Genes Controlling to Fe Uptake in Lentil Seed (Lens culinaris L.) Using Recombinant Inbred Lines. *Plant Anim. Genome*.

Allen, L., de Benoist, B., Dary, O. and Hurrell, R. (2006). Guidelines on food fortification with micronutrients.

Amarakoon, D., Thavarajah, D., McPhee, K. and Thavarajah, P. (2012). Iron-, zinc-, and magnesium-rich field peas (Pisum sativum L.) with naturally low phytic acid: A potential food-based solution to global micronutrient malnutrition. *J. Food Compos. Anal.* 27:8–13. doi:10.1016/j.jfca.2012.05.007

Ariza-Nieto, M., Blair, M. W., Welch, R. M. and Glahn, R. P. (2007). Screening of iron bioavailability patterns in eight bean (Phaseolus vulgaris L.) genotypes using the Caco-2 cell in vitro model. *J Agric. Food Chem.* 55:7950–6. doi:10.1021/jf070023y

Asare-Marfo, D., Herrington, C., Alwang, J., et al. (2016). Assessing the Adoption of High Iron Bean Varieties and Their Impact on Iron Intakes and Other Livelihood Outcomes in Rwanda.

Ates, D., Aldemir, S. B., Sever, T., et al. (2014). Identification of QTLs controlling genes to Mn and Zn uptake in lentil seed. *Plant Anim. Genome*.

Baloch, F. S., Karaköy, T., Demirbaş, A., et al. (2014). Variation of some seed mineral contents in open pollinated faba bean (Vicia faba L.) landraces from Turkey. *Turkish J. Agric. For.* 38:591–602. doi:10.3906/tar-1311-31

Basavarajeshwari, R., Rajashekar Reddy, B. H., Iranacolaivalu, M., et al. (2014). Development of mapping population for grain zinc content in pigeonpea. *Agric. Hortic. Sci.* 2:1–16.

Beebe, S., Gonzalez, A. V. and Rengifo, J. (2000). Research on trace minerals in the common bean. *Food Nutr. Bull.* 21:387–91.

Beebe, S. E. (2012). Common bean breeding in the tropics. *Plant Breed. Rev.* 36:357–426. doi:10.1002/9781118358566

Birol, E., Meenakshi, J. V., Oparinde, A., et al. (2015). Developing country consumers??? acceptance of biofortified foods: a synthesis. *Food Secur.* 7:555–68. doi:10.1007/s12571-015-0464-7

Blair, M. W. (2013). Mineral biofortification strategies for food staples: The example of common bean. *J. Agric. Food Chem.* 61:8287–94. doi:10.1021/jf400774y

Blair, M. W., Astudillo, C., Grusak, M. A., et al. (2009a). Inheritance of seed iron and zinc concentrations in common bean (Phaseolus vulgaris L.). *Mol. Breed.* 23:197–207. doi:10.1007/s11032-008-9225-z

Blair, M. W., Astudillo, C., Rengifo, J., et al. (2011). QTL analyses for seed iron and zinc concentrations in an intra-genepool population of Andean common beans (Phaseolus vulgaris L.). *Theor. Appl. Genet.* 122:511–21. doi:10.1007/s00122-010-1465-8

Blair, M. W., Herrera, A. L., Sandoval, T. A., et al. (2012). Inheritance of seed phytate and phosphorus levels in common bean (Phaseolus vulgaris L.) and association with newly-mapped candidate genes. *Mol. Breed.* 30:1265–77. doi:10.1007/s11032-012-9713-z

Blair, M. W. and Izquierdo, P. (2012). Use of the advanced backcross-QTL method to transfer seed mineral accumulation nutrition traits from wild to Andean cultivated common beans. *Theor. Appl. Genet.* 125:1015–31. doi:10.1007/s00122-012-1891-x

Blair, M. W., Izquierdo, P., Astudillo, C. and Grusak, M. A. (2013). A legume biofortification quandary: variability and genetic control of seed coat micronutrient accumulation in common beans. *Front Plant Sci.* 4:275. doi:10.3389/fpls.2013.00275

Blair, M. W., Knewtson, S. J., Astudillo, C., et al. (2010a). Variation and inheritance of iron reductase activity in the roots of common bean (Phaseolus vulgaris L.) and association with seed iron accumulation QTL. *BMC Plant Biol.* 10:215. doi:10.1186/1471-2229-10-215

Blair, M. W., Medina, J. I., Astudillo, C., et al. (2010b). QTL for seed iron and zinc concentration and content in a Mesoamerican common bean (Phaseolus vulgaris L.) population. *Theor. Appl. Genet.* 121:1059–70. doi:10.1007/s00122-010-1371-0

Blair, M. W., Monserrate, F., Beebe, S. E., et al. (2010c). Registration of High Mineral Common Bean Germplasm Lines NUA35 and NUA56 from the Red-Mottled Seed Class. *J. Plant Regist.* 4:55. doi:10.3198/jpr2008.09.0562crg

Blair, M. W., Sandoval, T. A., Caldas, G. V., et al. (2009b). Quantitative trait locus analysis of seed phosphorus and seed phytate content in a recombinant inbred line population of common bean. *Crop Sci.* 49:237. doi:10.2135/cropsci2008.05.0246

Bohra, A., Sahrawat, K. L., Kumar, S., et al. (2015). Genetics- and genomics-based interventions for nutritional enhancement of grain legume crops: status and outlook. *J. Appl. Genet.* 56:151–61. doi:10.1007/s13353-014-0268-z

Bouis, H. E. (1999). Economics of enhanced micronutrient density in food staples. *Field Crop Res.* 60:165–73.

Bouis, H. E. (2000). Enrichment of food staples through plant breeding: a new strategy for fighting micronutrient malnutrition. *Nutrition* 16:701–4. doi:10.1016/S0899-9007(00)00266-5

Bouis, H. E. (2003). Micronutrient fortification of plants through plant breeding: can it improve nutrition in man at low cost? Proceedings of the Nutritional Society, 62, 403–11.

Bouis, H. E., Hotz, C., McClafferty, B., et al. (2011). Biofortification: A new tool to reduce micronutrient malnutrition. *Food Nutr. Bull.* 32:31–40. doi:10.1177/15648265110321S105

Bouis, H. E. and Welch, R. M. (2010). Biofortification – A sustainable agricultural strategy for reducing micronutrient malnutrition in the global south. *Crop Sci*, 50:S-20-S-32. doi:10.2135/cropsci2009.09.0531

Boukar, O., Massawe, F., Muranaka, S., et al. (2011). Evaluation of cowpea germplasm lines for protein and mineral concentrations in grains. *Plant Genet. Resour. FirstView*:1–8. doi:10.1017/S1479262111000815

Caldas, G. V. and Blair, M. W. (2009). Inheritance of seed condensed tannins and their relationship with seed-coat color and pattern genes in common bean (Phaseolus vulgaris L.). *Theor. Appl. Genet.* 119:131–42. doi:10.1007/s00122-009-1023-4

Campion, B., Glahn, R. P., Tava, A., et al. (2013). Genetic reduction of antinutrients in common bean (Phaseolus vulgaris L.) seed, increases nutrients and in vitro iron bioavailability without depressing main agronomic traits. *Field Crop Res.* 141:27–37. doi:10.1016/j.fcr.2012.10.015

Campion, B., Sparvoli, F., Doria, E., et al. (2009). Isolation and characterisation of an lpa (low phytic acid) mutant in common bean (Phaseolus vulgaris L.). *Theor. Appl. Genet.* 118:1211–21. doi:10.1007/s00122-009-0975-8

Chiaiese, P., Ohkama-Ohtsu, N., Molvig, L., et al. (2004). Sulphur and nitrogen nutrition influence the response of chickpea seeds to an added, transgenic sink for organic sulphur. *J. Exp. Bot.* 55:1889–1901. doi:10.1093/jxb/erh198

CIAT (2016) High-iron beans: a new 'superfood' for Colombia | CIAT Blog. http://blog.ciat.cgiar.org/high-iron-beans-a-new-superfood-for-colombia/ Accessed 9 September 2016.

Cichy, K. A., Caldas, G. V., Snapp, S. S. and Blair, M. W. (2009). QTL analysis of seed iron, zinc, and phosphorus levels in an Andean bean population. *Crop Sci.* 49:1742–50. doi:10.2135/cropsci2008.10.0605

DellaValle, D. M., Thavarajah, D., Thavarajah, P., et al. (2013). Lentil (Lens culinaris L.) as a candidate crop for iron biofortification: Is there genetic potential for iron bioavailability? *Field Crop Res.* 144:119–25. doi:10.1016/j.fcr.2013.01.002

Diapari, M., Sindhu, A., Bett, K., et al. (2014). Genetic diversity and association mapping of iron and zinc concentrations in chickpea (Cicer arietinum L.). *Genome* 57:459–68. doi:http://dx.doi.org/10.1139/gen-2014-0108

Díaz, A. M., Caldas, G. V. and Blair, M. W. (2010). Concentrations of condensed tannins and anthocyanins in common bean seed coats. *Food Res. Int.* 43:595–601. doi:10.1016/j.foodres.2009.07.014

Díaz de la Garza, R. I., Gregory, J. F. and Hanson, A. D. (2007). Folate biofortification of tomato fruit. *Proc. Natl. Acad. Sci. U S A* 104:4218–22. doi:10.1073/pnas.0700409104

Drakakaki, G., Marcel, S., Glahn, R. P., et al. (2005). Endosperm-specific co-expression of recombinant soybean ferritin and Aspergillus phytase in maize results in significant increases in the levels of bioavailable iron. *Plant Mol. Biol.* 59:869–80. doi:10.1007/s11103-005-1537-3

Dwivedi, S. L., Sahrawat, K. L., Rai, K. N., et al. (2012). Nutritionally enhanced staple food crops. *Plant Breed. Rev.* 36:169–291. doi:10.1002/9781118358566

Falco, S. C., Guida, T., Locke, M., et al. (1995). Transgenic canola and soybean seeds with increased lysine. *Biotechnology* (N Y) 13:577–82.

FAO/WHO (1998). Vitamin and mineral requirements in human nutrition.

FAO, WHO (2001). Human Vitamin and Mineral Requirements.

Gelin, J. R., Forster, S., Grafton, K. F., et al. (2007). Analysis of seed zinc and other minerals in a recombinant inbred population of navy bean (Phaseolus vulgaris L.). *Crop Sci.* 47:1361–6. doi:10.2135/cropsci2006.08.0510

Glahn, R. P., Lee, O. A., Yeung, A., et al. (1998). Caco-2 cell ferritin formation predicts nonradiolabeled food iron availability in an in vitro digestion/Caco-2 cell culture model. *J. Nutr.* 128:1555–61.

Graham, R. D., Welch, R. M. and Bouis, H. E. (2001). Addressing micronutrient malnutrition through enhancing the nutritional quality of staple foods: Principles, perspectives and knowledge gaps. *Adv. Agron.* 70:77–142.

Gupta, R. K., Gangoliya, S. S. and Singh, N. K. (2015). Reduction of phytic acid and enhancement of bioavailable micronutrients in food grains. *J. Food Sci. Technol.* 52:676–84. doi:10.1007/s13197-013-0978-y

Guzman-Maldonado, S. H., Martinez, O., Acosta-Gallegos, J. A., et al. (2003). Putative quantitative trait loci for physical and chemical components of common bean. *Crop Sci.* 43:1029–35. doi:10.2135/cropsci2003.1029

Haas, J. D., Luna, S. V., Lung'aho, M. G., et al. (2016). Consuming iron biofortified beans increases iron status in Rwandan women after 128 days in a randomized controlled feeding trial. *J. Nutr.* doi:10.3945/jn.115.224741

HarvestPlus (2014a). HarvestPlus 2014 Annual Report. http://www.harvestplus.org/content/harvestplus-2014-annual-report. Accessed 22 October 2015.

HarvestPlus (2014b). Biofortification Progress Briefs. 82.

Hirschi KD (2009). Nutrient biofortification of food crops. *Annu. Rev. Nutr.* 29:401–21. doi:10.1146/annurev-nutr-080508-141143

Khalid, S., Asghar, H. N., Akhtar, M. J., et al. (2015). Biofortification of iron in chickpea by plant growth promoting rhizobacteria. *Pakistan J. Bot.* 47:1191–4.

Klein, M. A. and Grusak, M. A. (2009). Identification of nutrient and physical seed trait QTL in the model legume Lotus japonicus. *Genome* 52:677–91. doi:10.1139/G09-039

Klein, M. A., López-Millán, A. F. and Grusak, M. A. (2012). Quantitative trait locus analysis of root ferric reductase activity and leaf chlorosis in the model legume, Lotus japonicus. *Plant Soil* 351:363–76. doi:10.1007/s11104-011-0972-y

Kumar, A. (2012). Phosphorus and Rain-Harvested Water Economy Through Vesicular Arbuscular Mycorrhizae (VAM) In OKRA-PEA Sequence. http://shodhganga.inflibnet.ac.in/handle/10603/9677

Kumar, A., Choudhary, A. K., Pooniya, V., et al. (2016a). Soil factors associated with micronutrient acquisition in crops- biofortification perspective. In *Biofortification of Food Crops.* Springer India, New Delhi, pp. 159–76.

Kumar, J., Kant, R., Quatadah, S. M., et al. (2016b). Biofortification through breeding interventions in lentil. In Singh, U., Praharaj, C. S., Singh, S. S. and Singh, N. P. (eds), *Biofortification of Food Crops*. Springer India, New Delhi, pp. 147–56.
Liu, X., Glahn, R. P., Arganosa, G. C. and Warkentin, T. D. (2015). Iron bioavailability in low phytate pea. *Crop Sci.* 55:320. doi:10.2135/cropsci2014.06.0412
Mba, C., Guimaraes, E. P. and Ghosh, K. (2012). Re-orienting crop improvement for the changing climatic conditions of the 21st century. *Agric. Food Secur.* 1:7. doi:10.1186/2048-7010-1-7
Mishra, P. K., Bisht, S. C., Jeevanandan, K., et al. (2014). Synergistic effect of inoculating plant growth-promoting Pseudomonas spp. and Rhizobium leguminosarum -FB1 on growth and nutrient uptake of rajmash (Phaseolus vulgaris L.). *Arch. Agron. Soil Sci.* 60:799–815. doi:10.1080/0365 0340.2013.843773
Mishra, P. K., Bisht, S. C., Mishra, S., et al. (2012). coinoculation of rhizobium leguminosarum -pr1 with a cold tolerant pseudomonas sp. Improves iron acquisition, nutrient uptake and growth of field pea (Pisum Sativum L.). *J. Plant Nutr.* 35:243–56. doi:10.1080/01904167.2012.636127
Molvig, L., Tabe, L. M., Eggum, B. O., et al. (1997). Enhanced methionine levels and increased nutritive value of seeds of transgenic lupins (Lupinus angustifolius L.) expressing a sunflower seed albumin gene. *Proc. Natl. Acad. Sci U S A* 94:8393–8. doi:10.1073/pnas.94.16.8393
Moura, F. F., De Palmer, A. C., Finkelstein, J. L., et al. (2014). Are biofortified staple food crops improving vitamin A and iron status in women and children ? New evidence from Efficacy trials. *Adv. Nutr.* 568–70. doi:10.3945/an.114.006627.568
Murgia, I., De Gara, L. and Grusak, M. A. (2014). Biofortification: how can we exploit plant science and biotechnology to reduce micronutrient deficiencies? *Front. Plant Sci.* https://doi.org/10.3389/fpls.2013.00429
Nair, R. M., Yang, R. Y., Easdown, W. J., et al. (2013). Biofortification of mungbean (Vigna radiata) as a whole food to enhance human health. *J. Sci. Food Agric.* 93:1805–13. doi:10.1002/jsfa.6110
Nestel, P., Bouis, H. E., Meenakshi, J. V. and Pfeiffer, W. (2006). Biofortification of staple food crops. *J. Nutr.* 136:1064–7.
Norton, G. J., Douglas, A., Lahner, B., et al. (2014). Genome wide association mapping of grain arsenic, copper, molybdenum and zinc in rice (Oryza sativa L.) grown at four international field sites. *PLoS ONE*. doi:10.1371/journal.pone.0089685
Nunes, A. C. S., Kalkmann, D. C. and Aragão, F. J. L. (2009). Folate biofortification of lettuce by expression of a codon optimized chicken GTP cyclohydrolase I gene. *Transgenic Res.* 18:661–7. doi:10.1007/s11248-009-9256-1
PABRA (2016). PABRA Database (Pan African Bean Research Alliance). http://database.pabra-africa.org/. Accessed 22 September 2016.
PABRA-Africa (2016). PABRA-Africa. http://www.pabra-africa.org/. Accessed 4 November 2016.
Paine, J. A., Shipton, C. A., Chaggar, S., et al. (2005). Improving the nutritional value of Golden Rice through increased pro-vitamin A content. *Nat. Biotechnol.* 23:482–7. doi:10.1038/nbt1082
Panzeri, D., Cassani, E., Doria, E., et al. (2011). A defective ABC transporter of the MRP family, responsible for the bean lpa1 mutation, affects the regulation of the phytic acid pathway, reduces seed myo -inositol and alters ABA sensitivity. *New Phytol.* 191(1):70–83. doi: 10.1111/j.1469-8137.2011.03666.x. Epub 2011 Mar 11.
Petry, N., Egli, I., Campion, B., et al. (2013). Genetic reduction of phytate in common bean (Phaseolus vulgaris L .) seeds increases iron absorption in young women 1–4. *J. Nutr.* 143(8):1219–24. doi:10.3945/jn.113.175067.
Petry, N, Egli, I., Gahutu, J. B., et al. (2014). Phytic acid concentration influences iron bioavailability from biofortified beans in rwandese women with low iron status 1, 2. 1–8. *J. Nutr.* doi:10.3945/jn.114.192989.1
Petry, N., Rohner, F., Gahutu, J. B., et al. (2016). In rwandese women with low iron status, iron absorption from low-phytic acid beans and biofortified beans is comparable, but low-phytic acid beans cause adverse gastrointestinal symptoms. *J. Nutr.* 146:970–5. doi:10.3945/jn.115.223693
Pfeiffer, W. H. and McClafferty, B. (2007). HarvestPlus: Breeding crops for better nutrition. *Crop Sci.* doi:10.2135/cropsci2007.09.0020IPBS

Poonya, V. I. and Shivay, Y. S. (2013). Enrichment of basmati rice grain and straw with zinc and nitrogen through ferti-fortification and summer green manuring under Indo-Gangetic plains of India. *J. Plant Nutr.* 36:91–117. doi:10.1080/01904167.2012.733052

Prasanna, R., Triveni, S., Bidyarani, N., et al. (2014). Evaluating the efficacy of cyanobacterial formulations and biofilmed inoculants for leguminous crops. *Arch. Agron. Soil Sci.* 60:349–66. doi:10.1080/03650340.2013.792407

Ram, H., Rashid, A., Zhang, W., et al. (2016). Biofortification of wheat, rice and common bean by applying foliar zinc fertilizer along with pesticides in seven countries. *Plant Soil* 1–13. doi:10.1007/s11104-016-2815-3

Rana, A., Joshi, M., Prasanna, R., et al. (2012). Biofortification of wheat through inoculation of plant growth promoting rhizobacteria and cyanobacteria. *Eur. J. Soil Biol.* 50:118–26. doi:10.1016/j.ejsobi.2012.01.005

Ray, H., Bett, K., Tar'an, B., et al. (2014). Mineral micronutrient content of cultivars of field pea, chickpea, common bean, and lentil grown in Saskatchewan, Canada. *Crop Sci.* 54:1698–1708. doi:10.2135/cropsci2013.08.0568

Saltzman, A., Andersson, M. S., Asare-Marfo, D., et al. (2016). Biofortification techniques to improve food security. *Ref. Modul. Food Sci.* 1:1–9.

Saltzman, A., Birol, E., Oparinde, A., et al. (2017). Availability, production, and consumption of crops biofortified by plant breeding : current evidence and future potential. *Ann. N Y Acad. Sci.* doi:10.1111/nyas.13314

Sankaran, R. P., Huguet, T., Grusak, M. A. (2009). Identification of QTL affecting seed mineral concentrations and content in the model legume Medicago truncatula. *Theor. Appl. Genet.* 119:241–53. doi:10.1007/s00122-009-1033-2

Santos, C. A. F. and Boiteux, L. S. (2015). Genetic control and transgressive segregation of zinc, Iron, Potassium, Phosphorus, Calcium, And sodium accumulation in cowpea (Vigna unguiculata) seeds. *Genet. Mol. Res.* 14:259–68. doi:10.4238/2015.January.16.10

Santos, C. A. F. and Boiteux, L. S. (2013). Breeding biofortifed cowpea lines for semi-arid tropical areas by combining higher seed protein and mineral levels. *Genet. Mol. Res.* 12:6782–9. doi:10.4238/2013.December.16.4

Sharma, A., Johri, B., Sharma, A. and Glick, B. (2003). Plant growth-promoting bacterium Pseudomonas sp. strain GRP3 influences iron acquisition in mung bean (Vigna radiata L. Wilzeck). *Soil Biol. Biochem.* 35:887–94. doi:10.1016/S0038-0717(03)00119-6

Shivay, Y. S., Prasad, R. and Pal, M. (2014). Effect of variety and zinc application on yield, profitability, protein content and zinc and nitrogen uptake by chickpea (Cicer arietinum). *Indian J. of Agron.* 59(2):317–21. Print ISSN : 0537-197X. Online ISSN : 0974-4460.

Shunmugam, A. S., Liu, X., Stonehouse, R., et al. (2015). Mapping seed phytic acid concentration and iron bioavailability in a pea recombinant inbred line population. *Crop Sci.* 55:828–36. doi:10.2135/cropsci2014.08.0544

Singh, J. (2016). Biofortification of food legumes and bioavailability of nutrients. In Singh, U., Praharaj, C. S., Singh, S. S. and Singh, N. P. (eds). *Biofortification of Food Crops*. Springer India, New Delhi, pp. 51–60.

Singh, R., van Heusden, A. W., Kumar, R., et al. (2013). Genetic diversity of mungbean (Vigna radiata L.) in iron and zinc content as impacted by farmers' varietal selection in Northern India. *Ecol. Food Nutr.* 52:148–62. doi:10.1080/03670244.2012.706006

Singh, U., Praharaj, C. S., Chaturvedi, S. K. and Bohra, A. (2016a). Biofortification: Introduction, approaches, limitations, and challenges. In Singh, U., Praharaj, C. S., Singh, S. S. and Singh, N. P. (eds), *Biofortification of Food Crops*. Springer India, New Delhi, pp. 3–18.

Singh, U., Praharaj, C. S., Singh, S. S. and Singh, N. P. (2016b). *Biofortification of Food Crops*, 1st edn. Springer India, New Delhi; Heidelberg, New York, Dordrecht and London.

Singh Sindhu, A., Zheng, Z. and Murai, N. (1997). The pea seed storage protein legumin was synthesized, processed, and accumulated stably in transgenic rice endosperm. *Plant Sci.* 130:189–96. doi:10.1016/S0168-9452(97)00219-7

Sompong, U., Kaewprasit, C., Nakasathien, S. and Srinives, P. (2009). Inheritance of seed phytate in mungbean (Vigna radiata (L.) Wilczek). *Euphytica* 171:389–96. doi:10.1007/s10681-009-0053-y

Sompong, U., Somta, P., Raboy, V. and Srinives, P. (2012). Mapping of quantitative trait loci for phytic acid and phosphorus contents in seed and seedling of mungbean (Vigna radiata (L.) Wilczek). *Breed Sci.* 62:87–92. doi:10.1270/jsbbs.62.87

Song, G., Jia, M., Chen, K., et al. (2016). CRISPR/Cas9: A powerful tool for crop genome editing. *Crop J.* 4:75–82. doi:10.1016/j.cj.2015.12.002

Storozhenko, S., De Brouwer, V., Volckaert, M., et al. (2007). Folate fortification of rice by metabolic engineering. *Nat. Biotechnol.* 25:1277–9. doi:10.1038/nbt1351

Takahashi, M., Nakanishi, H., Kawasaki, S., et al. (2001). Enhanced tolerance of rice to low iron availability in alkaline soils using barley nicotianamine aminotransferase genes. *Nat. Biotechnol.* 19:466–9. doi:10.1038/88143

Tako, E., Blair, M. W., Glahn, R. P., et al. (2011). Biofortified red mottled beans (Phaseolus vulgaris L.) in a maize and bean diet provide more bioavailable iron than standard red mottled beans: Studies in poultry (Gallus gallus) and an in vitro digestion/Caco-2 model. *Nutr. J.* 10:113. doi:10.1186/1475-2891-10-113

Taunk, J. and Yadav, N. (2012). Genetic diversity among greengram genotypes varying in micronutrient(Fe and Zn) content using RAPD markers. *Indian J.* 11:48–53.

Tofiño-Rivera, A. P., Pastrana-vargas, I. J., Melo-ríos, A. E., et al. (2016). Rendimiento, estabilidad fenotípica y contenido de micronutrientes de fríjol biofortificado en el caribe seco colombiano. Yield, Phenotypical Stability and Micronutrients Contents in the Biofortified Bean in the Colombian Sub-humid Caribbean. *Corpoica Cienc Tecnol Agropecuaria* 17(3):309–29.

Tulchinsky, T. H. (2010). Micronutrient deficiency conditions : Global health issues. *Public Health* 32:1–255.

Vasconcelos, M., Eckert, H., Arahana, V., et al. (2006). Molecular and phenotypic characterization of transgenic soybean expressing the Arabidopsis ferric chelate reductase gene, FRO2. *Planta* 224:1116–28. doi:10.1007/s00425-006-0293-1

Warkentin, T. D., Delgerjav, O., Arganosa, G., et al. (2012). Development and characterization of low-phytate pea. *Crop Sci.* 52:74–8. doi:10.2135/cropsci2011.05.0285

Welch, R. M. and Graham, R. D. (1999). A new paradigm for world agriculture: Meeting human needs. Productive, sustainable, nutritious. *Filed Crop Res.* 60:1–10. doi:10.1016/S0378-4290(98)00129-4

Welch, R. M., House, W. A., Beebe, S. and Cheng, Z. (2000). Genetic selection for enhanced bioavailable levels of iron in bean (Phaseolus vulgaris L.) seeds. *J. Agric. Food Chem.* 48:3576–80. doi:10.1021/jf0000981

White, P. J. and Broadley, M. R. (2009). Biofortification of crops with seven mineral elements often lacking in human diets – iron, zinc, copper, calcium, magnesium, selenium and iodine. *New Phytol.* 182:49–84. doi:10.1111/j.1469-8137.2008.02738.x

WHO (2000). The role of food fortification in the control of micronutrient malnutrition. In *Geneva World Heal.* Organ.

Wolt, J. D., Wang, K. and Yang, B. (2016). The regulatory status of genome-edited crops. *Plant Biotechnol. J.* 14:510–18. doi:10.1111/pbi.12444

Ye. X., Al-Babili. S., A. Klöti, et al. (2000). Engineering the provitamin A (Beta-Carotene) biosynthetic pathway into (Carotenoid-Free) rice. *Science* 287:303. doi:10.1126/science.287.5451.303.

Part 2

Cultivation

Chapter 9

Variety selection and seed quality management in grain legume cultivation

Jean Claude Rubyogo, Pan-Africa Bean Research Alliance (PABRA), International Center for Tropical Agriculture, Tanzania; and Wilfred Odhiambo, Seed Systems Consultant, Kenya

1 Introduction

1.1 Variety selection: key issues and challenges

Smallholder farmers in sub-Saharan Africa (SSA) grow grain legumes. The majority of smallholder farmers farm an average of less than 2 ha of landholding (Buruchara et al. 2011; Belmain et al. 2013; CTA 2014) and rely on non-mechanized farming and family labour (FAO and ICRISAT 2015). As short-duration crops, legumes shorten hunger periods for farming households and provide a wide range of useful products for human and livestock consumption (Popelka et al. 2004; Buruchara et al. 2011). Legume leaves and green pods are consumed as vegetables in many communities while the dry grains are traded for cash and non-cash items, cooked and eaten as complements to other staples such as maize, with numerous health benefits (Venn 2004; Bouchenak and Lamri-Senhadji 2013; Rebello et al. 2014; Odhiambo et al. 2016), or processed into various products including canned and pre-cooked beans and sold as ready-to-eat products. Legume residues or by-products are used as feed for livestock as a way of improving the quality of grass-dominated feeding systems (Valbuena et al. 2012). Legumes also contribute to the sustainability of cropping systems through nitrogen fixation, hence reducing the need for the application of inorganic fertilizers (Odhiambo et al. 2016) and providing other crop rotation-related benefits especially in mixed cropping systems which are the dominant

http://dx.doi.org/10.19103/AS.2017.0023.07

systems used by smallholders (Lupwayi et al. 2011). These factors not only place legumes at the centre of food/nutrition, income security and agro-ecological health but also inform the decisions on variety selection by farmers.

The various uses of legume systems is a result of variation in the level of market demand for legumes, farmers' cropping systems, household use and those within the household who make decisions about production, marketing of produce or consumption of legumes. Consequently, legume variety selection depends on a range of factors including market, household consumption, agro-ecological adaptation, post-harvest aspects and gender.

As the market becomes more differentiated, preferences for various types of legumes become diversified based on factors such as palatability, grain shape, colour and size and cooking time, among others. An increase in demand for a certain legume type with preferred market factors would trigger the selection of such a variety by farmers in response to the market demand and household consumption needs. Previous studies have used a stated preference elicitation method to estimate the trade-offs acceptable to farmers and the associated gains to farmers if such attributes were to be changed (Katungi et al. 2015). Many smallholder farmers grow legumes for household food/nutrition security and increasingly for income and would prefer varieties that mature early and have broader, consumable leaves, fresh pods and grains, in order to shorten the 'hunger period' and ensure food security.

Gender-based choice experiment studies (Ogato et al. 2009) indicated that men and women elicited different preferences for different legume attributes such as yield, cooking time and early maturity (Katungi et al. 2015). Higher-yielding legume varieties were more preferred by men than by women, because marketing decisions were made by men and yield levels influenced the profitability of legumes as an enterprise (Katungi et al. 2011). Women-preferred traits (such as cooking time) played a critical role in variety selection. Easy-to-cook varieties were preferred to hard-to-cook types (Kinyanjui et al. 2014) affording women extra time for other activities as well as saving on fuel costs. Cooking time thus influenced legumes utilization with implications for gender equity, nutritional diets and energy utilization (Cichy et al. 2015). As stated by David et al. (2000), some other traits were gender neutral. For instance, early maturing/short growth periods were gender neutral as these traits contributed to shortening the hunger period (which was important to women) and reducing cash/income gaps (which was important to both men and women). Post-cooking storability was also critical because this reduced the households' frequency of cooking the legumes while saving costs in terms of energy and time. Legume varieties had to be adapted to farmers' agro-ecological conditions in terms of the seasonal growth period, tolerance to disease/pests/moisture stress and poor soils, use for the market and household food security. Finally, the gender aspect is critical as household decisions on whether to grow or sell legumes and the proportion to be sold or consumed are made by either women or men.

1.2 Quality management: key issues and challenges

Farmers depend on seed as a critical input to crop production in general and legumes in particular cannot be overemphasized (CTA 2014). While not underrating the vital roles of other components of agriculture including markets, credit supply and support institutions, the use of appropriate and good-quality seed is a major starting point in crop production (Pelmer 2005). Seed quality management is the process through which quality control is carried out by mandated institutions (e.g. national seed authorities) to ensure that set standards are met (Matthews et al. 2012) or farmers' perceived desirable seed attributes or characteristics are maintained or enhanced (Odhiambo et al. 2016) through farm-based seed quality control practices (Rubyogo et al. 2010a). The seed and varietal quality influences legume productivity and adaptability due to variations in agro-ecologies (Belmain et al. 2013), and is defined by multiple constraints including abiotic, for example, drought, excess moisture and biotic, for example, diseases and pests (FAO and ICRISAT 2015).

2 Seed quality

2.1 Mechanisms of seed quality control

Traditionally, seed quality has been associated with either formal or informal seed sources. The formal seed source result in certified seeds of released legume varieties through certification processes (Odhiambo et al. 2016) and guarantees a clear distinction between seed and grain as well as maintaining varietal identity and purity, with focus being on select released varieties (David and Sperling 1999). The informal seed source is unstructured and embedded in farmers' cropping systems and local grain markets (Teshale et al. 2006; Rubyogo et al. 2007). The seed of local landraces and improved varieties are recycled and maintained by farmers through continuous selection of varieties with preferred traits, which are used in the next growing seasons. Quality control is vested in the farmers (Trutmann et al. 1996). While there has been a distinction between formal and informal seed sources, studies have indicated that there is minimal difference in seed quality from these two sources (CIAT 1991; Otsyula et al. 2004; Rubyogo et al. 2010a), thus affirming farmers' active role in seed quality control (Buruchara and David 1994; Odhiambo et al. 2016).

2.2 Seed quality parameters

In legume breeding, considerable efforts are channelled towards genotypic development and varietal attributes, such as yield level and days to maturity are a proxy for relative varietal superiority (Beebe 2012). Achieving high productivity particularly increased yield is a major challenge in legume production. This requires application of complementary crop management practices to catalyse legumes' reproductive traits in order to attain higher yield levels. Seed yield is a critical factor in market access of grain legumes. However, the commercialization of market-demanded and farmer-preferred improved legume varieties depends on a sustainable supply of affordable, high-quality seeds (Boelt et al. 2015). The absence of functional systems that sustainably supply legumes seed in convenient and affordable quantities is viewed as a major factor in the limited use of improved legume varieties, especially among marginalized smallholder farmers. The success with which a new legume crop is established depends mainly on the physical, pathological, physiological and genetic (varietal) quality parameters (FAO 2006) and more importantly, the seed vigour (Hampton et al. 2013).

Physical quality is characterized by (i) minimum damaged seed – this includes broken, cracked or shrivelled seed or seed with weevils. Insect (weevil)-damaged seed increases the chances of further attacks by microorganisms and have lower germination rates; (ii) absence of weed seed or inert matter – good-quality seed should be free of noxious weed seeds, chaff, stones and dirt; (iii) the absence of diseased seed – discoloured or stained seed may host organisms that attack the seed at germination and spread to other plants; (iv) uniformity of seed size – small seed size normally reduces germination and growth vigour.

Physiological quality entails germination and vigour. High germination enables the seed to emerge and produce a plant under normal field conditions. The capacity of the seed to emerge and survive in potentially harsh conditions is enhanced by its growth vigour. Seed vigour precedes germination hence seed with high germination may exhibit low growth vigour.

Genetic quality refers to those features that differentiate the distinct kinds of a particular crop, otherwise known as varieties or cultivars. These features must be reproducible and consistently similar when the varieties are used in production over generations. A cultivar is a collection of cultivated plants distinguishable by morphological, physiological and other traits, which are retained when reproduced.

Phytosanitary quality includes seed health attributes assessed through determining the presence or absence of pathogens such as bacteria, viruses, fungi and pests, including insects and nematodes.

The complexity of user/farmer systems and differentiated markets determines seed quality requirements by users and hence the variations in seed source preferences. For instance, generally large-scale, commercial legume farmers prefer to obtain seed from formal sources, given the perceived high seed quality associated with these sources. Smallholder farmers mainly source legume seed from informal sources because of its affordability and proximity of outlets (Rubyogo et al. 2010a). While certain quality parameters, such as the level of infestation by insects, are considered to be critical across formal and informal sources, others are often overlooked, especially when farmers obtain seed from informal sources. A study conducted in western Kenya revealed that while 97% of smallholder farmers obtained seed from informal sources, nearly half of the farmers considered that insect-free seeds had the highest quality. In contrast, the presence of chaff and the seed colour were not considered to be factors that determined seed quality (Odhiambo et al. 2016).

3 Addressing key issues and challenges in variety selection and quality control

3.1 Participatory varietal selection

Participatory varietal selection (PVS) has offered an effective alternative for fast adoption and wider distribution of improved legumes varieties. While farmers are afforded the opportunity to assess large numbers of test varieties and evaluate them under local conditions using their usual management practices, traders and other actors in the legumes value chain can select market-demanded varieties to inform production decisions. Notwithstanding the genetic potential of the different varieties, actors can make observational judgements based on their preferred traits. From such PVS, participating farmers obtain the novel varieties after harvest, hence increasing the number of crop varieties in their prior sets (Dorward et al. 2007; Aw-Hassan et al. 2008). The PVS approach has played an important role in enhancing poor farmers' resilience, especially women, to micro-ecological, niche-adapted bean varieties that are locally preferred by farmers

(Almekinders et al. 2007). The actors along the value chain act interdependently. For instance, traders are crucial in building farmers' overall market intelligence (i.e. which varieties to promote, which markets to target) and help farmers understand the range of acceptance for specific market-needed traits (e.g. colour, size, etc.). They influence farmers' marketing/organizing skills, help to make farmers aware of seed quality needs and disseminate new varieties quickly. Farmers are increasingly becoming engaged in variety release committees and are accessing seed of improved varieties from local seed traders (Sperling et al. 2014).

3.2 On-farm identification of new farmer-preferred legume varieties

The approach involves encouraging farmers to experiment with the varieties using centralized and decentralized approaches. Several variants of testing crop varieties have been used including the 'mother–baby' approach (Islam et al. 2008). A centralized approach exploits communal linkages with a focus on individual farmers (identified by peers) who host 'mother' trials. The approach involves identification of focal farmers, who establish demonstrations for several legume varieties and allows individual farmers to select those varieties with preferred characteristics at a central point, for example, a farmer field school. Decentralized plots ('baby' trials) are where farmers are given small samples of start-up seed of several varieties and allowed them to plant in their farms and select the most preferred varieties based on individual criteria. The overarching principle is that the varieties so chosen must have adhered to farmers' local conditions and needs.

3.3 Linking farmers' seed demands and supply of improved varieties

At the household level, farmers are considered as seed secure if they can access sufficient quantities of seeds of their preferred varieties with adequate physical quality, at the right time of the year (i.e. planting) (Sperling and Cooper 2003) and within reasonable geographical reach (Odhiambo et al. 2016). As the majority of legume farmers source their seed from their own seed stocks or local markets, one way to accelerate the use of preferred varieties and create impact is to support decentralized seed multiplication and dissemination of farmer-preferred varieties by local entrepreneurs who produce quality declared seed (QDS) (FAO 2006). The breeding/crop research organizations can contribute to this approach by providing parental seed and technical support to community-based seed entrepreneurs such as organized farmer groups.

3.4 Establishment of innovative public–private partnerships

In many SSA countries, full private sector involvement in seed supply has been limited to just a few crops such as hybrid maize and vegetables. Seed companies have shown limited interest in market-driven legume seed production due to its low profitability and unreliable seed demand. This has been largely due to legumes exhibiting a fragmented one-off seed demand, as farmers tend to save their own seeds after accessing the initial germplasm. As a result, the diffusion of new legume varieties, seed multiplication and distribution has been mainly left in the hands of government institutions and sometimes Non-Government Organizations (NGOs) and farmers' organizations (Teshale et al. 2006) with negative consequence for effectiveness and efficiency of legume seed systems. However, recently, several efforts have been made to establish public–private partnerships (PPPs) to enhance the effective diffusion of legume varieties, increase seed multiplication and create sustainable market linkages for seed and grain legume producers (Rubyogo et al. 2016a). The PPPs encourage private seed companies/entrepreneurs and large farmer organizations to produce and market critical quantity and quality legume seed with the support of public institutions (breeding organizations and policy makers) and NGOs. The approach has been used on beans with widespread impacts (Rubyogo et al. 2010b; Buruchara et al. 2011) and is currently being expanded to other legumes (Rubyogo et al. 2016b). Within these PPP innovative platforms, efficient decentralized seed production enterprises have been introduced to create demand for new varieties and complement the seed companies. Notwithstanding the limited profitability of legumes seed to the private sector, a five-step process for developing and sustaining the PPP approach (Rubyogo et al. 2016a) can be adopted to encourage investment by the private sector in legume seed production:

i Creating public awareness about released varieties

This involves identifying marketable released varieties through PVS and creating public awareness through open-house events aiming at displaying and exposing the released varieties to interested (existing and potential) stakeholders including agricultural research organizations, extension services and government departments, NGOs, the private sector involved in agricultural inputs and outputs (traders) and farmers. Based on seed producers' interests and their specific bean market targets and partners, especially in the private sector are encouraged to select preferred varieties for multiplication and possible commercialization.

ii Building partnerships to scale up supply

This step involves selecting, building and sustaining partnerships for producing seed on a wider scale and to generate a sustainable supply. A key consideration is engaging partners with shared interests and commitments and ensuring the free flow of information while ensuring seed availability to all interested partners. Both public and private partners engage in demand creation for the varieties identified in step i. While this is conducted jointly, the private sector should take the lead and assure the public of sustained seed supply of the varieties promoted. This further enables the private sector to be the drivers and eliminate the notion that the public sector is the competitor. The intention here is to develop alternatives and learn together.

iii Availability of preferred varieties of seed to the private sector

Here we make basic seed of marketable improved legume varieties available to identified seed suppliers. Private seed companies should be supplied with basic seed for production of a number of seed grades. During the production period, there should be frequent interactions, skills and knowledge-sharing between public and private sector on legume seed production. The interactions among other legume value chain actors (e.g. seed suppliers, grain traders, extension agents, researchers and policy makers) should also be enhanced through an innovative platform. The platform will provide an ongoing opportunity to exchange information and knowledge, and the interactions will help seed entrepreneurs to identify new seed business opportunities and broaden their client base. The platform will provide an opportunity to solve institutional bottlenecks together.

Through this forum, the private sector can lobby for adjustments in unfavourable seed policies.

iv Participatory evaluation and performance tracking

Availing of seeds of market-demanded and farmer-preferred legume varieties is a huge task, especially for foundation seed. However, this step also provides an opportunity to create demand for seed of improved varieties. Ex-post (qualitative and quantitative) evaluations to assess the effects of the seed distributions should be jointly assessed by both the public and private sectors. These assessments provide insights for improving subsequent operations and for meeting emerging demands. The evaluations are achieved through multiple interactions with stakeholders such as seed suppliers, farmers, extension staff and policy makers. For instance, farmers' satisfaction can be assessed based on a range of criteria including farmers' preference for pack size, time of seed delivery, distance to seed source, overall germination, yield levels and type and sources of information, especially on markets and extension.

v Generating and disseminating lessons learned for scaling up

This step involves sharing the experiences and lessons learned from the PPP approach with the possibility of scaling out best practices to other legume value chain actors including seed companies and private seed producers. Regular interactions with seed companies should be organized and feedback mechanisms established. The feedback mechanisms may include sharing of lessons and knowledge gained through verbal (face to face as well as media) and written processes. Some of the key issues to be addressed include increasing the role of private seed producers in multiplying basic seed and expanding the geographical impact (i.e. enhancing access to certified legumes seed by more farmers).

4 Case studies: the use of research to improve legume cultivation in practice

4.1 The case of PABRA collaborative bean research

The Pan-Africa Bean Research Alliance (PABRA) is a consortium of African-owned regional bean networks with membership from 30 countries in SSA. The regional networks include the Eastern and Central Africa Bean Research Network, the Southern Africa Bean Research Network and the Western and Central Africa Bean Research Network. The goal of PABRA is to enhance the food security, income generation and health of poor communities in a gender-equitable manner (Buruchara et al., 2011).

PABRA facilitates collaborative research within, among and beyond the regional networks and operates within a common framework, which upholds partners' national interests, shared regional vision and objectives. In fostering collaborative research for development through PPPs, PABRA is guided by:

> the principles of the 'network paradigm' which include: reciprocity, a willingness to exchange information, know-how, proprietary knowledge and goods; trust, a willingness to risk placing faith in the reliability of others; learning, a recognition that knowledge develops and best practice should be learnt; partnership, a preparedness to solidify

> reciprocal relationships preferentially; and decentralism, a realization that centralized information and decision-processing is inefficient. (Buruchara et al. 2011)

PABRA provides a conceptual framework for other players/actors to enhance synergy. Its collaborative research has yielded several benefits.

First, less endowed national programmes have been able to select varieties and release them through PABRA networking. Countries without breeding programmes, for example, Angola, Burundi, Cameroon, south DR Congo, Lesotho, Mozambique, Swaziland and others have released bean varieties through collaborative evaluation and sharing of germplasm. Though PVS involves various bean value chain actors in selecting varieties as part of the variety development process, PABRA has moved bean technologies and approaches across countries and networks. For instance, from 2006 to 2015, the Institute of Agricultural Research for Development in Cameroon released 12 varieties.

Second, as a collaborative research avenue, PABRA has attained milestones in policy influence anchored on exploring the possibility of co-owning the released varieties, that is, managing/sharing the intellectual property, bringing the private sector on board before the release and granting of exclusive/non-exclusive rights, asking the private sector to pay some royalties to support breeding or nucleus seed production, integrating QDS and support to private companies (for synergies and impact), exploring sustainable institutional mixes for delivering and out-scaling seed and complementary technologies.

Third, at the regional level, PABRA has achieved seed system integration through its role in: strengthening multiple varieties release across countries, facilitating variety licensing by seed companies and wider commercialization of bean seed and fine-tuning and deploying embedded crop insurance in the seed price to increase the use of certified seeds.

4.2 The case of white pea beans in Ethiopia

Bean production and marketing constraints in Ethiopia before 2004

In Ethiopia, the common bean is important to the national economy as an export commodity but is also a source of employment, food and cash income to farming households, especially those in lowlands and mid-altitude zones (Negash 2007). Before 2004, increasing bean production to meet the upsurge in demand was a challenge. Constraints to bean yield and productivity included low levels of input use (e.g. fertilizers, pesticides and improved seeds) that were attributable to low farmers' incomes, low levels of irrigation, soil degradation, inadequate agricultural research and extension and constraints in market development.

Although extensive research conducted in Ethiopia over the years (in response to the needs of the increasingly differentiated bean grain market) has generated many improved bean varieties, the seed supply for these varieties has remained limited, especially for small-scale, resource-poor farmers. Farmers' access to seed of improved bean varieties was constrained by many factors including the underdeveloped seed industry (Gemeda et al. 2001).

For several decades, bean research in the Ethiopian Institute of Agricultural Research (EIAR), Regional Agricultural Research Institutes (RARIs) and universities, which was coordinated by the EIAR bean research programme based in Melkassa, aimed to respond to production constraints by generating production technologies targeting at increased bean productivity for enhanced household food security in Ethiopia. EIAR and RARIs relied on a small number of farmer research groups and the Ethiopia Seed Enterprise (ESE), a government agency to disseminate improved varieties to farmers (Alemu and Spielman 2010).

However, there was limited interaction among the bean value chain actors, despite the increasing interest in beans as a marketable commodity by traders and exporters. The link between the producers and the export markets was weak due to the large number of ineffective intermediaries operating in the value chain. The small-scale intermediaries operated in limited geographic areas were fragmented between the producer and consumer markets and created a lack of transparency in bean marketing. The large bean importing companies were unable to manage thousands of smallholder bean farmers who were operating under non-standardized production systems and who lacked market forecast information, adequate financial systems and rudimentary logistics.

In terms of the bean seed supply, there was limited access for farmers to fresh seed from either the research organizations or the seed suppliers. The main source of certified bean seed of improved varieties was a government parastatal organization, the ESE. For instance, in 2004, ESE supplied about 40 t (or less than 1% of the national seed requirement in that year). ESE supplied three varieties of white pea bean out of 21 that were released up to 2003 by EIAR, RARIs and Hallemaya University (Assefa et al. 2006). ESE seed was supplied primarily to government-supported seed intervention projects. The woreda (district) bureaux of agriculture were responsible for distribution of seed to the farmers. There was also limited decentralized seed production by farm research groups supported by research centres around their experimental stations. The rate and extent of adoption of the released varieties was quite low (Assefa et al. 2005).

Bean seed access, grain production and marketing in Ethiopia after 2004

The robust seed systems approach that was adopted in Ethiopia after 2004 yielded remarkable results. Between 2004 and 2014, sustained and pluralistic bean seed supply dramatically increased access by farmers to seed of improved bean varieties. The Ethiopian bean multi-stakeholders' annual review meetings were held every year. The impact assessment indicated impressive impacts on bean production, resulting from novel targeting of demand-led varieties and engaging with decentralized seed producers, who were supported by partner organizations (including NGOs, farmer cooperatives and district agricultural offices).

The adoption and use of affordable bean seed packs expanded the reach of new varieties, especially in remote areas where variety penetration was low. The approach proved efficient in enhancing access to high-quality bean seed by resource-poor farmers, particularly women (Rubyogo et al. 2016a). Access to seeds of market-demanded varieties increased from less than 20% to about 65% across major bean growing areas between 2004 and 2010 (Katungi and Gebeyehu 2011). The number of partners also increased from 15 to 53 between 2004 and 2014. In the same period, the amount of seed produced increased 500-fold and the number of varieties promoted by the partner organizations increased 200-fold.

The wider use of improved bean varieties was linked to improved bean pricing and higher market demand which provided incentives to farmers. The Ministry of Agriculture facilitated the exposure of farmers to market-demanded varieties and complementary technologies. Between 2004 and 2014, there was 33% increase in area under beans from 245 507 to 326 500 ha, while production increased by 115% from 211 347 to 455 115 t. Bean production increased as a result of both increased area under beans and increased grain productivity, with productivity increasing from 0.861 t per ha to 1.49 t per ha between 2004 and 2014.

Better access to quality seed of marketable varieties improved the livelihoods of farmers through increased and diversified incomes, which helped them to meet their financial needs (such as school fees and essential household items) (Gebeyehu and Dadi 2010). Some farmers invested their proceedings from bean sales to expand the area they had under bean production and/or started livestock production (Assefa et al. 2007).

The drivers of the impact of white pea bean in Ethiopia were (i) the right product (highly marketable bean varieties); (ii) impact-oriented NARS to engage other bean value chain actors; (iii) policy support for legumes as a cash and export crop, including resource mobilization, structured and competitive market through Ethiopia commodity exchange; (iv) price incentives to value chain actors, particularly farmers; (v) continuous partnership and engagement between EIAR bean programme and CIAT from variety development to deployment; and (vi) sustained government and donors funding to the bean programme, especially for breeding, breeder and basic seed production, bean value chain actors capacity skills and knowledge enhancement and engagement of other value chain actors.

5 Summary: how research can contribute to enhanced and sustainable crop production

- In Ethiopia between 2004 and 2014, an integrated, impact-driven seed systems approach resulted in a drastic increase in white pea bean: productivity (by 73%), cultivated area (by 33%) and production (by 115%).
- In Malawi, building a PPP helped to stimulate wider delivery of bean seed across regions of the country and leverage the government's bean seed subsidy programme, thus creating an opportunity to build a foundation for a demand-driven and sustainable bean seed system (Rubyogo et al. 2016a).
- In Rwanda, research into the viability of climbing beans and adoption of these bean varieties by farmers have led to a tremendous increase in production of the highly consumed pulse crop. With climbing beans, farmers are harvesting up to 5 t per hectare, which is three times what can be achieved with bush beans. Farmers use more air space than land thus becoming as effective as high-rise buildings in urban areas. The Rwanda Agricultural Board in collaboration with institutions such as CIAT-PABRA, CIAT-Harvest Plus and Alliance for a Green Revolution in Africa (AGRA) have intensified the breeding of more resilient, highly productive and farmer-preferred climbers. These new varieties have recorded adoption levels of up to 65% nationally and 100% in several regions (Katungi et al. 2016).
- Through the collaborative Tropical Legumes research project, the participation of several seed producers in decentralized systems across 14 countries of SSA and two in South Asia (SA) has increased the production of certified and QDS of legumes and ensured access to seed of farmer-preferred legume varieties in remote areas. Seed production more than doubled (i.e. increased by 253% or from 168 775 to 426 456 t)

while seed access increased by 164% (from 5 033 913 to 8 276 910 beneficiaries) between 2007 and 2014. In Ethiopia, more than 124 000 t of certified and quality declared chickpea seed was produced. In Eastern and Southern Africa, more than 3000 t of pigeon pea seed and 42 250 t of bean seed were produced, while in Nigeria 10 237 t of soybean seed were produced. In India, 268 585 t of chickpea seed, 24 560 t of groundnut seed and 3522 t of pigeon pea seed were produced (Monyo and Varshney 2016).

- Across several African countries, collaborative legume research and increased awareness of improved varieties through multimedia and user-friendly communication strategies and tools led to rapid adoption and use of newly released varieties by farmers. In Nigeria, the adoption of cowpea and soybean varieties increased by 49% and 17%, respectively, while in Mozambique for the same crops, the increase in adoption was 11% and 8%, respectively. In Malawi, the adoption of soybean varieties increased by 22% of the cropped area (Monyo and Varshney 2016). Additionally, the increase in number of channels and outlets that conveniently provided seed to farmers, combined with the availability of affordable small packs enhanced farmers' access to quality seed of improved legume varieties, especially for previously disadvantaged women farmers.

6 Future trends in legume research

6.1 Demand-led legume breeding

Legumes are increasingly becoming cash crops for smallholders, particularly women who are moving from subsistence systems to market systems. This will require new approaches beyond PVS including a market (demand)-led breeding approach which entails identification of industry needs through anticipation of future agro-ecological and market demands. Consultation with industry and public policy makers will determine the critical traits that breeding of new varieties will focus on. This will also ensure the wider adoption of new varieties and stimulate private investments in legume value chains (Persley and Antony 2014).

6.2 Variety characterization and disease diagnosis/identification through molecular tools

The current legume genetic quality identification is carried out by morphological observations using descriptors developed during the tests of distinctiveness, uniformity and stability before the variety is released. However, this process is time-consuming, therefore use of molecular markers will be accurate and cost/time efficient. Disease identification through pathogen isolation is time-consuming, usually the results come in after the planting season. Therefore, a wider use of disease diagnostic tools such as the enzyme-linked immunosorbent assay and polymerase chain reaction (PCR)-based techniques, including terminal restriction fragment length polymorphism (T-RFLP) can be used in the detection and identification of bean fungal and bacterial diseases (Rubyogo et al. 2010a).

6.3 Beyond variety selection

Using innovative seed delivery models to speed up the rate and extent of seed access to remote areas

Several approaches have been tested and have proved to be viable in reaching millions of smallholder farmers over a short time as well as enhancing the adoption of improved legume varieties. For instance, the use of small affordable seed packs (Phiri et al. 2000) has been tested and is being extensively used across several SSA countries and SA both for growing legumes and vegetables (Chakraborty 2014). In some countries, the approach has provided opportunities for private companies to expand their seed business to remote and poor, hard-to-reach women farmers. For instance, seed companies in Kenya and Uganda reached more small-scale farmers through the use of small bean seed packs at various outlets than the NARS-embedded formal system. Other successful seed system models include the 'one-village, one-variety' concept with pigeon pea open-pollinated varieties and hybrids in India. In this model, small-scale farmers with limited access to land consolidated small units of land to produce seed collectively and to avoid contamination of different varieties; all members of the village had to adopt the same variety. In Malawi, seed loans were successful in disseminating new groundnut varieties. Under the scheme, small-scale farmers produced new varieties and passed them on to other members of the community as a loan. In Mali, groundnut varieties were successfully disseminated at community level through women's groups. Similarly, commercially oriented mixed gender farmers' groups and farmer cooperative unions have become the backbone of common bean and chickpea seed production in Ethiopia. In West Africa, community seed production has been strengthened for cowpea and soybean seed production. This has ensured availability of cowpea seeds in target communities in Mali, Nigeria and Niger Republic and soybean seeds in Nigeria. Community-based seed production involving NGOs, CBOs and farmers associations have improved access to cowpea and soybean seeds in target communities in Mozambique.

Adequate variety demand creation

Intensified sensitization on the use of quality seed of improved varieties and their importance through user-friendly promotional tools and multimedia approaches have also helped to improve the rate and magnitude of legume varieties' diffusion. One of the reasons for suboptimal use of improved legume varieties is inadequate information about these varieties and their importance. The formal seed systems practised in many countries upheld a stipulated procedure for variety dissemination through approved institutions. This top-down approach had no regard for sensitization of the target farmers prior to or during seed distribution. With adequate information supporting demand-led breeding, PVS and value chain actors from the variety design to the release will allow farmers to make informed choices. The use of efficient and electronic mass communication tools, for example, social media and radio talks linked to local demonstrations, seed fairs, farmer field schools as well as open and field days will enable farmers to gain practical experience with the varieties being promoted and to access seed of improved varieties either as seed-for-cash or seed-for-seed exchange.

Exploit existing subsidy programmes to catalyse demand-driven seed systems

Subsidy programmes have been extensively used in some African countries to help avail of inputs including seed, with the aim of enabling smallholder farmers to increase production levels of certain crops, especially cereals. In countries such as Malawi, the merits of this subsidy programme, which distributes vouchers to farmers to buy seed at reduced cost from agro-dealers, have been debated over time (Xu et al. 2009). Essentially, through subsidy programmes, farmers gain access to new varieties and good-quality seed but the delivery mechanism is often unsustainable. The constraints associated with the distribution of seed through these subsidy programmes are a heavy financial burden that must be borne by the government and crowding out of the private seed sector by seed business opportunities. The inclusion of the private sector for complementarity could lay the foundation for a more sustainable legume seed systems in most African countries (Xu et al. 2009; Rubyogo et al. 2016a).

7 Where to look for further information

As legumes contribute to several uses (e.g. food, feed, soil fertility and cash), there is a lot of information available from both public and private sector information platforms (websites and offices). These sites are either owned by the GGIAR centres, NARS or universities working on legumes, for example, http://grainlegumes.cgiar.org and PABRA (http://www.pabra-africa.org/). However, increasingly, private sector companies are getting involved in both seed and grain markets/traders and they are showing increasing interested in communicating their products and interests in legumes to the wider public, for example, Kapitan Trading (http://www.kapitantrading.com/). Nutrition-oriented public departments and development partners may also provide information on the nutritional values of legumes, for example, the National Health Service in the United Kingdom (http://www.nhs.uk/Livewell/Goodfood/Pages/pulses.aspx). However, more needs to be done to intensify information in Africa and Asia where legume consumption is high.

8 References

Abate, T. (Ed.). (2012). *Four Seasons of Learning and Engaging Smallholder Farmers: Progress of phase 1*. Nairobi, Kenya: International Crops Research Institute for the Semi-Arid Tropics, p. 258.

Alemu Dawit, A. and Spielman, D. J. (2010). The Ethiopian seed system: Regulations, institutions and stakeholders. Paper submitted to ESSP Policy Conference 2006, Bridging, Balancing, and Scaling up: Advancing the Rural Growth Agenda in Ethiopia. 6–8 June 2006, Addis Ababa, Ethiopia.

Almekinders, C. J. M., Thiele, G. and Danial, D. L. (2007). Can cultivars from participatory plant breeding improve seed provision to small scale farmers? *Euphytica*, 153, 363–72.

Assefa, T., Rubyogo, J. C., Sperling, L., Amsalu, B., Abate, T., Deressa, A, Reda, R., Kirkby, R. and Buruchara, R. (2006). Creating partnerships for enhanced impact: Bean variety delivery in Ethiopia. *J. Crop Sci. Soc. Ethiopia*, 12, 1–19.

Aw-Hassan, A., Mazid, A. and Salahieh, H. (2008). The role of informal farmer-to-farmer seed distribution in diffusion of new barley varieties in Syria. *Exp. Agr.*, 44, 413–31.

Beebe, E. S. (2012). Common bean breeding in the tropics. In J. Janick (Ed.), *Plant Breeding Reviews*. Wiley-Blackwell: John Wiley and sons. Inc., pp. 357–426.

Belmain, S. R., Haggar, J., Holt, J. and Stevenson, P. C. (2013). *Managing Legume Pests in sub-Saharan Africa: Challenges and Prospects for Improving Food Security and Nutrition Through Agro-ecological Intensification*. Chatham Maritime, UK: Natural Resources Institute (NRI), University of Greenwich, p. 34.

Boelt, B., Julier, B., Karagić, D. and Hampton, J. (2015). Legume seed production meeting market requirements and economic impacts. *Crit. Rev. Plant Sci.*, 34(1–3), 412–27.

Bouchenak, M. and Lamri-Senhadji, M. (2013). Nutritional quality of legumes, and their role in cardiometabolic risk prevention: A review. *J. Med. Food*, 16(3), 185–98

Buruchara, R. A. and David, S. 1994. Seed quality: Issues in small-scale farmer bean production. Paper presented at the 4th SADC Regional Bean Workshop, Potchefstroom, South Africa, 2–4 October 1994. CIAT African Workshop Series No 31. Kampala, Uganda: CIAT.

Buruchara, R., Chirwa, R., Sperling, L., Mukankusi, C., Rubyogo, J. C., Muthoni, R. and Abang, M. M. (2011). Development and delivery of bean varieties in Africa: The Pan-Africa Bean Research Alliance (PABRA) model. *Afr. Crop Sci. J.* 19(4), 227–45.

Chakraborty, S. (2014). An intervention on 'distribution of quality vegetable seeds in mini-packets to remote farmers: Impact assessment report (http://www.enterprise-development.org/wp-content/uploads/KatalystCase-3.pdf;16.6.2017).

[CIAT] International Center for Tropical Agriculture. (1991). Comparisons of seed quality, from bean seed obtained from range of formal and informal sources annual report bean program, 1991. Cali: CIAT, pp. 310–12.

[CTA] (2014). Seed systems, science and policy in East and Central Africa. P.O. Box 380 6700 AJ Wageningen. The Netherlands: CTA.

Cichy, K. A., Wiesinger, J. A. and Mendoza, F. A. (2015). Genetic diversity and genome-wide association analysis of cooking time in dry bean (*Phaseolus vulgaris* L.). *Theor. Appl. Genet.*, 128, 1555–67.

[DARS] Department of Agricultural Research and Technical Services. (2010). Annual report. Seed Services. Lilongwe, Malawi: DARS.

David, S., Kirkby, R. and Kasozi, S. (2000). Assessing the impact of bush bean varieties on poverty reduction in sub-Saharan Africa: Evidence from Uganda. Network on Bean Research in Africa. Occasional Publications Series, No. 31. Kampala, Uganda: CIAT.

David, S. and Sperling, L. (1999). Improving the technology delivery mechanisms: Lessons from bean seed systems research in East and Central Africa. *Agr. Hum. Values*, 16, 381–8.

Dorward, P., Craufurd, P., Marfo, K., Dogbe, W. and Bam, R. (2007). Improving participatory variety selection processes: participatory varietal selection and the role of informal seed diffusion mechanisms for upland rice in Ghana. *Euphytica* 155, 315–27.

[FAO] Food and Agriculture Organization of the United Nations. (2006). Quality declared seed: Technical guidelines for standards and procedures. Plant Production and Protection Paper 185. FAO, Viale delle Temedi Carcalla 00100 Rome, Italy.

[FAO and ICRISAT] Food and Agriculture Organization of the United Nations and International Crops Research Institute for the Semi-Arid Tropics. (2015). Community seed production. C. O. Ojiewo, S. Kugbei, Z. Bishaw and J. C. Rubyogo (Eds), Workshop Proceedings, 9–11 December 2013. FAO, Rome and ICRISAT, Addis Ababa, p. 176.

FAOSTAT 1993–2004. Food and Agriculture Organization statistics at www.fao.org.

Gebeyehu and Dadi (2010). Improved common bean production and post-harvest handling practices in Ethiopia. Catholic Relief Service Ethiopia 2010. *A guideline that outlines improved common bean production, post-harvest and marketing practices in Ethiopia*. Addis Ababa, Ethiopia.

Gemeda, A., Aboma, G., Verkuijl, H. and Mwangi, W. 2001. *Farmers' Maize Seed Systems in Western Oromia, Ethiopia*. Mexico, D.F.: International Maize and Wheat Improvement Center (CIMMYT) and Ethiopian Agricultural Research Organisation (EARO), p. 32.

Hampton, J. G., Boelt, B., Rolston, M. P. and Chastain, T. G. (2013). Effects of elevated CO_2 and temperature on seed quality. *J. Agr. Sci.* 151, 154–62.

[IFPRI] International Food Policy Research Institute. (2010). Pulses value chain in Ethiopia: Constraints and opportunities for enhancing exports. Working Paper. July 2010.

Islam, M. M., Pandit, D. B., Hasanuzzaman, M., Hussain, A. and Farhad, M. (2008). Participatory variety selection for selecting and disseminating farm preferred wheat varieties. *Intl. J. Biores.* 4(5), 13–19

Katungi, E. and Gebeyehu, S. (2011) Enhanced seed accessibility and adoption of improved common bean varieties in Ethiopia: monitoring early adoption of improved varieties – Survey report.

Katungi, E., Sperling, L., Karanja, D., Wozemba, D., Mutuoki, T. and Rubyogo, J. C. (2011). A cost-benefit analysis of farmer based seed production for common bean in Kenya. *Afr. Crop Sci. J.* 19(4), 119–31.

Katungi, E., Kikulwe, E. and Emongor, R. (2015). Analysis of farmers' valuation of common bean attributes and preference heterogeneity under environmental stresses of Kenya. *Afr. J. Agr. Res.*, 10(30), 2889–901.

Katungi E., Larochelle C., Mugabo J. and Buruchara, R. (2016). Estimating the impact of climbing bean adoption on bean productivity in Rwanda: Endogenous Switching Regression. Invited paper presented at the 5th International Conference of the African Association of Agricultural Economists, September 23–26 2016, Addis Ababa, Ethiopia. (http://ageconsearch.umn.edu/bitstream/246456/2/218.%20Climbing%20bean%20productivity.pdf).

Kinyanjui, P. K., Njoroge, D. M., Makokha, A. O., Christiaens, S., Sila, D. N. and Hendrickx, M. (2014). Hydration properties and texture fingerprints of easy- and hard-to-cook bean varieties. *Food Sci. Nutr.*, 61, 275–82.

Lupwayi, N. Z., Kennedy, A. C. and Chirwa, R. M. (2011). Grain legume impacts on soil biological processes in sub-Saharan Africa. *Afr. J. Plant Sci.*, 5(1), 1–7.

Matthews, S., Noli, E., Demir, I., Khajeh-Hosseini, M. and Wagner, M.-H. (2012). Evaluation of seed quality: From physiology to international standardization. *Seed Sci. Res.*, 22, S69–S73.

Monyo, E. S. and Varshney, R. K. (Eds). (2016). *Seven Seasons of Learning and Engaging Smallholder Farmers in the Drought-prone Areas of sub-Saharan Africa and South Asia Through Tropical Legumes, 2007–2014*. Patancheru 502 324, Telangana, India: International Crops Research Institute for the Semi-Arid Tropics, p. 236. ISBN: 978-92-9066-568-7.

Negash, R. (2007). Determinants of adoption of improved haricot bean production package in Alaba Special Woreda, southern Ethiopia. [MSc thesis]. Haramaya, Ethiopia: Haramaya University.

Odhiambo, W., Ngigi, M., Lagat, J., Binswanger, H. P. and Rubyogo, J. C. (2016). Analysis of quality control in the informal seed sector: Case of smallholder bean farmers in Bondo sub-county, Kenya. *J. Econ. Sust. Dev.*, 8(7), 8–29.

Ogato G. S., Boon, E. K. and Subramani, J. (2009). Gender Roles in Crop Production and Management Practices: A Case Study of Three Rural Communities in Ambo District, Ethiopia. *J. Hum. Ecol.*, 27(1), 1–20.

Otsyula, R., Rachier, G., Ambitsi, N., Juma, R., Ndiya, C., Buruchara, A. R. and Sperling, L. (2004). The use of informal seed producer groups for diffusing root rot resistant varieties during period of acute stress. In L. Sperling, T. Remington, J. M. Haugen and S. Nagoda (Eds), *Addressing Seed Security in Disaster Response: Linking Relief with Development*. Cali, Colombia: International Center for Tropical Agriculture (CIAT), pp 66–89.

Pelmer, D. P. (2005). Agriculture in the developing world: Connecting innovation in plant breeding research to downstream applications. *PNAS,* 102(44): 15739–46.

Persley, G. and Antony, V. (2014). *Demand led plant variety design for emerging markets in Africa.* http://aciar.gov.au/aifsc/demand-led-plant-variety-design-emerging-markets-africa [15.04.2014 15:03:01].

Phiri, M. A. R., Chirwa, R., Kandoole, S. and Tripp, R. (2000). Introducing new bean varieties with small seed packs: Experience from Malawi. Network on Bean Research in Africa, Occasional Publications Series, No. 32. Kampala, Uganda: CIAT.

Popelka, J. C., Terryn, N. and Higgins, T. J. V. (2004). Gene technology for grain legumes: Can it contribute to the food challenge in developing countries? *Plant Sci.*, 167, 195–206.

Rebello, C. (2014). A review of the nutritional value of legumes and their effects on obesity and its related co-morbidities. *J. Food Nutr. Res.*, 15(5), 392–407.

Rubyogo, J. C., Dickinson, M., Mayes, S. and Assefa, T. (2010a). Seed-health assessment of different bean seed grades and sources from Ethiopia using molecular tools. *J. New Seeds*, 10, 293–310.

Rubyogo, J. C., Sperling, L., Muthoni, R. and Buruchara, R. (2010b). Bean seed delivery for small farmers in sub-Saharan Africa: The power of partnerships. *Soc. Natur. Resour.*, 23, 285–302.

Rubyogo, J. C., Sperling, L. and Assefa, T. (2007). A new approach for facilitating farmers' access to bean seed. *LEISA Magazine*, 23(2), 27–9.

Rubyogo, J. C., Magreta, R., Kambewa, D., Chirwa, R., Mazuma, E. and Andrews, M. (2016a). Using subsidized seed to catalyse demand-driven bean seed systems in Malawi. *Dev. Pract.*, 26(1), 15–26.

Rubyogo, J. C., Myers, M., Ajeigbe, H., Kamara, A., Boahen, S., Buruchara, R., Upadhyaya, H., Pasupuleti, J., Okori, P., Desmae, H. and et al. (2016b). Integrated seed systems delivering on the promise: Experiences from Tropical Legumes II. In E. S. Monyo and R. K. Varshney (Eds), *Seven Seasons of Learning and Engaging Smallholder Farmers in the Drought-Prone Areas of sub-Saharan Africa and South Asia Through Tropical Legumes*. Panthancheru, India: ICRISAT, pp. 167–79.

Shi, J. (2004). Saponins from edible legume: Chemistry, processing and health benefits. *J. Med. Food*, 7(1), 67–78.

Sperling, L., Boettiger, S. and Barker, I. (2014). Integrating seed systems. Planning for Scale Brief # 3. Ag Partner X Change (APXC).

Sperling, L. and Cooper, H. D. (2003). Understanding seed systems in seed security. In: *Improving the effectiveness and sustainability of seed relief*. Proceedings of a stakeholders workshop in Rome, 26–28 May 2003. Rome: Food and Agriculture Organization of the United Nations.

Teshale A., Rubyogo, J. C., Sperling, L., Amsalu, B., Abate, T., Deressa, A., Reda, F., Kirkby, R. and Buruchara, R. (2006). Creating partnerships for enhanced impact; bean variety delivery in Ethiopia. *J. Crop Soc. Ethiopia* 12, 27–30.

Tonapi, V. A., Ravinder Reddy, C., Prasad, S. R., Tomar, B. S., Singh, S., Pandey, S., Natarajan, S. and Lal, S. (2012). *Strategies to Build Viable Community Seed Systems in Dry Land Ecosystems for Sustainable Seed and Food Security in India*. Post Bag No. 11, Village: Kushmaur, PO: Kaithauli, Mau 275 101 (UP), India: Directorate of Seed Research, p. 43.

Trutmann, P., Voss, J. and Fairhead, J. (1996). Local knowledge and farmer perceptions of bean diseases in Central Africa Highlands. *Agric. Hum. Values* 13, 64–74.

Valbuena, D., Erenstein, O., Homann-Kee Tu, S., Abdoulay, T., Claassen, L., Duncan, A. J., Gérard, B., Rufino, C., Teufel, N., Van Rooyen, A. and Van Wijk, M. T. (2012). Conservation agriculture in mixed crop-livestock systems: Scoping crop residue trade-offs in sub-Saharan Africa and South Asia. *Field Crops Res.*, 132, 175–84.

Venn, B. (2004). Cereal Grains, Legumes and Diabetes. *Eur. J. Clin. Nutr.*, 58(11), 1443–61.

Xu, Z., Burkea, W. J., Jayne, T. S. and Govereh, J. (2009). Do input subsidy programs 'crowd in' or 'crowd out' commercial market development? Modeling fertilizer demand in a two-channel marketing system. *Agr. Econ.*, 40, 79–94.

Chapter 10

Grain legumes in integrated crop management systems

Chris Johansen and Kadambot H. M. Siddique, The University of Western Australia, Australia

1 Introduction

Grain legumes (GLs) are not only important food and feed sources; they also play a major role in the sustainability and productivity of cropping systems (Siddique et al., 2008, 2012). GLs contribute to cropping system diversity when grown with crops of other plant families (e.g. Gramineae), disrupting the pest and disease cycles that develop with continuous cropping of the same species. If effectively nodulated, they can significantly enhance the nitrogen (N) dynamics of the cropping system, firstly by fixing atmospheric N to meet their own needs, and thus not relying on inorganic N fertilizer or depleting soil N, and by leaving residual N for following crops (Kumar Rao et al., 1998). Decaying legume residue, above and below ground, can enhance soil organic matter (Barrios et al., 1996), but the extent of net gain or loss in available soil N for a following crop depends on the degree of N sequestration by that residue (Kumar and Goh, 2000). Further, some GLs can release sparingly available soil phosphorus (P) to increase the available P pool of the entire cropping system (Nuruzzaman et al., 2005).

GLs are usually grown as secondary components of cereal-dominated cropping systems. Consequently, but understandably, GLs receive less research attention than the cereals they share land with (Foyer et al., 2016). This is despite GLs generally being exposed to a

http://dx.doi.org/10.19103/AS.2017.0023.08

plethora of biotic and abiotic stresses. Further, their cultivation must adjust to the primary requirements of the cereal crop, as well as other components of the farming system (e.g. animal husbandry), often displacing GLs from their optimum growing environment. Nevertheless, there is a considerable bank of component knowledge on how to maximize GL productivity, but not enough of this knowledge is translated to farmers, especially resource-poor ones.

Improvement in GL production necessarily requires a holistic approach, not only to tackle the multiple stresses directly affecting the crop but also to integrate external factors ultimately affecting production, like competition with other crops, input availability and market opportunities. These factors can interact in ways ranging from synergism to antagonism. For example, application of a limiting nutrient may induce better root growth that would enhance soil water extraction, but the resultant improved vegetative growth may attract pests and diseases. Thus an integrated crop management (ICM) approach is required, which the National Resources Institute of the University of Greenwich, UK, usefully defines as 'a system of crop production which conserves and enhances natural resources while producing food on an economically viable and sustainable foundation' (NRI, 2003). This necessitates 'a good understanding of the interactions between biology, environment and land management systems' and 'is particularly appropriate for small farmers because it aims to minimize dependence on expensive inputs and to make the fullest possible use of indigenous technical knowledge and land use practices' (NRI, 2003).

Human population growth – and the subsequent increased demand for food and declining area of agricultural land which is often degrading in soil health – puts even more pressure on GLs (Foyer et al., 2016). There is ever-growing pressure to increase the production of staple cereal crops, thereby further threatening the area that can be allocated to GLs. Another factor of growing importance is climate change, which modifies local plant growth conditions, disrupts traditional cropping patterns and increases the riskiness of cropping. This presents further constraints but also some opportunities for GLs (Siddique et al., 2012).

This chapter will focus on resource-poor, smallholder farming systems, where most GLs are grown, the yields are low and unstable, and the need for GL consumption is greatest. Nevertheless, many considerations for improving GLs in these systems apply to large-scale commercialized production systems. We will first examine the nature of the yield gap for GLs and explore options for narrowing it. There is indeed a broad range of potential solutions, genetic and management based, to alleviate the yield constraints, but the challenge is to find solutions that will work for resource-poor farmers.

A top–down approach to disseminating the potential solutions for increasing GL cultivation and yields has had limited success, especially for resource-poor farmers in rainfed environments. Therefore, a more enhanced farmer-participatory approach than has so far been implemented is advocated, and suggestions of how best to achieve this are given.

2 Yield gaps

An understanding of yield gaps, the difference between the yields realized by farmers and the potential yield under 'ideal' cultural conditions, is necessary to assess the scope for improving on-farm yields. The potential yield of rainfed crops is usually limited by suboptimal soil moisture, mostly deficit but sometimes excess, at some time during the growing period. Therefore, knowledge of soil water content through the growing season is crucial as this sets the yield ceiling for that season. It is therefore evident that potential rainfed yield is a year-to-year moving target, depending on pre-season and growing season rainfall (Anderson et al., 2016).

Simple soil water balance models, such as those used by Slatyer (1956), Fitzpatrick and Nix (1969), and French and Schultz (1984), have been effective in defining potential crop yields in a given season (Anderson et al., 2016). Transpiration use efficiency and rainwater lost to the crop, through soil evaporation, surface run-off or drainage below the root zone, are the only two parameters required for these calculations (Siddique et al., 2001). More sophisticated crop simulation models – taking account of additional factors affecting crop growth and yield, including radiation use efficiency and crop phenology – can give more precise estimates of potential rainfed yield. One such model that has been widely evaluated is APSIM (Keating et al., 2003). Such sophisticated models have a higher data input requirement; data that is not always readily available in resource-poor farming environments.

In the northern grain region of Australia where data requirements for APSIM can be met, Chauhan and Rao (2014) used this model to characterize the seasonal soil water status for mungbean. From historical rainfall records, they simulated yields over time and estimated the risk of drought stress at different crop growth stages. Cluster analysis identified different target production environments within the region based on local climate and soil water-holding capacity. They proposed that this should help refine specific genotype and agronomic requirements for the different target production environments.

A comprehensive attempt at estimating yield gaps of the major GLs grown under rainfed conditions in India – chickpea, pigeonpea, soybean and groundnut – was made by Bhatia et al. (2006). Using CROPGRO and APSIM models, they simulated water-limited potential yields across the different production zones and compared these values with district average yields as representative of farmer yields. The range of yield gaps found, as compared with the national average yield, is shown in Table 1. In general, realized yields were around half of what they could be, but with considerable spatial and temporal variation. The authors partitioned the yield gap into two components: (1) the difference between achievable yields, obtained from on-farm demonstrations incorporating currently recommended production technology, and farmer yields, and (2) the difference between model-calculated potential and achievable yields. They found that the first component was usually greater than the second, suggesting that there is much scope for applying known technology.

Elsewhere in resource-poor agriculture, farmers' yields also only reach about half of potential rainfed yields. For example, in West Africa, respective farmer and potential yields

Table 1 The range of yield gaps derived by subtracting district average yields from simulated rainfed yields across Indian production zones, compared with national average yield for GLs. Data from Bhatia et al. (2006)

Grain legume	National average yield (kg/ha)	Yield gap (kg/ha)
Chickpea	800	610–1150
Pigeonpea	690	550–770
Soybean	1040	850–1320
Groundnut	1150	1180–2010

are 0.54 and 1.14 t/ha for beans, 0.83 and 1.35 t/ha for groundnut, and 0.79 and 1.5 t/ha for soybean (Nin-Pratt et al., 2011). In Sudan, on-farm faba bean yields average around half of those obtained on research stations (Mohammed et al., 2015). In developed country agriculture, however, farmers usually achieve yields closer to the rainfed potential, due to their greater use of established yield-enhancing technologies. For example, soybean yields in the United States and lupin yields in Australia reach 70% of their potential yields (Fischer et al., 2014).

Although in rainfed environments available soil water (mostly deficit but sometimes excess) determines realized and potential yields, assessments made across global production systems indicate that disease, insects and nutrient stress are also major contributors to the yield gap in GLs (Kelley et al., 1995; Waddington et al., 2010). For example, in chickpea foliar diseases such as ascochyta blight, root diseases such as wilt and root rots (Kelley et al., 1995; Allen and Lenné, 1997) and pod borers (Sharma, 2006) are major yield reducers.

Estimates of yield gaps in GLs are fraught with uncertainty due in part to their greater sensitivity than cereals to a range of biotic and abiotic stress factors, increasing the spatial and temporal variability of yield (Srivastava et al., 2010). This particularly applies in developing countries where resource-poor farmers are less able to implement alleviatory measures. Further, uncertainties occur in the 'crop cutting' methodology used to quantify yields and the reporting of national statistics, which possibly overestimate yields. Potential yields derived from research station trials or on-farm demonstrations are also likely to be overestimated due to small plot sizes and sampling bias (Gomez and Gomez, 1984). It can only be concluded that there is an unsatisfactorily large gap between realized and potential yields of GLs. Established technology is available to narrow that gap, but its implementation in resource-poor farming communities faces many constraints in addition to technical ones – local availability of information, timely input availability, risk management, economic, social and markets, to name a few.

The narrowing of yield gaps depends on an understanding of the causes of those gaps. Yield gap analyses for pigeonpea and chickpea in India indicate that many farmers are not practising the recommended agronomy, such as fertilizer application and plant protection, to alleviate known constraints (Nain et al., 2014). Meta-analyses based on surveys of opinions of researchers have been conducted to better identify yield constraints of chickpea, pigeonpea, groundnut, cowpea and other crops across global regions, with the aim of sharpening research priorities (Waddington et al., 2010; Kelley et al., 1995). This methodology does not account for spatial and temporal variability of particular constraints or the potential biases of those surveyed or the survey takers, so at best this approach can only provide a broad guideline for more precise diagnosis at a local level (i.e. farm, district). As discussed below it is recommended that future attempts at constraint analysis be focused at the local, farmer-field level, taking account of all possible factors contributing to yield formation and their interactions, and the spatial and temporal variability involved.

3 Better pinpointing of constraints

In the 1980s, concerns were raised that the top-down approach of relying on scientists' perceptions of yield constraints to design research of relevance to resource-poor farmers was largely ineffective, particularly in rainfed cropping systems (Chambers et al., 1989).

There was a growing realization that agricultural development could be more effective if a first step in the process was to '*ask the customers what actual problems they were facing and what they think could be done about it*' – to get a farmers' perspective of the constraints facing their cropping endeavours and how they thought improvements could be made. This led to concepts of 'participatory rural appraisal' (PRA) (Chambers et al., 1989), which directly gather the farmers' perspectives of production-constraining factors but do not necessarily permit specific identification and quantification of causal factors.

Even if the improvement of yield and yield stability of a particular GL is the prime focus of an agricultural development programme, it is necessary to gain a holistic view of the total farming situation in the target area. Not only must farmers' perspectives be understood but so must those of other persons and processes that would impinge, however remotely, on the cropping activity. This necessitates a value chain analysis, or similar, to assess the role of other participants in the value chain, such as input manufacturers and suppliers, local and international grain traders, grain processors, retailers and consumers. Only recently has value chain analysis been rigorously applied to assess research priorities for GLs, such as for bean, cowpea, groundnut and soybean in West Africa (Rusike et al., 2013). In hindsight, a project involving the present authors on the alleviation of botrytis grey mould in traditional chickpea-growing regions of Bangladesh would have benefited from a comprehensive *ex-ante* value chain analysis (Johansen et al., 2008a). Although we demonstrated that this disease could be effectively and economically managed in a traditional growing region, chickpea production in that region continued to decline after project completion. This was largely attributable to increased sowings of more reliable and remunerative crops – particularly vegetables for urban markets and, later, maize – which could be irrigated, competing with the winter-grown chickpea. Another factor was the shift in chickpea production to areas less prone to botrytis grey mould and with less irrigation capacity, and thus less opportunity for competing crops, in northwest Bangladesh. Therefore, being able to economically alleviate an overriding constraint of a GL does not necessarily ensure its increased production; a holistic understanding of the agricultural production system is required.

Following participatory rural appraisal and gaining an understanding of value chains, on-farm diagnostic trials are usually needed to pinpoint local causal factors and to suggest possible alleviatory measures. However, not all possible remedies would be feasible in resource-poor farming situations and thus close farmer involvement with researchers at all stages of on-farm experimentation is required to identify remedies that may, or may not, work in a given farmer's field. At this point, farmers and scientists can synergistically interact, with farmers describing symptoms through the growing season and assessing the extent of crop damage, and the significance of this to their overall farming enterprise, and with scientists providing technical knowledge perhaps not available to the farmer. An example was the general poor growth, with reddish-yellow leaf symptoms, of chickpea complained about by farmers in parts of northwest Bangladesh. Scientists interacting with them could attribute these symptoms to N deficiency and noted that crops were being grown on acid soils. The poor nodulation observed suggested to the scientists that inadequate rhizobial infection or molybdenum deficiency could be problems, topics about which these farmers knew little. Farmer-participatory experiments in farmers' fields confirmed positive responses to *Rhizobium* inoculation, combined with molybdenum application (Johansen et al., 2007). A method of alleviating these constraints by including *Rhizobium* inoculum and molybdenum in the seed-priming process was then subsequently evolved through a sequence of experiments on farmer's fields. Due to their involvement in the research process, farmers were ready to adopt this relatively complex technology, but the subsequent adoption was constrained by supply chain factors – unavailability of molybdenum and *Rhizobium* inoculum in the area – which pointed to business development opportunities in this regard.

A problem often raised about conducting on-farm experimentation is the inherent variability involved. In on-station research, it is argued that there would be better 'control' of yield-affecting factors other than treatment factors, and thus greater ability to differentiate treatment effects and interactions between treatments. However, it needs to be remembered that most research stations have been established for plant-breeding purposes, with the aim of minimizing environmental effects (e.g. spatial soil differences) to maximize the expression of genetic effects (Vanlauwe et al., 2016). Although research stations were originally meant to be representative of a target agricultural region, the on-station cropping environment inevitably becomes differentiated from the actual farming environment as a result of attempts to minimize on-station variability. Thus extrapolation of crop management results from on-station to on-farm becomes problematic, especially with respect to the spatial and temporal variability experienced by farmers.

Methods have been developed to account for variability in on-farm trials, and indeed using that variability to assess the risk associated with farmers adopting an improved technology. For single factor on-farm trials, Vanlauwe et al. (2016) proposed plotting the cumulative frequency against the magnitude of treatment differences and associated confidence intervals to assess the likelihood of the response at particular yield levels. These authors pointed out that the options, or treatments to be tested, need to be analysed in relation to the biophysical, social and economic context, and the interactions between options and context understood. It would then be possible to give more realistic farmer-specific recommendations, rather than those based on a mean treatment response for a region. This approach requires the conduct of many farmer replicates, at least 20 per option, repeated iteratively over time to allow an understanding of option x context interactions. This approach seems appropriate for the stepwise improvement of GL cultivation.

The above example involving seed priming with *Rhizobium* inoculum and molybdenum followed earlier on-farm research re-introducing priming technology to the region, which is discussed in the first case study below. Other examples of intensive farmer participation in the research process include the expansion of short-duration pigeonpea in India (case study 2) and legume diversification in Malawi (case study 3). These case studies are discussed in detail in the next section.

4 Case studies

4.1 Case study 1: seed priming of chickpea in Bangladesh

Chickpea is a traditional pulse crop of Bangladesh and an important source of protein for impoverished people with limited access to animal protein. It is mainly grown during winter in the mid-western region of the country (Rahman et al., 2000). The crop relies on residual soil moisture after the harvest of the rainy-season rice, supplemented by limited but erratic winter rainfall. However, over the last 30 years or so, chickpea area and production have been declining in this region, due mainly to the expansion of irrigated agriculture and increasing disease incidence, particularly botrytis grey mould and root rots (Rahman et al., 2000). However, chickpea cultivation has been increasing in the High Barind Tract (HBT), to the north of the traditional growing region. In the HBT, irrigation potential is limited, with extensive areas remaining fallow after rainy-season rice, and the drier atmosphere during winter less conducive to botrytis grey mould. However, the quick drying nature of the soil permits only a narrow window for sowing chickpea, before the topsoil becomes too dry for crop establishment. Consequently, chickpea yields in the HBT remained low (<1 t/ha) and erratic.

In other environments, seed priming – soaking the seed overnight then drying before sowing the next day – has improved crop establishment. Indeed, as related by some older community members, it had been a traditional practice for upland crops of the region but was no longer practised. Therefore, a project to evaluate seed priming of chickpea in the HBT was conducted during 1998–2000. It involved the University of Wales, ICRISAT, the Bangladesh Agricultural Research Institute, and a non-government organization, PROVA, and was funded by the UK Department for International Development (DFID). Paired plots, with and without seed priming, were grown on farmers' fields across the region from 1998 to 2000. Plot size was around 0.13 ha. The trials were farmer implemented, with guidance by researchers. In the 1998–9 season, the mean yield response to priming across 30 trials was 47%, and in the 130 trials of the HBT in 1999–2000 it ranged from 17 to 22% in different parts of the HBT (Musa et al., 2001).

The priming response was attributed mainly to rapid seedling establishment, with higher plant stand and earlier crop maturity allowing escape from end-of-season moisture and heat stress (Musa et al., 2001). Priming also reduced the incidence of collar rot and increased nodulation by native rhizobia.

An impact analysis conducted during the 2001–2 season estimated that the area of chickpea under seed priming due to project influence, implemented by farmers alone without researcher intervention, was at least 20% (Saha, 2002). The negligible input cost of implementing priming meant that any resultant yield increase would be the net benefit of priming. Funding constraints prevented further formal follow-up of the impact but informal observation over subsequent seasons suggested that priming became universal for chickpea, and some other rainfed winter crops, across the HBT. The ready uptake of this technology could be attributed to its simplicity and maximum farmer involvement in the evaluation process, such that they 'owned' the technology from its inception. The success of the priming evaluation increased the receptivity of HBT farmers to other possible technology improvements, such as combining molybdenum and *Rhizobium* inoculation with priming (Johansen et al., 2007) and the introduction of improved varieties (Johansen et al., 2008b).

4.2 Case study 2: short-duration pigeonpea in peninsular India

Pigeonpea traditionally grown through the monsoon season in the Indo-Gangetic Plain region of India (sub-tropics) was either of medium (150–250 days) or long (250–350 days) duration. Although medium-duration genotypes offered the option of growing a winter crop, their maturity usually extended into December, well beyond the optimum sowing time for winter crops, principally wheat, thereby limiting the yield potential of the winter crop. Therefore, a primary objective of ICRISAT plant-breeding efforts, and those in the associated Indian national programme, was to identify or breed pigeonpea genotypes that would mature in time for optimum sowing of wheat, by late November. Indeed, many such genotypes, of both indeterminate and determinate growth habit, were produced and subsequently released through the Indian national programme. Subsequent on-farm evaluation found that indeterminate genotypes that matured within 120–50 days, such as ICPL 88039, were most productive and best fitted into cropping systems in Uttar Pradesh, Haryana and Punjab states (Dahiya et al., 2001). In this case, the indeterminate character was an advantage as it allowed better recovery from pod borer (*Helicoverpa armigera*) damage than determinate types within the allowed growing period; this damage is a major constraint to pigeonpea production throughout India.

The diversity of the short-duration genotypes produced, even though originally targeted for northern India, were also evaluated on-station in southern India. One such genotype, ICPL 87, later released in India as Pragati, was studied in detail at the ICRISAT centre near Hyderabad in peninsular India. It was a determinate genotype producing a first flush yield 120–30 days after sowing. However, if soil water was adequate, usually requiring irrigation due to limited winter rainfall in this region, it could produce subsequent reproductive flushes. This was possible because the higher winter temperatures than in northern India allowed continued growth of this warm-season, semi-perennial plant. From three flushes, up to 5 t/ha grain could be harvested over a growing period of about 220 days (Chauhan et al., 1987). The ability of ICPL 87, and similar genotypes, to produce subsequent reproductive flushes provided a means of recovery from pod borer damage to the first flush, different to northern India where winter temperatures are too low to permit continued growth.

While elucidation of the yield potential in these new short-duration genotypes was scientifically stimulating and rewarding, some of the researchers directly involved began to realize this potential was far removed from the resource-poor farmers that could benefit from it. Therefore, in the mid-1980s, they initiated on-farm trials around the then state of Andhra Pradesh. Farmers were provided with seed of ICPL 87 and had the characteristics of the genotype explained to them. They were encouraged to find niches in their existing cropping systems where this genotype might fit, rather than following a handed-down prescription. However, there was limited subsequent adoption in Andhra Pradesh itself, probably due to the low water-holding capacity of the mainly Alfisol soils where it was introduced and the limited capacity for supplementary irrigation during winter.

Nevertheless, when on-farm evaluation spread to the neighbouring states of Maharashtra and Karnataka, substantial adoption occurred. For example, in western Maharashtra, the percentage of the pigeonpea crop as ICPL 87 rose from 0% in 1987 to 57% in 1994, with the area increasing to about 150 000 ha (Bantilan and Parthasarathy Rao, 1999). This occurred predominantly in irrigated areas, growing mainly sugarcane, cotton and horticultural crops. Farmers not only liked the flexibility offered by this short-duration pigeonpea crop, its higher average yield than traditional pigeonpea and its remunerative value, they also realized its value in improving soil fertility so as to reduce fertilizer requirement for subsequent crops (Bantilan and Parthasarathy Rao, 1999).

This is an example of scientific innovation, originally targeting another region, being found beneficial in a non-target region via on-station research. In a sense, it is an example of successful trickle-down technology but, in hindsight, it can be speculated as to how the process could have been more efficient and rapid with greater farmer involvement at earlier stages. This would include a better understanding of farmers' needs, expectations and problems of their traditional pigeonpea cultivation earlier in the process, and then experimenting with farmers in their fields to unlock the potential of short-duration pigeonpea in their particular farming systems.

4.3 Case study 3: Grain legume options for resource-poor farmers in Malawi

In sub-Saharan Africa, in particular, soils of low and declining fertility present a major challenge to the food security of resource-poor subsistence farmers. Not only do they limit the quantum of crop production but they also limit the nutritional quality of the produce. It has generally been recognized that greater cultivation of legumes could alleviate these problems, but increases in legume cultivation have been marginal. Impediments to further GL cultivation include their low and unstable yields and increasing competition for land by staple cereal crops.

Nevertheless, concerted holistic efforts can increase GL production, as demonstrated in northern Malawi (Kerr et al., 2007). The basic rationale of this study was the poor nutritional status of children in the target region, which was hypothesized to be most effectively addressed by increasing legume cultivation in the region. This would increase the supply of legume grain to improve human nutrition and soil fertility so as to increase the quantum of crop production. Thus a farmer-participatory programme was established in 2000, combining the evaluation of alternative legume cultivation options with farmer education on soil fertility, crop production and human nutrition. Farmer research teams were established to implement the field testing, interacting closely with researchers providing technical and scientific knowledge inputs.

The legume options evaluated, over five years with 3000 farmers in 80 communities, were: i) groundnut and pigeonpea intercropped in year 1 and rotated with maize in year 2; ii) soybean and pigeonpea intercropped in year 1 and rotated with maize in year 2; iii) maize and pigeonpea intercropped; iv) *Mucuna* spp. rotated with maize; and v) *Tephrosia vogelii* (a small tree legume) relay intercropped with maize. The mother and baby trial procedure (Snapp, 2002) was followed, with mother trials containing all treatments at a central village location and baby trials distributed among individual farmers whom each tested 1–2 legume options. The mother and baby trial method has since been adopted for participatory varietal selection studies elsewhere in the world (Witcombe et al., 2005).

The average increase in area of legume systems per farmer was 862 m^2 in 2005, with women farmers favouring cultivation of edible legumes and mainly male farmers including legume green manure-only options, but only where the availability of land area was less of a constraint. This confirms previous findings of the difficulties of inducing resource-poor farmers to grow legumes for the sole purpose of green manure, in the hope of future soil fertility improvement (Kumar Rao et al., 1998; Schulz et al., 2003). However, this project did result in increased GL contribution to soil fertility as it successfully increased the incidence of burial of GL residue, rather than its traditional removal or burning (Kerr et al., 2007). Pigeonpea/groundnut rotating with maize was the most popular cropping choice in this region. The regular knowledge sharing between the farmers and researchers also improved human nutritional knowledge, to the extent of increased edible legumes being fed to children by the end of the project. The centrally located mother trials provided a focus for this knowledge sharing.

5 Avenues for agronomic improvement: conservation agriculture

Conservation agriculture is one of the strategies with the most scope for agronomic improvement of GLs in smallholder agriculture. We discuss conservation agriculture (CA) in this section, and then cover two further strategies with high potential in the next two sections.

CA refers to cropping systems involving minimal soil disturbance (i.e. minimum or zero tillage), maintenance of soil mulch cover and crop species diversification. It evolved due to concerns about soil deterioration with excess tillage, the lower labour and fuel requirements possible with reduced tillage, the widespread availability of herbicides from the 1960s and the development of minimum tillage seed drills (Serraj and Siddique, 2012). In the early 1970s, CA was practised on 3 M ha, mainly in the Americas, and in 2013 on 157 M ha globally (Kassam, 2014). The GL most widely cultivated under CA is soybean, usually grown in rotation with maize, and predominantly in North and South America.

GLs are highly compatible with CA – providing soil cover as a growing crop or N-rich residue and as a rotation option adding fixed N to the system and breaking pest and disease cycles. Their sowing is readily accomplished by standard minimum tillage seed drills, originally developed primarily for cereal crops, due to their usual spherical seed shape and intermediate seed size. Modern minimum tillage seed drills were first developed to be drawn by four-wheel tractors used in large fields. They have been modified, with their cost reduced, for widespread use in the rice–wheat systems of the Indo-Gangetic Plain (Hobbs et al., 2008). However, for resource-poor farmers, access to four-wheel tractors is usually impractical, due to their local unavailability or cost of hiring and the small field sizes those farmers cultivate. Recently, low-cost minimum tillage implements fitted to two-wheel tractors have been developed, which are manoeuvrable in small fields and obviate the need for four-wheel tractors (Haque et al., 2011; Bell et al., 2016; Johansen et al., 2012). However, it must be remembered that resource-poor subsistence farmers have traditionally used minimum tillage techniques involving hand tools – such as dibble sticks, hoeing planting basins and jab planters – and simple animal-drawn implements in various parts of the world (Johansen et al., 2012).

Soybean is an example of a GL both contributing to and benefiting from CA practices. This GL is a major component of CA systems in Brazil, where there has been widespread adoption of this practice over the last three decades (Bolliger et al., 2006). There are Brazilian examples of CA resulting in continuous soil cover (minimizing erosion), gradual increases in soil organic matter, integration of livestock, movement of surface-applied lime through the acid soil profile, breakage of compact soil layers and reduced reliance on agrochemicals, all under a variety of climatic and edaphic conditions, farm sizes and degree of mechanization (Bolliger et al., 2006). Where GLs are grown in extensive commercial agricultural systems, as in the Americas and Australia, they are predominantly cultivated under CA, with measurable increases in profitability, sustainability and environmental impact (Kassam et al., 2012).

For other GLs grown by resource-poor farmers, however, there is much work to be done; first, in adapting GLs to CA techniques and then evaluating the advantages or otherwise of doing so. Broadcast sowing with full tillage remains the predominant practice of resource-poor farmers growing GLs. With the increasing availability of seed drills, it becomes possible to place seed at desired depths and spatial geometry, increasing the likelihood of successful emergence and seedling establishment for each seed sown. Greater seedling populations have been observed for chickpea in Bangladesh when sown using strip tillage, zero tillage or single-pass shallow tillage, as compared with broadcasting with full tillage, using the same seed rate (Vance, 2013). Thus, with drilling, lower seed rates are possible, which lowers the input costs. There is, therefore, much seed rate experimentation to be done for GLs sown with the recently developed minimum tillage implements to determine the optimum seed rates for particular soil types and environmental conditions.

Moving from the broadcasting of seed and fertilizer to drilling them also increases the scope for reducing the fertilizer requirement, due to the ability to place fertilizer in the seed row, thus increasing fertilizer use efficiency. For example, banding of P fertilizer increases its use efficiency in lupin (Bolland and Jarvis, 1996) and faba bean (Bolland et al., 2001), provided the proximity between fertilizer and seed is sufficient to avoid toxicity. Thus, reassessment of optimum fertilizer rates in resource-poor farming situations is required, as this is a particularly costly input for those farmers.

Row sowing of GLs allows greater farmer control over crop canopy structure. Wider row spacing can effectively reduce the incidence of foliar disease (Gan et al., 2006; Reddy et al., 2012) and insect pests (Sithanantham and Reed, 1979) as well as allowing more efficient use of limiting soil water resources (Ritchie, 1983).

Caution is required in the type of CA used for GL cultivation. In Ethiopian vertisols, Erkossa et al. (2006) found that while lentil growth and yield was considerably advantaged by broad bed and furrow cultivation, compared with the traditional practice of multiple tillage to form ridges and furrows, a reduced tillage system was best for the cereals, wheat and teff (*Eragrostis tef*). This was because of the greater waterlogging susceptibility of lentil than the cereals, with the broad bed and furrow system allowing more effective drainage from the waterlogging-prone vertisols.

Nevertheless, the benefits of CA for smallholder farmers can be questionable due to the lack of local and timely availability of key inputs, such as appropriate herbicides, inadequate access to technical guidance, alternative needs for crop residues such as for animal feed, fuel and so on and a disparity between farmer-preferred crops and those most likely to improve soil health in a crop rotation (Giller et al., 2009, 2011). Perhaps the most problematic factor mitigating against smallholder farmers adopting CA practices is the use of herbicides. In large-scale commercial agriculture, these are considered essential, to replace tillage as the main means of weed control. For smallholders, although row sowing would enhance manual weeding by allowing inter-row cultivation with manual implements, increased use of herbicides seems inevitable. This is despite problems of local availability of appropriate herbicides, concerns about product integrity, and the lack of instruction about their effective and safe use (Vogel, 1994). A major problem arising due to the inappropriate use of herbicides, in all forms of agriculture, is the development of herbicide resistance among major weed species (Powles and Yu, 2010). Other non-chemical options for integrated weed management in smallholder cropping systems rely on a holistic understanding of crop and weed ecology. They include using crop rotations unfavourable to major weed species, using cover crops, adjusting sowing time and procedure, using competitive crop genotypes, arranging the planting pattern, minimizing the contamination of crop seed with weed seeds and adjusting the fertilizer strategy to minimize weed competition (Vogel, 1995; Bàrberi, 2002).

It should be noted that CA developed through farmer initiatives on their land, rather than through formal research approaches initiated on research stations (Kassam et al., 2009). If CA is to evolve productively in smallholder agriculture, then it needs to continue as a farmer-led process rather than as a researcher-directed one.

6 Avenues for agronomic improvement: crop intensification

6.1 Short-duration grain legumes

The development of short-duration GLs has increased the scope for fitting them into cropping patterns where traditional longer-duration varieties would be disadvantaged. This permits cropping on land that would otherwise remain fallow, or present a high risk to a crop not properly adapted to those particular conditions. An example is short-duration chickpea, which has permitted the expansion of chickpea production in peninsular India, where abrupt increases in temperature and accelerated depletion of stored soil moisture from February limit productivity and the reliability of traditional longer-duration varieties derived from higher latitudes (Bantilan et al., 2014; Joshi et al., 2007). The development of short-duration pigeonpea has also permitted the expansion of this crop into new cropping systems in peninsular India (Case study 2).

Short-duration mungbean offers an opportunity to fit this crop into a traditionally fallow period between a winter crop, such as wheat, rapeseed, mustard or pulse, and the rainy-season rice crop, in the spring–summer period in South Asia. Cultivation of mungbean has been expanding in the Indo-Gangetic Plain region, thereby increasing the cropping intensity and overall productivity of rice-based or rice–wheat cropping systems (Chadha, 2010).

6.2 Relay sowing

Late harvest of rainy-season rice can delay the planting of winter GLs, and other winter-grown crops, thereby reducing their yield or even preventing their sowing. In this case, cropping intensity can be maintained or improved by relay sowing of the legume crop into rice, that is, broadcasting seed into a rice crop before its harvest often into standing water. This has been a traditional practice in parts of South Asia, particularly for lentil and grasspea which are relatively well adapted to establishing under saturated soil conditions (Johansen et al., 2000). There is scope for promoting this practice and further researching how to improve lentil and grasspea performance under relay sowing, and perhaps adapting other GLs less able to establish under saturated soil conditions, such as chickpea, to the practice. Another opportunity to expand relay sowing is in double-cropped rice systems in South Asia. Where there is a gap of about three months between the harvest of rainy-season rice and transplanting of irrigated spring–summer rice, Malik et al. (2016) showed that relay sowing could be viable.

6.3 Intercropping

Legumes are particularly suited to intercropping with cereal crops, due to the different phenology and resource use of the cereal and the legume (Ofori and Stern, 1987; Willey, 1990). This minimizes competition between them and can thus increase crop productivity per unit of land. A prime example is sorghum–pigeonpea intercropping common in India where the fast-growing sorghum makes first use of light, water and soil nutrients while the initial growth of pigeonpea is slow. After harvesting sorghum, the pigeonpea canopy develops, and its deep-rooting capability allows it to access water and nutrients not used by the sorghum crop (Willey, 1990).

Intercropping can also be advantageous for temperate crops. Intercropping of barley or wheat with faba bean increased total yield and revenue, reduced weed and disease pressure, increased land-use efficiency and thereby enhanced the sustainability of crop production in the Ethiopian highlands (Agegnehu et al., 2006, 2008).

Most farmers who practise cereal–legume intercropping would not want to sacrifice yield of the cereal, which is usually the staple crop, from what they would obtain from a sole cereal crop. They, therefore, adjust the legume population accordingly such that, due to competition from the cereal and suboptimal plant population, intercropped legume yields are below what could be achieved by cropping with a sole legume (ICRISAT, 1981). For example, the greatest productivity and financial return of wheat–faba bean intercropping in Ethiopia was obtained when wheat was sown at its full seed rate and faba bean at 37.5% of its optimum seed rate (Agegnehu et al., 2008). However, intercropping provides stability to the cropping system; if the cereal crop is damaged by biotic or abiotic stress factors, then total crop yield can be compensated for by the legume component.

The principles of intercropping and theory of optimizing agronomy for cereal–GL intercropping have been well researched (Ofori and Stern, 1987; Willey, 1990; Fukai, 1993). There are infinite combinations of species/genotypes, plant populations, spatial geometry, environments and so on to examine through formal agronomic research, but we suggest that all future study be done with farmers on their fields. The challenge is to apply known theory to improve existing farmer practice, build upon their existing intercropping know-how, and encourage their experimentation with alternative options. One such option is to identify GL genotypes better adapted to cereal competition in an intercropping situation, considering that most available genotypes were bred for sole crop situations.

A factor limiting the further deployment of intercropping is the mechanization of harvesting, usually, to harvest the cereal crop without damaging the still-developing GL crop. Indeed, for resource-poor farming in general, mechanization of harvesting of prostrate-habit GLs remains a challenge, particularly to reduce labour costs of manual harvesting. ICARDA research has developed low-cost machine harvest systems for lentil, primarily for use in the West Asia–North Africa region (Sarker and Kumar, 2011). For smallholder farmers generally, there is scope for adapting two-wheel tractors for this purpose.

Many resource-poor farmers still practise traditional mixed-cropping practices, with random broadcasting of the seed of component crops. This retains the row-sown intercropping advantages such as differential resource use, total yield stability, pest and disease suppression (Ramert et al., 2002) and so on but presents problems for weeding and harvesting without damaging the crop. There is little scope for mechanical harvesting of any crop component except the one maturing last. The only scope for improvement in such systems is farmer experimentation with the genotypes used and ratios of the different crop components.

7 Avenues for agronomic improvement: adaptation to climate change

Rainfed GLs have traditionally been at the mercy of variable weather, but existing effects are likely to be exacerbated by climate change, manifesting in increasing weather extremes and gradually increasing mean temperatures and CO_2 levels. There are many obvious threats to GLs, in view of the range of abiotic and biotic stresses to which they are

Table 2 Possible disadvantages and advantages of climate change to the cultivation of grain legumes, summarized from Cooper et al. (2009), Siddique et al. (2012) and Vadez et al. (2012)

Disadvantages	Advantages
Extremes of rainfall and temperature reducing GL yields and increasing cultivation risk	Where competing cereals or other crops are more affected by water and temperature stress, possible new niches arise for GLs
High temperature stress-inducing reproductive damage and end-of-season water deficit	Higher temperatures accelerating phenology to escape end-of-season stress (but with normally lower yield potential of shorter duration crops)
Lower vernalization opportunity for temperate legumes and reduced opportunity for their cultivation at lower latitudes	Greater opportunities for tropical legumes at higher latitudes
Appearance of new pests, diseases and weeds	Reductions in existing pests, diseases and weeds
	Higher atmospheric CO_2 increasing photosynthesis
	Indeterminate growth habit of most GLs allowing better adjustment to weather variability than determinate cereals

prone, but there are also opportunities for increasing GL production. Some of the possible constraints and opportunities are listed in Table 2.

While controlled environment experiments can unravel effects of temperature, soil water availability, CO_2 and their interactions, a more realistic way of assessing possible effects of climate change on GLs is to use holistic crop simulation models. For example, Cooper et al. (2009) used APSIM to simulate groundnut yields as affected by likely climate change at Bulawayo, Zimbabwe. They found that a 10% decrease in rainfall resulted in a median yield reduction of 9%, a 3°C temperature rise reduced yield by 35% and an increase in CO_2 concentration to 700 ppm increased yield by 7%, resulting in an overall climate-change effect of a 31% yield reduction.

Cooper et al. (2009) pointed out that the overall effect of climate change depends on the extent of the yield gap. Where the gap is large, implying the existence of substantial biotic and/or abiotic constraints, the positive effect on yield of directly addressing these constraints is likely to well exceed any yield reduction due to climate change *per se*. Thus climate change is likely to have a larger effect on high input systems with crops yielding near their potential. It is not that resource-poor, smallholder farmers do not need to worry about climate change, but suggests that addressing their existing crop yield constraints is the most effective way of managing any climate-change consequences for them. Thus we need to approach the problem by referring back to Liebig's law of the minimum.

Specific adaptation of GLs to a changing climate over the longer term will largely depend on genetic enhancement of drought resistance, high temperature tolerance during reproductive growth and frost tolerance in temperate zones where climate change increases frequency of frost events (Gu et al., 2008). It will also require genetic alteration of phenology to better fit crops to changed climate patterns (Vadez et al., 2012).

Over the shorter term, however, there are a range of agronomic measures that can be used to buffer existing cultivars against effects of climate change (Siddique et al., 2012).

These include planting density, the judicious use of fertilizers, appropriate rotations, time of planting and supplementary irrigation. Where soil water availability is likely to be lower, planting density could be reduced. In low-input agriculture, alleviation of nutrient deficiencies can compensate effects of temperature rise (Cooper et al., 2009). More diversified crop rotations would provide increased opportunity for introducing crops that may better suit a changing climate, as well as improving soil fertility and breaking pest, disease and weed cycles. Adjustment of time of planting can alter crop phenology to better fit a changing climate. For rainfed production of GLs it becomes ever more important to develop catchment systems to supply supplementary irrigation to assist high-value GL crops through the increased drought periods likely with climate change.

While climate change poses a threat to most crops, some already possessing drought and heat-resistance characteristics may be favoured in replacing more susceptible crops. Chickpea and pigeonpea would be advantaged in this regard (Saxena, 2003). Pigeonpea and mung bean are relatively tolerant of high temperatures, as in northern India they thrive during seasons where maximum temperatures regularly are in the range of 35–45°C, provided soil water is adequate.

Thus climate change is opening up a host of possible interactions, whereby it is impossible to predict the ultimate consequence on GL production at a particular location. To attempt to unpick these interactions and examine possible future scenarios clearly, intensive simulation modelling is required. However, because climate change will induce greater site-to-site variability, a widespread distribution of experimentation is required to test model extrapolations and evolve location-specific best options. This inevitably requires greater emphasis on on-farm research. Although implementation of existing agronomic best options can substantially raise yields even though the climate is changing, ongoing field-based studies are required to adjust that agronomy to climate-change consequences.

8 Refocusing agronomy for grain legumes

Although it has generally been realized, for the last 25 years at least, that meaningful research to improve the agronomy of GLs grown by resource-poor farmers in rainfed environments needs to be more precisely focused on farmers' fields, the institutional and financial commitment to this goal has been inadequate (Nelson et al., 2016). On-farm orientation has largely been treated as peripheral to core agronomic research under 'controlled' conditions and supported by only short-term project funding. Project duration and funding have been inadequate to effectively accomplish the baseline work, including target area characterization and seeking stakeholder opinion, and outcome and impact analysis.

It is proposed that agronomic improvement of GLs grown by resource-poor farmers in rainfed environments needs to follow the same procedure as developed for participatory varietal selection (PVS) and participatory plant breeding (PPB) in targeting resource-poor farmers in rainfed agriculture (Joshi et al., 2007; Ceccarelli et al., 2009). The entire genetic improvement process can be conducted with farmers, mainly on their farms, and has been demonstrated to be more effective in delivering improved varieties to resource-poor farmers than traditional top–down models based on research-station breeding and selection (Ceccarelli et al., 2009). It is suggested that the approach to agronomic improvement of GLs for resource-poor farming situations be modelled on PVS and PPB, as depicted in Table 3.

Table 3 Comparison of on-farm participatory varietal selection (PVS) and participatory plant breeding (PPB) procedures with a suggested procedure for on-farm agronomic improvement of GLs for smallholder farming in rainfed environments

PVS/PPB	On-farm agronomic improvement
Defining and understanding the target environment; baseline analysis	Ditto
PRA and value chain analysis – getting stakeholder opinions	Ditto
PVS with promising genotypes introduced in farmers' fields under their agronomy – mainly mother trials	On-farm diagnostic trials quantifying major limiting and interacting factors
Parental selection and making crosses – with farmers, and can be done on-farm	Evaluation of agronomic solutions feasible for target farmers – on-farm with farmers
Selection in segregating generations PVS of selections – mother and baby trials	Mother and baby trial approach (Snapp, 2002) particularly applicable
Community-based seed production and dissemination	Business development for required inputs for evolved agronomic solutions
Outcome assessment and impact analysis	Ditto

Indeed, it is suggested that there be a more holistic approach to simultaneously targeting both genetic and agronomic options to improving performance of GLs in the fields of resource-poor farmers. Wide distribution of simple treatment comparisons involving both genetic and agronomic options across farmers' fields provides increased understanding of genotype x environment interactions, a key to breeding relevant genotypes (Sarker and Kumar, 2011), as well as to understanding the spatial efficacy of agronomic options. Seed production and distribution of best locally adapted genotypes, around which to develop optimum agronomy, has remained a major constraint to GL improvement in resource-poor farming situations. Seed production strategies premised on a widespread network of farmers in the target areas seems to be most effective, rather than traditional centralized schemes. For example, in the first six months of a project to increase production of quality seed of improved chickpea varieties in Ethiopia, the project expanded its reach from 45 farmers in the previous research phase to 794, of which 20% were women (Woldeyes et al., 2014).

An important aspect of the on-farm approach to agronomic improvement is to focus on feasible solutions within the resource limitations of the farmer. This requires the inclusion of farmers throughout the experimental process. The testing of identified solutions requires comparison with controls on multiple farmers' fields, under farmer management but with researcher guidance in layout and data collection. There are existing and evolving ways of statistically analysing these types of multi-location trials (Riley and Alexander, 1997; Vanlauwe et al., 2016; Reddy et al., 2012), to provide more information about the usefulness of an improved technology than could be obtained from traditional, multi-replicate randomized block trials at a few locations. The involvement of farmers in all stages of the process creates a sense of their ownership of any improvements identified and therefore increases the likelihood of widespread adoption.

Much more needs to be done in moving from an on-station to an on-farm focus for GL agronomic research. Nelson et al. (2016) have recently suggested that this could be best achieved under the framework of Farmer Research Networks. These would be, essentially, groupings of farmers and researchers using emerging developments in ICT to scale up on-farm participatory practices that have so far evolved. This would facilitate more widespread collection of data at the farm level, including that from iterative on-farm trials, its more rigorous analysis and interpretation, and effective communication of outcomes back to farmers. It would permit more rigorous analysis of 'option x context' interactions, which ultimately determine whether or not a new technology is adopted. 'Option' refers to features such as technological innovation and resource requirements and 'context' to biophysical conditions, farming system and social, policy and market aspects (Nelson et al., 2016).

Implementation of the approach proposed by Nelson et al. (2016) would require a massive reorientation of current agricultural research and development focus for resource-poor agriculture. This is now still biased towards replication of simple, comprehensive recommendations rather than confronting the realities of local adaptation in diverse environments and investigating this complexity. Roles of agricultural researchers and agronomists in particular would need to change, from being assessed primarily according to their academic publication record to their contribution to effective outcomes for farmer groups. Funding amounts and project lengths would need to be increased to permit adequate baseline characterization, multi-location and multi-year on-farm trials and built-in outcome and impact analysis. More enduring partnerships need to be established between national and international research bodies, non-government organizations, community-based organizations and commercial entities interacting with farmers. Only then can we expect that smallholder farmers will practically implement the extensive knowledge we have so far to increase the production of GLs.

9 Where to look for further information

A classical publication highlighting the shortcomings of top–down research and development for resource-poor, smallholder farming communities and the need for a farmer-participatory approach was that of Chambers et al. (1989). This approach has been most advanced for participatory plant breeding, with the outcome summarized in Ceccarelli et al. (2009), and the present chapter argues that much of the methodology here can also be applied to agronomic aspects of improving grain legume production.

A recently published comprehensive source for developing sustainable farming systems, of which grain legumes would be an important component, is Snapp, S. and Pound, B. (eds). 2017. *Agricultural Systems: Agroecology and Rural Innovation for Development.* 2nd Edition. Academic Press. ISBN: 9780128020708. 558p.

Detailed information on research progress for individual grain legumes is archived with institutes of the Consultative Group on International Agricultural Research (CGIAR) (http://www.cgiar.org/), as follows:

International Center for Agricultural Research in Dry Areas (ICARDA) (http://www.icarda.org/) – chickpea, lentil, faba bean, grasspea

International Crops Research Institute for the Semi-arid Tropics (ICRISAT) (http://www.icrisat.org/) – chickpea, pigeonpea, groundnut
International Institute of Tropical Agriculture (IITA)(http://newint.iita.org/) – cowpea, soybean
International Centre for Tropical Agriculture (CIAT)(https://ciat.cgiar.org/) – common beans
World Vegetable Center (AVRDC)(https://avrdc.org/)(affiliated with but not a member of CGIAR) – mungbean, soybean.

The CGIAR coordinates grain legume research among its institutes through the CGIAR Research Program (CRP) on Grain Legumes (http://grainlegumes.cgiar.org/).

Other sources for information on recent developments in grain legume research include the International Legume Society (http://ils.nsseme.com/) and the regularly held International Food Legumes Research Conferences (IFLRC), the latest of which was IFLRC VI, held along with the 7th International Conference on Legume Genetics and Genomics (ICLGG VII), at Saskatoon, Canada, during 7–11 July 2014.

10 References

Agegnehu, G., Ghizaw, A. and Sinebo, W. 2006. Yield performance and land-use efficiency of barley and faba bean mixed cropping in Ethiopian highlands. *European Journal of Agronomy*, 25, 202–7. https://doi.org/10.1016/j.eja.2006.05.002

Agegnehu, G., Ghizaw, A. and Sinebo, W. 2008. Yield potential and land-use efficiency of wheat and faba bean mixed intercropping. *Agronomy for Sustainable Development*, 28, 257–63. https://doi.org/10.1051/agro:2008012

Allen, D. J. and Lenné, J. M. 1997. *The Pathology of Food and Pasture Legumes*, Wallingford, UK, CAB International.

Anderson, W., Johansen, C. and Siddique, K. H. M. 2016. Addressing the yield gap in rainfed crops: a review. *Agronomy for Sustainable Development*, 36, 1–13. https://doi.org/10.1007/s13593-015-0341-y

Bantilan, M. C. S., Charyulu, D. K., Gaur, P., Shyam, M. and Davis, J. 2014. Short-duration chickpea technology: Enabling legumes revolution in Andhra Pradesh. ICRISAT Research Program Markets, Institutions and Policies Research. Report No. 23. Patancheru, Telengana, India: International Crops Research Institute for the Semi-Arid Tropics.

Bantilan, M. C. S. and Parthasarathy Rao, D. 1999. Efficiency and sustainability gains from adoption of short-duration pigeonpea in nonlegume-based cropping systems. *Impact Series No. 5.* Patencheru, India: International Crops Research Institute for the Semi-Arid Tropics.

Bàrberi, P. 2002. Weed management in organic agriculture: are we addressing the right issues? *Weed Research*, 42, 177–93. https://doi.org/10.1046/j.1365-3180.2002.00277.x

Barrios, E., Buresh, R. J. and Sprent, J. I. 1996. Organic matter in soil particle size and density fractions from maize and legume cropping systems. *Soil Biology and Biochemistry*, 28, 185–93. https://doi.org/10.1016/0038-0717(95)00110-7

Bell, R. W., Haque, M. E., Johansen, C., Vance, W., Kabir, M. E., Musa, M. A., Mia, M. N. N., Neogi, M. G. and Islam, M. A. 2016. Mechanized minimum soil disturbance, establishment and yield of diverse crops in paddy fields using a two-wheel tractor-mounted planter suitable for smallholder cropping. *Experimental Agriculture*, Submitted.

Bhatia, V. S., Singh, P., Wani, S. P., Rao, A. V. R. K. and Srinivas, K. 2006. Yield Gap Analysis of Soybean, Groundnut, Pigeonpea and Chickpea in India Using Simulation Modeling: Global Theme on Agroecosystems. *Global Theme on Agroecosystems Report no. 31*. Hyderabad, India: International Crops Research Institute for the Semi-Arid Tropics.

Bolland, M. D. A. and Jarvis, R. J. 1996. Effectiveness of different methods of applying superphosphate for lupins grown on sandplain soils. *Australian Journal of Experimental Agriculture*, 36, 707–15. https://doi.org/10.1071/ea9960707

Bolland, M. D. A., Riethmuller, G. P., Siddique, K. H. M. and Loss, S. P. 2001. Method of phosphorus fertiliser application and row spacing on grain yield of faba bean (*Vicia faba* L.). *Australian Journal of Experimental Agriculture*, 41, 227–34.

Bolliger, A., Magid, J., Amado, J. C. T., Neto, F. S., Dos Santos Ribeiro, M. D. F., Calegari, A., Ralisch, R. and De Neergaard, A. 2006. Taking stock of the Brazilian 'Zero-Till Revolution': a review of landmark research and farmers' practice. *Advances in Agronomy*, 91, 47–110. https://doi.org/10.1016/s0065-2113(06)91002-5

Ceccarelli, S., Guimarães, E. P. and Weltzien, E., eds. 2009. *Plant Breeding and Farmer Participation*, FAO, Rome (Italy).

Chadha, M. L. 2010. *Short Duration Mungbean: A New Success in South Asia*, Bangkok, Thailand, APAARI.

Chambers, R., Pacey, A. and Thrupp, L. A. 1989. *Farmer First. Farmer Innovation and Agricultural Research*, London, UK, Intermediate Technology Publications.

Chauhan, Y. S. and Rao, R. C. N. 2014. Defining agro-ecological regions for field crops in variable target production environments: a case study on mungbean in the northern grains region of Australia. *Agricultural and Forest Meteorology*, 194, 207–17. https://doi.org/10.1016/j.agrformet.2014.04.007

Chauhan, Y. S., Venkataratnam, N. and Sheldrake, A. R. 1987. Factors affecting growth and yield of short-duration pigeonpea and its potential for multiple harvests. *The Journal of Agricultural Science*, 109, 519–29. https://doi.org/10.1017/s0021859600081739

Cooper, P., Rao, K. P. C., Singh, P., Dimes, J. P., Traore, P. S., Rao, K., Dixit, P. and Twomlow, S. J. 2009. Farming with current and future climate risk: advancing a 'Hypothesis of Hope' for rainfed agriculture in the semi-arid tropics. *Journal of SAT Agricultural Research*, 7, 1–19.

Dahiya, S. S., Chauhan, Y. S., Srinvastava, S. K., Sekhon, H. S., Waldia, R. S., Gowda, C. L. L. and Johansen, C. 2001. Growing extra-short duration pigeonpea in rotation with wheat in the Indo-Gangetic Plain. *Natural Resouce Management Program – Report*. Patencheru, India: International Crops Research Institute for the Semi-Arid Tropics.

Erkossa, T., Stahr, K. and Gaiser, T. 2006. Soil tillage and crop productivity on a Vertisol in Ethiopian highlands. *Soil and Tillage Research*, 85, 200–11. https://doi.org/10.1016/j.still.2005.01.009

Fischer, R. A., Byerlee, D. and Edmeades, G. 2014. *Crop Yields and Global Food Security*, Canberra, Australia, Australian Centre for International Agricultural Research.

Fitzpatrick, E. A. and Nix, H. A. 1969. A model for simulating soil water regime in alternating fallow-crop systems. *Agricultural Meteorology*, 6, 303–19. https://doi.org/10.1016/0002-1571(69)90023-5

Foyer, C. H., Lam, H.-M., Nguyen, H. T., Siddique, K. H. M, Varshney, R. K., Colmer, T. D., Cowling, W., Bramley, H., Mori, T. A. and Hodgson, J. M. 2016. Neglecting legumes has compromised human health and sustainable food production. *Nature Plants*, 2, 16112. https://doi.org/10.1038/nplants.2016.112

French, R. J. and Schultz, J. E. 1984. Water use efficiency of wheat in a Mediterranean-type environment. II. some limitations to efficiency. *Australian Journal of Agricultural Research*, 35, 765–75. https://doi.org/10.1071/ar9840765

Fukai, S. 1993. Intercropping – bases of productivity. *Field Crops Research*, 34, 239–45. https://doi.org/10.1016/0378-4290(93)90116-5

Gan, Y. T., Siddique, K. H. M., Macleod, W. J. and Jayakumar, P. 2006. Management options for minimizing the damage by ascochyta blight (*Ascochyta rabiei*) in chickpea (*Cicer arietinum* L.). *Field Crops Research*, 97, 121–34. https://doi.org/10.1016/j.fcr.2005.10.002

Giller, K. E., Corbeels, M., Nyamangara, J., Triomphe, B., Affholder, F., Scopel, E. and Tittonell, P. 2011. A research agenda to explore the role of conservation agriculture in African smallholder farming systems. *Field Crops Research*, 124, 468–72. https://doi.org/10.1016/j.fcr.2011.04.010

Giller, K. E., Witter, E., Corbeels, M. and Tittonell, P. 2009. Conservation agriculture and smallholder farming in Africa: the heretics' view. *Field Crops Research*, 114, 23–34. https://doi.org/10.1016/j.fcr.2009.06.017

Gomez, K. A. and Gomez, A. A. 1984. *Statistical Procedures for Agricultural Research*, New York, John Wiley & Sons.

Gu, L., Hanson, P. J., Mac Post, W., Kaiser, D. P., Yang, B., Nemani, R., Pallardy, S. G. and Meyers, T. 2008. The 2007 eastern US spring freeze: increased cold damage in a warming world? *BioScience*, 58, 253–62. https://doi.org/10.1641/b580311

Haque, M. E., Bell, R. W., Islam, A. K. M. S., Sayre, K. and Hossain, M. M. 2011. Versatile Multicrop Planter for two-wheel tractors: An innovative option for smallholders. *5th World Congress of Conservation Agriculture Incorporating 3rd Farming Systems Design Conference*. Brisbane, Australia.

Hobbs, P. R., Sayre, K. and Gupta, R. 2008. The role of conservation agriculture in sustainable agriculture. *Philosophical Transactions of the Royal Society B: Biological Sciences*, 363, 543–55.

ICRISAT. 1981. Proceedings of the International Workshop on Intercropping. 10–13 January 1979 Hyderabad, India. Hyderabad, India: ICRISAT.

Johansen, C., Bakr, M., Islam, M. S., Mondal, N., Afzal, A., Macleod, W., Pande, S. and Siddique, K. H. 2008a. Integrated crop management of chickpea in environments of Bangladesh prone to Botrytis grey mould. *Field Crops Research*, 108, 238–49. https://doi.org/10.1016/j.fcr.2008.05.008

Johansen, C., Duxbury, J. M., Virmani, S. M., Gowda, C. L. L., Pande, S. and Joshi, P. K. 2000. *Legumes in Rice and Wheat Cropping Systems of the Indo-Gangetic Plain-Constraints* and *Opportunities*., Patancheru 502 324, Andhra Pradesh, India; Ithica, New York, USA, International Crops Research institute for the Semi-Arid Tropics; Cornell University.

Johansen, C., Haque, M. E., Bell, R. W., Thierfelder, C. and Esdaile, R. J. 2012. Conservation agriculture for small holder rainfed farming: opportunities and constraints of new mechanized seeding systems. *Field Crops Research*, 132, 18–32. https://doi.org/10.1016/j.fcr.2011.11.026

Johansen, C., Musa, A. M., Harris, D., Islam, M. S. and Ali, M. O. 2008. Integration of chickpea and other rabi crops into rainfed rice-based systems of the High Barind Tract. In Riches, C., Harris, D., Johnson, D. E. and Hardy, B., eds. *Improving Agricultural Productivity in Rice-Based Systems of the High Barind Tract of Bangladesh*, 2008b Los Banos, Philippines. International Rice Research Institute, 135–46.

Johansen, C., Musa, A. M., Kumar Rao, J. V. D. K., Harris, D., Yusuf Ali, M., Shahidullah, A. K. M. and Lauren, J. G. 2007. Correcting molybdenum deficiency of chickpea in the High Barind Tract of Bangladesh. *Journal of Plant Nutrition and Soil Science*, 170, 752–61. https://doi.org/10.1002/jpln.200700249

Joshi, K. D., Musa, A. M., Johansen, C., Gyawali, S., Harris, D. and Witcombe, J. R. 2007. Highly client-oriented breeding, using local preferences and selection, produces widely adapted rice varieties. *Field Crops Research*, 100, 107–16. https://doi.org/10.1016/j.fcr.2006.05.011

Kassam, A. 2014. Overview of the current status of Conservation Agriculture globally and challenges with designing and adapting CA to the circumstances of the smallholders. In Vance, W. H., Bell, R. W. and Haque, M. E. (eds), *Proceedings of the Conference on Conservation Agriculture for Smallholders in Asia and Africa* Mymensingh, Bangladesh: Murdoch University.

Kassam, A., Friedrich, T., Derpsch, R., Lahmar, R., Mrabet, R., Basch, G., González-Sánchez, E. J. and Serraj, R. 2012. Conservation agriculture in the dry Mediterranean climate. *Field Crops Research*, 132, 7–17. https://doi.org/10.1016/j.fcr.2012.02.023

Kassam, A., Friedrich, T., Shaxson, F. and Pretty, J. 2009. The spread of conservation agriculture: justification, sustainability and uptake. *International Journal of Agricultural Sustainability*, 7, 292–320. https://doi.org/10.3763/ijas.2009.0477

Keating, B. A., Carberry, P. S., Hammer, G. L., Probert, M. E., Robertson, M. J., Holzworth, D., Huth, N. I., Hargreaves, J. N. G., Meinke, H., Hochman, Z., Mclean, G., Verburg, K., Snow, V., Dimes, J. P., Silburn, M., Wang, E., Brown, S., Bristow, K. L., Asseng, S., Chapman, S., McCown, R. L.,

Freebairn, D. M. and Smith, C. J. 2003. An overview of APSIM, a model designed for farming systems simulation. *European Journal of Agronomy*, 18, 267–88. https://doi.org/10.1016/s1161-0301(02)00108-9

Kelley, T. G., Ryan, J. G. and Patel, B. K. 1995. Applied participatory priority setting in international agricultural research: making trade-offs transparent and explicit. *Agricultural Systems*, 49, 177–216. https://doi.org/10.1016/0308-521x(94)00030-u

Kerr, R. B., Snapp, S., Chirwa, M., Shumba, L. and Msachi, R. 2007. Participatory research on legume diversification with Malawian smallholder farmers for improved human nutrition and soil fertility. *Experimental Agriculture*, 43, 437–53. https://doi.org/10.1017/s0014479707005339

Kumar, K. and Goh, K. M. 2000. Crop residues and management practices: effects on soil quality, soil nitrogen dynamics, crop yield, and nitrogen recovery. *Advances in Agronomy*, 68, 197–319. https://doi.org/10.1016/s0065-2113(08)60846-9

Kumar Rao, J. V. D. K., Johansen, C. and Rego, T. J. 1998. *Residual Effects of Legumes in Rice and Wheat Cropping Systems of the Indo-Gangetic Plain*, New Delhi, Oxford and IBH Publishing Co. Pvt. Ltd.

Malik, A. I., Ali, M. O., Zaman, M. S., Flower, K., Rahman, M. M. and Erskine, W. 2016. Relay sowing of lentil (*Lens culinaris* subsp. *culinaris*) to intensify rice-based cropping. *The Journal of Agricultural Science*, 154, 850–7. https://doi.org/10.1017/s0021859614001324

Mohammed, H. S. A., Khalifa, G. E. and Ibrahim, O. H. 2015. Economy of faba bean production in the River Nile State: yield gap between research and on-farm *NAF International Working Paper Series*. Pavia, Italy: Natural Resources, Agricultural Development and Food Security International Research Network.

Musa, A. M., Harris, D., Johansen, C. and Kumar, J. V. D. K. 2001. Short duration chickpea to replace fallow after aman rice: the role of on-farm seed priming in the High Barind Tract of Bangladesh. *Experimental Agriculture*, 37, 509–21. https://doi.org/10.1017/s0014479701000448

Nain, M. S., Dubey, S. K., Kumbhare, N. V. and Bahal, R. 2014. Adoption gap as the determinant of instability in Indian legume production. *Journal of Food Legumes*, 27, 146–50.

Nelson, R., Coe, R. and Haussmann, B. I. G. 2016. Farmer research networks as a strategy for matching diverse options and contexts in smallholder agriculture. *Experimental Agriculture*, doi:10.1017/S0014479716000454, 1–20.

Nin-Pratt, A., Johnson, M., Magalhaes, E., You, L., Diao, X. and Chamberlin, J. 2011. *Yield Gaps and Potential Agricultural Growth in West and Central Africa*. Washington DC, USA: International Food Policy Research Institute.

NRI. 2003. *Small Producers in Export Horticulture: A Guide to Best Practice* [Online]. University of Greenwich, UK: Natural Resources Institute. Available: http://projects.nri.org/nret/SPCDR/Chapter4/agriculture-4-4-1.htm [Accessed 5 September 2016].

Nuruzzaman, M., Lambers, H., Bolland, M. D. A. and Veneklaas, E. J. 2005. Phosphorus benefits of different legume crops to subsequent wheat grown in different soils of Western Australia. *Plant and Soil*, 271, 175–87. https://doi.org/10.1007/s11104-004-2386-6

Ofori, F. and Stern, W. R. 1987. Cereal–legume intercropping systems. *Advances in Agronomy*, 41, 41–90. https://doi.org/10.1016/s0065-2113(08)60802-0

Powles, S. B. and Yu, Q. 2010. Evolution in action: plants resistant to herbicides. *Annual Review of Plant Biology*, 61, 317–47. https://doi.org/10.1146/annurev-arplant-042809-112119

Rahman, M. M., Bakr, M. A., Mia, M. F., Idris, K. M., Gowda, C. L. L., Kumar, J., Deb, U. K., Malek, M. A. and Sobhan, A. 2000. Legumes in Bangladesh. In Johansen, C., Duxbury, J. M., Virmani, S. M., Gowda, C. L. L., Pande, S. and Joshi, P. K. (eds), *Legumes in Rice and Wheat Cropping Systems of the Indo-Gangetic Plain: Constraints and Opportunities*. Patancheru 502 324, Andhra Pradesh, India; Ithica, New York, USA: International Crops Research institute for the Semi-Arid Tropics; Cornell University.

Ramert, B., Lennartsson, M. and Davies, G. The use of mixed species cropping to manage pests and diseases–theory and practice. Proceedings of the UK Organic Research 2002 Conference, 2002. Organic Centre Wales, Institute of Rural Studies, University of Wales Aberystwyth, 207–10.

Reddy, M. V., Raju, T. N., Sharma, S. B., Nene, Y. L., Mcdonald, D., Pande, S. and Sharma, M. 2012. *Handbook of Pigeonpea Diseases (Revised). Information Bulletin No. 42*, Patencheru, India, International Crops Research Institute for the Semi-Arid Tropics.

Riley, J. and Alexander, C. J. 1997. Statistical literature for participatory on-farm research. *Experimental Agriculture*, 33, 73–82. https://doi.org/10.1017/s0014479797000185

Ritchie, J. T. 1983. Efficient water use in crop production: discussion on the generality of relations between biomass production and evapotranspiration. In Taylor, H. M., Jordan, W. R. and Sinclair, T. R. (eds), *Limitations to Efficient Water use in Crop Production*. Madison, Wisconsin, USA: American Society of Agronomy.

Rusike, J., Van Den Brand, G. J., Boahen, S., Dashiell, K., Katengwa, S., Ongoma, J., Mongane, D. M., Kasongo, G., Jamagani, Z. B. and Aidoo, R. 2013. Value chain analyses of grain legumes in N2Africa: Kenya, Rwanda, eastern DRC, Ghana, Nigeria, Mozambique, Malawi and Zimbabwe. Wageningen University: N2Africa.

Saha, A. K. 2002. Final Report on Impact Assessment Study for the DFID-funded Project R 7540, Promotion of Chickpea following Rainfed Rice in the Barind Area of Bangladesh. Dhaka, Bangladesh: DFID Plant Science Research Programme, Centre for Arid Zone Studies (CAZS), University of Wales, Bangor, U.K.

Sarker, A. and Kumar, S. 2011. Lentils in production and food systems in West Asia and Africa. *Grain Legumes*, 56, 46–8.

Saxena, N. P. 2003. Management of Agricultural Drought: Agronomic and Genetic Options, New Delhi, Oxford, IBH Publishing Co. Pvt. Ltd.

Schulz, S., Honlonkou, A. N., Carsky, R. J., Manyong, V. M. and Oyewole, B. D. 2003. Alternatives to mucuna for soil fertility management in southern Benin: Farmer perception and use of traditional and exotic grain legumes. *Experimental Agriculture*, 39, 267–78. https://doi.org/10.1017/s0014479703001273

Serraj, R. and Siddique, K. H. M. 2012. Conservation agriculture in dry areas. *Field Crops Research*, 132, 1–6. https://doi.org/10.1016/j.fcr.2012.03.002

Sharma, H. C. 2006. Integrated pest management research at ICRISAT: present status and future priorities. Monograph. Hyderabad, India: International Crops Research Institute for the Semi-Arid Tropics.

Siddique, K. H. M., Johansen, C., Kumar Rao, J. V. D. K. and Ali, M. 2008. Legumes in sustainable cropping systems. In Kharkwal, M. C. (ed.), *Food Legumes for Nutritional Security and Sustainable Agriculture* New Delhi, India: Indian Society of Genetics and Plant Breeding.

Siddique, K. H. M., Johansen, C., Turner, N. C., Jeuffroy, M.-H., Hashem, A., Sakar, D., Gan, Y. and Alghamdi, S. S. 2012. Innovations in agronomy for food legumes. A review. *Agronomy for Sustainable Development*, 32, 45–64. https://doi.org/10.1007/s13593-011-0021-5

Siddique, K. H. M., Regan, K. L., Tennant, D. and Thomson, B. D. 2001. Water use and water use efficiency of cool season grian legumes in low rainfall Mediterranean-type environments. *European Journal of Agronomy*, 15, 267–80. https://doi.org/10.1016/s1161-0301(01)00106-x

Sithanantham, S. and Reed, W. 1979. Plant density and pest damage in chickpea. *International Chickpea Newsletter*, 1, 9–10.

Slatyer, R. O. 1956. Absorption of water from atmospheres of different humidity and its transport through plants. *Australian Journal of Biological Sciences*, 9, 552–8. https://doi.org/10.1071/bi9560552

Snapp, S. 2002. Quantifying farmer evaluation of technologies: the mother and baby trial design. In Bellon, M. R. and Reeves, J. (eds), *Quantitative Analysis of Data from Participatory Methods in Plant Breeding*. Mexico, DF: CIMMYT.

Srivastava, S. K., Sivaramane, N. and Mathura, V. C. 2010. Diagnosis of pulses performance of India. *Agricultural Economics Research Review*, 23, 137–48.

Vadez, V., Berger, J. D., Warkentin, T., Asseng, S., Ratnakumar, P., Rao, K. P. C., Gaur, P. M., Munier-Jolain, N., Larmure, A. and Voisin, A.-S. 2012. Adaptation of grain legumes to climate change: a review. *Agronomy for Sustainable Development*, 32, 31–44.

Vance, W. 2013. *Overcoming Soil Water Constraints to Chickpea Yield in Rainfed Environments of Western Australia and Bangladesh*. PhD, Murdoch University.

Vanlauwe, B., Coe, R. and Giller, K. E. 2016. Beyond averages: new approaches to understand heterogeneity and risk of technology success or failure in smallholder farming. *Experimental Agriculture*, doi:10.1017/S0014479716000193, 1–23.

Vogel, H. 1994. Weeds in single-crop conservation farming in Zimbabwe. *Soil and Tillage Research*, 31, 169–85. https://doi.org/10.1016/0167-1987(94)90078-7

Vogel, H. 1995. The need for integrated weed management systems in smallholder conservation farming in Zimbabwe. *Der Tropenlandwirt-Journal of Agriculture in the Tropics and Subtropics*, 96, 35–56.

Waddington, S. R., Li, X., Dixon, J., Hyman, G. and De Vicente, M. C. 2010. Getting the focus right: production constraints for six major food crops in Asian and African farming systems. *Food Security*, 2, 27–48. https://doi.org/10.1007/s12571-010-0053-8

Willey, R. W. 1990. Resource use in intercropping systems. *Agricultural Water Management*, 17, 215–31. https://doi.org/10.1016/0378-3774(90)90069-b

Witcombe, J. R., Joshi, K. D., Gyawali, S., Musa, A. M., Johansen, C., Virk, D. S. and Sthapit, B. R. 2005. Participatory plant breeding is better described as highly client-oriented plant breeding. I. Four indicators of client-orientation in plant breeding. *Experimental Agriculture*, 41, 299–319. https://doi.org/10.1017/s0014479705002656

Woldeyes, Z. W., Ameda, T. T. and Jiru, S. B. 2014. *Bringing Benefits of Chickpea to more Men and Women Farmers in Southern Ethiopia*. Ottawa, Canada: International Development Research Centre, Canada.

Chapter 11

Grain legume–cereal intercropping systems

L. Bedoussac, ENSFEA, INRA AGIR, France; E-P. Journet, CNRS LIPM, INRA AGIR, France; H. Hauggaard-Nielsen, Roskilde University, Denmark; C. Naudin and G. Corre Hellou, Ecole Supérieure d'Agricultures, France; E. S. Jensen, Swedish University of Agricultural Sciences, Sweden; and E. Justes, INRA AGIR, France[1]

1 Introduction

Intercropping involves simultaneously growing two or more crops in the same field for a significant period of time that could be the whole period of cultural season for some arable crops such as in cereal–legume intercrops (Willey 1979a; Vandermeer et al. 1998; Malézieux et al. 2008; Jensen et al. 2015). The practice is ancient as early records from many human societies all over the world have shown (Willey 1979a). According to Altieri (1999), intercropping systems are estimated to still provide as much as 15–20% of the world's food supply. In Latin America farmers grow 70–90% of their beans with maize, potatoes and other crops, and maize is intercropped on 60% of the region's maize-growing area (Francis, 1986). In rural sub-Saharan Africa, intercropping is considered as a traditional cropping system with the predominant crop combinations being maize, bean/cowpea and pumpkin (Matusso et al. 2014). Intercropping has also been practised in China for thousands of years and it has been estimated that intercropping surfaces are about 30 million ha representing 20–25% of the Chinese arable land (Li, 2001). The practice was widespread in some European farming systems up until the 1950s – before the so-called fossilisation of agriculture. At that time as much as 50% of all available nitrogen (N) may have originated from symbiotic N_2 fixation by leguminous food, forage and green manure crops used in rotation which limited the reactive N received from fertiliser-N and then the negative impact

[1] Full author affiliations: *L. Bedoussac, AGIR, Université de Toulouse, INRA, ENSFEA, Castanet-Tolosan, France; E-P. Journet, LIPM, Université de Toulouse, INRA, CNRS, Castanet-Tolosan, France; AGIR, Université de Toulouse, INRA, Castanet-Tolosan, France; H. Hauggaard-Nielsen, Department of Environmental, Social and Spatial Change, Roskilde University, Denmark; C. Naudin, USC LEVA, INRA, Ecole Supérieure d'Agricultures, Angers, France; G. Corre Hellou, USC LEVA, INRA, Ecole Supérieure d'Agricultures, Angers, France; E. Justes, AGIR, Université de Toulouse, INRA, Castanet-Tolosan, France*

http://dx.doi.org/10.19103/AS.2017.0023.14

of NO_3, NH_3 and N_2O on the environment (Peoples et al. 2009). In those systems, land was dedicated to fertility-generating legume rotations, which potentially also contributed to other ecosystem services such as carbon sequestration and biodiversity (Peoples et al. 2009). Despite these advantages, grain legume cropping is less favoured now, even in organic crop rotations, because of a reputation of low yield and instability related to several factors like intolerance to water stress, harvest difficulties because of lodging, pathogen attacks, sensitivity to insect pests and weed competition. Aiming at higher crop diversity, intercropping is an interesting option to introduce legumes in cropping systems in a more efficient way as compared to sole cropping rotations. Therefore, grain legumes were often grown in different cereal intercrop combinations to secure yield stability and soil fertility, lowering nutrient losses and reducing weeds, diseases and pests (Hauggaard-Nielsen et al. 2001b, 2007). This suggests that intercropping is a way to improve adaptability to climate change taking into account both biotic stresses (Padulosi et al. 2002) and abiotic stresses (e.g. of more and more unpredictable weather patterns (IAASTD 2009)). These intercrops were found all the more efficient and productive when grown in low-input systems, and in particular in organic farming (Bedoussac et al. 2015). Since the 1950s, intercropping declined in Europe and some other parts of the world in the post-war period due to the intensification of agriculture focusing on maximising yields of sole crops using external inputs (artificial fertilisers and pesticides) together with the optimisation of mechanical management (Crews and Peoples 2004; Anil et al. 1998; Malézieux et al. 2008).

There is now renewed interest in intercropping in Europe in achieving 'sustainable, ecological or eco-functional intensification' of agricultural production, particularly in organic farming (Anil et al. 1998; Hauggaard-Nielsen and Jensen 2005; Malézieux et al. 2008; Bedoussac et al. 2015). Exploiting the ability of legumes to fix free atmospheric N_2 means potentially less reliance on the use of N fertiliser input (Fustec et al. 2010), thus reducing equivalent CO_2 emissions (Nieder and Benbi 2008) and the carbon footprints of agricultural products (Gan et al. 2011). Because intercrops could be an efficient way to reduce the damages due to some pests (e.g. aphids), weeds and shoot diseases in comparison to legume sole crops, intercropping is of particular interest in organic farming systems (Jensen, 1996a; Corre-Hellou et al. 2011; Ndzana et al. 2014; Bedoussac and Justes 2010a,b; Naudin et al. 2010). Intercropping is also known as a method to boost crop productivity (Qin et al. 2013), improve land utilisation (Agegnehu et al. 2008), reduce reliance on fertilisers and risks of nitrate leaching compared to sole cropping (Hauggaard-Nielsen et al. 2003; Adad et al. 2004; Corre-Hellou and Crozat 2005) and reduce greenhouse gas emissions compared to the use of sole cropping strategies (Oelhermann et al. 2009; Naudin et al. 2014). Intercropping is of particular interest in temperate regions where organic arable crop rotations often comprise sole crops, that is, pure stands, including annual legumes (Hauggaard-Nielsen et al. 2001b) with a number of abiotic and biotic factors influencing yields. Moreover, Jensen et al. (2015) proposed the concept that intercropping could be an efficient way of precision ecological farming allowing the improvement of resource use efficiency due to the adjustment of the mixed cover to resource spatial variability.

This chapter summarises data from over 50 field experiments undertaken since 2001 on cereal–grain legume intercropping in 13 sites in Southern and Western France as well as in Denmark using spring and winter cereal–grain legume intercrops (Bedoussac et al. 2014, 2015). More detailed information concerning these experiments can be found in Hauggaard-Nielsen et al. (2007, 2001a,b), Knudsen et al. (2004), Naudin et al. (2009), Bedoussac and Justes (2010a,b) and Bedoussac et al. (2014, 2015). The following combinations of intercrops were evaluated:

- spring barley (*Hordeum vulgare*)–spring pea (*Pisum sativum*);
- spring barley–spring faba bean (*Vicia faba*);
- soft wheat (*Triticum aestivum*)–winter pea;
- soft wheat–spring faba bean;
- durum wheat (*Triticum turgidum*)–winter pea;
- durum wheat–winter faba bean.

The experiments covered a range of intercropping methods mostly in substitutive design such as:

- sowing in separate rows or mixing within the same row;
- sowing different proportions of cereals and legumes in a field.

They also tested the effects of using organic fertiliser-N on the interspecific competitive interactions taking place in the crop stand. Yields of intercrops were compared to yields of the corresponding sole crops sown on the same date, with the same level of fertiliser-N and harvested at crop maturity (i.e. of the later crop in the case of intercrops).

2 Effects on yields and quality

Intercropping has been shown to increase crop yields in low-N input systems in particular (Hauggaard-Nielsen et al. 2009b; Lithourgidis et al. 2006). In the French and Danish intercropping studies, the total grain yield of the intercrop (cereal plus legume) was on average 3.3 ± 1.0 Mg ha^{-1}. This was typically more than the mean yield of the comparable sole cereal crop (2.9 ± 0.9 Mg ha^{-1}) and the comparable sole legume crop (2.4 ± 1.4 Mg ha^{-1}) (Bedoussac et al. 2014, 2015). These results confirm others analysing both conventional and organic farming which also show a higher grain yield from intercrops compared to sole crops, particularly for cereal–legume combinations (Jensen 1996a; Bedoussac and Justes 2010a,b; Hauggaard-Nielsen et al. 2009b).

Numerous studies suggest that intercrops are particularly suited to low-N-input systems (Willey 1979a,b; Ofori and Stern 1987; Vandermeer 1989; Willey 1990; Fukai 1993; Jensen 1996a; Hauggaard-Nielsen et al. 2003; Corre-Hellou et al. 2006; Bedoussac and Justes 2010a,b; Bedoussac et al. 2015). A study of durum wheat–winter pea intercrops by Bedoussac and Justes (2010a) showed that when fertiliser-N is applied, whilst wheat yields increased, intercropped legume growth and yield were reduced, leading to a lower yield of the intercrop compared to the fertilised sole crop wheat. This implies that intercropping may be advantageous when N availability (soil N plus fertiliser-N) is below a determined threshold (12 g N m^{-2} in the French and Danish experiments). Similar results have been reported for several cereal–legume intercrops grown in arid, semi-arid, tropical and temperate climates (Fujita et al. 1992; Ofori and Stern 1987; Jensen 1996a; Naudin et al. 2010).

The land equivalent ratio (LER) is a widely used indicator to compare the efficiency of sole crops and intercrops. The LER is defined as the relative land area required when growing sole crops to produce, for example, the yield achieved in an intercrop with the same species proportion (Willey and Osiru 1972). LER>1 indicates a per-area advantage of intercropping compared to sole cropping in terms of improved use of resources such as light, water and N. The French and Danish studies show that grain-yield-based LER values are greater than 1 for almost all the experiments (1.27 on average), indicating an advantage of intercropping compared to sole cropping (Bedoussac et al. 2015). It is important to note that the LER is dependent on the conditions relating to the sole crop (Mead and Willey 1980; Jolliffe 2000) and that relative and absolute production performances are not necessarily linked (Bedoussac and Justes 2011). Species mixtures with the highest LER values do not necessarily have the highest absolute productivity (Garnier et al. 1997; Jolliffe and Wanjau 1999). Other indices can also be useful for evaluating species interactions and intercrop efficiency (Weigelt and Jolliffe 2003). A comparison of commonly used indices has been done by Bedoussac and Justes (2011) on durum wheat–winter pea intercrops demonstrating the interests and limits of LER that must be analysed in the framework of level of yield produced.

To obtain profitable yield and grain protein concentration, cereals are generally fertilised with high levels of nitrogen not only in conventional cropping systems but also in organic systems. On the contrary, in lower-nitrogen-input systems, limiting nitrogen resource makes it difficult to reach a sufficient grain protein concentration as required by agrofood industries either for soft wheat to make bread or for durum wheat to make semolina and pasta. Yet, in addition to interesting increases in yield and productivity (global performance of both species), intercropping has also been shown to increase cereal grain protein concentration and baking quality (Gooding et al. 2007). Results from the French and Danish studies show that the protein concentration of the intercropped cereal is almost always greater than that of the respective cereal sole crop (11.1 ± 1.7% and 9.8 ± 1.7%, respectively), although no significant difference was observed in the grain protein concentration between the intercrop and sole crop legume (Bedoussac et al. 2014). A number of studies have shown that the effect of intercropping on wheat grain protein concentration is due to a higher mineral soil N availability for the cereal on a per-plant or a per-grain basis in intercropping compared to sole crops (Gooding et al. 2007; Bedoussac and Justes 2010a,b). In a review of wheat–faba bean intercrops in five regions across Europe, Gooding et al. (2007) concluded that despite a 25–30% reduction in wheat yield compared to conventional sole cropping cereal with high levels of N fertilisation, intercropping could still have an overall economic benefit resulting from the higher protein concentration of the intercropped wheat, combined with the added value of the legume crop. There is therefore an economic benefit in intercropping in sectors such as organic agriculture. Intercropping can also offset problems of low and variable yields in organic grain legume cropping associated with water stress, lodging, diseases and competition from weeds.

3 Agronomical performance of intercropping

3.1 Ecological processes involved in intercrops

One key pathway to ecological intensification is the exploitation of diversity in cropping systems by increasing the number of cultivated species and varieties cropped in species mixtures, potentially including a larger proportion of legumes in cropping systems. The interactions between species taking place in species mixtures are complex and can be described by the 'four Cs': 1) *Competition* for resources, 2) *Complementarity*, allowing a more efficient use of resources by the mixtures compared with sole cropping, 3) *Cooperation* through facilitation, when the modification of the environment by one species is beneficial to the other(s), for example, by reducing disease attack (Finckh and Wolfe 2015), weed competition (Hauggaard-Nielsen et al. 2001b) or by increasing mineral N and P availability (Hinsinger et al. 2011) and 4) *Compensation*, when species differ in their sensitivity to abiotic and/or biotic stress and the demise of one is compensated by the other(s) through release from competition. The relative contribution of these processes to yield advantages of mixtures depends on the species composition and varies in time and space.

There are a number of factors explaining the performances of intercropping and the processes involved that can boost the agronomical intercrop performance.

3.2 Nitrogen use efficiency

Some studies have shown a higher quantity of N accumulated per kg of grain in intercropped cereals compared to sole cropped cereals. A number of studies have demonstrated the way the intercrop legume facilitates the absorption of soil mineral N by the intercrop cereal (Stern 1993; Xiao et al. 2004) together with the transfer of N from the legume to the cereal which represent in annual barley/peas intercrops only up to 19% (Jensen 1996b). The amount of N exchanged between plants is higher when soil mineral N is low and when roots are intermingled (Jensen 1996a; Xiao et al. 2004; Fustec et al. 2014). However, within the growth time of annual intercrops, Fustec et al. (2014) showed that the net balance between the legume and the companion crop was negligible because the transfers from pea into wheat were not significantly different from those from wheat towards pea. Then, this higher efficiency in intercrops is mostly the result of the complementary way intercropped species use available N from the soil or atmosphere (Jensen 1996a; Bedoussac and Justes 2010a; Corre-Hellou et al. 2006). Legumes in an intercrop are 'forced' to rely more on atmospheric N because cereals strongly compete for soil mineral N (Hauggaard-Nielsen et al. 2001a; Bellostas et al. 2003). This increases the legume N proportion from N_2 fixation and expands niche complementarity increasing the global N availability for the intercrop (Jensen 1996a; Corre-Hellou et al. 2006; Hauggaard-Nielsen et al. 2009b; Naudin et al. 2010).

3.3 Light interception and utilisation

Intercrops are known to be more efficient users of light compared to sole crops (Jahansooz et al. 2007; Bedoussac and Justes 2010b). This reflects differences in their shoot architecture and crop life cycles which allow them to intercept light in ways that are more complementary and less competitive, supporting higher overall biomass and yield (Trenbath 1986; Tsubo et al. 2001; Tsubo and Walker 2002). There is a general agreement that the partition of radiation when intercropping different species is primarily influenced by vertical competition (Spitters and Aerts 1983; Caldwell 1987; Cudney et al. 1991; Cenpukdee and Fukai 1992a,b) and secondly by the crop row orientation and the light extinction coefficient of the leaves of each species. As an illustration, modern pea varieties differ in morphology, primarily in stem length and leaf type. Then, semi-leafless varieties with short stems and all the leaflets transformed into tendrils could be more adapted for intercropping. Indeed, compared to the normal-leafed varieties with a large leaflet area, semi-leafless varieties allow more light available for the initial growth of the companion crop (Mikić et al. 2015). Similar results can be expected when considering different legume growth habits (indeterminate, semi-determinate and determinate), which could have an important influence on competition and complementarity.

3.4 Weed control

Intercropping reduces the weed infestation level, in particular compared to sole crop of legume (Hauggaard-Nielsen et al. 2007) reducing the needs for chemical or mechanical weed control (Vasilakoglou et al. 2005; Banik et al. 2006; Corre-Hellou et al. 2011). It has been shown that more competitive cereals especially for soil mineral N uptake such as barley can help reduce weed growth in grain legumes such as peas which are less able to take up soil mineral N and thus compete with weeds (Corre-Hellou et al. 2011). These crops could strongly control or suppress weed growth by competing for resources such as light and nutrients (Bedoussac and Justes 2010b; Anil et al. 1998). Results from the French and Danish studies show that weed biomass within the intercrops or the cereal sole crops at harvest are comparable (0.04 kg m^{-2}) but significantly lower compared to the legume sole crops (0.14 kg m^{-2}). These results are consistent with those obtained by Corre-Hellou et al. (2011) on pea–barley intercrops where weed control was high and consistent even with a low percentage of barley in the total biomass, whereas it was lower and more variable in pea sole crops.

3.5 Pests and diseases control

Intercropping has been found to reduce some pests and diseases (Trenbath 1993; Altieri 1999; Hauggaard-Nielsen et al. 2007; Corre-Hellou and Crozat 2005; Ratnadass et al. 2012). Andow (1991) analysed 209 studies on crop mixtures involving 287 different species of parasitic insects. The insects were significantly fewer in 52% of the cases (149 species) compared with monocultures, and greater in 15% of the cases (44 species). This has been linked to physical barrier effects and possible chemical effects against insect and disease spread from the use of particular intercrops (Vandermeer 1989; Hauggaard-Nielsen and Jensen 2005; Ndzana et al. 2014). As an illustration, *Acyrthosiphon pisum* Harris (Aphididae: Hemiptera), the pea aphid, can be significantly decreased when intercropping winter pea with durum wheat (Ndzana et al. 2014). Ndzana et al. (2014) suggested that a mechanism related to the resource concentration hypothesis may explain the associational resistance of the intercrop towards *A. pisum* since substitutive systems were less infested than additive systems and row mixtures were less infested than mixtures on the row.

4 Cultivation practices in intercropping

In multispecies mixtures such as intercrop, the interactions between species can be represented as the effect of one species on the environment and the response of the other species to this change (Vandermeer 1989; Goldberg 1990). The interactions are complex, occur dynamically over time and space (Connolly et al. 1990) and depend, amongst other things, on the availability of nutrients, soil–climatic conditions and the companion species and cultivars. Despite this complexity which makes generalisation difficult, it is possible to identify cultivation practices which will improve the overall performance.

4.1 Agricultural management practices adapted for intercropping

Naudin et al. (2010) have suggested cultivation needs to match the farmer's overall production objective, whether

- improving the quality of the cereal by maximising the availability of soil mineral N and by increasing the symbiotic fixation rate of the legume; or
- producing legumes using intercrops by reducing weed pressure and spread of diseases and pests because of a cereal competition or physical barrier effect, respectively.

In the former case, one would favour an early developing cereal to rapidly deplete the mineral N pool available to the legume. For the legume, one would choose species and varieties with a root development able to cover its early N needs, including an early start of leguminous symbiotic N_2 fixation. If the focus is on legumes, one would choose high legume densities close to those of sole cropping and lower densities for the companion cereal without the use of N fertilisation. The main role of the cereal in such intercrop is to reduce weed pressure and the spread of diseases and pests and, in the case of peas, to provide mechanical support to avoid lodging.

4.2 Management practices to optimise the use of light and nitrogen

Whatever the intercrop species mixture is, two general principles in improving output are to:

- improve the use of light energy; and
- improve the use of N sources.

Looking at the use of light, the dominant species should have a shoot architecture and biomass production that allows a reasonable amount of light to reach the understorey (Berntsen et al. 2004; Jahansooz et al. 2007). As an example, in the case of durum wheat–winter pea intercrops, a short-strawed durum wheat variety would be selected for intercropping with winter peas, and a long-strawed variety for intercropping with faba beans (Bedoussac 2009).

Nitrogen availability strongly affects species complementarity. Increased availability of soil mineral N in early growth stages will result in reduced amounts of fixed N, a reduced legume yield and an increased cereal yield (Hauggaard-Nielsen and Jensen 2001b; Bedoussac and Justes 2010a; Naudin et al. 2010). As an example, Naudin et al. (2010) showed that mineral N fertilisation applied after the beginning of pea flowering stops its symbiotic fixation activity compared to the unfertilised treatment. This was confirmed by the experiments on pea plants grown under greenhouse conditions (Naudin et al. 2011). Conversely, late availability of soil N will have little or no effect on overall symbiotic fixation and yield of the legume but will improve the protein concentration of the cereal. Unlike mineral N, which is immediately available, organic manures undergo soil microbial processes including mineralisation. Consequently, only early applications of organic N from animal manure, green manuring, etc., can have an effect on the behaviour of the intercrop and, in particular, on the proportions of the two species at harvest (Andersen et al. 2007).

4.3 Sowing designs

Hauggaard-Nielsen et al (2006) showed in a pea–barley intercrop field study how changes in plant density and relative crop frequency had a marked effect on the interspecies dynamics, obviously influencing the functioning of intercrops and their potential benefits. In intercrops, the optimal total plant density can be greater than that of each of the sole crops because of the complementarity between species (Willey and Osiru 1972). The increase in plant density increases the competition between the components of the mixture which tends to favour the dominant species (Willey 1979a). This suggests a higher density of the dominated species (more than 50% of that in sole cropping) and a lower density of the dominant species (less than 50% of that in sole cropping) to manage competitive effects. As an example, Bedoussac (2009) has shown that in mixtures with durum wheat, the density of peas could be the same as that of the sole crop with a proportion of cereal grain in the intercrop of about 50%.

Variations in the spatial structure of intercrops (such as mixtures within the row, alternate rows or strips of varying width) and row orientation will also modify the distribution of radiation, water and nutrients. Such effects have been reported on maize–pigeon pea mixtures (Dalal 1974), maize–soya and sorghum–soya mixtures (Mohta and De 1980) or barley–pea intercrops (Chen et al. 2003). This suggests that densities should be chosen according to the spatial arrangement of the species and how they compete as well as production objectives. The sowing pattern also needs to take mechanical weeding operations into account, for example, using a tine harrow (an effective tool widely used in organic farming).

4.4 Harvesting intercrops

Intercrops can only be marketed if the grains of each species can be correctly sorted. One of the main obstacles in intercropping is the capacity for sorting large volumes efficiently, quickly and cheaply. On the basis of the French studies, it is possible to correctly separate the grains of the two species, that is, wheat and pea or faba bean, provided that they sufficiently differ in size and/or shape and that the mixture does not contain too many broken grains (Bedoussac et al. 2015). To achieve this requires both species of intercrop reach maturity at similar dates and that adjustments are made to combined harvesters to suit the more fragile species (at the risk of losing some of the grains of the other species). This practical question thus poses various difficulties not only in terms of choice of machinery and its adjustment but also from the logistic point of view of the companies collecting and storing seeds. Indeed, their organisational structure can play the role of a self-reinforcement mechanism that reduces the incentives to adopt new practices (Fares et al. 2012).

5 Future trends and conclusion

This chapter has shown the potential for increasing the use of grain legume–cereal intercropping reviewing primarily the work conducted under low-N input conditions such as organic agriculture. However, the complexities of different species combinations and production objectives of intercrops make generalisation difficult. As an example, varietal selection criteria typically used for sole crops are not necessarily appropriate for intercrops (Carr et al. 1998). This highlights the limits of field experiments and the value of modelling multispecies cropping systems to optimise decisions on combinations to use as well as cultivation practice and production outcomes (Brisson et al. 2004; Corre-Hellou et al. 2009; Barillot et al. 2012, 2014a,b). This requires a better mechanistic understanding of the behaviour of multispecies cropping systems and the integration of this knowledge into current crop models or the development of new models that better represent inter- and intraspecific competitions (Launay et al. 2009). Furthermore, increased focus on farmer-driven objectives for intercropping use and stakeholder needs for markets is required to challenge perceptions, routines, rules and regulations at the level of the cropping system and the social and economic context in which farmers and actors of the agrofood chain operate. Reintroduction of intercropping strategies has to be negotiated and locally adjusted by relevant stakeholders to be effective. Such dedicated intercropping systems are not restricted to cereal/grain legume mixtures, so it would certainly be appropriate to establish linkages with other strategies such as perennial grasslands and agroforestry by sharing knowledge and tools to create a generic model of the behaviour of multispecies canopies. Such an approach should be a way to extend this knowledge to more traditional intercropping systems practised by farmers notably in Africa and Asia. Indeed, research works dedicated to these intercrop systems are limited compared to those focusing on cereal–legume intercrops in Europe. Finally, intercropping is one option with crop rotation to increase diversification which must be combined with the spatial distribution of crop at the farm scale in order to increase the delivery of ecological services.

6 References

Abad, A., Lloveras, J. and Michelena, A. (2004) Nitrogen fertilization and foliar urea effects on durum wheat yield and quality and on residual soil nitrate in irrigated Mediterranean conditions. *Field Crops Res.* 87:257–69.

Agegnehu, G., Ghizaw, A. and Sinebo, W. (2008) Yield potential and land-use efficiency of wheat and faba bean mixed intercropping. *Agron. Sustain. Dev.* 28:257–63.

Altieri, M. (1999). The ecological role of biodiversity in agroecosystems. *Agric. Ecosyst. Environ.* 74:19–31.

Andersen, M. K., Hauggaard-Nielsen, H., Weiner, J. and Jensen, E. S. (2007) Evaluating competitive dynamics in two and three component intercrops of *Pisum sativum*, *Hordeum vulgare* and *Brassica napus*. *J. Appl. Ecol.* 44:545–51.

Andow, D. A. (1991). Yield loss to arthropods in vegetationally diverse ecosystems. *Environ. Entomol.* 20:1228–35.

Anil, L., Park, J., Phipps, R. H. and Miller, F. A. (1998). Temperate intercropping of cereals for forage: A review of the potential for growth and utilization with particular reference to the UK. *Grass Forage. Sci.* 53:301–17.

Banik, P., Midya, A., Sarkar, B. K. and Ghose, S. S. (2006) Wheat and chickpea intercropping systems in an additive experiment. Advantages and weed smothering. *Europ. J. Agron.* 24:325–32.

Barillot, R., Combes, D., Pineau, S., Huynh, P. and Escobar-Gutiérrez, A. (2014a) Comparison of the morphogenesis of three genotypes of pea (*Pisum sativum*) grown in pure stands and wheat-based intercrops. *AoB Plants,* 6. doi:10.1093/aobpla/plu006.

Barillot, R., Escobar-Gutiérrez, A. J., Fournier, C., Huynh, P. and Combes, D. (2014b) Assessing the effects of architectural variations on light partitioning within virtual wheat-pea mixtures. *Ann. Bot.* 114(4):725–37.

Barillot, R., Combes, D., Chevalier, V., Fournier, C. and Escobar-Gutierrez, A. (2012). How does pea architecture influence light sharing in virtual wheat–pea mixtures? A simulation study based on pea genotypes with contrasting architectures. *AoB Plants*. 2012:1–15.

Bedoussac, L., Journet, E. P., Hauggaard-Nielsen, H., Naudin, C., Corre Hellou, G., Jensen, E. S., Prieur, L. and Justes, E. (2015). Ecological principles underlying the increase of productivity achieved by cereal-grain legume intercrops in organic farming: A review. *Agron. Sustain. Dev.*, 35(3):911–35.

Bedoussac, L., Journet, E. P., Hauggaard-Nielsen, H., Naudin, C., Corre Hellou, G., Prieur, L., Jensen, E. S. and Justes, E. (2014). 'Eco-functional intensification by cereal-grain legume intercropping in organic farming systems for increased yields, reduced weeds and improved grain protein concentrations', In: Bellon, S. and Penvern, S. (Eds), *Organic Farming: Prototype for Sustainable Agricultures*. Springer, Berlin, Germany. pp. 47–63.

Bedoussac, L. and Justes, E. (2011). Comparison of commonly used indices for evaluating species interactions and intercrop efficiency: Application to durum wheat-winter pea intercrops. *Field Crops Res.* 124:25–36.

Bedoussac, L. and Justes, E. (2010a). The efficiency of a durum wheat-winter pea intercrop to improve yield and wheat grain protein concentration depends on N availability during early growth. *Plant Soil* 330:19–35.

Bedoussac, L. and Justes, E. (2010b). Dynamic analysis of competition and complementarity for light and N use to understand the yield and the protein content of a durum wheat-winter pea intercrop. *Plant Soil* 330:37–54.

Bedoussac, L. (2009). *Analyse du fonctionnement des performances des associations blé dur-pois d'hiver et blé dur-féverole d'hiver pour la conception d'itinéraires techniques adaptés à différents objectifs de production en systèmes bas-intrants.* Mémoire de thèse, Thèse de doctorat, Institut National Polytechnique de Toulouse, Toulouse, FRA, 220p.

Bellostas, N., Hauggaard-Nielsen, H., Andersen, M. and Jensen, E. S. (2003). Early interference dynamics in intercrops of pea, barley and oilseed rape. *Biol. Agric. Hort.* 21:337–48.

Berntsen, J., Hauggaard-Nielsen, H., Olesen, J. E., Petersen, B. M., Jensen, E. S. and Thomsen, A. (2004) Modelling dry matter production and resource use in intercrops of pea and barley. *Field Crops Res.* 88:69–83.

Brisson, N., Bussière, F., Ozier-Lafontaine, H., Tournebize, R. and Sinoquet, H. (2004). Adaptation of the crop model STICS to intercropping. Theoretical basis and parameterisation. *Agronomie* 24:409–21.

Caldwell, R. M. (1987). Plant architecture and ressource competition. *Ecol. Stud.* 61:164–79.

Carr, P. M., Martin, G. B., Caton, J. S. and Poland, W. (1998). Forage and nitrogen yield of barley-pea and oat-pea intercrops. *J. Agron.* 90:79–84.

Cenpukdee, U. and Fukai, S. (1992a). Cassava/legume intercropping with contrasting cassava cultivars. 1. Competition between component crops under three intercropping conditions. *Field Crops Res.* 29:113–33.

Cenpukdee, U. and Fukai, S. (1992b). Cassava/legume intercropping with contrasting cassava cultivars. 2. Selection criteria for cassava genotypes in intercropping with two contrasting legume crops. *Field Crops Res.* 29:135–49.

Chen, Y., Yu, S. and Yu, Z. (2003). Relationship between amount or distribution of PAR interception and grain output of wheat communities. *Acta Agronomica Sinica* 29:730–4.

Connolly, J., Wayne, P. and Murray ,R. (1990). Time course of plant-plant interactions in experimental mixtures of annuals – Density, frequency and nutrient effects. *Oecologia* 82:513–26.

Corre-Hellou, G., Dibet, A., Hauggaard-Nielsen, H., Crozat, Y., Gooding, M., Ambus, P., Dahlmann, C., von Fragstein, P., Pristeri, A., Monti, M. and Jensen, E. S. (2011). Competitive ability of pea-barley intercrops against weeds and interactions with crop productivity and soil N availability. *Field Crops Res.* 122:264–72.

Corre-Hellou, G., Faure, M., Launay, M., Brisson, N. and Crozat, Y. (2009). Adaptation of the STICS intercrop model to simulate crop growth and N accumulation in pea-barley intercrops. *Field Crops Res.* 113:72–81.

Corre-Hellou, G., Fustec, J. and Crozat, Y. (2006). Interspecific competition for soil N and its interaction with N2 fixation, leaf expansion and crop growth in pea-barley intercrops. *Plant Soil* 282:195–208.

Corre-Hellou, G. and Crozat, Y. (2005). N2 fixation and N supply in organic pea (*Pisum sativum L.*) cropping systems as affected by weeds and peaweevil (*Sitona lineatus L.*). *Eur. J. Agron.* 22:449–58.

Crews, T. E., Peoples, M. B. (2004). Legume versus fertilizer sources of nitrogen: Ecological tradeoffs and human needs. *Agric. Ecosyst. Environ.* 102:279–97.

Cudney, D., Jordan, L. and Hall, A. (1991). Effect of wild oat (*Avena fatua*) infestations on light interception and growth rate of wheat (*Triticum aestivum*). *Weed Sci.* 39:175–9.

Dalal, R. C. (1974). Effects of intercropping maize with pigeon peas on grain yield and nutrient uptake. *Exp. Agric.* 10:219–24.

Fares, M., Magrini, M. B. and Triboulet, P. (2012). Transition agroécologique, innovation et effets de verrouillage: Le rôle de la structure organisationnelle des filières. Le cas de la filière blé dur française. *Cah. Agric.* 21:34–45.

Finckh, M. and Wolfe, M. (2015). 'Biodiversity enhancement', In: Finckh, M., van Bruggen, A. and Tamm, L. (Eds), *Plant Diseases and Their Management in Organic Agriculture*.

Francis, C. A., 1986. *Multiple Cropping Systems*. MacMillan, New York, 383p.

Fujita, K., Ofosubudu, K. G. and Ogata, S. (1992). Biological nitrogen fixation in mixed legume-cereal cropping systems. *Plant Soil* 141:155–75.

Fukai, S. (1993). Intercropping-bases of productivity. *Field Crops Res.* 34:239–45.

Fustec, J., Jamont, M., Joly, F. X., Piva, G., Cannavacciuolo, M., Cassagne, N. (2014). Measuring nitrogen transfer in annual intercrops. In: *Proceedings of the European Society for Agronomy Congress*. ESA2014 25–29 August 2014, Debrecen, Hungary, 2 p.

Fustec, J., Lesuffleur, F., Mahieu, S. and Cliquet, J. B. (2010) Nitrogen rhizodeposition of legumes. *Rev. Agron. Sustain. Dev.* 30:57–66.

Gan, Y. T., Liang, C., Hamel, C., Cutforth, H. and Wang, H. (2011). Strategies for reducing the carbon footprint of field crops for semiarid areas - a review. *Agron. Sustain. Dev.* 31:643–56.

Garnier, E., Navas, M. L., Austin, M. P., Lilley, J. M. and Gifford, R. M. (1997). A problem for biodiversity-productivity studies: How to compare the productivity ofmultispecific plant mixtures to that ofmonocultures? *Acta Oecologica* 18(6):657–70.

Goldberg, D. (1990). *Components of Resource Competition in Plant Communities. Perspectives on Plant Competition*. Academic Press, San Diego, pp. 27–50.

Gooding, M. J., Kasynova, E., Ruske, R., Hauggaard-Nielsen, H., Jensen, E. S., Dahlmann, C., von Fragstein, P., Dibet, A., Corre Hellou, G., Crozat, Y., Pristeri, A., Romeo, M., Monti, M. and Launay, M. (2007). Intercropping with pulses to concentrate nitrogen and sulphur in wheat. *J. Agric. Sci.* 145:469–79.

Hauggaard-Nielsen, H., Gooding, M., Ambus, P., Corre-Hellou, G., Crozat, Y., Dahlmann, C., Dibet, A., von Fragstein, P., Pristeri, A., Monti, M. and Jensen, E. S. (2009b). Pea-barley intercropping for efficient symbiotic N2-fixation, soil N acquisition and use of other nutrients in European organic cropping systems. *Field Crops Res.* 113:64–71.

Hauggaard-Nielsen, H., Jørnsgard, B., Kinane, J., Jensen, E. S. (2007). Grain legume-cereal intercropping: The practical application of diversity, competition and facilitation in arable and organic cropping systems. *Renew. Agric. Food Syst.* 23:3–12.

Hauggaard-Nielsen, H., Andersen, M. K., Jørnsgard, B. and Jensen, E. S. (2006). Density and relative frequency effects on competitive interactions and resource use in pea-barley intercrops. *Field Crops Res.* 95:256–67.

Hauggaard-Nielsen, H. and Jensen, E. S. (2005). Facilitative root interactions in intercrops. *Plant Soil* 274:237–50.

Hauggaard-Nielsen, H., Ambus, P. and Jensen, E. S. (2003). The comparison of nitrogen use and leaching in sole cropped versus intercropped pea and barley. *Nutr. Cycl. Agroecosyst.* 65:289–300.

Hauggaard-Nielsen, H., Ambus, P., and Jensen, E. S. (2001a). Temporal and spatial distribution of roots and competition for nitrogen in pea-barley intercrops - a field study employing P-32 technique. *Plant Soil.* 236:63–74.

Hauggaard-Nielsen, H., Ambus, P. and Jensen, E. S. (2001b). Interspecific competition, *N* use and interference with weeds in pea-barley intercropping. *Field Crops Res.* 70:101–9.

Hinsinger, P., Betencourt, E., Bernard, L., Brauman, A., Plassard, C., Shen, J. ... Zhang, F. (2011). P for two, sharing a scarce resource: Soil phosphorus acquisition in the rhizosphere of intercropped species. *Plant Physiol.*, 156:1078–86.

Jahansooz, M. R., Yunusa, I. A. M., Coventry, D. R., Palmer, A. R. and Eamus, D. (2007). Radiation and wateruse associated with growth and yields of wheat and chickpea in sole and mixed crops. *Eur. J. Agron.* 26:275–82.

Jensen, E. S. (1996a) Grain yield, symbiotic N2 fixation and interspecific competition for inorganic N in pea-barley intercrops. *Plant Soil* 182:25–38.

Jensen, E. S. (1996b). Barley uptake of N deposited in the rhizosphere of associated field pea. *Soil Biol. Biochem.* 28:159–68.

Jensen, E. S., Bedoussac, L., Carlsson, G., Journet, E.-P., Justes, E. and Hauggaard-Nielsen, H. (2015). Enhancing yields in organic crop production by eco-functional intensification. *Sustain. Agric. Res.* 4(3), 42–50.

Jolliffe, P. A. (2000). The replacement series. *J. Ecol.* 88:371–85.

Jolliffe, P. A. and Wanjau, F. M. (1999). Competition and productivity in crop mixtures: Some properties of productive intercrops. *J. Agric. Sci.* 132: 425–35.

Knudsen, T., Hauggaard-Nielsen, H., Jørnsgard, B. and Jensen, E. S. (2004). Comparison of interspecific competition and *N* use in pea-barley, faba bean-barley and lupin-barley intercrops grown at two temperate locations. *J. Agric. Sci.* 142:617–27.

Launay, M., Brisson, N., Satger, S., Hauggaard-Nielsen, H., Corre-Hellou, G., Kasynova, E., Ruske, R., Jensen, E. S. and Gooding, M. J. (2009). Exploring options for managing strategies for pea-barley intercropping using a modeling approach. *Eur. J. Agron.* 31:85–98.

Li, W. (2001) *Agro-ecological Farming Systems in China*, Man and the biosphere series, J. N. R. Jeffers (Ed.), Vol. 26, Paris.

Lithourgidis, A. S., Vasilakoglou, I. B., Dhima, K. V., Dordas, C. A. and Yiakoulaki, M. D. (2006). Forage yield and quality of common vetch mixtures with oat and triticale in two seeding ratios. *Field Crops Res.* 99:106–13.

Malézieux, E., Crozat, Y., Dupraz, C., Laurans, M., Makowski, D., Ozier-Lafontaine, H., Rapidel, B., de Tourdonnet, S. and Valantin-Morison, M. (2008). Mixing plant species in cropping systems: Concepts, tools and models. A review. *Agron. Sustain Dev.* 29:43–62.

Matusso, J. M. M., Mugwe, J. N. and Mucheru-Muna, M. 2014. Effects of different maize (Zea mays L) – soybean (Glycine max (L) Merrill) intercropping patterns on soil mineral-N, N-uptake and soil properties. *Afr. J. Agric. Res.* 9:42–55.

Mead, R. and Willey, R. W. (1980). The concept of a 'land equivalent ratio' and advantages on yields from intercropping. *Exp. Agric.* 16:217–28.

Mikić, A., Ćupina, B., Rubiales, D., Mihailović, V., Šarūnaitė, L., Fustec, J., Antanasović, S., Krstić, Đ., Bedoussac, L., Zorić, L., Đorđević, V., Perić, V. and Srebrić, M. (2015). Models, developments, and perspectives of mutual legume intercropping. In: Sparks, D. L. (Ed.), *Advances in Agronomy*, Elsevier Academic Press Inc., San Diego, USA, pp. 337–419.

Mohta, N. K. and De, R. (1980). Intercropping maize and sorghum with soya beans. *J. Agric. Sci.* 95:117–22.

Naudin, C., van derWerf, H. M. G., Jeuffroy, M. H. and Corre-Hellou, G. (2014). Life cycle assessment applied to pea-wheat intercrops: A new method for handling the impacts of co-products. *J. Clean Prod.* 73:80–7.

Naudin, C., Corre-Hellou, G., Voisin, A. S., Oury, V., Salon, C., Crozat, Y. and Jeuffroy, M. H. (2011). Inhibition and recovery of symbiotic N_2 fixation by peas (*Pisum sativum* L.) in response to short-term nitrate exposure. *Plant Soil* 346:275–87.

Naudin, C., Corre-Hellou, G., Pineau, S., Crozat, Y. and Jeuffroy, M. H. (2010). The effect of various dynamics of N availability on winter pea-wheat intercrops: Crop growth, N partitioning and symbiotic N2 fixation. *Field Crops Res.* 119:2–11.

Naudin, C., Aveline, A., Corre-Hellou, G., Dibet, A., Jeuffroy, M. H. and Crozat, Y. (2009). Agronomic analysis of the performance of spring and winter cereal-legume intercrops in organic farming. *J. Agric. Sci. Technol.* 3:17–28.

Ndzana, R. A., Magro, A., Bedoussac, L., Justes, E., Journet, E. P. and Hemptinne, J. L. (2014). Is there an associational resistance of winter pea–durum wheat intercrops towards Acyrthosiphon pisum Harris? *J. Appl. Entomol.* 138 (8), 577–85.

Nieder, R. and Benbi, D. K. (2008). *Carbon and Nitrogen in the Terrestrial Environment.* Springer, Heidelberg.

Oelhermann, M., Echarte ,L., , Vachon ,K. and Dubois, C. (2009). The role of complex agroecosystems in sequestering carbon and mitigating global warming. *Earth Environ. Sci.* 6:20–31.

Ofori, F. and Stern, W. R. (1987). Cereal - legume intercropping systems. *Adv. Agron.* 41:41–90.

Padulosi, S., Hodgkin, T., Williams, J. T. and Haq, N. (2002). Underutilized crops: Trends, challenges and opportunities in the 21st century. In: Engels, J. M., Ramanatha, M., Rao, V., Brown, A. H. D. and Jackson, M. T. (Eds), *Managing Plant Genetic Resources.* CAB International, Wallingford, pp. 323–38.

Peoples, M. B., Hauggaard-Nielsen, H. and Jensen, E. S. (2009). The potential environmental benefits and risks derived from legumes in rotations. In: Emerich, D. W. and Krishnan, H. B. (Eds), *Nitrogen Fixation in Crop Production.* American Society of Agronomy, Madison, pp. 349–85.

Qin, A. Z., Huang G. B., Chai, Q., Yu, A. Z. and Huang, P. (2013) Grain yield and soil respiratory response to intercropping systems on arid land. *Field Crops Res.* 144:1–10.

Ratnadass, A., Fernandes, P., Avelino, J. and Habib, R. (2012). Plant species diversity for sustainable management of crop pests and diseases in agroecosystems: A review. *Agron. Sustain. Dev.* 32:273–303.

Spitters, C. and Aerts, R. (1983). Simulation of competition for light and water in crop-weed associations. *Aspects Appl. Biol.* 4:467–83.

Stern, W. (1993). Nitrogen fixation and transfer in intercrop systems. *Field Crops Res.* 34:335–56.

Trenbath, B. R. (1993). Intercropping for the management of pests and diseases. *Field Crops Res.* 34:381–405.

Trenbath, B. R. (1986) *Resource use by intercrops.* In: Francis, C. A. (Ed.) *Multiple Cropping Systems.* MacMillan, New York, pp. 57–81.

Tsubo, M. and Walker, S. (2002). A model of radiation interception and use by a maize-bean intercrop canopy. *Agric. Forest. Meteorol.* 110:203–15.

Tsubo, M., Walker, S. and Mukhala, E. (2001). Comparisons of radiation use efficiency of mono-/ intercropping systems with different row orientations. *Field Crops Res.* 71:17–29.

Vandermeer, J., van Noordwijk, M., Anderson, J., Ong, C. and Perfecto, I. (1998). Global change and multispecies agroecosystems: Concepts and issues. *Agric. Ecosyst. Environ.* 67:1–22.

Vandermeer, J. (1989). *The Ecology of Intercropping.* Cambridge University Press, Cambridge.

Vasilakoglou, I. B., Lithourgidis, A. S. and Dhima, K. V. (2005). Assessing common vetch-cereal intercrops for suppression of wild oat. In: *Proceedings of 13th International Symposium*, Session S5, European Weed Research Society, Bari, Italy.

Weigelt, A. and Jolliffe, P. (2003). Indices of plant competition. *J. Ecol.* 91:707–20.

Willey, R. (1979a). Intercropping-its importance and research needs. 1. Competition and yield advantages. *Field Crop Abstr.* 32:1-10.

Willey, R. W. (1979b). Intercropping-its importance and research needs. 2. Agronomy and research needs. *Field Crop Abstr.* 32:73–85.

Willey, R. and Osiru, D. (1972). Studies on mixtures of maize and beans (*Phaseolus vulgaris*) with particular reference to plant population. *J. Agric. Sci.* 79:517–29.

Willey, R. W. (1990). Resource use in intercropping systems. *Agric. Water Manag.* 1:215–31.

Xiao, Y., Li, L. and Zhang, F. (2004). Effect of root contact on interspecific competition and N transfer between wheat and fababean using direct and indirect 15N techniques. *Plant Soil* 262:45–54.

Chapter 12

Soil and nutrient management in grain legume cultivation

S. Adjei-Nsiah, International Institute of Tropical Agriculture (IITA), Ghana; and B. D. K. Ahiabor, CSIR-Savanna Agricultural Research Institute, Ghana

1 Introduction

The population of the world is projected to reach 9.7 billion by 2050 from the current 7.3 billion. Most of this increase is expected to occur in the developing world (UN, 2015) and this requires increase in food production (FAO, 2014). Sustainable intensification is crucial in areas of high population densities where land expansion is limited (Pretty et al., 2011; Vanlauwe et al., 2014; Cook et al., 2015). Integration of grain legumes in farming systems in the context of integrated soil fertility management offers a potential pathway for sustainable intensification as they are vital for improving soil fertility.

The contribution of grain legumes to soil fertility in the tropics has extensively been reported by several researchers (Sanginga, 2003; Ahiabor et al., 2007; Adjei-Nsiah et al., 2008; Yusif et al., 2009). For instance, in the southern Guinea Savanna zone of Nigeria, Sanginga (2003) reported residual nitrogen values of 10–24 kg N ha^{-1} (14–36% of the total N in maize) in a soybean–maize rotation. This resulted in a significantly higher maize grain yield (20–23% increase) with maize grown immediately after soybean compared to maize after maize. Similarly, in the Forest/Savanna transitional agro-ecological zone of Ghana, Adjei-Nsiah et al. (2008) also reported significantly higher maize grain yield increases of 100–150% when maize followed erect cowpea varieties and 275% increase when maize followed an indeterminate creeping cowpea variety compared to maize after maize. The N fertilizer equivalent values for the cowpeas were 18–23 kg N ha^{-1} for the erect varieties and 60 kg N ha^{-1} for the indeterminate creeping variety. In a field trial in the Guinea Savanna zone of Ghana, Ahiabor et al. (2007) compared the N_2-fixing capacities of six

http://dx.doi.org/10.19103/AS.2017.0023.09

legume species, namely cowpea (*Vigna unguiculata*), devil bean (*Crotalaria retusa*), black and white varieties of *Mucuna* (*Mucuna pruriens* var. *utilis*) and *Canavalia ensiformis*. They reported that the amount of total N fixed by the legumes ranged between 23 kg ha^{-1} for *C. ensiformis* and 145 kg ha^{-1} for cowpea.

Globally, about 40 million tons of N is fixed every year by agricultural legumes (Herridge et al., 2008) with an estimated fertilizer equivalence of about US$60 billion assuming a cost of US$1.25 kg^{-1} of fertilizer N (O'Hara et al., 2012). Chianu et al. (2010) estimated the financial value of nitrogen fixation in soybean within smallholder farming systems in Africa to be US$375 million, assuming a fertilizer price of US$795 t^{-1}.

In addition, grain legumes have excellent nutritional values in terms of protein, amino acids and micronutrients (Philips et al., 2003; Gibson and Ferguson, 2008; Belane and Dakora, 2011). However, despite their importance, yields of grain legumes are far below their potential due to several factors. These factors include use of low-yielding varieties, high cost or limited availability of good quality inputs, poor soil fertility, particularly low soil phosphorus and poor nutrient management (Kamara et al., 2007; Kolawole, 2012). Grain legumes are often considered as secondary crops to cereal crops and are commonly promoted as crops that require no fertilizer application. The soils in most parts of the tropics where grain legumes are cultivated are, however, inherently low in fertility. This situation is aggravated by nutrient depletion through continuous cultivation coupled with inadequate replenishment.

Grain legumes can access atmospheric N through symbiosis with soil-inhabiting bacteria collectively known as rhizobia through a process called biological nitrogen fixation (BNF) and therefore require minimal N fertilizer input. These rhizobia and the process can, however, be limited by the availability of other nutrients particularly phosphorus and unfavourable pH ranges as well as tillage practices (Peoples et al., 1995). For grain legumes, nodulation and nitrogen fixation are dependent upon adequate supply of essential micro (Mo, Zn, B, Cu and Mn) and macro (P, K, Mg, Ca and S) nutrients within a favourable pH range.

Nutrient deficiencies involving P, K, S, Mo and Zn are quite common and widespread in the tropics (Dakora and Keya, 1997). Several studies have shown that legume yields can be improved through the use of high-yielding crop varieties (Buruchara et al., 2011), phosphate (P)-based fertilizers (Kamanga et al., 2010; Kamara et al., 2007), rhizobia inoculants (Ahiabor et al., 2014; Boddey et al., 2016) or their combinations (Zerihun et al., 2015; Ronner et al., 2015).

This chapter will limit itself to how to maximize BNF and grain yields of grain legumes through soil and nutrient management options. We will first review existing literature on optimal conditions for legumes cultivation and BNF in grain legumes. This will be followed by discussions on recent studies in BNF in the Guinea Savanna zone of Ghana within the framework of N2Africa-Putting Nitrogen Fixation to work for smallholder farmers in Africa and other projects. We will conclude this chapter with future trend in grain legumes research.

2 Biological nitrogen fixation (BNF) in grain legumes

BNF is a process through which the stable inert atmospheric nitrogen (N_2) is reduced to a form (NH_4^+) that can be assimilated by plants through symbiosis between a legume host and a soil-inhabiting bacterium generally known as rhizobia. The symbiosis between grain legumes and rhizobia is a well-organized system that involves a number of steps. This symbiotic association begins when rhizobia infect the roots of the host legume and form nodules. In the nodules, the rhizobia convert gaseous atmospheric N_2 into ammonia (NH_3) which is largely assimilated by the legume for growth. In return, the legume host provides the rhizobia with carbon as a source of energy and also serves as a habitat for the rhizobia. The efficiency of the symbiotic BNF is dependent on the mutual compatibility of both partners and is influenced by environmental factors.

2.1 Optimal conditions for nitrogen fixations

Bacterial strains

Most agricultural soils contain root nodule bacteria which are a diverse group of soil-inhabiting Gram-negative bacteria (O'Hara, 2001). However, particular root nodule bacteria present in a particular soil may not be able to nodulate the host legume crop being grown or if they can may not be able to form effective symbiosis that can contribute nitrogen to the legume host in particular or to the farming system in general.

The symbiotic relationship between root nodule bacteria and legume host is characterized by a very high degree of specificity for effective nodulation of a particular host legume by a strain of root nodule bacteria (O'Hara et al., 2002). Specificity between host legumes and rhizobia occurs at three levels – the ability to form nodules, the ability to fix nitrogen and the extent to which symbiotic nitrogen fixation meets the plant's nitrogen requirement (Gibson et al., 1982).

When a new legume crop is introduced to a land for the first time, one of the three things will happen: (i) the uninoculated legume may form abundant effective nodules indicating the presence of a large population of effective root nodule bacteria in the soil, (ii) no nodules may form on the uninoculated legumes which suggests very little or, in some cases, no background population of root nodule bacteria in the soil that are capable of nodulating the particular host legume and (iii) the uninoculated legumes may form variably effective nodules, indicating the presence of a population of ineffective or variably effective root nodule bacteria (O'Hara et al., 2002).

Studies by Singleton et al. (1992) and Ronner et al. (2015) have shown clear benefits of rhizobium inoculation for major grain legumes. Response of grain legumes to rhizobium inoculation is, however, strong when they are introduced to new areas where the soil lacks the appropriate bacteria (van Kessel and Hartley, 2000).

According to Gibson et al. (1982), a successful inoculant strain of rhizobium should have a greater nitrogen-fixing ability with the selected legume host, being competitive in nodule formation on a wide range of the legume host and in many situations be able to persist in the soil indefinitely. In addition, the strain should possess characteristics that include genetic stability and ability to tolerate a wide range of environmental stress including moisture stress and aluminium toxicity.

One of the major constraints in legume rhizobium inoculation is the failure of the inoculant strain to occupy a significant proportion of the legume nodules (Vlassak and Vanderleyden, 1997). Factors affecting survival and competition for nodule occupancy by rhizobia have been reviewed by Vlassak and Vanderleyden (1997). To be able to survive and compete for nodule occupancy in the root of the legume, the introduced rhizobia in the inoculant must be able to compete with the indigenous often ineffective indigenous rhizobium population in the soil. This requires that they first have to adapt to the prevailing environmental conditions and rapidly colonize the legume rhizosphere and the soil and again compete for infection sites and nodulate the host roots. If the introduced strains are to persist in the soil, they have to live while not being associated with legumes (i.e. saprophytically) in the soil, multiply and produce effective nodules that are able to nodulate the next season of cultivated legume host (Al-Falih, 2002).

P nutrition

Phosphorus is one of the most important nutrients which affects nitrogen fixation. Grain legumes in general have high P requirements. Phosphorus is known to stimulate root growth, initiate nodulation as well as influence the efficiency of legume–rhizobia symbiosis (Yakubu et al., 2010). It is also known to be involved in energy transfer, more specifically as ATP in nitrogenase activity (Beegle and Durst, 2002). Nodules themselves are strong sinks for P, and P can strongly influence nodulation and nitrogen fixation. In white clover, Hart (1989) reported that nodules may contain 0.72–1.2% P. Hence when legumes which depend on symbiotic nitrogen receive an inadequate supply of phosphorus they may suffer from nitrogen deficiency. Almendras and Bottomley (1987) reported of a significant increase in the per cent nodule occupancy of *Trifolium subterranean* by *R. leguminosarum* bv. *trifolii* when they applied 25 mg of P kg^{-1} of soil in the form of KH_2PO_4 in an acid soil. In their studies to investigate the effects of cropping history and phosphorus source on nitrogen fixation in cowpeas, Vesterager et al. (2008) reported that the amount of nitrogen fixed was increased by 30–40% when P was applied as either TSP or rock phosphate, although grain yield was not affected. In the Sudano-Sahelian zone of north eastern Nigeria, Yakubu et al. (2010) also reported that P application resulted in 378, 169 and 138% increases in N fixation in cowpea, groundnut and Bambara groundnut, respectively.

In the savannas of West Africa, significant legume responses to phosphorus applied either as TSP or SSP have been obtained in several studies (Kamara et al., 2007; Yakubu et al., 2010; Kalewole, 2012; Ronner et al., 2015). In a two-year multi-location trial in northern Nigeria to determine the response of soybean to inoculation and phosphorus application, the additive effect of P and inoculation resulted in about 777 kg ha^{-1} increase in soybean grain yield, while P application alone resulted in about 444 kg ha^{-1} increase in grain yield (Ronner et al., 2015). High P fertilizer application could, however, induce Zn deficiency with concomitant decrease in yield when adequate supplementation is not provided (Murphy et al., 1981; Alloway, 2008; Nyoki and Ndakidemi, 2014).

Aluminium toxicity

Aluminium (Al) toxicity is a major constraint to agricultural production in the tropics, and grain legumes are no exception. Acid soils with pH of less than 5 often contain levels of Al and Mn that may be injurious to nodulation or growth of rhizobia (O'Hara et al., 2012). Soil pH affects both the survival of the rhizobia and the formation of nodules. Legume species also differ widely in their tolerance to Al and Mn ions and some plants may be more strongly affected than the rhizobia (Graham, 1992). Different legume symbioses have different tolerance to extremes of soil pH. However, most legume symbionts have an optimal pH range of 5–8, although a few host legumes such as Lupin and groundnut have pH preferences of 4.5 to 7.5 (O'Hara et al., 2012), whereas the optimal soil pH range for effective symbiosis by this group of bacteria is 6.5–7.0. Soil acidity limits symbiotic nitrogen fixation, affecting rhizobium survival and persistence in the soil and reducing nodulation. In acidic soils particularly with pH of less than 5, grain legumes may fail to nodulate, and the lack of persistence in the soil of certain rhizobium strains has been attributed to lack of nodulation (Al-Fali, 2002). Response to aluminium toxicity in acid soils and cultures by rhizobia varies widely. Selection of rhizobium strain tolerant to Al has been suggested as a possible strategy to overcome the problem of Al toxicity in acid soils (Al-Fali, 2002). While slow-growing rhizobia are in general more tolerant of low pH than the fast-growing rhizobia, differences exist among strains. Some legume species such as *Stylosanthes humilis* also grow better at pH 4 than in less acid soils (Andrew, 1976).

2.2 Nitrogen sensing

Nitrogen fixation in agricultural systems is sometimes very low. It has been reported that nodulation and nitrogen fixation in legumes are suppressed by high levels of soil nitrogen sometimes caused by high nitrogen inputs into the environment (Miller et al., 2007). This suppression occurs through a homeostatic mechanism which allows the plant to balance the high cost of nitrogen fixation with its nitrogen requirement (Gibson and Harper, 1985). Legumes have evolved to use soil nitrogen in the form of ammonia or nitrate instead of fixing their own N by nodulation that will require the use of their own energy. To choose between using soil nitrogen and fixing its own nitrogen by nodulation, it must be able to sense soil nitrogen. This mechanism is called nitrogen sensing and depends on several factors including the legume species, the host rhizobia and environmental conditions. Legumes often have specific N-containing molecules such as amino acids, ureides and amides that are found in the phloem and xylem. When these molecules are transported to the phloem and xylem, they may send a feed back signal to regulate the uptake of soil N, fixation and nodulation possibly through the phloem concentration. The sensing mechanism sometimes mediates the root of plants to adjust their root architecture to changes in N supply in the soil. The mechanisms involved in nitrogen sensing by legumes have been extensively reviewed by Murray et al. (2017).

2.3 Crop diversification in intercropping and crop rotation

Strategies for enhancing nitrogen fixation and grain yields in grain legumes have been reviewed extensively by van Kessel and Hartely (2000). According to van Kessel and Hartely, nitrogen fixation and grain yields of grain legumes could be improved through strategies such as application of starter N, intercropping, rotation with cereals and conservation agriculture. Intercropping grain legumes within non-N_2-fixing cereals is a common practice, especially in the dry savannas in the tropics. According to Willey (1979), reasons for intercropping legumes with cereals are varied and include the following:

1 Increasing total yield by increasing the land equivalent ratio,
2 Reducing the need for nitrogen fertilizer, either by increasing the available soil N or N transfer,
3 Increasing the utilization efficiency of available nutrients and water,
4 Diversification of the risk associated with crop failure, and
5 Reducing pest problems.

While studies (Danso et al., 1987) have shown increased nitrogen fixation in legumes when intercropped with non-cereals, other studies have produced mixed results. It is estimated that the total amount of N fixed per unit area in the intercropped system is often lower due to reduced legume population densities as well as increased competition for light and nutrients by the non-legume (van Kessel and Hartley, 2000).

The amount of soil-available N can also be influenced by recent cropping history or cropping sequence (Peoples et al., 1995). Studies (Bergersen et al., 1989; Herridge and Holland, 1992; Peoples and Craswell, 1992) have demonstrated that legumes can fix more nitrogen and contribute more fixed nitrogen to the soil when it is grown after cereals than when grown on previously fallowed plot or immediately after legume. Thus, with proper choice of crops in a rotational system, N fixation could be enhanced.

2.4 Conservation agriculture

The amount of N fixed by legumes depends on how well the legume grows and the level of soil nitrate. The lower the soil nitrate and greater the biomass produced, the greater the amount of N fixed. Therefore, management practices that maximize water use and provide the legume with ideal, stress-free growing conditions including low soil nitrate will improve N fixation. One management option that can optimize growing conditions and lower soil nitrate level is no-tillage or minimum tillage. No-tillage may lead to increased soil water and decreased soil nitrate accumulation through lower rates of N mineralization coupled with increased N immobilization (van Kessel and Hartley, 2000). Several studies (Herridge and Brockwell, 1988; Hughes and Herridge, 1989; Herridge and Holland, 1992) have shown that nodulation and N fixation by grain legumes can be markedly improved under no-tillage than under conventional tillage.

In a Mediterranean semi-arid rainfed environment, Stringi et al. (2004) showed that the no-tillage technique was most effective in enhancing grain yields of legumes in dry years. In their two-year study, mean grain yield under no-tillage increased over the conventional tillage by as much as 21.5%. In Western Kenya, Okoth et al. (2013) showed that the amount of nitrogen fixed in soybean was larger under zero tillage than under conventional tillage, although the tillage method did not influence grain yield. In their studies on the effect of tillage and crop residue management on nitrogen fixation in soybean in Western Kenya, Kihara et al. (2011) also reported that the total amount of N fixed under reduced tillage plus crop residue application was more than that in the other treatments by at least 55% in intercropping and 34% in rotation systems. They also reported that reduced tillage plus crop residue application also resulted in grain yield increase of 50–160 kg ha^{-1} in rotation and 70–360 kg ha^{-1} in intercropping with a better N balance realized than under conventional tillage.

3 Nutrient and soil management practices and grain legume production in the savanna agro-ecologies of Ghana

3.1 Grain legume production environment

The Guinea Savanna agro-ecological zones of Ghana have a great potential for the production of grain legumes, particularly soybean, cowpea and groundnut, and these crops are major components of the farming systems. The savanna agro-ecologies have a shorter growing period of 120–150 days, and rainfall in the zone is unimodal and occurs between May and October with a dry period between November and April. Total annual mean rainfall amounts range between 900 mm in the northern Guinea Savanna zone and 1163 mm in the southern Guinea Savanna zone. The soils are classified as Savanna Ochrosol and Groundwater Laterites in the Interim Ghana Soil Classification System (Adjei-Gyapong and Asiamah, 2002) and as Plinthosols in the World Reference Base for soil resources (WRB, 2015). One striking feature of the zone is the inherently low fertility of its soils (Braimoh and Vlek, 2004) expressed in low levels of organic carbon, total nitrogen and available phosphorus (Table 1). The savanna zone is characterized by relatively low vegetative cover resulting in reduced quantities of plant residues being added to the soil.

Soil-available P, the most limiting nutrient for grain legume production is very low and P deficiencies are widespread in the Guinea Savanna zone. While farmers may apply nitrogen fertilizers to maize, grain legumes rarely receive phosphorus fertilizers. Grain legumes are an important option in providing nutrition and cash and in rebuilding soil fertility but scanty knowledge by farmers on P fertilizer use and its application across fertility gradients has led to limited expected returns. Determining how to effectively increase grain legume yields in degraded soils and maintaining the fertility of still productive lands are of paramount importance to all farmers, especially those in degraded savanna zones. It is even hypothesized that identifying suitable grain legume varieties alongside improving soil P nutrient capital leads to increased mean plant survival, nutrient and dry matter accumulation and yields which ultimately reduces crop production risks and benefits system productivity, food security and human nutrition. Chien et al. (1993) and Sanginga et al. (1996) reported that the application of P to soybean increases the amount of N derived from the atmosphere (Ndfa) by the soybean–rhizobium symbiotic system.

Table 1 Means and ranges of selected physical and chemical properties of soils of the Guinea Savanna of northern Ghana from 110 locations

Parameter	Range	Mean
pH (H_2O 2:1)	5.48–7.38	6.17
Clay (%)	9.52–18.06	12.10
Sand (%)	40.32–74.62	61.55
Silt (%)	13.99–42.82	25.95
OM (%)	0.23–1.57	0.84
Total N (%)	0.02–0.12	0.05
Avail P (Bray-1) (mg/kg)	2.52–16.38	5.92
Exchangeable cations (me/100g)		
K	0.09–0.31	0.16
Ca	1.88–11.79	4.09
Mg	0.52–5.55	1.57
Na	0.06–0.34	0.11
ECEC	3.35–18.13	6.10

Therefore, the application of P should increase the N content of roots, haulms and fallen leaves. When these residues are not exported from the field, their mineralization should add to the soil part of the N derived from the atmosphere. This may enable farmers to reduce the amount of N fertilizer applied to cereals when grown after soybean.

3.2 Effect of nutrient and soil fertility management on grain legume production

Research results in the Guinea Savanna zone of Ghana have shown the importance of soil fertility improvement in grain legume production (Ahiabor et al., 2014; Asei et al., 2015; Adjei-Nsiah, unpublished). In the Guinea Savanna zone of Ghana, phosphorus appears to be the most limiting nutrient and the use of phosphorus fertilizers has been found to increase both fodder and grain yields of legumes (Aziz et al., 2016; Atakora et al., 2014).

Investigations into the effect of rhizobia inoculation on grain legume production in Ghana have received a lot of attention in recent times (Ahiabor et al., 2014; Asei et al., 2015; Aziz et al., 2016). Widespread response of soybean and cowpea to rhizobia inoculation has been obtained in northern Ghana in several studies (Ahiabor et al., 2014; Boddey et al., 2016; Lamptey et al., 2014a,b). Only few studies in Ghana (Aziz et al., 2016) have reported lack of response of soybean to rhizobia inoculation and this was suspected to be probably due to a possible large population of native rhizobia as a result of the soil having been planted to soybean on a few occasions in the past.

In separate studies to investigate the response of cowpea to rhizobia inoculation in northern Ghana, Boddey et al. (2016) reported that inoculation with rhizobia in an on-station trial increased nodule weight in cowpea by between 78 and 145% in one trial and between 75 and 212% in another trial. The corresponding increases in grain yields were 93–111% in the first trial and 40–57% in the second trial. In the same year, similar trials conducted on-farm with farmers yielded similar results. In these on-farm trials, inoculation enhanced nodule weight by 48% to 62%, while grain yield was enhanced by 37% and 64%. In their studies to evaluate the effect of rhizobium inoculation and phosphorus application on the performance of soybean and their economic implications to farmers in northern Ghana, Ahiabor et al. (2014) reported that rhizobium inoculation alone resulted in increases of 43% and 22% in nodule mass and grain yield, respectively.

Several studies (Ahiabor et al., 2014; Lamptey et al., 2014a; Adjei-Nsiah, unpublished) in the Guinea Savanna zone of Ghana have demonstrated that grain legumes respond better to inoculation when it is applied in combination with phosphorus. A similar observation has also been made in northern Nigeria under similar growing conditions by Ronner et al. (2015).

In the work of Ahiabor et al. (2014), inoculating soybean with rhizobium inoculant in the presence of 30 kg ha^{-1} P enhanced nodule biomass production by 105% while increasing grain yield by 122% compared with only 43% and 22% increase in nodule biomass and grain yield respectively when only rhizobium inoculant was applied to the soybean. This could be explained by the fact that P is known to stimulate root growth, initiate nodulation as well as influence the efficiency of legume–rhizobia symbiosis (Yakubu et al., 2010). P is also known to be involved in reactions with energy transfer, more specifically as ATP in nitrogenase activity (Israel, 1987).

4 Summary

This chapter summarizes the role of soil and nutrient management in enhancing BNF within the smallholder farming systems in the tropics. It also summarizes contributions of research in grain legumes to sustainable intensification of grain legumes in the savanna agro-ecologies of northern Ghana. In spite of the enormous contributions of grain legumes to soil fertility improvement and food and nutritional security within the predominantly cereal-based farming systems in the tropics, their productivity and nitrogen-fixing capabilities have been hampered by declining soil fertility and poor soil and nutrient management practices. Review of the available literature indicates that grain yield and BNF in grain legumes could be hampered by poor soil fertility, particularly phosphorus, Al toxicity, high soil nitrogen levels and lack of symbiotic compatibility between rhizobium and host.

Management options such as improved phosphorus nutrition and selection of rhizobium inoculant strains that have a greater nitrogen-fixing ability and are competitive in nodule formation with a wide range of the legume host could enhance nitrogen fixation. The amount of N fixed is also significantly influenced by the growth of the legume and the level of soil nitrate. Hence management practices that maximize water use and provide the legume with stress-free growing environment including low soil nitrate such as zero or minimum tillage will enhance N fixation.

Grain legume yields within smallholder farming systems hardly exceed 0.5 t ha^{-1} but recent studies have demonstrated that grain legume yields could be as high as 2 t ha^{-1} if appropriate crop and soil management practices are employed in legume cultivation. Recent studies in northern Ghana within the framework of the N2Africa project, Africa-Brazil Agricultural Innovation MarketPlace project and other projects have shown that grain legume yields could be enhanced through the application of phosphorus and or rhizobium inoculants. Research has also demonstrated the contribution of rhizobium inoculation to cowpea grain yield contrary to earlier belief that cowpea was a promiscuous legume and therefore could freely nodulate with any rhizobium in the soil and so would not respond to rhizobium inoculation. Yield increases of between 37 and 111% as a result of rhizobium inoculation have been obtained from different studies.

While cowpea and soybean have shown widespread responses to rhizobium inoculation in northern Ghana, no such responses have been published on groundnut yet, although preliminary studies at the Savanna Agricultural Research Institute (SARI) of the Council for Scientific and Industrial Research (CSIR) have shown that groundnut yield could be increased by as much as 75% with the combined application of TSP and rhizobium application and by 57% with the application of commercial composted manure and rhizobium application.

Several studies have shown that soybean yields could be enhanced through phosphorus application and yield increases of as much as 124% have been obtained through phosphorus application. Recent studies in rhizobiology have also demonstrated widespread response of promiscuous soybean to rhizobium inoculation contrary to earlier studies in Nigeria that reported no significant increases in grain yield of promiscuous soybean varieties with rhizobium inoculation (Okungu and Sanginga, 2003; Osunde et al., 2003).

A large majority of farmers could benefit from application of rhizobium inoculants to grain legumes both agronomically and economically as they are cheaper. Thus rhizobium inoculation is effective in increasing soybean yields with little financial cost provided farmers get access to good quality inoculants and good quality seed. Unfortunately, these inputs are not widely used in sub-Saharan Africa due to low levels of awareness and also due to inaccessibility of good quality inoculants in rural areas. There are also no proper public–private partnerships to commercialize the technology.

Although application of P fertilizers to grain legumes is agronomically effective, it may not benefit a large majority of farmers in economic terms due to poor output/input price ratios of some grain legumes. However, continuous cultivation of grain legumes without P fertilizer application will lead to depletion of soil P reserves, especially considering the already low soil-available P in northern Ghana. Besides the direct benefit of P to grain legumes, subsequent crops may also benefit from the residual effects of the P. In northern Ghana where grain legumes are often rotated with maize, maize grown after legumes may benefit from such residual P from the phosphorus applied to the legume in the preceding season. Furthermore, repeated application of P will reduce the need for P application in subsequent crops and thus improve the profitability of P fertilizers.

5 Future trends in grain legume research

In most parts of the savanna agro-ecologies of West Africa, grain legumes are either rotated or intercropped with cereals (Steiner, 1984). However, grain legume yields in the intercropped systems are very low, hardly exceeding 500 kg ha^{-1} compared with over 2000 kg ha^{-1} under optimized growing conditions in sole cropping systems. Thus, any research programme should focus on designing an intercropping system that will optimize the system in terms of resource use and yield.

Phosphorus appears to be the most limiting plant nutrient for grain legume production in the tropics, and recent studies in the savanna agro-ecologies of Ghana give credence to this. Studies have, however, shown that grain legume genotypes show marked differences in the uptake of P due to the ability of plants to (i) secrete enzymes and acids in the legume rhizosphere that solubilize soil P, (ii) explore a large volume of soil for soil-available P and (iii) absorb P from low P solution. Future studies should exploit grain legume crop genotypes with these characteristics that would help increase grain legume productivity and yield.

Cereal/legume rotation plays a major role in the farming systems in many parts of the savanna agro-ecologies of West Africa. Most of the increased yields in cereals in cereal/legume rotations have been attributed to improved N availability by grain legumes as well as non-nitrogen effects such as the breaking of cereal pests and disease cycles, enhanced P availability through secretion of enzymes and acids in legumes rhizosphere and arbuscular mycorrhizal colonization. While most studies have quantified the above-ground N fixed by different legume crop varieties, very little is known about the contribution of the below-ground N fixed by the legumes to soil fertility improvement. In the northern Guinea Savanna zone of West Africa, most of the grain legume crop residues are used for animal feed and are not recycled into the soil to maintain soil fertility, thereby leading to nutrient depletion when the nutrients are not replenished. Further research is therefore needed to quantify the below-ground contribution of grain legume cultivars to soil fertility improvement.

Despite the widespread response of grain legumes to rhizobium inoculation and P fertilizer application, the availability of these inputs in rural areas still remains a challenge because agro-input dealers are unwilling to import them to regions where grain legumes are produced. Research is therefore required to ascertain the willingness of smallholder farmers to purchase these inputs. In one such study, Martey et al. (2016) reported that willingness to invest in legume inoculants was relatively high among agro-input dealers and that agro-input dealers who had access to financial institutions were 36% more willing to intensively invest in legume inoculants as compared to those without access to financial institutions.

Currently most of the rhizobium inoculants being promoted and used in Africa in general and Ghana in particular are being imported and may become expensive in the future beyond the reach of the smallholder farmers. Research should therefore be intensified in the isolation and identification of effective rhizobia strains from African soils for the production of legume inoculants locally.

6 Where to look for further information

Giller, K. E. (1991). *Nitrogen fixation in tropical cropping systems.* CAB International, Wallingford.

Van Kessel, C., and Hartley, C. (2000). Agricultural management of grain legumes: has it led to an increase in nitrogen fixation? *Field Crops Research* 65, 165–181.

7 References

Adjei–Gyapong, T. and Asiamah, R. D. (2002). The interim Ghana soil classification system and its relation with the world reference base for soil resources, pp. 51–76. In: *Quatorzième réunion du Sous-Comité ouest et centre africain de corrélation des sols, Abomey, Benin, 9–13 October 2000.* World Soil Resources Report No. 98. FAO, Rome.

Adjei-Nsiah, S., Kuyper, T. W., Leeuwis, C., Abekoe, M. K., Cobbina, J., Sakyi-Dawson, O. and Giller, K. E. (2008). Farmers' agronomic and social evaluation of productivity, yield and N2-fixation in different cowpea varieties and their subsequent residual N effects on a succeeding maize crop. *Nutrient Cycling in Agroecosystems*, 80, 199–209.

Ahiabor, B. D. K., Fosu, M., Tibo, I. and Sumaila, I. (2007). Comparative nitrogen fixation, native arbuscular mycorrhiza formation and biomass production potentials of some grain legume species grown in the field in the Guinea Savanna Zone of Ghana. *West African Journal of Applied Ecology*, 11, 89–98.

Ahiabor, B. D. K., Lamptey, S., Yeboah, S. and Bahari, V. (2014). Application of phosphorus fertilizer on soybean (Gycine max L.) inoculated with rhizobium and its economic implication to farmers. *America Journal Experimental Agriculture*, 4, 1420–34.

Al-Falih, A. M. K. (2002). Factors affecting the efficiency of symbiotic nitrogen fixation by rhizobium. *Pakistan Journal of Biological Sciences*, 5(11), 1277–93.

Alloway, B. J. (2008). *Zinc in Soils and Crop Nutrition.* 2nd ed., IZA and IFA, Brussels, Belgium and Paris, France, 135p.

Almendras, A. S. and Bottomley, P. J. (1987). Influence of lime and phosphoate on nodulation of soil-grown *Trifolium subterraneum* L., by indigenous *Rhizobium trifolii*, App. *Environmental Microbiology*, 53, 2090–7.

Andrew, C. S. (1976). Effect of calcium, pH and nitrogen fixation on the growth and chemical composition of some tropical and temperate pasture legumes. I. Nodulation and growth. *Australian Journal of Agricultural Research*, 27, 611–23.

Asei, R., Ewusi-Mensah, N. and Abaidoo, R. C. (2015). Response of soybean (Glycine max L.) to Rhizobia inoculation and Molybdenum Application in the Northern Savanna zones of Ghana. *Journal of Plant Sciences*, 3, 64–70.

Atakora, W. K., Oppong, E., Mahama, R., Tuffour, H. O. and Adu Gyamfi, S. (2014). Cowpea yield as affected by level and time of mineral phosphorus fertilizer application in the Guinea Savanna agro-ecological zone of Ghana. *Global Journal of Science Frontier Research: D Agriculture and Veterinary*, 14(8), 12–18.

Aziz, A. L. A., Ahiabor, B. D. K., Opoku, A. and Abaidoo, R. C. (2016). Contributions of Rhizobium inoculants and phosphorus fertilizer to biological nitrogen fixation, growth and yields of three soybean varieties on a fluvic luxisol. *American Journal of Experimental Agriculture*, 10(2), 1–11.

Beegle, B. and Durst, P. T. (2002). Managing phosphorus for crop production. Agronomy Facts 13, Penn State Extension, Pennsylvania State University, USA.

Belane, A. K. and Dakora, F. D. (2011). Levels of nutritionally-important trace elements and macro-nutrients in edible leaves and grain of 27 nodulated cowpea genotypes grown in the Upper West Region of Ghana. *Food Chemistry*, 125, 99–105.

Bergersen, F. T., Blockwell, J., Gault, R. R., Morthotpe, L., Peoples, M. B. and Turner, G. L. (1989). Effects of available nitrogen and rates of inoculation on nitrogen fixation by irrigated soybeans and evaluation of ä^{15}N methods for measurements. *Australian Journal of Agricultural Research*, 40, 763–80.

Boddey, R. M., Fosu, M., Atakora, W. K., Miranda, C. H. B., Boddey, L.H., Guimaraes, A. P. and Ahiabor, B. D. K. (2016). Cowpea (Vigna unguiculata) crops in Africa can respond to inoculation with Rhizobium. *Experimental Agriculture*, doi:10.1017/S0014479716000594.

Bohloo, B.B., Ladha, J. K., George, T and Garrity, D.P. (1992). *Biological Nitrogen Fixation for Sustainable Agriculture*: A perspective. *Plant and Soil*, 141, 1–12.

Braimoh, A. K. and Vlek, P. L. G. (2004). The impact of land-cover change on soil properties in northern Ghana. *Land Degradation & Development*, 15, 65–74.

Buruchara, R., Chirwa, R., Sperling, L., Mukankusi, C., Rubyogo, J. C., Mutonhi, R. and Abang, M. (2011). Development and delivery of bean varieties in Africa, the Pan-Africa bean research alliance (PABRA) model. *African Crop Science Journal*, 19, 227–45.

Chianu, J. N., Huising, J., Danso, S., Okoth, P., Chianu, J. N. and Sanginga, N (2010). Financial value of Nitrogen fixation in soybean in Africa: Increasing benefits for smallholder farmers. *Journal of Life Science*, 4(6), 50–9.

Chien, S. H., Carmona, G., Menon, R. G. and Hellums, D. T. (1993). Effect of phosphate rock sources on biological nitrogen fixation by soybean. *Fertility Reserach and Practice*, 34, 153–9.

Cook, S., Silica, L., Adolph, B. and Walker, S. (2015). Sustainable intensification revisited. IIED Issue Paper, IIED, London.

Dakora, F. D. and Keya, S. O. (1997). Contribution of legume nitrogen fixation to sustainable agriculture in Sub-Saharan Africa. *Soil Biology and Biochemistry*, 29(5/6), 809–17.

Danso, S. K. A., Zapata, F. and Hardarson, G. and Fried, M. (1987). Nitrogen fixation in faba beans as affected by intercropping in sole or intercropped systems with barley. *Soil Biology and Biochemistry*, 19, 411–15.

FAO (2014). *Sustainable Crop Production Intensification*. FAO, Rome, http://www.fao.org/agriculture/crops/thematicsitemap/theme/Spi-home/framework/en/.

Gibson, A. H., Dreyfus, B. L. and Dommergues, Y. R. (1982). Nitrogen fixation by legumes in the tropics. In Y. R. Dommrgues and H. G. Diem (Eds), *Microbiology of Tropical Soils and Plant Productivity*. Martinus Nihoff/Dr W. Junk Publishers, The Hagues/Boston/London.

Gibson, R. and Ferguson, E. (2008). An interactive 24-hour Recall for Assessing the Adequacy of Iron and Zinc intakes of Developing Countries. Technical Monograph 8, HarvestPlus, Washington, USA.

Gibson, A. H. and Harper, J. E. (1985). Nitrate effect on nodulation of soybean by Bradyrhizobium japonicum. *Crop Science*, 25, 497–501.

Graham, P. H. (1992). Tress tolerance in Rhizobium and Bradyrhizobium, and nodulation under adverse soil conditions. *Canadian Journal of Microbiology*, 38, 475–84.

Hart, A. L. (1989). Nodule phosphorus and nodule activity in white clover. *New Zealand Journal of Agricultural Research*, 32, 145–9.
Herridge, D. F. and Brochwell, J. (1988). Contribution of fixed nitrogen and soil nitrate to the nitrogen economy of irrigated soybean. *Soil Biology and Biochemistry*, 20, 711–17.
Herridge, D. F. and Holland, J. F. (1992). Production of summer crops in nort hern New southwales I. Effects of tillage and double cropping on growth, grain and n yields of six crops. *Australian Journal of Agricultural Research*, 44, 137–49.
Herridge, D. E., Peoples, M. B. and Boddey, R. M. (2008). Global inputs of biological nitrogen fixation in agricultural systems. *Plant and Soil*, 3111, 1–8.
Hughes, R. M and Herridge, D. F. (1989). Effect tillage on yield, nodulation and nitrogen fixation of soybean in far north-coastal New South Wales. *Australian Journal of Experimental Agriculture*, 29, 671–7.
Israel, D. N. 1987. Investigation of the role of phosphorus in symbiotic nitrogen fixation. *Plant Physiology*, 84, 835–40.
Kamanga, B. C. G., Whitbread, A., Wall P., Waddington, S. R., Almekinders, C. and Giller, K. E. (2010). Farmer evaluation of P fertilizer application to annual legumes in Chisepo, Central Malawi. *African Journal of Agricultural Research*, 8, 668–80.
Kamara, A. Y., Abaidoo, R. C., Kwari, J. and Omoigui, L. (2007). Influence of phosphorus application on growth and yield of soybean genotypes in the tropical savannas of northeast Nigeria. *Archieves Of Agronomy and Soil Science*, 53, 539–52.
Kihara, J., Martius, C., Bationo, A. and Vlek, P. L. G. (2011). Effects of tillage and crop residue application on soybean nitrogen fixation in a tropical ferralsol. *Agriculture*, 1, 22–37. doi:10.3390/agriculture 1010022.
Kolawole, G. O (2012). Effect of phosphorus fertilizer application on the performance of maize/ soybean intercrop in the Southern Guinea Savanna of Nigeria. *Archieves of Agronomy and Soil Science*, 58, 180–98.
Lamptey, S., Ahiabor, B. D. K., Yeboah, S. and Osei, D. (2014a). Effect of rhizobium inoculants and reproductive growth stages on shoot biomass and yield of soybean (Glycine max (L.) Merril). *Journal of Agricultural Science*, 6(5), 44–54.
Lamptey, S., Ahiabor, B. D. K., Yeboah, S. and Asiama, C (2014b). Response of soybean (Glycine max) to rhizobium inoculation and phosphorus application. *Journal of Experimental Biology and Agricultural Sciences*, 2(1), 72–7.
Martey, E., Ahiabor, B. D. K., Buah, S. S. J. and Kusi, F. 2016. Assessment of agro-input dealers' willingness to Invest in legume inoculants in northern Ghana. *Asian Journal of Agricultural Extension, Economics & Sociology*, 8(3), 1–13.
Miller, A. J., Fan, X, Orsel, M. and et al. (2007). Nitrate transport and signalling. *Journal of Experimental Botany*, 58, 2297–306.
Murphy, L. S., Ellis Jr., R. and Adriano, D. C. (1981). Phosphorus and micro-nutrient interaction effects on crop production. *Journal of Plant Nutrition*, 3, 1–4, 593–613.
Murray, J. D., Liu, C. W., Chen, Y and Miller, A. J. (2017). Nitrogen sensing in legumes. *Journal of Experimental Botany*, 68 (8), 1919–1926. doi:10.1093/jxb/erw405.
Nyoki, D. and Ndakidemi, P. A. (2014). Influence of Bradyrhizobium japonicum and Phosphorus on micronutrient uptake in cowpea. A case study of zinc (Zn), iron (Fe), copper (Cu) and manganese (Mn). *American Journal of Plant Sciences*, 5, 427–35.
O'Hara, G. W. (2001). Nutritional constraints on root nodule bacteria affecting symbiotic nitrogen fixation: A review. *Australian Journal of Experimental Agriculture*, 41, 417–33.
O'Hara, G. W., Yates, R. and Howieson, J. (2002). Selection of strains of root nodule bacteria to improve inoculant performance and increase legume productivity. In D. Herridge (Ed.) *Inoculants and Nitrogen Fixation of Legumes in Vietnam*, pp. 75–80. ACIAR Proceedings 109e.
O'Hara, G., Howieson, J., Drew, E., Ballard, R., Herridge, D., Gemmell, G., Hartley, E., Phillips, L., Deaker, R., Denten, M., Yates, R., Seymour, N. and Ballard, N. (2012). *Inoculating Legumes: A Practical Guide*. Grains Research and Development Corporation, Australia.
Okoth, J. O., Mungai, N. W., Ouma, J. P. and Baijukya, F. P. (2013). Tillage effects on biological nitrogen fixation and soybean grain yields in western Kenya. Transforming rural livelihoods in Africa: How

can land and water management contribute to enhanced food security and address climate change adaptation and mitigation? Nakuru, Kenya, 20–25 October, 2013. *Joint proceedings of the 27th Soil Science Society of East Africa and the 6th Soil Science Society of Africa.*

Osunde, A. O. and Gwan, S., Bala, A., Sanginga, N., Okogun, J. A. (2003). Response to rhizobial inoculation by two promiscuous soybean cultivars in soils of the Southern Guinea savannah zone of Nigeria. *Biology and Fertility of Soils*, 37, 274–9.

Peoples, M. B. and Craswel, T. (1992). Nitrogen fixation in Agriculture: investments, expectations and actual contribution to agriculture.*Plant and Soil*, 141, 13–39 .

Peoples, M. B., Ladha, J. K. and Herridge, D. F. (1995). Enhancing legume N_2 fixation through soil and plant management. *Plant and Soil*, 174, 83–101.

Philips, R. D., McWatters, K. H., Chinnan, M. S., Hung, Y. C., Beuchat, L. R., Sefa-Dedeh, S., Sakyi-Dawson, E., Ngoddy, P., Nnanyelugo, D., Enwere, J., Komey, N. S., Lia, K., Mensa-Wilmot, Y., Nnanna, I. A., Okeke, C., Prinyawiwatkat, W and Saalia, F. K. (2003). Utilization of cowpeas for human food. *Field Crops Research*, 82, 193–213.

Pretty, J., Toulmin, C. and Williams, S. (2011). SustainableIntensification in African agriculture. *International Journal of Agricultural Sustainability*, 9, 5–24.

Ronner, E., Franke, A. C., Vanlauwe, B., Dianda, M., Edeh, E., Ukem, B., Bala, A., van Heerwaarden, J. and Giller, K. E. (2015). Understanding variability in soybean yield and response to P-fertilizer and rhizobium inoculants on farmers' fields in northern Nigeria. *Field Crops Research*, 186, 133–45.

Sanginga, N., Okpogun, J. A., Akobundu, I. O. and Kang, B. T. (1996). Phosphorus requirement and nodulation of herbaceous and shrub legumes in low P soils of a Guinea savanna in Nigeria. *Applied Soil Ecology*, 3, 247–55.

Sanginga, N. (2003). Role of biological nitrogen fixation in legume-based cropping systems: A case study of West Africa farming systems. *Plant and Soil*, 252, 25–39.

Singleton, P. W., Bohlool, B. B., Nakao, P. L., 1992. Legume response to rhizobial inoculation in the tropics: myths and realities. In: R. Lal and P. A. Sanchez (Eds), *Myths and Science of Soils of the Tropics*. Soil Sci. Soc. Am. and Am. Soc. Agron. Spec. Publ., Vol. 29, pp. 135–55.

Steiner, K. G. (1984). *Intercropping in the Tropical Smallholder Agriculture with Special Reference to West Africa Soils*. Germany Agency for Technical Cooperation, West Germany, 304p.

Stringi, L., Giambalvo, D., Traponi, P., Scarpello, C. and Attardo, C. (2004). Effect of soil management techniques on performance of different grain legumes in a Mediterranean environment. In: C. Cantero-Martinez and D. Gabina (Eds), *Mediterranean Rainfed Agriculture: Strategies for Sustainability*. CIHEAM, Zaragoza, pp. 79–83.

United Nations, Department of Economic and Social Affairs, Population Division (2015). World Population Prospects. The 2015 Revision, Key Findings and Advance Tables. Working Paper 40. ESA/P/WP 241.

Van Kessel, C. and Hartley, C. (2000). Agricultural management of grain legumes: Has it led to an increase in nitrogen fixation? *Field Crops Research*, 65, 165–81

Vanlauwe, B., Descheemaeker, K., Giller, K. E., Huising, J., Merckx, R., Nziguheba, G., Wendt, J. and Zingore, S. (2014). Integrated soil fertility management in sub-Saharan Africa: Unravelling local adaptation. *Soil*, 1, 1239–86.

Vesterager, J. M., Nielson, N. E. and Hogh-Jenson, H. (2008). Effects of cropping history and phosphorus source on yield and nitrogen fixation in sole and intercropped cowpea-maize systems. *Nutrient Cycling Agroecosystems*, 80, 61–73.

Vlassak, K. M. and Vandurleyden, J. (1997). Factors influencing nodule occupancy by inoculant rhizobia. *Critical Reviews in Plant Sciences*, 16, 163–229.

Whitbread, A., Blair, G., Konboon, Y., Lefroy, R. and Naklang, K. (2003). Managing crop residues, fertilizers and leaf litters to improve soil C, nutrient balances, and the grain yield of rice and wheat cropping systems in Thailand and Australia. *Agriculture Ecosystems & Environment*, 100(2–3), 251–63. In: O. Lyasse, N. Sanginga, B. Vanlauwe, J. Diels and R. Merckx (Eds), *Special Issue: Balanced Nutrient Management Systems for Cropping Systems in the Tropics: From Concept to Practice.*

Wiley, R. W. (1979). Intercropping – its importance and research needs. Part 1. Competition and yield advantages. *Field Crops Abstr.* 32, 2–10.
WRB. 2015. World Reference Base for Soil Resources 2014, update 2015. International soil classification system for naming soils and creating legends for soil maps. World Soil Resources Reports No. 106, FAO, Rome.
Yacubu, H., Kwari, J. D. and Sandabe, M. K. (2010). Effect of phosphorus fertilization on nitrogen fixation by some grain legume varieties in Sudano-Sahelian zone of North Eastern Nigeria. *Nigeria Journal of Basic and Applied Sciences*, 18(1), 19–26.
Yanni, Y. G. (1992). Performance of chickpea, lentil and lupin nodulated with indigenous or inoculated rhizobia micropartners under nitrogen, boron, cobalt and molybdenum fertilization schedules. *World Journal of Microbiology and Biotechnology*, 8, 607–13.
Yusif, A. A., Iwuafor, E. N. D., Abaidoo, R. C., Olufajo, O. O. and Sanginga, N. (2009). Grain legume rotation benefits to maize in the northern Guinea Savanna of Nigeria: Fixed –nitrogen versus other rotation effects. *Nutrient Cycling in Agroecosystems*, 84, 124–39.
Zerihun, A., Alemayo, D. and Wolde-Meskel, E. (2015). On-farm yield responses of soybean (Glycine max L.) to fertilizer sources under different soil acidity status in Gobu Sayo District Western Ethiopia. *Journal of Agronomy*, 14(1), 30–6.

Chapter 13

Diseases affecting grain legumes and their management

Keith Thomas, University of Sunderland, UK

1 Introduction

All field crops suffer from diseases of some sort, often in combination. This is increasingly so in monoculture growth where pathogen evolution may be out of synchrony with crop development and new crop varieties push mutualistic microorganisms towards parasitism.

Moreover, as modern agriculture is increasingly pressured to improve productivity, breeding is incentivised to select for traits associated with yield and character. Traits to resist disease may be judged of lower importance if other means of controlling disease are available. For many years, the proliferation and effectiveness of chemical control allowed this, but with recent moves to limit exposure to chemicals, the priorities of other means of disease resistance have increased. This is, of course, a difficult balance. Pressures for yield are motivated by the justifiable need to satisfy food demand, but in recent years awareness of pesticide hazards have tempered their use. Calculations that pesticide poisoning can cause more deaths than infectious diseases in the developing world (Eddleston et al., 2002) have indicated that uncontrolled means of achieving food security cannot justify the target.

Increases in disease may be seen as a consequence of changes in pathogen or host physiology or through a trade-off between physiological systems where resistance against one disease reduces resistance to others (McGrann et al., 2014; Brown and Rant 2013). Progressive legume cultivation is no different but has specific features relevant to disease development as well as a range of adapted pathogens with both biotrophic and necrotrophic consequences.

http://dx.doi.org/10.19103/AS.2017.0023.10

Climate change contributes additional variabilities to the anticipation of disease and is a further element in food insecurity (Ghini et al., 2008; Chakroborty et al., 2000) and financial cost. Losses to soya bean crops by rust are estimated to have cost up to $2 billion between 1996 and 2007 (Wrather and Koenning, 2009). Losses in other continents are similarly severe (Murithi et al., 2015), with up to 15% of food legume production in India lost to disease (Pandey et al., 2009).

In some cases, disease is a local concern often in small-scale production, possibly due to poor management of soils, infected seeds or lack of crop rotation. In other cases, large-scale production may be threatened by regional or epidemic disease spread as when a virulent strain of a pathogen arises and spreads without warning. Obviously, these two overlap and may cross infect, thus requiring a broad approach as advocated by integrated management systems (Davidson and Kimber, 2007).

2 Grain legume diseases

Grain legumes incur infection by broad pathogen groups such as *Fusarium* and *Botrytis* as well as more species-specific infections from *Colletrotrichum* and *Ascochyta* fungi. Bacterial blight from *Pseudomonas syringae* and a range of viral infections are also found in many crop areas. Table 1 lists the range of these diseases, but it is by no means exhaustive. The incidence and severity of each disease depends, though, on local climate and circumstance and on the host legume. Nevertheless, some diseases stand out for their prevalence. For example, *Ascochyta* blight is perhaps the most important worldwide constraint on chickpea production (Siddique et al., 2012).

This basic listing, although extensive, is an incomplete view of a dynamic crop condition with plants susceptible not only to different types of diseases but also to different species in each disease. *Ascochyta* blight has a range of infective species, including *Ascochyta lentis* on lentil, *A. rabiei* on chickpea and *A. fabae* on faba bean, while *Mycosphaerella pinodes* infects field pea. Crossover between these species is likely and *Mycosphaerella* is advised to be a suitable terminology for the *Ascochyta* blight complex (Gossen et al., 2011).

These distinctions themselves are increasingly incomplete as varieties of virulence continually appear, making protection an increasingly mobile process dependent on rapid monitoring and identification. Each group of pathogens has their own elements of this, with viruses mutating faster than bacteria and fungi, bacteria multiplying rapidly in suitable conditions but less dependent on airborne transmission. Fungi grow more slowly but have extensive sporulation and greater resistance to chemical treatment because of their physiological similarities to plant hosts.

3 Traditional vs. integrated disease management

3.1 Traditional disease management

Developing a disease management system requires both broad protection and, when circumstances arise, specific responses to localised epidemics. Disease management involves an integrated approach ranging from selection of resistant varieties to soil management

Table 1 Major diseases of grain legumes

Pathogen	Disease	Crop
Ascochyta fabae	*Ascochyta* blight	Faba bean
Ascochyta lentis	*Ascochyta* blight	Lentil
Ascochyta rabei	*Ascochyta* blight	Chickpea
Aphanomyces euteiches	Common root rot	Pea, faba bean, lentil
Colletotrichum lindemuthianum	Anthracnose	Common bean, chickpea
Erysiphe polygoni	Powdery mildew	Pea
Fusarium solani	*Fusarium* root rot	Pea, chickpea
Fusarium oxysporum f. sp.	*Fusarium* wilt	Chickpea, pea, lupin, faba bean, lentil
Macrophomina phaseolina	Dry root rot	Chickpea
Mycosphaerella pinodes	*Ascochyta* blight	Pea
Peronospora viciae	Downy mildew	Pea
Phoma medicaginis var. *pinodella*	Foot rot	Pea
Pythium spp.	Seed, seedling and root rot	Pea, chickpea, lentil, common bean, lupin
Rhizoctonia solani	Seedling blight	Lentil, pea, lupin
Sclerontinia sclerotiorum	Stem rot	Lentil, common bean, lupin
Uromyces viciae-fabae	Rust	Pea, faba bean, lentil
Uromyces ciceris-arietini	Rust	Chickpea
Bacteria diseases		
Botrytis cinerea	Botrytis grey mould	Chickpea, lupin
Botrytis fabae, B. cinerea	Chocolate spot	Faba bean
Viral diseases		
Cowpea golden mosaic	Cowpea golden mosaic virus	Cowpea
Sterility mosaic	Pigeonpea sterility mosaic virus	Pigeonpea
Yellow vein mosaic	Yellow vein mosaic	Mung bean

in advance of planting to pesticide application during cultivation. Certain critical disease sources such as seed infection (Chang et al., 2007) are well recognised and can be addressed centrally, while others, such as climate dependency, vary according to location.

For some time pesticides provided unqualified support to the aim of increasing crop yields and allowed extensive industrialisation of production (Clark, 2006). In a parallel to the effectiveness of clinical antibiotics, disease resistance in plant pathogens developed rapidly, leading to a progressive increase in pesticide production – from a negligible global level in 1945 to three million tons per year in 1985 (Tilman et al., 2002). Because of the selection pressure exerted by extensive antibiotic applications, microorganisms can develop resistance to antibiotics within one to three years of their application requiring continual refinement of treatments.

The disease process in plants is a mirror of clinical disease development with progression stages from transmission, exposure, adherence, infection, colonisation and growth, and finally symptoms. Each of these stages has the potential for control and a constant battle between pathogen and host has driven the evolution of increasing virulence and concomitant host response for millennia.

3.2 Integrated disease management

It has become increasingly appreciated that no one solution can address the complexities of multiple disease incidence on crops (Makkouk and Kumari, 2009), even with the development of resistant varieties (Pande et al., 2006; Pande et al., 2005). Pesticide treatment alone is increasingly infeasible.

Integrated disease management (IDM) does not seek to remove pesticides entirely but to optimise their effectiveness and minimise their application (Stoddard et al., 2010; Davidson and Kimber 2007). This may be conducted in a number of ways, some direct and individual, others as part of a complex matrix of actions. IDM is also an essential tool for farming in resource-poor locations where pesticide purchase, advanced testing technology or even mechanisation may be limited. Moreover, legumes are more sensitive to stress factors than other crops because of the complex relationship with *Rhizobium* bacteria (Bordeleau and Prevost, 1994) and require more attention on a continual basis.

Similar approaches to disease prevention have been developed in other areas, for example, hurdle technology in food preservation, where a series of barriers provides challenges to microbial growth or degenerative processes (Sing and Shalini, 2016; Leistner, 2000). For example, in beer stabilisation, the combined effect of low pH, high acidity, alcohol, low oxygen and hop compounds provides very effective protection against the development of pathogens (Vriesekoop et al., 2012). In other foods, elements of these hurdles such as citric acid or salt are added in formulation to assist intrinsic factors.

IDM need not be complex as a single intervention may incorporate a number of effects. An example of a simple initiative to minimising pesticide application is reported from Yunnan Province in China using the co-culturing of rice varieties susceptible and resistant to rice blast disease (*Magnaporthe grisea*). Planting alternative rows of the two varieties resulted in a 94% reduction in disease severity and 89% greater yield of the susceptible variety than when in monoculture (Zhu et al., 2000). This programme was so successful that fungicide applications were stopped after two years. A variety of mechanisms may explain these results which do need further analysis before extrapolation. Barrier exclusion may be a simple option but also incorporate effects on microclimate. Possible transmission of endophytes between plants may be a more complex possibility. Other mechanisms to explain the impact of intercropping plants include the production of volatile organic compounds by trap plants to attract predators carrying pathogens (Stenberg et al., 2015). Such volatiles may also be considered as markers to indicate plant disease and feasibly could be bioengineered into crops as indicator systems (Lucas, 2011). In many cases of IDM, however, more complex management systems are developed and applied, often targeted to specific crops or diseases.

A. rabiei blight is a major crop infection in chickpea and other legume crops attacking leaves, stems and pods. Infection can develop at all stages, particularly during flowering, resulting in up to 100% losses in susceptible varieties. Cereals are a major host of many *Ascochyta* species, but some are found to be pathogenic only on specific legume species (Ahmed and Beniwal, 1991). Control is possible with targeted fungicide treatment (Chongo et al., 2003) but benefits greatly from a multilayer approach to integrate prediction and active prevention. This is increasingly desirable as increasing use of systemic fungicides appears to carry a risk of increased fungicide insensitivity (Gossen et al., 2014).

Dispersal of inoculum by rainsplash from infected field debris (Pedersen et al, 1994) and active foliage infection are major sources of disease spread and greatly aided by persistent wetness followed by dry periods with optimal infection temperatures of 20°C for young seedlings. Spores can persist for up to two years in low humidity conditions below 65% relative humidity (RH), but viability declines if diseased tissue is buried below a 10-cm depth (Kaiser, 1973). Ascospores are, however, more infective than conidiospores under stress conditions, thus making physiology of infection on plant residues an important feature (Trapero-Casas and Kaiser 2007). Pathogen genetics of *A. rabiei* in Canada indicates high levels of diversity, suggesting that variations in virulence may pose challenges to static management practices as more virulent strains of pathogen arise (Chongo et al., 2004). A similar extent of variation was also noted in isolates from the western provinces of Iran (Vafaei et al., 2016).

With specific knowledge of a pathogen physiology and virulence, it is increasingly possible to develop an IDM to both predict disease potential and act on its presence. In the next section, we review the components which are potential elements of an integrated approach to any disease control, using *Ascochyta* as an example.

4 Components of IDM

4.1 Selection of variety

Selection of seed variety has a major impact on disease incidence (Chang et al., 2007), but the mechanism for this is difficult to elucidate without details of pathogen virulence and host resistance (Rubiales and Fondevilla, 2012).

Identification of potential resistant varieties has traditionally depended on identifying resistant plants in the field, but this has difficulties as disease virulence changes (Tivoli et al., 2006). Advanced techniques have been reported to identify potential resistance genes (Ohm et al., 2012; Skiba et al., 2005; Satovic et al., 2003) but more extensive genetic information is required to apply this information into a practical context, particularly as many genes code for partial resistance (Rubiales and Fondevilla, 2012; Jayakumar et al., 2005; Tar'an et al., 2003).

High-throughput phenotyping of plant features is also being developed to help selection of varieties with improved growth and resistance traits. These traits may be biochemical but also physiological such as root architecture (Burridge et al., 2016).

In addition, it is relevant that as disease incidence increases with plant stress (Bostock et al., 2014; Schoeneweiss, 1975), resistant varieties are tested to cover such growing conditions, particularly as, for example, chickpea yield is improved if plants are drought stressed in later growth stages to induce increased root growth (Kashiwagi et al., 2014).

It is further evident that infection of plants with one disease may increase vulnerability of seeds to other diseases (Hepperly, 1979). Developing disease resistance to one pathogen need not necessarily provide protection against another, possibly because of the disease trade-off noted earlier (McGrann et al., 2014).

A further dimension of pathogen control and variety selection is the presence of endophyte microorganisms providing intrinsic protection. Endophyte microorganisms have been noted to be closely related to plant pathogens and may have derived a commensal relationship to the benefit of both partners (De Meyer et al., 2015: Schardl et al., 1997). Endophytes may have many potential modes of action (Dedeja et al., 2012), but specific antagonism to pathogenic microorganisms has been documented for *Pseudomonas* species (O'Sullivan and O'Gara, 1992). Endophytes have also shown action as growth promoters (Zhao et al., 2013). Beneficial inoculation of chickpea with endophyte bacteria has been reported and the potential for including endophytes in addition to *Rhizobia* in legume seeds is a valuable area of development (Maltzahn et al., 2016: Rajendran et al., 2015).

4.2 Seed quality

Infected seeds are a fundamental source of plant pathogenesis (Burgess et al., 1997; Bretag et al., 1995) and, if based in a major production facility, can disseminate disease to a broad area. Locating seed production away from a known infected area may help reduce the potential for infection, but hygienic production facilities are critical to achieve a target of less than 0.3% infection. Management of production is also important as damage to seeds in processing will make seeds vulnerable to infection. Large-seeded legumes are particularly susceptible to damage in handling due to their thin testa which are easily cracked and having embryo parts close to the seed coat. Harvesting seed crops, thus, require mechanical surfaces to be protected and impact to be minimised.

Additional controls for seed hygiene commonly include addition of fungicides and fumigation (Kaiser, 1982) but also heat treatment to reduce pathogen vigour (Ahmed and Beniwal, 1991; Zinnen, 1982). The use of agents to stimulate natural defence activators or elicitors has also been developed and patented (Eulgem et al., 2016; Melander and Rogers, 2016) as has been the application of small-interfering RNAs (siRNAs) (Huang et al., 2016). A caution on the use of fungicides on seeds is to ensure that they are compatible with the addition of *Rhizobium* inoculant.

4.3 Location

Choice of field location for cultivation has broad and narrow options of control. Areas with moist warm climates are more likely to induce *Ascochyta* growth (Casas and Kaiser, 1992), while fields with a history of disease will clearly have a reservoir of infection. A three or four species crop rotation may minimise the latter but will have limited impact if wind dispersal is likely from more distant locations (Salam et al., 2011; Madden, 1997; Peddersen and Morrall, 1995). Ideally, a separation of 500 metres from fields with an infection history is desirable along with bordering fields hosting non-infective crops.

4.4 Field hygiene

Where growing location options are constrained, an active management of crop residues and soil quality can be important factors in limiting infection. Removal or combustion of infected residues will reduce infection sources but is difficult to achieve fully as soil will remain a reservoir. Soil itself is not inert and, as with the current awareness of human intestinal biome, active microorganisms in soil have been shown to have suppressive effects (Camsos et al., 2016; Mazzola, 2002). Management practices to enhance commensal microbes with disease suppression indicate that no-tillage systems had the greatest effect in reducing disease by *Fusarium graminearum* (Campos et al., 2016).

4.5 Biocontrol

Fluorescent *Pseudomonas* species have been suggested as being a facilitator in the suppression of plant diseases by the production of the antibiotic 2,4-diacetylphloroglucinol (Mazzola, 2002), and antibiotics have long been noted as viable disease control agents (Handelsman and Stabb, 1996). *Rhizobium* inoculation has also been noted as reducing the development of disease in legumes (Hemissi et al., 2013). Antibiotic-mediated suppression of pathogens has been elucidated to specific mechanisms such as interfering with zoospore-homing events in *Pythium* infection of pea (Heungens and Parke, 2000).

More intimate measures to control infection include modifying the plant genome to increase resistance. This, for example, can include increasing intrinsic barriers such as root cell walls and phytoalexin production (Lozavaya et al., 2005). One issue in altering intrinsic resistance is the balance within the plant genome between resistance against biotrophic and necrotrophic pathogens and the possibility of disrupting a trade-off between these (Erb et al., 2011). Nevertheless, a number of promising avenues are being explored (Chen et al., 2012), particularly control of hormone production, for example, abscisic acid, defence modulators such as NPR1 (Kim and Delaney, 2002) and transcription factors controlling gene activation (Gurr and Rushton, 2005).

Of course, plant disease is not all microbial, parasites and pests are equally effective in depleting growth and yields. Nematodes are well-known soil parasites of legumes (Davis et al., 2005), and the possibility of identifying mutants with enhanced resistance has been developed using Target-induced Local Lesions in Genomes (TILLING) technologies (Perry et al., 2003).

4.6 Plant density and arrangement

Ascochyta germination and infection is greatly affected by moisture and can easily be enhanced by canopy humidity (Pedersen and Morrall, 1994). Planting density and positioning can thus be important controls to limit disease initiation and spread both from direct infection and via insect vectors (Hema et al., 2014). *Ascochyta* incidence is noted to correlate with plant density (Chang et al., 2007). Density of planting is a well-established factor for crop yield with, for example, between 35 and 70 plants per m^2 recommended for chickpea in Canada (Gan et al., 2003) and 40 plants per m^2 in Queensland, Australia (Beech and Leach, 1989). Because chickpea initially grows slowly compared to other plants, a moderate to high density is advantageous in reducing weed competition. However, impacts on disease spread may conflict with yield targets. For example, wide row spacing has been noted to reduce *Ascochyta* incidence (Chang et al., 2007) but will also reduce yields, making a careful calculation necessary to achieve optimal planting for final yield.

Studies also indicate that the time of sowing can affect disease incidence, with later sowing dates producing less *Ascochyta* disease in field pea in Australia (Bretag et al., 2000), although yields may be reduced, making it a difficult factor to balance without local experience.

4.7 Pesticide application

Pesticide use has been curtailed in the face of increased awareness of toxin effects (Eddleston et al., 2002) and in the light of knowledge of how more varied applications may be effective (Madonald and Richardson, 2006). Early fungicides were profoundly broad spectrum and highly toxic but with increasing appearance of pathogen resistance have become more targeted and formulated to allow mixed application at different times and seasons of application (Gossen et al., 2014).

Preventative sprays at early stages of disease development are recommended, but this does require effective crop monitoring and attention to patterns of disease in local areas. Attention to two additional factors, timing and dosing, can help achieve a successful outcome (Banniza et al., 2011).

4.8 Education

A wealth of support services is available to provide information and guidance to farmers to minimise and treat legume diseases, on both local and international bases.

However, past experience and traditional beliefs may be difficult to change. A belief in the power of pesticides or lack of knowledge of disease features may limit IDM (Schreinemachers et al., 2015: Abtew et al., 2016).

4.9 Examples

The above components are all potential elements of an integrated approach to any disease control. They can be applied at local and international levels and are the basis of many collaborative initiatives to predict and manage food security.

Local initiatives often couple with agronomic advice for local areas, for example, Saskatchewan Pulse Growers (http://saskpulse.com/growing/faba-beans/disease-management/). Broader initiatives such as the US Government's Feed the Future programme have a long-term international perspective incorporating research institutes and universities (https://feedthefuture.gov/sites/default/files/resource/files/ftf_factsheet_fsiclegumes_may2015.pdf) and which has a targeted legume programme.

The CGIAR Systemwide Program on Integrated Pest Management (SP-IPM) is a further example of an international collaboration between specialists with broad aims, including anticipating climate change effects and sustainable agriculture (http://www.spipm.cgiar.org/home).

5 Practical developments: modelling, sampling and identification

Looking at future prospects for disease control raises the questions of how IDM may develop and how it may take advantage of advanced technical inputs. As the term is defined, it incorporates sound practical management of existing techniques but can greatly benefit from additional elements of monitoring and prediction, both of which have specific technologies to apply – modelling and genetic analysis.

5.1 Modelling

Modelling has both academic relevance and practical application in disease control. Academic studies using experimental data are essential to identify elements in disease development and can include fundamental investigations of pathogen physiology and field exposures.

Pathogen exposure is not essential for model development. Some modelling is based on virtual pathogenesis where a plant's features are adjusted to determine disease dynamics and epidemiology and are a good theoretical basis for field studies (Wilson and Chakrabrty, 1998). Similarly, features of infection such as germination speed can be based on laboratory experimentation (Balkaya, 2004).

In practice, models can allow for direct disease forecasting using field observations of mesoclimatic conditions and daily infection values (Schoeny et al., 2007) and also for broader-scale predictions of continental epidemics (Isard et al., 2007).

Climatic analysis can involve different data sets where inputs can be provided from direct measurement or from regional meteorological information (Palmeri et al., 2006). A study of both of these inputs applied to the same system by Palmeri et al. (2006) indicated similar predictability from both of these sources for fungal disease in turfgrass, suggesting that on-site assessment is not essential, at least for this application. Daily and prior 24-h weather were noted to be the most successful predictors in a study of models for anthracnose disease using data from Australia, Brazil and Colombia when analysed by neural networks (Chakraborty et al., 2004).

Weather data are important factors in model development and, while temperature and humidity are major inputs, other associated data can also feature. Mean RH, number of hours of RH >95%, leaf wetness duration and precipitation during the 48 h prior to sunrise, and minimum and maximum soil temperatures can all be incorporated. The value of timing of leaf wetness is particularly important as these infective periods are strongly correlated to disease incidence by providing free water for germination (Roger et al., 1999; Kaiser, 1973). Although difficult to measure, leaf wetness may be indicated and predicted by high RH.

Direct observation of disease can also be a feature of modelling where threshold levels trigger action to treat a crop or initiate warnings for adjacent cultivation (Adomou et al., 2005; Sikora et al., 2014).

A broader predictability based on accumulating conditions is also possible. Degree-days predictions based on the accumulation of heat units during a growing season are used to predict flowering in crops, and similar approaches can be applied to estimating the potential appearance of disease (Schoeny et al., 2007).

5.2 Sampling

Predication of disease may be based on weather modelling, but on-site sampling allows a more direct and definitive assessment of a location. Visual assessment of crops can indicate the beginnings of an outbreak but does require experienced and skilled observation as well as a rigorous sampling plan to cover the target areas. Too few samples and a localised outbreak may be missed. Too many samples and assessors may be overwhelmed.

In this regard, it is perhaps fortunate that many legume diseases are wind borne and may be detected by air sampling. The target point for air sampling is not exclusively the crop area itself but may be some distance away and most usefully part of a continual monitoring programme (Kennedy and Wakeham, 2008).

Air sampling is typically conducted with mechanised air sampling devices passing a set volume of air over an agar plate, a sticky tape or into tubes which can be processed for molecular analysis (West et al., 2008). Tests may be conducted for fungal spores, bacteria or viruses and with rapid analysis provide a result within 24 h but potentially less (West and Kimber, 2015). Automation can schedule sampling at regular intervals allowing changes in contaminants to be plotted (Kaczmarek et al., 2008).

5.3 Identification

Application of models has considerable potential to predict and manage crop diseases. However, models do depend on accurate knowledge of the presence of disease and this relies on obtaining rapid and correct identification. In addition, being able to detect further features of virulence or strain changes require both advanced technologies and rapid response, ideally as close to field conditions as possible rather than relying on distant, and delayed, laboratory testing.

This latter demand is difficult to achieve with traditional techniques. Classical identification of field crop diseases has relied initially on observation by growers and on evaluations by experienced practitioners. Unfortunately, visual symptoms are frequently obtained late in the disease cycle and in many cases too late to avoid crop losses. They do, nevertheless, form part of a broader system to alert nearby and more distant locations to the hazard as well as help direct action on the crop itself.

Difficulties in crop observation are evident in the need for experienced practitioners to make assessments and, in many cases, for access to technical resources for confirmation. Developments to overcome this include using machine learning and neural networks to provide identification of images forwarded from field observations. Public-access archives of photographs of healthy and diseased plants can be used for assessment by appropriate algorithms to provide direct feedback, for example PlantVillage developed by Hughes and Salathé from Penn State College of Agricultural Sciences in 2012 (https://www.plantvillage.org/). Remote surveillance is an extension to this where airborne drones may provide the input from a distance (Lucas, 2011).

Disease symptoms may overlap with stress symptoms compounding confusion. The acceleration of infection by stress is a further element enhancing disease progression (Schoenweiss, 1975) and making it essential to act on the presence of symptoms.

Confirmation of field identification relies on practitioners having appropriate training and experience, but subjective elements are inevitably involved and may vary between regions and even between countries due to differences in training and access to contemporary literature. In cases of bacterial disease, a series of staining and biochemical tests may provide identification, particularly if these tests involve a means of rapid development of reagents such as those contained in test strip systems or microtitre plate growth monitoring.

Laboratory assessments are particularly problematic for fungal diseases where fungal species can have complex life cycles often involving multiple differentiation of spores and tissues. Microscope observation requires access to these differentiated features as hyphae of different species may be similar and indistinguishable.

With suitable expertise, classical analysis may provide essential information for both direct feedback and epidemiological study, and standard techniques remain in use by analysis services (Montana Regional Pulse Crop Diagnostic Laboratory 2016).

Classical identification procedures have further difficulties in that not all species are easily cultured and so cannot provide material for observation. Some pathogens may have very specific nutrient needs, some of which may be unknown, making the pathogen unculturable and impossible to identify without advanced techniques.

Extra demands of increased identification and analysis include the need for speed and also to provide data which may be used in broad epidemiological analysis of a disease outbreak. Advanced techniques are not always rapid, particularly if specialist skills are required. These advanced techniques are increasingly prevalent in both investigative and routine studies of phytopathogens. Many are more suitable for academic investigations into specific disease mechanisms, but some have promise of translating into systems suitable for on-site field or local laboratory use.

An additional application is in food safety assessment where legume products are released for consumption and require confirmation of the absence of contamination. These tests are also relevant to assess imported shipments to ensure that they do not carry disease to novel regions.

As well as providing identification of a disease, methods are also required to provide an assessment of its severity and whether mutations may have produced or activated specific virulence factors (Ohm et al., 2012). Quantification of disease may be conducted on soils and plant material as well as on air and water samples. An integrated plan of monitoring can not just identify the locations and sources of disease but profile its development and anticipate potential effect as well. Techniques are available for assessing all of these and so allow rapid response and effective control.

6 Advanced and rapid analysis techniques

The value of advanced techniques in clinical disease management is well recognised and is a major focus for clinical disease prevention. While the specific benefits of these techniques are clear, it is relevant to assess which of the many different approaches are relevant and suitable for long-term application. Given the pace of technical developments, it is likely that many practitioners and disease managers will have relevant instruments in their hands in the not too distant future.

Advances in the past two decades have taken place with particular focus on molecular biology but also with targets to address speed of processing and automation. The ideal for future applications is to have portable equipment which can process small volumes of samples within the time of a field visit and provide a direct diagnosis with a high level of accuracy.

Because legume diseases are easily encountered by airborne distribution, methods to integrate molecular biology with air sampling are particularly relevant (West and Kimber, 2015: Kaczmarek et al., 2008; West et al., 2008; Fraaije et al., 2005).

Additional targets of analysis are to obtain discrete data such as DNA sequences which can be used for future reference and to allow processing in epidemiological studies.

Disease success depends on the presence and activation of virulence factors. Identifying the presence of these, irrespective of identification of a disease agent, is a further benefit, particularly in the light of horizontal gene transfer where pathogenesis may move between species.

6.1 Sequence-based methods

Molecular methods have certainly addressed the targets of obtaining discrete data along with speed and automation, albeit with varying degrees of success. Molecular biology based on DNA, RNA or protein is typically the basis of many contemporary procedures, with the molecular sequence of these being the fingerprint providing relevant information. The cornerstone of nucleic acid methods is the polymerase chain reaction (PCR) catalysing amplification of DNA sections corresponding to specific gene sequences. Short primer sequences of DNA are designed to initiate this amplification and are key to making the process specific for target organisms or genes. As such, the design of primers is a critical factor in the success of PCR assays (Robertson and Walsh-Weller, 1998). For species identification, primer sequences based on the ribosomal genes are typically used.

Identification is, inevitably, a major requirement to allow relevant and specific control measures to be enacted. However, this is not always guaranteed. For successful identification, a database is necessary and the sequence of a novel pathogen or one with limited prior study may not be available for comparison. In addition, strains of a pathogen species may have different virulence factors and identification of these may require more specific targeting of genes rather than the universal 16S ribosomal gene sequences commonly used for identification.

Success in identification also varies according to the type of pathogen. Bacteria and viruses may be more efficiently identified than fungi due to their single chromosome character. This is much more applicable to whole-genome sequencing (WGS) which will provide much more information than a single gene profile.

Fungi, being eukaryotic organisms with multiple chromosomes, are less amenable to WGS and require more complex approaches. It is, perhaps, not simple chance that fungi pose more pathogenic risks to crops than viruses and bacteria do. However, the advantage of their more complex genome provides an ability to differentiate and more extensively integrate with the host. Despite this, whole-genome sequences have been obtained for major legume diseases (Dash et al., 2016), with a genomic data portal accessible at http://legumeinfo.org.

While WGS allows high-level identification, it opens additional avenues of analysis which molecular methods may further develop. In particular, it permits extensive comparison of sequences from different samples and therefore allows progress of a disease to be tracked and its location to be specified. It further allows changes to be identified and, perhaps, correlated to selection stresses present in different environments. However, despite many applications in clinical diseases (Berg et al., 2011), costs are likely to make WGS more suitable for academic and international interest than for local legume disease management.

More applicable methods of monitoring legume diseases centre on amplification of short sequences of DNA or RNA, patterns of amplified DNA, immunological reactions or protein mass spectroscopy. These approaches will be considered in comparison, but there are some general issues which molecular tests must address in order to provide reliable use.

Despite the potential of molecular methods and their advantages over classical observation and laboratory study, a number of limitations must be addressed for specific tests to be validated. These centre particularly on sensitivity and specificity. Sensitivity is less of an issue when a sample is obtained from a purified laboratory culture but is critical when field and environmental samples are processed. A small pick of a plate culture will provide enough DNA, RNA or protein for a direct analysis by genotyping, PCR or protein MALDI TOF.

In contrast, an environmental sample may contain only a few microorganisms and moreover be mixed with other species and with potential inhibitors of the analysis method. Soil is a particularly difficult medium to process due to levels of humus-inhibiting PCR polymerase enzymes (Fatima et al., 2014), but plant material can be similarly challenging (Maropolo et al., 2015). Extracting and purifying the target species becomes a limiting factor, particularly when hard cell walls or spores require digestion. Early classical extraction techniques using phenol and alcohol precipitation (Yeates et al., 1997) have been superseded by optimised enzyme-based kits from many commercial suppliers, but low microbial levels may require concentration of samples or bulking.

While sensitivity addresses the limits of detection, specificity is needed to distinguish one species from another but is limited by the uniqueness of the features chosen to be assessed. These may vary from the whole genome as in WGS to short sections of interspecific spacer areas of the ribosomal gene to the pattern of amplified DNA bands produced by primers amplifying random areas of the genome. In all cases for identification, a database is required to match the produced fingerprint to ones with an agreed taxonomic identification.

Databases are required for comparison to previously identified samples deposited as type samples. Some databases such as GenBank at the National Centre for Biotechnology Information (http://blast.ncbi.nlm.nih.gov/blast/Blast.cgi) are extensive, covering the full extent of prokaryotes and eukaryotes for DNA and protein sequences. Some are focused on disease organisms in general, such as the International Society for Human and Animal Mycology (http://its.mycologylab.org). Others may be more targeted to specific genera such as the *Aspergillus* website (http://www.aspergillus.org.uk/) and thus contain more detailed information on strains as well as information on recent developments, methods and pathogenesis.

Sites dedicated to specific diseases such as *Aspergillus* or *Fusarium* (http://www.fusariumdb.org/index.php) provide genomic database libraries with search and compare features to allow such identifications and also to log entries for future comparisons. More specific sites such as the Pathogen Host Interactions (PHI-base, www.phi-base.org) focus on disease genes to assist targeting anti-infective chemistries. The Legume Information System (http://legumeinfo.org/genomes) hosts the genome data of seven legume species at present (2016) and can be a resource of information for resistance genes.

An essential caution of databases is, of course, that the reference and type samples are accurately identified by classical means. In addition, skills are required in interpreting the results of database returns to avoid false positives and false negatives (Foldfeder et al., 2016). Difficulties with accuracy were noted in early use of databases (Clayton et al 1995) and in surveys since (Ashelford et al., 2005; Janda and Abbott, 2007), but more rigorous screening will increase reliability.

Sensitivity and specificity in molecular biology analysis depends on careful development of methods. Many species have similar sequences in their genes and identifying which area is most discriminatory requires *in silico* analysis and careful primer design. The internal transcribed spacer (ITS) region of the ribosomal genes in fungi are routinely targeted for primer design and provide suitable sources of variation to identify species (Manter and Vivanco, 2007). Sensitivity is also affected by primer design and amplification method, thus requiring validation to determine minimum copy number for a reliable response. This can make a difference if low numbers of cells are to be detected, such as on air sample filters (West et al., 2008).

Bacteria identification also uses ribosomal genes, typically those of the 16S ribosome. The value of this location was recognised early for clinical use (Clarridge 2004) and has since been applied to most areas of bacterial analysis (Chakravorty et al., 2007). As with fungi, the choice of ribosomal genes has the typical advantage of the presence of multiple copies, thus making the test more sensitive than if a specific genome were chosen. A similar rationale applies to targeting mitochondrial genes. Ribosomal RNA may also be amplified and has advantages of being applicable to environmental samples with multiple organisms. The higher copy numbers of the RNA providing a higher starting concentration of nucleic acid and thus a greater sensitivity (Moreno-Paz and Parro, 2006).

Extensions to the basic PCR procedure provide greater precision (Srinivasa et al., 2012) and it is likely that further developments will be applied to more specific genera as genome information becomes more extensively studied. Other developments in methodology have focused on optimising amplification of microbial DNA in the presence of plant tissue as is required in disease analysis of live samples (Tsoktouridis et al., 2014).

6.2 Genotyping methods

Sequencing is not the only means by which a DNA sample may be assessed. Amplification of target parts of the genome is directed by the choice of primer, but the amplified DNA may be compared directly. For this the DNA may be separated and viewed as band patterns on an agarose or on a polyacrylamide gel. The size of a band is indicative of the target gene and can be taken as indicative of a suspect organism if the primer is totally specific for a gene present only in that organism.

In a less definitive version of standard PCR, the use of more random primers may produce a range of amplified DNA products of different size. This random amplified polymorphic DNA (RAPD) technique uses short primers of 9–11 base pairs to scan whole genomes. When run on a gel, the amplified DNA produce a definitive pattern which may be compared to database sets for identification (Irinyi et al., 2016: Micheli et al., 1994). A major limitation of this approach is that the sample must be pure when isolating DNA as DNA of other organisms, including operators, will confuse the banding.

In more complex analysis, a purified sample of amplified DNA can be digested by nuclease enzymes and the digest profiled on a gel to obtain a fingerprint. Amplified fragment length polymorphism (AFLP) has been widely used to genotype fungi species (Bensch and Akesson, 2005) and also to identify resistance markers in host plants (Ajay et al., 2015).

Identification of individual pathogens is, inevitably, a simplification of the microbial community present at any one time in a plant's life cycle and similarly of the microbiome in the soil and environment. Next-generation sequencing can provide a broader view of these communities, not always at species level but specific enough to provide data for trend analyses. Its use in fingerprinting genomes of pathogens is particularly valuable (Barba et al., 2014) and has been applied to legume viruses (Kehoe et al., 2014).

6.3 Quantitative techniques

Determining the levels of a disease organism can be equally important as detecting its presence. Routine PCR will produce amplified DNA, but some techniques can quantify the amount and relate this to the numbers of copy number and microbial cells initially present.

Real-time PCR (RT PCR) is perhaps the most prevalent method to do this and is extensively applied to many quality assurance processes for food safety as well as clinical analysis. RT PCR follows the same protocols as standard PCR but is able to measure the increase in DNA amplification with time. The amplification system involves the inclusion of a component containing fluorescent probes and quenchers. Because of the action of the TAQ polymerase enzyme, the fluorescent probe is activated during the amplification in proportion to the DNA produced. This will in turn depend on the amount of DNA present at the start of the reaction and so allow a quantification, either in comparison with another product or for absolute measure against a standard curve of purified target DNA.

Applications of RT PCR are increasingly evident in disease analysis both for quantification of a pathogen (Kaczmarek et al., 2009) and for detection of the presence of a mutation in a fungicide target gene (McCartney et al., 2003).

RT PCR also allows a melt curve of the amplified DNA to be calculated. The temperature and profile of the melt curve will depend on the DNA sequence and can be an indicator for identification. It is becoming increasingly common to combine a number of primers into one multiplex RT PCR reaction so that more than one disease may be assessed at the same time, thus reducing costs. In providing results directly, RT PCR allows for a rapid analysis. Coupled with the development of small-scale instruments, the potential of the technique for field use is increasingly close.

While developments of RT PCR lie in the use of increasingly complex reagents, an alternative approach relies on the use of increasingly complex primers. Loop-mediated isothermal amplification (LAMP) is based on conducting PCR assays at a constant temperature and analysing the result in a simple spectrophotometer to detect an increase in optical density at 600nm as the amplified DNA precipitates pyrophosphate salts (Notomi et al., 2000). LAMP uses a mix of four primers targeting six target sequences on the DNA template and acting sequentially to amplify the DNA into complex three-dimensional structures with multiple loops. Detection by simple optical density allows the technique to be used in more basic conditions than standard PCR and so has the potential for extensive application. The technique has been used as a rapid test for a range of bacterial pathogens and more recently for fungi (Dai et al., 2012) including *Asochyta* in chickpeas (Chen et al., 2016) and *Erwinia* species (Temple and Johnson, 2011).

PCR is not the only approach to rapid analysis. Immunochemical tests such as ELISA have been used for decades with increasing effectiveness. As noted earlier, applications to disease monitoring include coupling automatic air sample systems impacting air samples on microtitre strips to collect samples at regular intervals dictated by the sampling control systems. Subsequent processing allows a rapid development of response to indicate amounts of a target antibody (Rogers et al., 2008; Kaczmarek et al., 2009).

7 Conclusion

From the range of considerations outlined here, it is evident that legume disease management has many elements and a strong potential to achieve success. This is, however, assuming the availability of strong and effective management and access to skilled practitioners on farms and in supporting laboratories. It requires local engagement and international initiatives with regional applications.

In particular, it requires the ability to act rapidly and effectively and increasingly so with climate change effects. Speed is often on the side of the disease with ability to germinate in hours, penetrate tissue rapidly and multiply extensively. Legume varieties may have partial resistance but require additional hurdles to suppress pathogen growth. It is the task of all practitioners to keep these hurdles in place and develop more when the track of infection changes course.

8 Where to look for further information

Africa Soil Health Consortium. Legume Crop Pests and Diseases. https://africasoilhealth.cabi.org/materials/legumes-crop-pests-and-diseases/

CGIAR Research Program on Grain Legumes. Annual Report 2015. http://library.cgiar.org/bitstream/handle/10947/4458/Grain-Legumes-Annual-Report-2015.pdf?sequence=1

Food and Agriculture Organization of the United Nations. http://agris.fao.org/agris-search/search.do?recordID=ES2003002116

Legume Federation. https://legumefederation.org/

United States Department of Agriculture. Agricultural Research Service. Grain Legume Research Physiology Research Publications. https://www.ars.usda.gov/research/publications/publications-at-this-location/?modecode=20-90-20-00

9 References

Abtew, A., Niassy, S., Affognon, H., Subramanian, S., Kreiter, S., Garzia, G. T. and Martin, T., 2016. Farmers' knowledge and perception of grain legume pests and their management in the Eastern province of Kenya. *Crop Protect.* 87, 90–7.

Adomou, M., Prasad, P. V. V., Boote, K. J. and Detongnon, J., 2005. Disease assessment methods and their use in simulating growth and yield of peanut crops affected by leafspot disease. *Ann. App. Biol.* 146, 469–79.

Ahmed, S. and Beniwal, S. P. S., 1991. Ascochyta blight of lentil and its control in Ethiopia. *Tropical Pest Manage.* 37, 368–73.

Ajay, B. C., Gowda, M. B., Prasad, P. S., Veerakumar, G. N., Babu, P. H., Ganesh, B. N., Venkatesh, S. C., Ganapathy, K. N., Fiyaz, R. A. and Ramya, K. T., 2015. Identification of AFLP markers linked to Fusarium wilt disease in pigeonpea [*Cajanus cajan* (L.) Millsp.]. *Legume Res.* 38, 126–30.

Ashelford, K. E., Chuzhanova, N. A., Fry, J. C., Jones, A. J. and Weightman, A. J., 2005. At least 1 in 20 16S rRNA sequence records currently held in public repositories is estimated to contain substantial anomalies. *Appl. Environ. Microbiol.* 71, 7724–36.

Barba, M., Czosnek, H. and Hadidi, A., 2014. Historical perspective, development and applications of next-generation sequencing in plant virology. *Viruses* 6, 106–36.

Balkaja, A., 2004. Modelling the effect of temperature on the germination speed in some legume crops. *J. Agron.* 3, 179–83.

Banniza, S., Armstrong-Cho, C. L., Gan, Y. and Chongo, G., 2011. Evaluation of fungicide efficacy and application frequency for the control of ascochyta blight in chickpea. *Can. J. Plant Pathol.* 33, 135–49.

Beech, D. and Leach, G., 1989. Effect of plant density and row spacing on the yield of chickpea (cv. Tyson) grown on the Darling Downs, south-eastern Queensland. *Aust. J. Expt. Agric.* 29, 241.

Bensch, S. and Akesson, M., 2005. Ten years of AFLP in ecology and evolution: why so few animals? *Mol. Ecol.* 14, 2899–2914.

Berg, J. S., Khoury, M. J. and Evans, J. P., 2011. Deploying whole genome sequencing in clinical practice and public health: Meeting the challenge one bin at a time. *Genet. Med.* 13, 499–504.

Bordeleau, L. M. and Prévost, D., 1994. Nodulation and nitrogen fixation in extreme environments, in Graham, P. H., Sadowsky, M. J. and Vance, C. P. (eds), *Symbiotic Nitrogen Fixation, Developments in Plant and Soil Sciences*. Springer: Netherlands, pp. 115–25.

Bos, L. and Makkouk, K. M., 1994. Insects in relation to virus epidemiology in cool season food legumes, in: Muehlbauer, F. J. and Kaiser, W. J. (eds), *Expanding the Production and Use of Cool Season Food Legumes, Current Plant Science and Biotechnology in Agriculture*. Springer Netherlands, pp. 305–32.

Bostock, R. M., Pye, M. F. and Roubtsova, T. V., 2014. Predisposition in plant disease: Exploiting the nexus in abiotic and biotic stress perception and response. *Annu. Rev. Phytopathol.* 52, 517–49.

Bouràoui, M., Abbes, A., Abdi, N., Hmissi, I. and Sifi, B., 2012, Evaluation of efficient *Rhizobium* isolates as biological control agents of *Orobanche foetida* Poir. Parasitizing *Vicia Faba* L. minor in Tunisia. *Bulg. J. Agric. Sci.* 18, 557–64.

Bretag, T., Price, T. and Keane, P., 1995. Importance of seed-borne inoculum in the etiology of the Ascochyta blight complex of field peas (*Pisum sativum* L.) grown in Victoria. *Aust. J. Exp. Agric.* 35, 525.

Bretag, T. W., Keane, P. J. and Price, T. V., 2000. Effect of sowing date on the severity of ascochyta blight in field peas (*Pisum sativuum* L) grown in the Wimmera region of Victoria. *Aust. J. Exp. Agric.* 40, 1113–19.

Brown, J. K. M. and Rant, J. C., 2013. Fitness costs and trade-offs of disease resistance and their consequences for breeding arable crops. *Plant Pathol.* 62, 83–95.

Burgess, D. R., Bretag, T. and Keane, P. J., 1997. Seed-to-seedling transmission of *Botrytis cinerea* in chickpea and disinfestation of seed with moist heat. *Aust. J. Exp. Agric.* 37, 223.

Burridge, J., Jochua, C. N., Bucksch, A. and Lynch, J. P., 2016. Legume shovelomics: High – Throughput phenotyping of common bean (*Phaseolus vulgaris* L.) and cowpea (*Vigna unguiculata* subsp, unguiculata) root architecture in the field. *Field Crops Res.* 192, 21–32.

Campos, S. B., Lisboa, B. B., Camargo, F. A. O., Bayer, C., Sczyrba, A., Dirksen, P., Albersmeier, A., Kalinowski, J., Beneduzi, A., Costa, P. B., Passaglia, L. M. P., Vargas, L. K. and Wendisch, V. F., 2016. Soil suppressiveness and its relations with the microbial community in a Brazilian subtropical agroecosystem under different management systems. *Soil Biol. Biochem.* 96, 191–7.

Casas, A. T., 1992. Development of *Didymella rabiei*, the Telomorph of *Ascochyta rabiei* on Chickpea straw. *Phytopathology* 82, 1261–6.

Chakraborty, S., Ghosh, R., Ghosh, M., Fernandes, C. D., Charchar, M. J. and Kelemu, S., 2004. Weather-based prediction of anthracnose severity using artificial neural network models. *Plant Pathol.* 53, 375–86.

Chakravorty, S., Helb, D., Burday, M., Connell, N. and Alland, D., 2007. A detailed analysis of 16S ribosomal RNA gene segments for the diagnosis of pathogenic bacteria. *J. Microbiol. Methods* 69, 330–9.

Chang, K. F., Ahmed, H. U., Hwang, S. F., Gossen, B. D., Howard, R. J., Warkentin, T. D., Strelkov, S. E. and Blade, S. F., 2007. Impact of cultivar, row spacing and seeding rate on ascochyta blight severity and yield of chickpea. *Can. J. Plant Sci.* 87, 395–403.

Chen, X., Ma, L., Qiang, S. and Ma, D., 2016. Development of a loop-mediated isothermal amplification method for the rapid diagnosis of *Ascochyta rabiei* L. in chickpeas. *Sci. Rep.* 6, 25688.

Chen, Y.-J., Lyngkjær, M. F. and Collinge, D. B., 2012. Future prospects for genetically engineering disease-resistant plants, in: Sessa, G. (ed.), *Molecular Plant Immunity*. Wiley-Blackwell, New York, pp. 251–75.

Chongo, G., Buchwaldt, L., Gossen, B. D., Lafond, G. P., May, W. E., Johnson, E. N. and Hogg, T., 2003. Foliar fungicides to manage ascochyta blight [*Ascochyta rabiei*] of chickpea in Canada. *Can. J. Plant Pathol.* 25, 135–42.

Chongo, G., Gossen, B. D., Buchwaldt, L., Adhikari, T. and Rimmer, S. R., 2004. Genetic diversity of *Ascochyta rabiei* in Canada. *Plant Dis.* 88, 4–10.

Clark, W. S., 2006. Fungicide resistance: are we winning the battle but losing the war. *Aspects Appl. Biol.* 78, 119–26.

Clarridge, J. E., 2004. Impact of 16S rRNA gene sequence analysis for identification of Bacteria on clinical microbiology and infectious diseases. *Clin. Microbiol. Rev.* 17, 840–62.

Clayton, R. A., Sutton, G., Hinkle, P. S., Bult, C. and Fields, C., 1995. Intraspecific variation in small-subunit rRNA sequences in GenBank: why single sequences may not adequately represent prokaryotic taxa. *Int. J. Syst. Bacteriol.* 45, 595–9.

Dai, T.-T., Lu, C.-C., Lu, J., Dong, S., Ye, W., Wang, Y. and Zheng, X., 2012. Development of a loop-mediated isothermal amplification assay for detection of *Phytophthora sojae*. *FEMS Microbiol. Lett.* 334, 27–34.

Dash, S., Campbell, J. D., Cannon, E. K. S., Cleary, A. M., Huang, W., Kalberer, S. R., Karingula, V., Rice, A. G., Singh, J., Umale, P. E., Weeks, N. T., Wilkey, A. P., Farmer, A. D. and Cannon, S. B., 2016. Legume information system (LegumeInfo.org): a key component of a set of federated data resources for the legume family. *Nucleic Acids Res.* 44, D1181–8.

Davidson, J. A. and Kimber, R. B. E., 2007. Integrated disease management of ascochyta blight in pulse crops. *Eur. J. Plant Pathol.* 119, 99–110.

Davis, E. L. and Mitchum, M. G., 2005. Nematodes. Sophisticated parasites of legumes. *Plant Physiol.* 137, 1182–8.

De Meyer, S. E., De Beuf, K., Vekeman, B. and Willems, A., 2015. A large diversity of non-rhizobial endophytes found in legume root nodules in Flanders (Belgium). *Soil Biol. Biochem.* 83, 1–11.

Denton, M. D., Pearce, D. J., Ballard, R. A., Hannah, M. C., Mutch, L. A., Norng, S. and Slattery, J. F., 2009. A multi-site field evaluation of granular inoculants for legume nodulation. *Soil Biol. Biochem.* 41, 2508–16.

Dudeja, S. S., 2016. Beneficial effects and molecular diversity of endophytic Bacteria in legume and Nonlegumes, in: Singh, D. P., Singh, H. B. and Prabha, R. (eds), *Microbial Inoculants in Sustainable Agricultural Productivity*. Springer India, New Delhi, pp. 245–56.

Dudeja, S. S., Giri, R., Saini, R., Suneja-Madan, P. and Kothe, E., 2012. Interaction of endophytic microbes with legumes. *J. Basic Microbiol.* 52, 248–60.

Eddleston, M., Karalliedde, L., Buckley, N., Fernando, R., Hutchinson, G., Isbister, G., Konradsen, F., Murray, D., Piola, J. C., Senanayake, N., Sheriff, R., Singh, S., Siwach, S. B. and Smit, L., 2002. Pesticide poisoning in the developing world – a minimum pesticides list. *Lancet* 360, 1163–7.

Erb, M., Balmer, D., De Lange, E. S., Von Merey, G., Planchamp, C., Robert, C. a. M., Röder, G., Sobhy, I., Zwahlen, C., Mauch-Mani, B. and Turlings, T. C. J., 2011. Synergies and trade-offs between insect and pathogen resistance in maize leaves and roots. *Plant, Cell Environ.* 34, 1088–1103.

Eulgem, T., Rodriguez-Salus, M. and Knoth, C. M., 2016. United States Patent: 9420790 – Molecules that induce disease resistance and improve growth in plants. 9420790.

Fatima, F., Pathak, N. and Rastogi Verma, S., 2014. An improved method for soil DNA extraction to study the microbial assortment within Rhizospheric region. *Mol. Biol. Int.* 2014.

Fondong, V. N., Thresh, J. M. and Zok, S., 2002. Spatial and temporal spread of cassava mosaic virus disease in cassava grown alone and when intercropped with maize and/or cowpea. *J. Phytopathol.* 150, 365–74.

Fraaije, B. A., Cools, H. J., Fountaine, J., Lovell, D. J., Motteram, J., West, J. S. and Lucas, J. A., 2005. Role of Ascospores in further spread of QoI-resistant cytochrome b alleles (G143A) in field populations of Mycosphaerella graminicola. *Phytopathology* 95, 933–41.

Gan, Y. T., Miller, P. R., McConkey, B. G., Zentner, R. P., Liu, P. H. and McDonald, C. L., 2003. Optimum plant population density for chickpea and dry pea in a semiarid environment. *Can. J. Plant Sci.* 83, 1–9.

Ghini, R., Hamada, E. and Bettiol, W., 2008. Climate change and plant diseases. *Sci. Agricola* 65, 98–107.

Goldfeder, R. L., Priest, J. R., Zook, J. M., Grove, M. E., Waggott, D., Wheeler, M. T., Salit, M., Ashley, E. A., 2016. Medical implications of technical accuracy in genome sequencing. *Genome Med* 8, 24.

Gossen, B. D., Hwang, S. F., Conner, R. L. and Chang, K. F., 2011, Managing the ascochyta blight complex on field pea in western Canada. *Prairie Soils Crops J.* 4, 135–41.

Gossen, B. D., Carisse, O., Kawchuk, L. M., Heyden, H. V. D. and McDonald, M. R., 2014. Recent changes in fungicide use and the fungicide insensitivity of plant pathogens in Canada. *Can. J. Plant Pathol.* 36, 327–40.

Gurr, S. J. and Rushton, P. J., 2005. Engineering plants with increased disease resistance: how are we going to express it? *Trends Biotechnol.* 23, 283–90.

Handelsman, J. and Stabb, E. V., 1996. Biocontrol of Soilborne plant pathogens. *Plant Cell* 8, 1855–69.

Hema, M., Sreenivasulu, P., Patil, B. L., Kumar, P. L. and Reddy, D. V. R., 2014. Tropical food legumes: virus diseases of economic importance and their control. *Adv. Virus Res.* 90, 431–505.

Hemissi, I., Mabrouk, Y., Abdi, N., Bouraoui, M., Saidi, M. and Sifi, B., 2013. Growth promotion and protection against *Orobanche foetida* of chickpea (*Cicer aerietinum*) by two *Rhizobium* strains under greenhouse conditions. *Afr. J. Biotechnol.* 12.

Hepperly, P. R., 1979. Predisposition to Seed Infection by *Phomopsis sojae* in Soybean plants infected by Soybean Mosaic virus. *Phytopathology* 69, 846.

Heungens, K. and Parke, J. L., 2000. Zoospore homing and infection events: effects of the biocontrol Bacterium *Burkholderia cepacia* AMMDR1 on two Oomycete pathogens of pea (*Pisum sativum* L.). *Appl. Environ. Microbiol.* 66, 5192–5200.

Huang, X., McNeill, T. and Schweiner, M., 2016. United States Patent: 9434942 – Small interfering RNAS with target-specific seed sequences. 9434942.

Irinyi, L., Lackner, M., de Hoog, G. S. and Meyer, W., 2016. DNA barcoding of fungi causing infections in humans and animals. *Fungal Biol.* 120, 125–36.

Isard, S. A., Russo, J. M. and Ariatti, A., 2007. The Integrated Aerobiology Modeling System applied to the spread of soybean rust into the Ohio River valley during September 2006. *Aerobiologia* 23, 271–82.

Janda, J. M. and Abbott, S. L., 2007. 16S rRNA gene sequencing for Bacterial identification in the diagnostic laboratory: pluses, perils, and pitfalls. *J. Clin Microbiol.* 45, 2761–4.

Jayakumar, P., Gan, Y. T., Gossen, B. D., Warkentin, T. D. and Banniza, S., 2005. Ascochyta blight of chickpea: infection and host resistance mechanisms. *Can. J. Plant Pathol.* 27, 499–509.

Kaczmarek, J., Fitt, B. F., Jedryczka, M. and Latunde-Dada, A. O., 2008. Detection by real-time PCR and quantification of *Leptospaeria maculans* and *L. biglobosa* in air samples from north Poland. *Aspects Appl. Biol.* 89, 71–6.

Kaiser, W. J., 1983. Etiology and control of seed decay and preemergence damping-off of chickpea by *Pythium ultimum. Plant Dis.* 67, 77.

Kaiser, W. J., 1973. Factors affecting growth, sporulation, pathogenicity, and survival of *Ascochyta rabiei. Mycologia* 65, 444–57.

Kashiwagi, J., Krishnamurthy, L., Crouch, J. H. and Serraj, R., 2006. Variability of root length density and its contributions to seed yield in chickpea (*Cicer arietinum* L.) under terminal drought stress. *Field Crops Res.* 95, 171–81.

Kehoe, M. A., Coutts, B. A., Buirchell, B. J. and Jones, R. A. C., 2014. Plant virology and next generation sequencing: experiences with a Potyvirus. *PLoS ONE* 9, e104580.
Kennedy, R. and Wakeham, A. J., 2008. Development and use of detection systems for the sporangia of *Peronospora destructor. Aspects Appl. Biol.* 89, 55–6.
Kim, H. S. and Delaney, T. P., 2002. Over-expression of TGA5, which encodes a bZIP transcription factor that interacts with NIM1/NPR1, confers SAR-independent resistance in *Arabidopsis thaliana* to *Peronospora parasitica*. *Plant J.* 32, 151–63.
Koenning, S. R. and Wrather, J. A., 2010. Suppression of Soybean Yield Potential in the Continental United States by Plant Diseases from 2006 to 2009. *Plant Health Progress*. http://www.plantmanagementnetwork.org/sub/php/research/2010/yield/yield.pdf doi:10.1094/PHP-2010-1122-01-RS
Leistner, L., 2000. Basic aspects of food preservation by hurdle technology. *Int. J. Food Microbiol.* 55, 181–6.
Lozovaya, V., Lygin, A., Zernova, O. and Widholm, J., 2005. Genetic engineering of plant root disease resistance by modification of the phenylpropanoid pathway. *Plant Biosyst.* 139, 20–3.
Lucas, J. A., 2011. Advances in plant disease and pest management. *J. Agric. Sci.* 149, 91–114.
Macdonald, O. and Richardson, D., 2006, Regulatory challenges from fungicide resistance. *Aspects Appl. Biol.* 78, 105–12.
Madden, L. V., 1997. Effects of rain on splash dispersal of fungal pathogens. *Can. J. Plant Pathol.* 19, 225–30.
Makkouk, K. M. and Kumari, S. G., 2009. Epidemiology and integrated management of persistently transmitted aphid-borne viruses of legume and cereal crops in West Asia and North Africa. Virus Research, Plant Virus Epidemiology: Controlling epidemics of emerging and established plant viruses – the way forward 141, 209–18.
Manter, D. K. and Vivanco, J. M., 2007. Use of the ITS primers, ITS1F and ITS4, to characterize fungal abundance and diversity in mixed-template samples by qPCR and length heterogeneity analysis. *J. Microbiol. Methods* 71, 7–14.
Maropola, M. K. A., Ramond, J.-B. and Trindade, M., 2015. Impact of metagenomic DNA extraction procedures on the identifiable endophytic bacterial diversity in *Sorghum bicolor* (L. Moench). *J. Microbiol. Methods* 112, 104–17.
Mazzola, M., 2002. Mechanisms of natural soil suppressiveness to soilborne diseases. An*tonie Van Leeuwenhoek* 81, 557–64.
McGrann, G. R. D., Stavrinides, A., Russell, J., Corbitt, M. M., Booth, A., Chartrain, L., Thomas, W. T. B. and Brown, J. K. M., 2014. A trade off between mlo resistance to powdery mildew and increased susceptibility of barley to a newly important disease, Ramularia leaf spot. *J. Exp. Bot.* 65, 1025–37.
Melander, C. and Rogers, S. A., 2016. United States Patent: 9439436 – Use of aryl carbamates in agriculture and other plant-related areas. 9439436.
Micheli, M. R., Bova, R., Pascale, E. and D'Ambrosio, E., 1994. Reproducible DNA fingerprinting with the random amplified polymorphic DNA (RAPD) method. *Nucleic Acids Res.* 22, 1921–2.
Moreno-Paz, M. and Parro, V., 2006. Amplification of low quantity bacterial RNA for microarray studies: time-course analysis of *Leptospirillum ferrooxidans* under nitrogen-fixing conditions. *Environ. Microbiol.* 8, 1064–73.
Murithi, H. M., Beed, F., Tukamuhabwa, P., Thomma, B. P. H. J. and Joosten, M. H. A. J., 2016. Soybean production in eastern and southern Africa and threat of yield loss due to soybean rust caused by *Phakopsora pachyrhizi*. *Plant Pathol.* 65, 176–88.
Narayanasamy, P., 2008. *Molecular Biology in Plant Pathogenesis and Disease Management: Microbial Plant Pathogens*. Springer Science & Business Media, Berlin.
Notomi, T., Okayama, H., Masubuchi, H., Yonekawa, T., Watanabe, K., Amino, N. and Hase, T., 2000. Loop-mediated isothermal amplification of DNA. *Nucleic Acids Res.* 28, E63.
Ohm, R. A., Feau, N., Henrissat, B., Schoch, C. L., Horwitz, B. A., Barry, K. W., Condon, B. J., Copeland, A. C., Dhillon, B., Glaser, F., Hesse, C. N., Kosti, I., LaButti, K., Lindquist, E. A.,

Lucas, S., Salamov, A. A., Bradshaw, R. E., Ciuffetti, L., Hamelin, R. C., Kema, G. H. J., Lawrence, C., Scott, J. A., Spatafora, J. W., Turgeon, B. G., de Wit, P. J. G. M., Zhong, S., Goodwin, S. B. and Grigoriev, I. V., 2012. Diverse lifestyles and strategies of plant pathogenesis encoded in the genomes of eighteen *Dothideomycetes* fungi. *PLoS Pathog.* 8, e1003037.

O'Sullivan, D. J. and O'Gara, F., 1992. Traits of fluorescent *Pseudomonas* spp. involved in suppression of plant root pathogens. *Microbiol. Rev.* 56, 662–76.

Palmieri, R., Tredway, L., Niyogi, D. and Lackmann, G. M., 2006. Development and evaluation of a forecasting system for fungal disease in turfgrass. *Met. Apps.* 13, 405–16.

Pande, S., Galloway, J., Gaur, P. M., Siddique, K. H. M., Tripathi, H. S., Taylor, P., MacLeod, M. W. J., Basandrai, A. K., Bakr, A., Joshi, S., Kishore, G. K., Isenegger, D. A., Rao, J. N. and Sharma, M., 2006a. Botrytis grey mould of chickpea: a review of biology, epidemiology, and disease management. *Aust. J. Agric. Res.* 57, 1137–50.

Pande, S., Siddique, K. H. M., Kishore, G. K., Bayaa, B., Gaur, P. M., Gowda, C. L. L., Bretag, T. W. and Crouch, J. H., 2005. Ascochyta blight of chickpea (*Cicer arietinum* L.): a review of biology, pathogenicity, and disease management. *Aust. J. Agric. Res.* 56, 317–32.

Pappa, V. A., Rees, R. M., Walker, R. L., Baddeley, J. A. and Watson, C. A., 2011. Intercropping of legumes and cereals: effect on yield and N uptake in a three year low input crop rotation.

Pedersen, E. A., Morrall, R. a. A., McCartney, H. A. and Fitt, B. D. L., 1994. Dispersal of conidia of *Ascochyta fabae* f. sp. *lentis* from infected lentil plants by simulated wind and rain. *Plant Pathol.* 43, 50–5.

Pedersen, E. A. and Morrall, R. A. A., 1995. Effect of wind speed and direction on horizontal spread of ascochyta blight of lentil. *Can. J. Plant Pathol.* 17, 223–32.

Perry, J. A., Wang, T. L., Welham, T. J., Gardner, S., Pike, J. M., Yoshida, S. and Parniske, M., 2003. A TILLING reverse genetics tool and a web-accessible collection of mutants of the legume *Lotus japonicus*. *Plant Physiol.* 131, 866–71.

Rajendran, G., Sing, F., Desai, A. J. and Archana, G., 2008. Enhanced growth and nodulation of pigeon pea by co-inoculation of Bacillus strains with *Rhizobium* spp. *Bioresource Technology, exploring horizons in biotechnology. Global Venture* 99, 4544–50.

Robertson, J. M. and Walsh-Weller, J., 1998. An introduction to PCR primer design and optimization of amplification reactions. *Methods Mol. Biol.* 98, 121–54.

Roger, C., Tivoli, B. and Huber, L., 1999. Effects of temperature and moisture on disease and fruit body development of Mycosphaerella pinodes on pea (*Pisum sativum*). *Plant Pathol.* 48, 1–9.

Rubiales, D. and Fondevilla, S., 2012. Future prospects for Ascochyta blight resistance breeding in cool season food legumes. *Front Plant Sci.* 3, 1–5.

Salam, M. U., Galloway, J., Diggle, A. J., MacLeod, W. J. and Maling, T., 2011. Predicting regional-scale spread of ascospores of *Didymella pinodes* causing ascochyta blight disease on field pea. *Aust. Plant Pathol.* 40, 640–7.

Satovic, Z., Avila, C. M., Cruz-Izquierdo, S., Díaz-Ruíz, R., García-Ruíz, G. M., Palomino, C., Gutiérrez, N., Vitale, S., Ocaña-Moral, S., Gutiérrez, M. V., Cubero, J. I. and Torres, A. M., 2013. A reference consensus genetic map for molecular markers and economically important traits in faba bean (*Vicia faba* L.). *BMC Genomics* 14, 932.

Schardl, C. L., Leuchtmann, A., Chung, K.-R., Penny, D. and Siegel, M. R., 1997. Coevolution by common descent of fungal symbionts (Epichloe spp.) and grass hosts. *Mol. Biol. Evol.* 14, 133–43.

Schoeneweiss, D. F., 1975. Predisposition, stress, and plant disease. *Ann. Rev. Phytopathol.* 13, 193–211.

Schoeny, A., Jumel, S., Rouault, F., May, C. L. and Tivoli, B., 2007. Assessment of airborne primary inoculum availability and modelling of disease onset of ascochyta blight in field peas. *Eur. J. Plant Pathol.* 119, 87–97.

Schreinemachers, P., Balasubramaniam, S., Boopathi, N. M., Ha, C. V., Kenyon, L., Praneetvatakul, S., Sirijinda, A., Le, N. T., Srinivasan, R. and Wu, M.-H., 2015. Farmers' perceptions and management of plant viruses in vegetables and legumes in tropical and subtropical Asia. *Crop Prot.* 75, 115–23.

Siddique, K. H. M., Johansen, C., Turner, N. C., Jeuffroy, M.-H., Hashem, A., Sakar, D., Gan, Y. and Alghamdi, S. S., 2012. Innovations in agronomy for food legumes. A review. *Agron. Sustain. Dev.* 32, 45–64.

Sikora, E. J., Allen, T. W., Wise, K. A., Bergstrom, G., Bradley, C. A., Bond, J., Brown-Rytlewski, D., Chilvers, M., Damicone, J., DeWolf, E., Dorrance, A., Dufault, N., Esker, P., Faske, T. R., Giesler, L., Goldberg, N., Golod, J., Gómez, I. R. G., Grau, C., Grybauskas, A., Franc, G., Hammerschmidt, R., Hartman, G. L., Henn, R. A., Hershman, D., Hollier, C., Isakeit, T., Isard, S., Jacobsen, B., Jardine, D., Kemerait, R., Koenning, S., Langham, M., Malvick, D., Markell, S., Marois, J. J., Monfort, S., Mueller, D., Mueller, J., Mulrooney, R., Newman, M., Osborne, L., Padgett, G. B., Ruden, B. E., Rupe, J., Schneider, R., Schwartz, H., Shaner, G., Singh, S., Stromberg, E., Sweets, L., Tenuta, A., Vaiciunas, S., Yang, X. B., Young-Kelly, H. and Zidek, J., 2014. A coordinated effort to manage Soybean rust in North America: a success story in Soybean disease monitoring. *Plant Dis.* 98, 864–75.

Singh, S. and Shalini, R., 2016. Effect of hurdle technology in food preservation: a review. *Crit. Rev. Food Sci. Nutr.* 56, 641–9.

Skiba, B., Ford, R. and Pang, E. C. K., 2005. Construction of a cDNA library of Lathyrus sativus inoculated with *Mycosphaerella pinodes* and the expression of potential defence-related expressed sequence tags (ESTs). *Physiol. Mol. Plant Pathol.* 66, 55–67.

Srinivasa, C., Sharanaiah, U. and Shivamallu, C., 2012. Molecular detection of plant pathogenic bacteria using polymerase chain reaction single-strand conformation polymorphism. *Acta Biochim. Biophys. Sin.* (Shanghai) 44, 217–23.

Stenberg, J. A., Heil, M., Åhman, I. and Björkman, C., 2015. Optimizing crops for biocontrol of pests and disease. *Trends Plant Sci.* 20, 698–712.

Stoddard, F. L., Nicholas, A. H., Rubiales, D., Thomas, J. and Villegas-Fernández, A. M., 2010. Integrated pest management in faba bean. *Field Crops Res. Faba Beans Sustain. Agric.* 115, 308–18.

Tar'an, B., Warkentin, T., Somers, D. J., Miranda, D., Vandenberg, A., Blade, S., Woods, S., Bing, D., Xue, A., DeKoeyer, D. and Penner, G., 2003. Quantitative trait loci for lodging resistance, plant height and partial resistance to mycosphaerella blight in field pea (*Pisum sativum* L.). *Theor. Appl. Genet.* 107, 1482–91.

Temple, T. N. and Johnson, K. B., 2010. Evaluation of loop-mediated isothermal amplification for rapid detection of *Erwinia amylovora* on pear and apple fruit flowers. *Plant Dis.* 95, 423–30.

Tilman, D., Cassman, K. G., Matson, P. A., Naylor, R. and Polasky, S., 2002. Agricultural sustainability and intensive production practices. *Nature* 418, 671–7.

Tivoli, B., Baranger, A., Avila, C. M., Banniza, S., Barbetti, M., Chen, W., Davidson, J., Lindeck, K., Kharrat, M., Rubiales, D., Sadiki, M., Sillero, J. C., Sweetingham, M. and Muehlbauer, F. J., 2006. Screening techniques and sources of resistance to foliar diseases caused by major necrotrophic fungi in grain legumes. *Euphytica* 147, 223–53.

Trapero-Casas, A. and Kaiser, W. J., 2007. Differences between Ascospores and Conidia of *Didymella rabiei* in spore germination and infection of Chickpea. *Phytopathology* 97, 1600–7.

Tsoktouridis, G., Tsiamis, G., Koutinas, N. and Mantell, S., 2014. Molecular detection of bacteria in plant tissues, using universal 16S ribosomal DNA degenerated primers. *Biotechnol Biotechnol Equip* 28, 583–91.

Uzokwe, V. N. E., Mlay, D. P., Masunga, H. R., Kanju, E., Odeh, I. O. A. and Onyeka, J., 2016. Combating viral mosaic disease of cassava in the Lake Zone of Tanzania by intercropping with legumes. *Crop Prot.* 84, 69–80.

Vafaei, S. H., Rezaee, S., Moghadam, A. A. and Zamanizadeh, H. R., 2015. Virulence diversity of *Ascochyta rabiei* the causal agent of Ascochyta blight of chickpea in the western provinces of Iran. *Arch. Phytopathol. Plant Prot.* 48, 921–30.

von Maltzahn, G., Flavell, R. B., Toledo, G. V., Leff, J. W., Samayoa, P., Marquez, L. M., Johnston, D. M., Djonovic, S., Millet, Y. A., Sadowski, C., Lyford, J., Ambrose, K. V. and Zhang, X., 2016. United States Patent: 9408394 – Endophytes, associated compositions, and methods of use thereof. 9408394.

Vriesekoop, F., Krahl, M., Hucker, B. and Menz, G., 2012. 125th Anniversary review: Bacteria in brewing: the good, the bad and the ugly. *J. Inst. Brew.* 118, 335–45. doi:10.1002/jib.49

West, J. and Kimber, R., 2015. Innovations in air sampling to detect plant pathogens. *Ann. Appl. Biol.* 166, 4–17.

West, J. S., Atkins, S. D., Emberlin, J. and Fitt, B. D. L., 2008a. PCR to predict risk of airborne disease. *Trends Microbiol.* 16, 380–7.

Wilson, P. A. and Chakrabrty, S., 1998. The virtual plant: a new tool for the study and management of plant diseases. *Crop Prot.* 17, 231–9.

Wrather, A. and Koenning, S., 2009. Effects of diseases on Soybean yields in the United States 1996 to 2007. *Plant Health Progress*.

Yeates, C., Gillings, M. r., Davison, A. d., Altavilla, N. and Veal, D. a., 1997. PCR amplification of crude microbial DNA extracted from soil. *Lett. Appl. Microbiol.* 25, 303–7.

Zhao, L. F., Xu, Y. J., Ma, Z. Q., Deng, Z. S., Shan, C. J. and Wei, G. H., 2013. Colonization and plant growth promoting characterization of endophytic *Pseudomonas chlororaphis* strain Zong1 isolated from *Sophora alopecuroides* root nodules. *Braz. J. Microbiol.* 44, 623–31.

Zhu, Y., Chen, H., Fan, J., Wang, Y., Li, Y., Chen, J., Fan, J., Yang, S., Hu, L., Leung, H., Mew, T. W., Teng, P. S., Wang, Z. and Mundt, C. C., 2000. Genetic diversity and disease control in rice. *Nature* 406, 718–22.

Zinnen, T. M., 1982. Thermotherapy of Soybean seeds to control Seedborne Fungi. *Phytopathology* 72, 831.

Chapter 14

Insect pests and integrated pest management techniques in grain legume cultivation

Tolulope A. Agunbiade, Yale University, USA; Weilin Sun, Michigan State University, USA; Brad S. Coates, USDA-ARS, USA; Fousséni Traore, Institut de l'Environnement et de Recherches Agricoles, Burkina Faso; James A. Ojo, Kwara State University, Nigeria; Anne N. Lutomia, University of Illinois at Urbana-Champaign, USA; Julia Bello-Bravo, Michigan State University, USA; Saber Miresmailli, Ecoation Innovative Solutions Inc., Canada; Joseph E. Huesing, USAID, USA; Michael Agyekum, Department of Agricultural, Food and Resource Economics, Michigan State University, USA; Manuele Tamò, International Institute of Tropical Agriculture (IITA), Benin; and Barry R. Pittendrigh, Michigan State University, USA

1 Introduction

Cowpea, *Vigna unguiculata* (L.) Walp. (Leguminosae: Papilionaceae), is an important grain legume crop in the semi-arid and dry savannah areas of the tropics (Singh and van Emden, 1979). The grain is a major nutritional source of protein for about 200 million people in sub-Saharan Africa. Cowpea production is highest in the West African countries of Nigeria, Niger and Burkina Faso. Insect-feeding damage caused by over one hundred pest species comprises a major biotic constraint on field production and post-harvest grain storage (Cork et al., 2009), and can dramatically reduce yields for subsistence farmers in West Africa. The insect pests feed on and damage the cowpea at virtually every crop developmental stage; major species include the legume pod borer, *Maruca vitrata*

http://dx.doi.org/10.19103/AS.2017.0023.11

Fabricius (Lepidoptera: Crambidae); the coreid pod bugs, *Clavigralla tomentosicollis* Stål (Hemiptera: Coreidae) and *Anoplocnemis curvipes* Fabricius (Hemiptera: Coreidae); the cowpea aphid, *Aphis craccivora* Koch (Hemiptera: Aphididae); and thrips, *Megalurothrips sjostedti* Trybom (Thysanoptera: Thripidae). The annual level of crop damage inflicted by these insect pests is estimated to be as high as 60–100% in the field (Singh et al., 1990).

Efforts to control these pests have remained a challenge. Despite decades of capacity building and educational efforts, the tenets of basic integrated pest management (IPM) are poorly implemented across sub-Saharan Africa. The control of cowpea pests still relies largely on the timely application of chemical insecticides, but the efficacy of this tactic remains variable because of evolving insect population resistances, which afford survival when exposed to one or more classes of insecticides (Ekesi, 1999). Also, the insecticides are prohibitively expensive, often counterfeit or simply otherwise unavailable to low-income farmers in West Africa, who are in most cases usually inadequately trained in their safe and effective use (Giga and Biscoe, 1989; Alghali, 1991; Bottenberg, 1995; Van Alebeek, 1996). Effective control of this insect pest complex and protection against yield loss in cowpea crops are likely to occur only subsequent to the development and successful implementation of IPM strategies.

IPM describes a pest control strategy whereby a variety of complementary approaches – for example, cultural control, biological control, pesticides and host plant resistance(s) – economically minimize the negative effects of pests on a given crop or cropping system (Oyewale and Bamaiyi, 2013). The goal of IPM is to suppress pest populations to below the economic injury level (EIL) using techniques that also minimize harm to the environment, including people. The UN's Food and Agriculture Organization (AGP, 2016 - http://www.fao.org/agriculture/crops/core-themes/theme/pests/ipm/en/) defines IPM as:

> The careful consideration of all available pest control techniques and subsequent integration of appropriate measures that discourage the development of pest populations and keep pesticides and other interventions to levels that are economically justified and reduce or minimize risks to human health and the environment. IPM emphasizes the growth of a healthy crop with the least possible disruption to agro-ecosystems and encourages natural pest control mechanisms.

Cowpea IPM programmes in West Africa will likely use a combination of locally produced biopesticides and/or conventional pesticides along with biological control agents and host plant resistance traits, for example, traditional breeding and biotechnologies such as marker-assisted breeding, genome-edited crops and genetic engineering (Wolt et al., 2016). Given the central role of smallholder farmers in cowpea production, it is also critical to understand their needs in order to facilitate and ensure successful implementation of any innovative IPM strategies. The creation and deployment of supporting educational content in local languages using scale-neutral approaches that will be appropriate for farmers of diverse literacy levels will then be crucial for the widespread adoption of improved IPM practices.

Although biocontrol agents, pesticides and both classical and biotechnology-based host plant resistance are all proven methods for controlling the pests of cultivated cowpea, increasing the basic understanding of pest population dynamics will better afford developing the most effective IPM strategies through combinations of the available approaches. Yet, the successful application of IPM strategies has remained elusive for cowpea insect pests in part due to (1) an overall lack of population demographic data, (2) the existence of cryptic species complexes within pest populations or among different populations of the same pests and (3) the seasonal migration patterns of some insects such as *M. vitrata*.

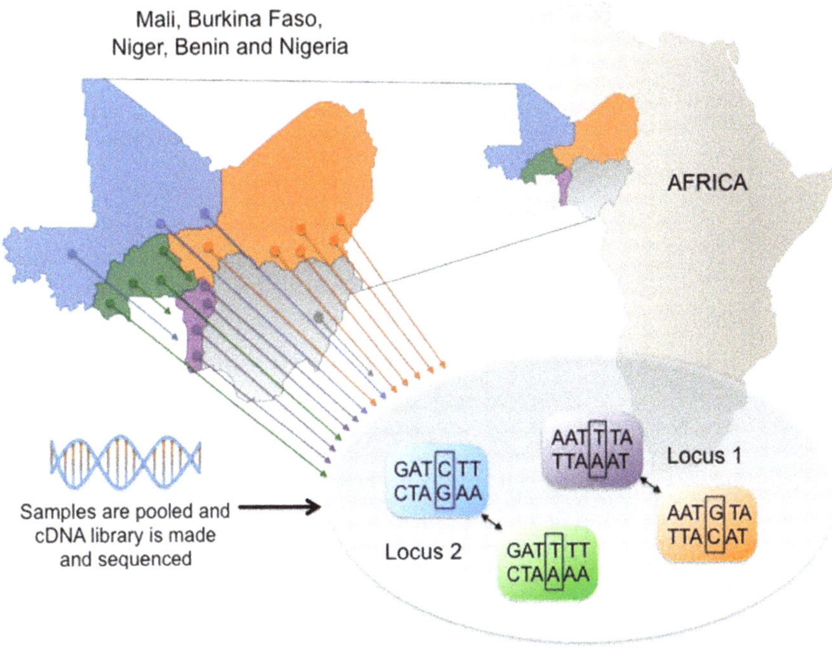

Figure 1 Hypothetical representation of the construction of a cDNA library by pooling insect samples from different cowpea-growing areas of West Africa (the arrows do not represent exact current sampling locations for any given insect species). The insects are selected from a diverse set of geographical locations, with mRNA pooled to create the cDNA library. Alignment of subsequent DNA sequence forms a set of expressed sequence tags potentially derived from the same locus and thus representative of alleles at a locus. The presence of polymorphisms among alleles in the form of SNPs can readily be adapted for genotyping individuals with high-throughput systems. Reprinted with permission from Agunbiade et al. (2012a).

These factors will continue to hinder any new application of control efforts against these pests if not taken into account. As such, IPM strategies are developed, refined and deployed to better characterize the life histories and population-level hierarchies of these insect pests in order to understand their relationships to the environment. To that end, we have developed a comprehensive toolset for the major pests of cowpea that combines traditional field studies with genomic techniques. This improved toolset enables the study of the population dynamics of cowpea pests (see Fig. 1 and 2) in order to craft those results into decision-making around the most effective control measure(s) for this complex of insect pests. This is outlined in the following section.

Population genetic tools can be used to determine the temporal and spatial distribution of pest insect populations and then leverage these insights for developing and deploying efficacious, cost-effective and environmentally sustainable strategies for controlling these pests (see Fig. 2 and 3). To date, the use of these tools has disclosed a more in-depth understanding of genetic diversity within *M. vitrata* populations (Margam et al., 2011b,c), and that *M. vitrata* living on cowpea have a great diversity of alternative host plants and common populations (Agunbiade et al., 2012b and 2014). From these results, the hypothesis that the

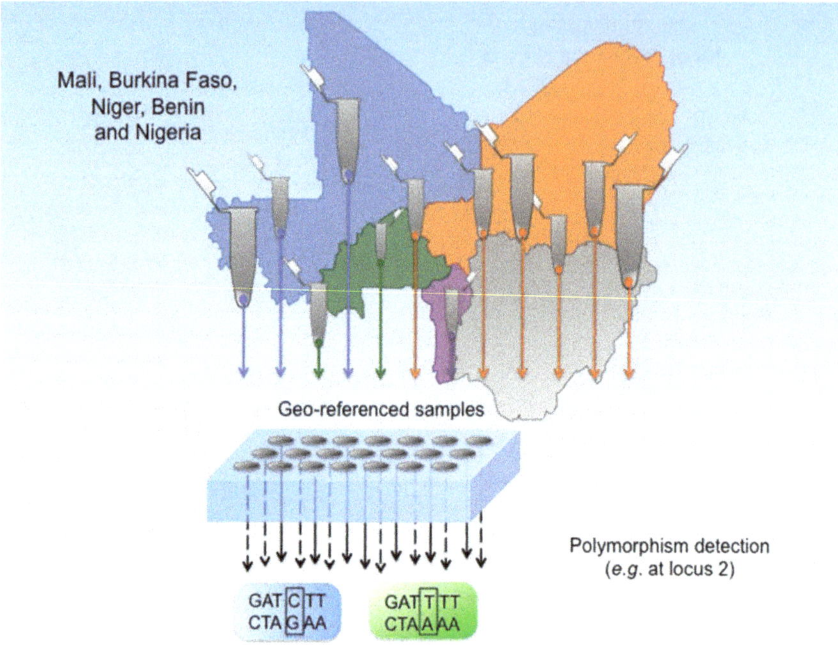

Figure 2 Representation of a hypothetical population genomics study in which individual insects from different locations (shown by the tubes) are collected from West Africa. Polymorphisms are detected among individual insect samples with high-throughput system(s) (shown as a blue plate) and are used to differentiate and correlate genotypes back to geographic locations, host plants, ecotype or other ecological characteristics described at the initial collection. Polymorphisms detected between individual insects are indicated as the C/G to T/A differences at the bottom of the figure. Reprinted with permission from Agunbiade et al. (2012a).

Maruca population living on cowpeas were interbreeding with *Maruca* on alternative hosts suggests that alternative host plants could likely act as a natural susceptible insect reservoir for transgenic cowpea expressing *Bacillus thuringiensis* (*Bt*) toxins. If so, then insects developing on natural refuge plants could serve as an overwhelming pool of susceptible individuals capable of mating with rarer, resistant individuals emerging from the transgenic crop.

Alternative host plants may also serve as a release point for biocontrol agents. Having developed a large transcriptome resource for *Megalurothrips sjostedti*, *Aphis craccivora*, *Clavigralla tomentosicollis* and *Anoplocnemis curvipes* that were used to detect polymorphisms among transcripts from the different populations (Agunbiade et al., 2013), this work potentially establishes a foundation for the future development of molecular markers and subsequent population dynamics studies within these insect populations. Use of such molecular tools may lead to the elucidation of biotypes that respond differently to host plant resistance traits, to the population structures of the pest populations or even to a better understanding of migratory patterns of the pests between wet and dry seasons.

To improve adoption and scalability of novel IPM technologies across West Africa, several important questions concerning the population of smallholder farmers must also

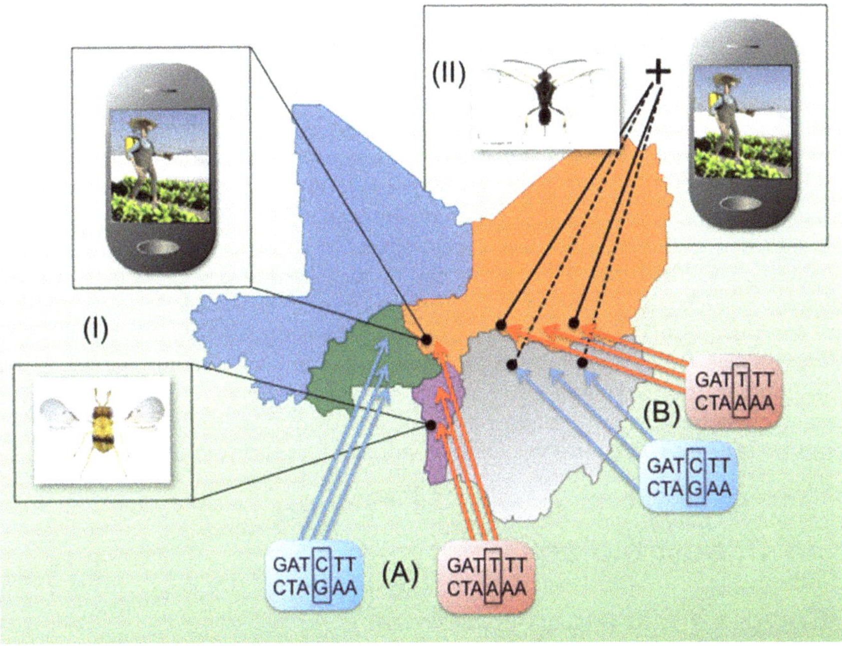

Figure 3 Hypothetical use of information gained from insect population genomic studies and how such information could be used in decisions associated with deployment of IPM strategies in two scenarios. In scenario A, the insects move from the south to the north in the rainy season, with biocontrol agents released in the south (in endemic regions) prior to migration for a long-term suppression of the pest population; temporary control measures (e.g. neem sprays) may be required in non-endemic regions if pest levels become too high. *Apanteles taragamae* is shown as an example of a biocontrol agent. In scenario B, the pest is endemic in both regions, with no significant longitudinal movement patterns. A possible pest control strategy in scenario B could involve the release of parasitoids (in both the south and north regions) along with implementation of an education programme concerning temporary control (e.g. with neem sprays) if pest levels become damaging. *Trichogrammatoidea eldanae* is shown as an example of a parasitoid. The farmer in the figure is applying a plant-based extract spray and thus does not require personal protection equipment; if a synthetic pesticide were used in this extension material, the farmer would have to be shown with personal protection equipment. Reprinted with permission from Agunbiade et al. (2012a).

be addressed; in particular, (a) What are the main drivers of farmers' pest control decisions? (b) Are farmers aware of the unintended negative health and environmental consequences that may be caused by persistent misuse of chemical pesticides? (c) What are the socio-economic characteristics of these smallholder farmers? and (d) How do farming systems compare across countries? We address these important questions by drawing not only on farmer survey results from Benin (Agyekum et al., 2016a) and Burkina Faso (Reyes et al., 2013) but also from insights from a choice experiment that sought to evaluate cowpea farmers' preferences for pest control options in Benin (Agyekum et al., 2016b).

Cowpea technologies developed by scientists, including novel control methods, can be utilized by low-resource farmers if the information is accessible to them. Thus, we have also collaborated with Scientific Animations Without Borders (SAWBO) to create educational animated videos specifically designed to serve individuals in highly multilingual African countries who have previously not had, or were not given adequate, access to the dominant, typically ex-colonial, national/administrative language in a given country. These educational videos can be transmitted from cell phone to cell phone via Bluetooth® technology, now widely available almost everywhere in Africa. To date, we have developed videos for improved cowpea production practices in West Africa, including solar drying, triple bagging of cowpea seeds to eliminate storage insect pests and the use of neem seed extract as a bioinsecticide against major insect pests of cowpea.

Developing and distributing insect pest control tactics that are accessible to indigenous populations is crucial for stabilizing food resources in many developing nations. Below, we outline specific research progress made towards enhancing cowpea IPM efforts in terms of both production and control of the pest complex. But it is by coupling these efforts to the development of novel video educational resources tailored in local languages that significant gains have already been made for improving the welfare of people across regions of West Africa. We term this coupled approach as 'precision-IPM'.

2 Known pest biology

2.1 Cowpea aphid (*Aphis craccivora*)

Aphis craccivora is a medium-sized, shiny black insect which feeds on the undersurface of young cowpea leaves, stem tissue and on pods of mature plants with a varied life cycle across Africa (Singh and Allen, 1980). An adult may live for 6–15 days and may produce more than one hundred progeny in its lifetime. They undergo incomplete metamorphosis (Whitworth and Ahmad, 2009) and reproduce both sexually and asexually (Trionnaire et al., 2008). The insect passes through four nymphal instars before reaching adulthood. Development time from the first instar to adulthood takes on average 3–5 days. Their typical annual life cycle is characterized by a succession of parthenogenetic generations and one sexual generation (Moran, 1992; Simon et al., 2002). A generation can be completed within 10–20 days. The growth, development, fecundity and longevity of *A. craccivora* varies with weather conditions, soil fertility, soil moisture and host plant (Ofuya, 1997). Winged adults are easily disseminated from one plant to another (Blackman and Eastop, 2008).

Damage to cowpea is caused by both adults and nymphs. Attacks cause plant stunting, leaf distortion, premature defoliation and the death of seedlings (Singh and Allen, 1980). They are known to cause significant damage even at low population densities due to the ability to transmit at least fourteen viruses (Rossel and Thottappilly, 1985; Thottapilly et al., 1990), including the cowpea aphid-borne virus (Raheja and Leleji, 1974; Atiri, 1984; Atiri et al., 1984, 1986, 1987; Atiri and Thottappilly, 1985). These viruses produce severe cowpea mottling, chlorosis and seed shrivelling, which can dramatically reduce yields.

2.2 Flower thrips (*Megalurothrips sjostedti*)

Adult thrips are shiny, black, minute insects that feed on the reproductive structures (flower buds and flowers) of cowpea. Infestations cause flower discolouration, distortion, abortion and abscission, thereby preventing pod formation, with estimated yield losses from 20–100% (Singh and Allen, 1980; Rusoke and Rubaihayo, 1994; Edema and Adipala, 1996). The entire life cycle takes 10–15 days (Alabi et al., 2004; Dabiré-Binso et al., 2008). Eggs are laid in the flower buds, and larvae feed voraciously on the plant, causing extensive damage (Singh and Allen, 1980). They have a rapid development, comprising the egg, two larval instars, prepupal, pupal, and adult stages. The larval stage is the most destructive. Pupation takes place in the soil. Males are haploid, and are produced only by unfertilized eggs, while females are diploid.

2.3 Legume pod borer (*Maruca vitrata*)

Maruca vitrata is endemic in all cowpea-growing regions of West Africa (Jackai and Adalla, 1997; Baoua et al., 2011; Ba et al., 2009). It is one of the most serious pests of cowpea due to its extensive host range, high damage potential and cosmopolitan distribution (Taylor, 1967; Raheja, 1974; Katayama and Suzuki, 1984; Ke et al., 1985; Sharma et al., 1999). Being a nocturnal insect, females fly at night and deposit eggs singly or in batches of two to six on flower buds, vegetative buds, flowers and leaf axils (Taylor, 1967, 1978; Traoré et al., 2013). Eggs hatch in about five days. The larvae undergo five larval developmental stages. The larvae are very active and tend to fall off the webbed flowers and pods with the slightest disturbance, by spinning a silken thread. The larvae feed on flowers, stems, peduncles and pods of cowpea, thereby causing damage at all developmental stages from seedling to podding stages (Singh and Taylor, 1978). However, the greatest damage occurs at flowering (Singh and Jackai, 1988). The larval stage is the most destructive one, causing yield losses of as much as 80% (Singh et al., 1990) depending on location, management and the cultural practices employed. The prepupal period lasts for 2 days. Pupation occurs in a silken cocoon among webbed leaves/pods or in the soil. The life cycle is completed in 18–35 days (Taylor, 1967; Akinfenwa, 1975). During the rainy season, *M. vitrata* population overlaps on cultivated cowpea and different alternative host plants especially in the family Leguminosae (Bottenberg et al., 1997; Arodokoun et al., 2006; Traoré et al., 2014).

2.4 Pod-sucking bugs (*Clavigralla tomentosicollis* and *Anoplocnemis curvipes*)

Pod-sucking bugs can cause a total loss of productivity to cowpea production, especially in Sudan zones where rainfall is between 600 and 1400 mm of water. After thrips, they are the second most damaging insect to cowpea pods during the seed-filling stage. In Africa, a complex of pod-sucking bugs predate cowpea pods and seeds, including *A. curvipes* and *C. tomentosicollis* (Singh and Allen, 1980; Singh and Jackai, 1985). In West Africa, *C. tomentosicollis* is the most widespread and damaging species of these. Females lay eggs on the underside of the leaves to avoid their exposure to sunlight and prevent washing off by rain splash. Hatched larvae undergo five instars before reaching adult stage. Both adults and larvae of *C. tomentosicollis* feed on all stages of the pods (including mature pods) by sucking the sap (Suh, 1986), causing premature pod drying and abnormal seed formation. The tolerable temperature range for a complete developmental cycle is 25°C–31°C (Dabiré, 2001). Like *C. tomentosicollis*, *A. curvipes* larvae undergo five nymphal instars. The adults are strong fliers and suck the sap from green pods, causing them to shrivel and dry prematurely with a resulting loss of seed (Singh and Allen, 1980).

3 Applying molecular biology tools to cowpea insect pests

Next-generation sequencing (NGS) platforms have the capacity to produce previously unparalleled amounts of DNA sequence data and offer a rapid entry point into genomic research (Coates et al., 2011). Applications involving NGS data have been especially useful for genomic-level research on non-model organisms (McCombie et al., 1992; Ellegren, 2008; Sloan et al., 2012; Harismendy et al., 2009) and have facilitated hypothesis-driven research into a diversity of research questions (Harismendy et al., 2009; Rothberg and Leamon, 2008). NGS technologies use massively parallel pyrosequencing to collect millions of independent nucleotide sequence reads in very short time frames at a very low cost per nucleotide (Margulies et al., 2005; Moore et al., 2006; Wicker et al., 2006; Huse et al., 2007; Weber et al., 2007), such that researchers can acquire large amounts of DNA sequence data from organisms where no prior information had previously existed (Margam et al., 2011b,c; Agunbiade et al., 2013).

While this genomics revolution has ushered in a fundamental change in how agricultural crop pest control is investigated (Coates et al., 2015), and is evident within many recent research avenues aimed at understanding the cowpea pest complex. This has also occurred concurrently with a general trend in crop protection towards a reduction in chemical pesticide use due to heightened conscientiousness towards human health and environmental stewardship (Tedeschi, 2012). Thus, NGS data are sought to facilitate the development of novel genetics-based and species-specific methods of controlling insect pest populations that will improve the efficacy, while reducing adverse environmental impacts from current methods (Alphey and Bonsall, 2014).

In the case of most cowpea pests, a lack of sufficient population genetic data has hindered the estimation of gene flow across these insect populations within their endemic range, such that the regional scales necessary for effective insect resistance management (IRM) and IPM programmes remain difficult to assess. This scenario, however, is quickly changing with the merging and application of previous ecological data and NGS tools to the control of these insect pests (Poelchau et al., 2016). Cowpea researchers are currently developing molecular genetic markers from genomic sequence data that afford identification of not only genetic variation within and between populations of a cowpea pest species and their associated natural enemies but also geographic distributions and patterns of gene flow – hence, determining where the pest populations in the wet season originate and where the source population is endemic in the dry season. These data are critical for targeting and timing IPM-based control measures in the field, as well as enhancing biological control efforts by determining the coincidence of endogenous natural enemy populations. High-throughput single-nucleotide polymorphism (SNP) genotyping (Margam et al., 2011b), as well as microsatellite marker assays (Agunbiade et al., 2012 and 2014) developed from NGS data, was used to answer questions of temporal and spatial scale for *M. vitrata* migration, while the above-mentioned molecular marker work, combined with light-trapping data, helped to support the South–North migratory hypothesis: that insects observed in the north in the rainy season were most closely related to those found directly south in the endemic zone in the dry season. From these data, then, the most logical place to deploy a biocontrol agent would be in the endemic zones, with the long-term potential to minimize populations that may be migrating north.

Findings by Margam et al. (2011a) further suggest that *M. vitrata* from West Africa (Old World) and Puerto Rico (New World) represent two distinct mitochondrial haplotype groups, and proposed that the samples are possibly a complex of sibling species. Studies that refine this prior haplotype divergence estimate and provide an estimated time of divergence between the New and Old World strains of *M. vitrata* are ongoing (Margam et al., 2011c; Agunbiade et al., 2016). Further characterization of *M. vitrata* populations collected from different host plants, especially in the more humid south of the West African country of Benin, suggests that host plants do not significantly influence the genetic structure of *M. vitrata* (Agunbiade et al., 2014). These prior studies demonstrate that genome-based NGS data can be applied to devise tools that increase our collective understanding of *M. vitrata* populations, some of which may be of practical applications for IPM and IRM; other components may be of only academic interest.

In the absence of whole-genome sequence assemblies, *de novo* transcriptome analysis has proven to be valuable for the identification of gene sequences, allelic variants and their expression levels in different developmental and metabolic pathways, for insecticide resistance traits and also to discover SNPs in model and non-model organisms (Sloan et al., 2012; Xue et al., 2010; Mittapalli et al., 2010; Poelchau et al., 2011). These polymorphisms can be developed into genetic markers for population genetic studies (Margam et al., 2011b). Much transcriptome research has been directed towards developing strategies to control the feeding of *M. vitrata* on cowpea crops, including understanding genetic mechanisms involved in insecticide resistance evolution and spread within insect populations, along with efforts to identify potential insecticidal RNA interference (RNAi) targets.

Recent shifts towards the use of more affordable and sustainable biocontrol measures have been initiated within West Africa (Adati et al., 2007; Amevoin et al., 2007). Some potential *M. vitrata* control strategies include the deployment of biopesticides (Oparaeke, 2010; Tamò et al., 2012) and the development of a transgenic cowpea that expresses *Bt* toxins (Cowpea CRSP, 2001). Transgenic cowpea that expresses a *Bt* Cry1Ab toxin is highly specific for lepidopteran insects such as *Maruca*. As such, these cowpea crops still remain susceptible to feeding damage and plant disease transmission by thrips, aphids and pod-sucking pests, including *A. craccivora*, *A. curvipes*, *C. tomentosicollis* and *M. sjostedti*.

The only method widely available to indigenous farmers in West Africa for controlling these non-lepidopteran pests is the application of chemical insecticides. Thus, the establishment of effective IRM plans for *A. craccivora*, *A. curvipes*, *C. tomentosicollis* and *M. sjostedti* may be critical for delaying the evolution of chemical insecticide resistance, as well as the possible evolution of *Bt* toxin resistance by *M. vitrata*. To partially address these concerns, Agunbiade et al. (2013) applied NGS technology to generate independent transcriptomes from normalized cDNA libraries in *A. curvipes*, *A. craccivora*, *C. tomentosicollis* and *M. sjostedti*. Annotation of individual gene transcripts identified candidate genes putatively involved in insecticide resistance, regulation of insect growth and response to disease transmission. The study was the first to report genomic data for these insect pests and provides an initial catalogue of molecular gene functions from several major field insect pests in cowpea cropping systems of West Africa. The future application of these genomics datasets might ultimately lead to a better understanding of the pest populations, with the long-term potential to improve the effectiveness of IPM programmes by defining pest–pathogen interactions and pest population dynamics that enhance the efficacy of biocontrol agents.

4 Biological control approaches to sustainable IPM

IPM has been the strategy of choice for developing plant protection options in cowpea, relying on commonly implemented practices worldwide (Cook, 2000) and focusing on plant health in a holistic manner (Tamò et al., 2012). In an ideal world, the first line of defence against crop pests is host plant resistance. This approach has been explored extensively in the past with regard to cowpea insect pests (Singh et al., 1990; Jackai and Adalla, 1997). There have been promising sources of resistance/tolerance to aphids (Souleymane et al., 2013) and flower thrips (Abudulai et al., 2006), which are currently the focus of ongoing screening efforts. However, for some of the key pests such as the pod borer, *M. vitrata,* and pod-sucking bugs such as *C. tomentosicollis*, there is virtually no useful source of resistance in the cultivated germplasm. Nevertheless, some wild relatives of cowpea, for example, *Vigna vexillata,* do possess useful levels of resistance to these recalcitrant pests. Unfortunately, sustained efforts to cross these wild cowpea species with cultivated lines have remained largely unsuccessful (Fatokun, 2002).

As stated above, cowpea has also been transformed with a gene construct encoding a Cry1Ab endotoxin, targeting the legume pod borer. Not yet published laboratory assays and confined field trials from the African Agricultural Technology Foundation have confirmed the technical efficacy of this approach. A second-generation product expressing *Maruca* control toxins with two independent modes of action is currently under development as well. Both the first- and second-generation products will be used within the standard IPM stewardship framework to address IRM concerns. At the same time, work is underway to establish a more robust cowpea seed value chain in order to be able to distribute large amounts of *Bt* cowpea under the auspices of a robust stewardship model.

As a normal and accepted component of agricultural production, farmers very often apply chemical pesticides, if and where available and affordable. However, handling and utilizing chemical insecticides in the context of West African subsistence agriculture is challenging (Jackai and Adalla, 1997). Arguably, synthetic pesticides in the short term continue to provide good control for most of the pest problems, with the added advantage of their knock-down effect, which can be easily appreciated by the low-literate farmer (Singh et al., 1990). However, in the long term, synthetic pesticides need to be carefully evaluated for their possible adverse side effects which are starting to emerge even in rural areas of West Africa (e.g. Diabate et al., 2002). As described in Coulibaly and Lowenberg-DeBoer (2002), the first issues to be addressed while applying pesticides are those related to human health. In the developed world, pesticides are applied following strict regulations, including protective equipment and according to the instructions written on the accompanying documentation. In West Africa, in general, farmers perceive the use of protective equipment as an unnecessary burden because of the hot and humid climatic conditions. Moreover, resource-poor farmers very often do not consider the purchase of protective implements as a priority. This leads to the practice of spraying with no protective equipment at all. From an environmental perspective, the inappropriate application of pesticides can have serious long-term effects as well, including the evolution of insecticide resistance in target pests as is already documented for *M. vitrata* in Nigeria (Ekesi, 1999). Indiscriminate use of broad-spectrum insecticides, particularly at the beginning of the cropping season, very often results in elimination of natural enemies as well, which limits the control impact on pest populations later in the season, leading to pest outbreaks (Heong et al., 1992).

Over the last 10 years, alternative control methods relying on both biological control and biopesticides have been given serious consideration and have now reached the mainstream of cowpea plant protection (Tamò et al., 2003). Biological control has been considered for many years as having marginal importance in limiting cowpea pest populations (Singh et al., 1990) but has regained interest recently. Biological control easily integrates with other pest control options, such as host plant resistance, and is generally only negatively impacted by the inappropriate use of pesticides. Biological control, in fact, complements the limited levels of insect resistance, which have been put into improved varieties during years of hard work by the breeders.

Biological control involves putting agrobiodiversity to work, restoring disturbed ecosystem services or establishing new ones (Moonen and Barberi, 2008), whereby specific parasitoids and predators that target specific life history stages for optimal control are deployed. Hence, the first step in determining the potential for biological control of a pest in a given system starts with an assessment of its life history, including studies of trophic-level interactions with beneficial organisms (Hoelmer and Kirk, 2005). Our biodiversity assessment studies have revealed the presence of hitherto unexploited natural enemies for most of the cowpea pests, some of which are already in an advanced testing stage and are starting to see deployment.

An example involves the East/Central African strain of the hymenopteran thrips parasitoid, *Ceranisus femoratus*, which was released and became successfully established on nursery plots of the leguminous shrub, *Tephrosia candida*, in Benin, Southern Ghana and Southwestern Nigeria (Tamò et al., 2003). In Benin, we intensified our post-release assessment sampling to not only measure the area covered by the parasitoid but also to obtain quantitative data on its ecological impact in reducing the population of the target host. So far, we continue to observe significant differences in thrips larval abundance – up to a 43% reduction on two major host plants for *M. sjostedti*, *Pterocarpus santalinoides* and *Lonchocarpus sericeus* (Tamò et al., 2012). This substantial reduction has led to a limiting of damage to cowpea in the surrounding areas, for example, where these host plants can provide a reservoir of parasitoids during the off-season. If combined with improved host plant resistance in a good agronomic background and punctual applications of biopesticides, for example, the newly developed emulsifiable neem oil mixture (see Sokame et al., 2015), this will provide a solution for the thrips problem in cowpea by lowering the pest populations below EILs.

The other major pest, the pod borer *M. vitrata*, has also been the object of intensive studies focusing on locally available natural enemies in Africa (Taylor, 1967; Agyen-Sampong, 1978; Usua and Singh, 1978). However, Arodokoun et al. (2006) were the first to assess the presence and impact of hymenopteran parasitoids on widespread alternative host plants over a three-year period. From this work, it appeared that the locally available parasitoids were location-specific rather than host-specific and were therefore not able to control high levels of *M. vitrata* populations. Because the presumed area of origin of *M. vitrata* was suspected to be in the Tonkin area in South East Asia (Tamò et al., 2003), inter-regional collaborative studies were undertaken to scout for more adapted natural enemies, including hymenopteran parasitoids.

Consequently, biodiversity studies carried out in Taiwan revealed the presence of seven parasitoid species, all unknown from Africa (Huang et al., 2003). The first one to be tested in West Africa was *Apanteles taragamae* Viereck (Hymenoptera: Braconidae) due to its impressive field parasitism rates of up to 63% on *M. vitrata* feeding on *Sesbania cannabina* (Retz.) Pers. (Leguminosae: Fabaceae) leaves, a common green manure crop in tropical Asia. However, extensive studies on this parasitoids ecology and biology (Dannon et al., 2010a,b and 2012) revealed it as poorly adapted to West African conditions, particularly because it was not capable of recognizing major wild host plants for *M. vitrata* present in West Africa.

Subsequent studies from Lao PDR, Taiwan and Vietnam indicated the presence of a larval parasitoid, *Therophilus javanus* (Bhat and Gupta, 1977), together with an egg-larval parasitoid, *Phanerotoma syleptae* Zettel (Hymenoptera: Braconidae), as important antagonists of *M. vitrata*. In fact, very recently, they have replaced *A. taragamae* as key parasitoids for *M. vitrata* in Taiwan (R. Srinivasan, AVRDC, pers. comm.). It is noteworthy to mention that these two organisms, albeit under a different nomenclature, had already been identified by Kalshoven (1950–1) as the possible cause for the low incidence of *M. vitrata* in Indonesia.

Both parasitoids have subsequently been introduced to the IITA laboratories in Benin, and, after clearance from the regulatory authorities, have been the object of experimental releases on patches of wild vegetation across Western Benin. So far, both organisms have been recovered from various host plants, with a clear preference of *P. syleptae* for legume trees and shrubs, and of *T. javanus* for herbaceous legumes including cultivated cowpea (B. Datinon, IITA, pers. comm.).

With a better grasp now of what needs to be considered for the development and deployment of a 'biocontrol pipeline' with regard to cowpea pests, rather than a reactive pest control approach using chemical pesticides as the main – and in many cases the sole – control option, we can propose a science-driven, proactive approach to IPM. Coupled with an educational component to elicit buy-in by farmers, we propose the term 'precision-IPM' to designate this approach. Given the future of pest control in light of climate change and changing societal domains, the need to involve information and communication technology (ICT) systems to understand and respond to pest patterns in real time, while remaining environmentally sustainable and economically viable, becomes mandatory.

In fact, one of the most challenging issues has been how to deploy IPM tactical knowledge to farmers with varying literacy levels, who typically live in rural areas, and generally speak a great diversity of non-national languages more fluently than the national language. To address this critical issue, a programme to format these knowledge strategies and concepts in an accessible form was needed, one that particularly allowed translation into local fluencies, while being appealing and electronically highly portable. The SAWBO video animations met all of these criteria.

5 Adoption of precision-IPM strategy by cowpea farmers

As primary stakeholders and implementers of agricultural technologies, it is important to study smallholder farmers and understand how they value alternative pest control strategies in their crop production decisions. This section discusses socio-economic research results from two separate farmer surveys conducted on cowpea producers in Benin and Burkina Faso. The discussion focuses on existing pest control characteristics and implications for sustainable agriculture, farmers' preferences for alternative pest control strategies, and other issues critical for adoption and scalability of the precision-IPM strategy across West Africa.

5.1 Existing pest control characteristics and farmers' awareness

In recent years, comprehensive farmer surveys have been carried out in two of the major cowpea-producing countries in West Africa, namely Benin (Agyekum et al., 2016a) and

Table 1 Characteristics of smallholder cowpea production in Benin and Burkina Faso

Characteristics	Benin[1]	Burkina Faso[2]
Farmer/household		
Gender of person responsible for cowpea production decisions (% male)	69	99
Number of people living in household	7	10.5
Use of agricultural credit for cowpea production (% yes)	17	3
Household received remittances from relatives (% yes)	14	18
Income from non-crop sources (% yes)	59	89
Cowpea farms		
Land size for cowpea production (ha)	0.6	0.8
No. of cowpea fields grown per season	1	1.3
Was cowpea inter-cropped? (% yes)	9	37
Cowpea grain yield (kg/ha)	316	345
Portion of cowpea grains sold (%)	67	59
Did you harvest fodder? (% yes)	10	64
Existing pest management characteristics		
Application of chemical pesticides on cowpea crops (% yes)	88	79
Chemicals on other crops? (% yes)	14	3
How volume of pesticide use compares to previous years (% lower)	5	26
Satisfaction with effectiveness of pesticides (% yes)	65	70
No. of chemical applications per cowpea season	3	2.5
Chemical hazard awareness, and exposure to chemical pesticides		
Do you think pesticides are harmful to people? (% yes)	93	91
Knows colour label identifying most toxic pesticides (% yes)	11	18
Has anyone you know been sick due to pesticide poisoning? (% yes)	45	33
Hired labour for pesticide application? (% yes)	0	16
Does household use gloves during spraying? (% yes)	25	27
Use face mask to apply pesticides? (% yes)	24	27
Clothes/skin wet with pesticides after spraying? (% yes)	71	46
Reported chemical exposure symptoms, e.g. skin/eye issues (% yes)	57	52
Awareness of beneficial insects		
Farmer's awareness of beneficial insects (% yes)	9	8
Sample size (no. of households)	505	560

Sources: [1] Benin information is from Agyekum et al. (2016a).
[2] Information on Burkina Faso is from Reyes et al. (2013).

Burkina Faso (Reyes et al., 2013). Each of the two field surveys interviewed over five hundred randomly selected farm households across several farming communities in the respective countries. Cowpea production data collected in the farmer surveys include pest pressure, pest control strategies, self-assessments of existing pest control methods, production costs, yield levels, prices, sources of household income, access to agricultural credit, household cash expenditure, assets owned by the household and socio-demographic characteristics. Table 1 is a cross-country comparison of both survey results. The surveys show that cowpea farmers in Benin and Burkina Faso have remarkable similarities, particularly on their current crop defence methods and awareness of associated risks to health.

On crop protection strategy, the majority of farmers apply chemical pesticides to control pest pressure on their cowpea fields. However, only a small fraction of farm households apply chemical pesticides on crops aside from cowpeas. Relative to other crops grown on farmers' fields, the widespread application of chemical pesticides on cowpeas is both a manifestation of the incessant pest attacks during cropping seasons and the importance of the crop to household food and income security. Furthermore, it is interesting to note that nearly all cowpea farmers believe chemical pesticides are harmful to human health. In spite of this level of awareness among smallholder farmers, survey results indicate limited use of personal protective equipment (PPE) such as gloves and protective masks; only a quarter of the farmers use some form of protection during chemical pesticide application.

The apparent lack of recommended PPE means that cowpea farmers may be at high risk of exposure to toxic active ingredients in chemical pesticides. Not surprisingly, more than half of the farmers reported cases of skin irritation and eye problems suffered after spraying chemical pesticides. Also, a quarter or more of the farm households reported knowing someone in their community who fell sick because of poisoning from chemical pesticide application on their farms.

The studies also assessed the farmers' awareness of environmental damage caused by misuse of chemical pesticides. Given the prevalence of broad-spectrum chemical pesticides in cowpea production across West Africa, farmers were asked if they were aware of any beneficial insects in their farming environment. More than 90% of farmers indicated an unawareness regarding potential disruptions to the ecosystem balance in the farming areas as a result of unintentionally killing natural enemies by an inappropriate use of broad-spectrum chemical pesticides.

5.2 Farmer pest control needs and adoption of precision-IPM strategy

To predict uptake of precision-IPM strategies, Agyekum et al. (2016b) implemented a choice experiment for cowpea farmers in Benin during the survey mentioned above (for details on choice experiments, see Kanninen, 2007; Bennett and Birol, 2010; Johnson et al., 2013). A conservative approach was employed in the choice experiment such that precision-IPM was deliberately presented as relatively worse in performance compared to a farmer's own chemical control method (in terms of cost, associated yield loss and labour input). Even with the relatively poor performance from the precision-IPM strategy, the majority of farmers indicated willingness to switch from their current chemical pest control method in favour of the innovative IPM method (see Fig. 4). The choice experiment results help reveal the trade-offs farmers are willing to make in their choice of pest management strategies.

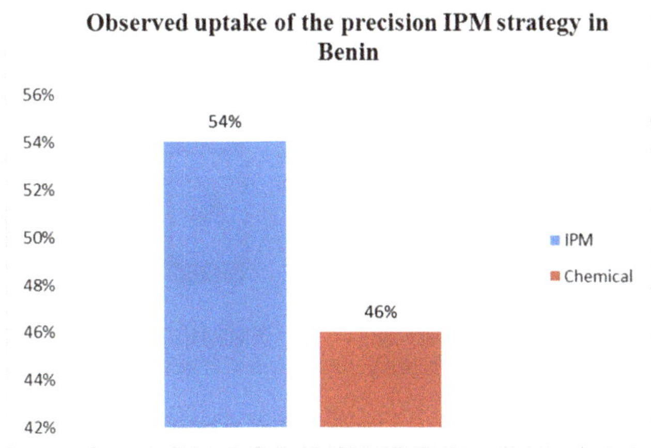

Figure 4 A choice experiment on preferences for alternative pest control methods conducted on more than five hundred cowpea farmers randomly selected across Benin. This chart shows the distribution of farmers' preferences when presented with the decision to choose between their own chemical pest control methods and the precision-IPM strategy. Note that the majority of farmers opted for the precision-IPM strategy even though the IPM scenarios presented in the experiment were intentionally described as costly (i.e. higher cost of biopesticides and more labour input) and less effective compared to existing chemical methods. Results show farmers desire to try alternative pest control methods that are safer for their health.

Agyekum et al. (2016b) show that cost, potential yield loss and labour requirements for the IPM strategy are important to farmers such that high values of these attributes would likely reduce uptake. Conversely, the socio-economic factors of age, gender, level of formal education and household income had no effect on potential adoption of the IPM method (Agyekum et al., 2016b). Farmer preference for the IPM strategy, then appears to be uniform, with a high potential for widespread adoption across West Africa.

Agyekum et al. (2016b) also explored the possible relationship between social pressure – in terms of the proportion of farmers in a community who choose to participate in an innovative IPM programme – and how this communal behaviour may impact other farmers' decision to adopt the precision-IPM strategy. Agyekum et al. (2016b) show a strong positive correlation between community participation rate and its influence on farmers' decision to choose the IPM method. Thus, farmers recognize that the success of the precision-IPM programme depends on cooperation from other community members in order to avoid the broad-spectrum chemical pesticides that would otherwise undermine the establishment of the biological control agents deployed in their farm environments.

6 SAWBO and mobile learning experiences

Founded in 2010 and publicly released in 2011, SAWBO's mission is to create educational animation materials presented in fluent local languages. To date, SAWBO has produced more than seventy videos in more than one hundred different languages. These videos

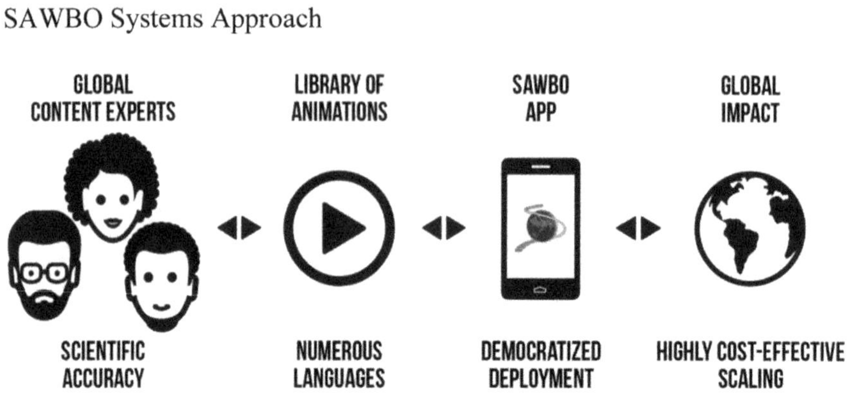

Figure 5 SAWBO systems approach.

span the agricultural, health and women's empowerment sectors and can be accessed from a variety of SAWBO channels for use on mobile phones, tablets, televisions, computers and overhead projectors. The programme relies on volunteers from all over the world to translate, record, voice overlay, share, deploy and train others. SAWBO values and has conducted research on indigenous knowledge (Bello-Bravo and Pittendrigh, 2012), aiming particularly to give voice to the most impoverished communities, to preserve their local knowledges and languages, and to share them globally.

The SAWBO systems approach entails the following: (1) to continually create educational content with experts on diverse topics, (2) to place content into diverse local languages and (3) to make it accessible to educators who can then (4) deploy such content in their communities as part of their own educational programmes. This approach allows organizations of any size – from small, community-based organizations to large government and international organizations – to access educational content in local languages that is scientifically accurate and accessible at no cost to these groups. This approach is illustrated in Fig. 5. Pre- and post-testing learning gain studies have shown that SAWBO animations are more popular and are also effective or more effective for increasing farmers' knowledge of IPM topics such as preparation of neem sprays as compared to traditional extension agent presentations of the same information (Bello-Bravo, unpublished results).

6.1 Information resources

As a next step in this ICT process, we are developing a prototype cell phone app to both draw upon data from the field and then deliver back pest control solutions to farmers in their own languages in a format that will be understandable by them across a great diversity of literacy levels. To ensure the effectiveness of educational content, a two-way communication mechanism, the Farmer Interface Application (FIA), is used to both deploy the educational material and then collect users' reaction to it while they are using it in the field. The FIA is a communication and notification portal that (1) enables farmers to have access to educational materials on proper pest scouting and management in cowpea,

(2) guides farmers and records their observations in the field, (3) provides notification with proper action suggestions on the basis of the field collected data, (4) transmits the observations to a central database and an expert system (ES) along with time/date stamp and geo-location and (5) receives further notifications generated by the ES that warns farmers about the upcoming pest outbreaks.

FIA is a user-friendly application that allows farmers to reach the action point in three simple steps. The app is designed to convey information with simple graphics and easy-to-follow visual and auditory instructions. Once farmers learn about proper scouting techniques via a SAWBO animation, they are asked to go to the field and monitor a prescribed number of plants and record the level of infestation. The application then utilizes an internal algorithm to determine whether the farmers should conduct further scouting on a later date or should manage the pests now.

Relatedly, the ES itself is a cloud-based platform that collects information generated by FIA users and matches that information with other data sources to generate notifications using predictive models and algorithms. The system is designed to connect to weather and geographic information system (GIS)-based application programme interfaces (APIs) that provide updated information about climate variables, like temperature, moisture, rainfall and cloud cover as well as geographical variables such as elevation, rivers, mountains and so on. Updated periodically, this system can also take into consideration in-depth information about the genetic make-up of pest populations, on the basis of data input from researchers, and factor this into the models used for assessing insect pest threats. The ES provides a platform where experts can set certain thresholds for each variable to design predictive models that can generate and send specific notifications to FIA users on the basis of their geo-location. The ES platform is extremely flexible and scalable and can use multiple decision-making algorithms simultaneously.

7 Conclusion and future trends

Over the past decades, changes in the available pest control technologies used by crop producers have led to changes in IPM strategies. Conceptually, IPM strategies recognize that many environmental factors interact and work in conjunction to affect the abundance, distribution, movement and phenotype of insect pest populations. The systems approach of IPM leverages an in-depth knowledge of how management practices affect an insect's biology and other trophic interactions. It also involves the monitoring of pest populations to identify appropriate control measures as well as to aid in making decisions about how, when and where to apply them. Advances in genomics make it possible to better understand the dynamics of pest populations (e.g. movement patterns or biotypes that respond differently to host plant resistance traits), which may provide the requisite knowledge for developing new approaches for managing insects.

However, these more recent IPM strategies will have large-scale impact only if they can be coupled with effective educational and deployment tools that can be rapidly updated when new or other practical control approaches emerge. A mechanism will be needed to facilitate the flow of information between scientists and the end users who apply the strategies. In turn, scientists will need to rely on these kinds of real-time information sharing systems to obtain data on insect populations from local users, which can then be leveraged to continually and efficiently build and remodel solutions. Extension agents will also require a centralized system for information exchange so that they can access information about available practical pest control strategies.

This is facilitated by the advent of the Internet and cell phones, and increasingly so in Africa as cell phone availability and use continues to expand dramatically. These widely available technologies can be used not only to communicate information on pest populations between farmers and researchers but also to convey practical pest management recommendations back to farmers. The strategies developed in the preceding sections have already been applied within cowpea cropping systems in most West African countries, but such research and communication tactics can be applied to other cropping systems in other countries, both developing and developed.

8 Where to look for further information

Bello-Bravo, J., Gemechu, O. W., and Pittendrigh, B. (2015). 'A pilot study using educational animations as a way to improve farmers' agricultural practices and health around Adama, Ethiopia'. *Information Technologies and International Development*, 11: 3, 23–37.

Bello-Bravo, J., Dannon, E., Agunbiade, T., Tamò, M., and Pittendrigh, B. (2013). 'The prospect of animated videos in agriculture and health: a case study in Benin'. *International Journal of Education and Development using Information and Communication Technology*, 9: 3, 4–16.

9 References

Abudulai, M., Salifu, A. and Haruna, M. (2006). 'Screening of cowpeas for resistance to the flower bud thrips, Megalurothrips sjostedti Trybom (Thysanoptera: Thripidae)'. *Journal of Applied Sciences*, 6, 1621–4.

Adati, T., Tamò, M., Yusuf, S., Downham, M., Singh, B. and Hammond, W. (2007). 'Integrated pest management for cowpea–cereal cropping systems in the West African savannah'. *International Journal of Tropical Insect Science*, 27, 123–37.

AGP. (2016). Integrated Pest Management (IPM). Available: http://tinyurl.com/z57r7ng [Accessed 9 December 2016].

Agunbiade, T. A., Coates, B. S., Datinon, B., Djouaka, R., Sun, W., Tamò, M. and Pittendrigh, B R. (2014). 'Genetic differentiation among Maruca vitrata F. (Lepidoptera: Crambidae) populations on cultivated cowpea and wild host plants: implications for insect resistance management and biological control strategies'. *PLoS ONE*, 9, e92072.

Agunbiade, T., Steele, L., Coates, B., Gassmann, A., Margam, V., Ba, M., Dabire-Binso, C., Baoua, I., Bello-Bravo, J., Seufferheld, F., Sun, W., Tamò, M., and Pittendrigh, B. (2012a). IPM-omics: from genomics to extension for integrated pest management of cowpea. In: Boukar, O., Coulibaly, O., Fatokun, C., Lopez, K., and Tamò, M., editors. Improving livelihoods in the cowpea value chain through advancements in science. Proceedings of the 5th World Cowpea Research Conference pp. 231–248.

Agunbiade, T. A., Coates, B. S., Kim, K. S., Forgacs, D., Margam, V., Murdock, L. L., Ba, M. N., Binso-Dabiré, C. L., Baoua, I. and Ishiyaku, M. F. (2012b). 'The spatial genetic differentiation of the legume pod borer, Maruca vitrata F. (Lepidoptera: Crambidae) populations in West Africa'. *Bulletin of Entomological Research*, 102, 589–99.

Agunbiade, T. A., Coates, B. S., Sun, W., Tsai, M.-R., Valero, M. C., Tamò, M. and Pittendrigh, B. R. (2016). 'Comparison of the mitochondrial genomes of the Old and New World strains of the legume pod borer, Maruca vitrata (Lepidoptera: Crambidae)'. *International Journal of Tropical Insect Science*, [in press].

Agunbiade, T. A., Sun, W., Coates, B. S., Djouaka, R., Tamò, M., Ba, M. N., Binso-Dabiré, C. Baoua, I., Olds, B. P. and Pittendrigh, B. R. (2013). 'Development of reference transcriptomes for the major field insect pests of cowpea: a toolbox for insect pest management approaches in West Africa'. *PLoS ONE*, 8, e79929.

Agyekum, M., Donovan, C., Pittendrigh, B. and Tamò. M. (2016a). 'Farmer baseline survey report on smallholder cowpea production and existing pest management practices in Benin, West Africa (Draft Technical Report)'. Michigan State University.

Agyekum, M., Donovan, C. and Lupi, F. (2016b). Novel IPM intervention for West Africa: smallholder farmers' preferences for biological versus synthetic control strategies for cowpea pests. Agricultural & Applied Economics Association Annual Meeting, July 31 to 2 August 2016, Boston, MA.

Agyen-Sampong, M. (1978). 'Pests of cowpea and their control in Ghana'. In: Singh, S. R., Van Emden, H. F. and Taylor, T. A. (eds), *Pests of Grain Legumes: Ecology and Control*. London: Academic Press, pp. 85–92.

Akinfenwa, S. (1975). *Biological Study of Maruca Testulalis (Geyer) (Lepidoptera: Pyralidae) in the Zaria Area of Northern Nigeria*. MSc, Ahmadu Bello University.

Alabi, O., Odebiyi, J. and Tamò, M. (2004). 'Effect of host plant resistance in some cowpea (Vigna unguiculata {L.} Walp.) cultivars on growth and developmental parameters of the flower bud thrips, Megalurothrips sjostedti (Trybom)'. *Crop Protection*, 23, 83–8.

Alghali, A. M. (1991). 'Studies on cowpea farming practices in Nigeria, with emphasis on insect pest control'. *International Journal of Pest Management*, 37, 71–4.

Alphey, N. and Bonsall, M. B. (2014). 'Interplay of population genetics and dynamics in the genetic control of mosquitoes'. *Journal of The Royal Society Interface*, 11, 20131071.

Amevoin, K., Sanon, A., Apossaba, M. and Glitho, I. (2007). 'Biological control of bruchids infesting cowpea by the introduction of Dinarmus basalis (Rondani)(Hymenoptera: Pteromalidae) adults into farmers' stores in West Africa'. *Journal of Stored Products Research*, 43, 240–7.

Arodokoun, D., Tamò, M., Cloutier, C. and Brodeur, J. (2006). 'Larval parasitoids occurring on Maruca vitrata Fabricius (Lepidoptera: Pyralidae) in Benin, West Africa'. *Agriculture, Ecosystems and Environment*, 113, 320–5.

Atiri, G. I. (1984). 'The occurrence and importance of okra mosaic virus in Nigerian weeds'. *Annals of Applied Biology*, 104, 261–5.

Atiri, G. I., Ekpo, E. and Thottappilly, G. (1984). 'The effect of aphid-resistance in cowpea on infestation and development of Aphis craccivora and the transmission of cowpea aphid-borne mosaic virus1'. *Annals of Applied Biology*, 104, 339–46.

Atiri, G. I., Enobakhare, D. and Thottappilly, G. (1986). 'The importance of colonizing and non-colonizing aphid vectors in the spread of cowpea aphid-borne mosaic virus in cowpea'. *Crop Protection*, 5, 406–10.

Atiri, G. I., Thottappilly, G. and Ligan, D. (1987). 'Effects of cypermethrin and deltamethrin on the feeding behaviour of Aphis craccivora and transmission of cowpea aphid-borne mosaic virus'. *Annals of Applied Biology*, 110, 455–61.

Atiri, G. I. and Thottappilly, G. (1985). 'Aphis craccivora settling behaviour and acquisition of cowpea aphid-borne mosaic virus in aphid-resistant cowpea lines'. *Entomologia experimentalis et applicata*, 39, 241–5.

Ba, N. M., Margam, V. M., Binso-Dabire, C. L., Sanon, A., McNeil, J. N., Murdock, L. L. and Pittendrigh, B. R. (2009). 'Seasonal and regional distribution of the cowpea pod borer Maruca vitrata (Lepidoptera: Crambidae) in Burkina Faso'. *International Journal of Tropical Insect Science*, 29, 109–13.

Baoua, I., Ba, N. M., Agunbiade, T. A., Margam, V., Binso-Dabiré, C. L., Antoine, S. and Pittendrigh, B. R. (2011). 'Potential use of Sesbania pachycarpa (Fabace ae: Papilionoideae) as a refugia for the legume pod borer Maruca vitrata (Lepidoptera: Crambidae)'. *International Journal of Tropical Insect Science*, 31, 212–18.

Bello-Bravo, J. and Pittendrigh, B. (2012). 'Scientific Animations Without Borders: A new approach to capture, preserve and share indigenous knowledge'. *The Journal of World Universities Forum*, 5, 11–20.

Bennett, J. and Birol, E. (2010). *Choice Experiments in Developing Countries: Implementation, Challenges and Policy Implications*, Cheltenham, Glos, UK, Edward Elgar.

Bhat, S. and Gupta, V. K. (1977). *Ichneumonologia Orientalis, or, A Monographic Study of Oriental Ichneumonidae (Part VI)*, Delhi, Association for the Study of Oriental Insects.

Blackman, R. and Eastop, V. (2008). *Aphids on the World's Herbaceous Plants and Shrubs*, Chichester, England, John Wiley & Sons.

Bottenberg, H. (1995). 'Farmers' perceptions of crop pests and pest control practices in rainfed cowpea cropping systems in Kano, Nigeria'. *International Journal of Pest Management*, 41, 195–200.

Bottenberg, H., Tamò, M., Arodokoun, D., Jackai, L. E. N., Singh, B. B. and Youm, O. (1997). 'Population dynamics and migration of cowpea pests in northern Nigeria: implications for integrated pest management'. In: Singh, B. B., Mohan Raj, D. R., Dashiell, K. E. and Jackai, L. E. N. (eds), *Advances in Cowpea Research*. Ibadan, Nigeria: International Institute of Tropical Agriculture; Japan International Research Center for Agricultural Sciences, pp. 271–84.

Coates, B. S., Kroemer, J., Sumerford, D. V. and Hellmich, R. L. (2011). 'A novel class of miniature inverted repeat transposable elements (MITEs) that contain hitchhiking (GTCY) n microsatellites'. *Insect Molecular Biology*, 20, 15–27.

Coates, B. S., Poelchau, M., Childers, C., Evans, J. D., Handler, A., Guerrero, F., Skoda, S., Hopper, K., Wintermantel, W. M. and Ling, K. (2015). 'Arthropod genomics research in the United States Department of Agriculture – Agricultural Research Service: current impacts and future prospects'. *Trends in Entomology*, 11, 1–27.

Cook, R. J. (2000). 'Advances in plant health management in the twentieth century'. *Annual Review of Phytopathology*, 38, 95–116.

Cork, A., Dobson, H., Grzywacz, D., Hodges, R., Orr, A. and Stevenson, P. (2009). 'Review of pre- and post-harvest pest management for pulses with special reference to Eastern and Southern Africa'. Chatham Maritime, Kent, UK: McKnight Foundation. Collaborative Crops Research Program.

Coulibaly, O. and Lowenberg-DeBoer, J. (2002). 'The economics of cowpea in West Africa'. In: Fatokun, C. A., Tarawali, S. A., Singh, B. B., Kormawa, P. M. and Tamò, M. (eds), *Challenges and Opportunities for Enhancing Sustainable Cowpea Production*. Proceedings of the World Cowpea Conference III, 4–8 September 2000, Ibadan, Nigeria. Ibadan, Nigeria: International Institute of Tropical Agriculture (Ibadan), Nigeria, pp. 351–66.

Cowpea, C. R. S. P. (2001). 'The Dakar Symposium/Workshop on the Genetic Improvement of Cowpea'.

Dabiré-Binso, C., Ba, N. and Sanon, A (2008). 'Laboratoire Entomologie Kamboinsé (Document Technique)'. INERA.

Dabiré, C. (2001). *Etude de quelques paramètres biologiques et écologiques de Clavigralla tomentosicollis STAL., 1855 (Hemiptera: Coreidae), punaises suceuses de gousses du niébé Vigna unguiculata (L) WALP. Dans une perspective de lutte durable contre l'insecte au Burkina Faso*. PhD, Université de Cocody.

Dannon, E. A., Tamò, M., Agboton, C., van Huis, A. and Dicke, M. (2012). 'Effect of Maruca vitrata (Lepidoptera: Crambidae) host plants on life-history parameters of the parasitoid Apanteles taragamae (Hymenoptera: Braconidae)'. *Insect Science*, 19, 518–28.

Dannon, E. A., Tamò, M., Van Huis, A. and Dicke, M. (2010a). 'Effects of volatiles from Maruca vitrata larvae and caterpillar-infested flowers of their host plant Vigna unguiculata on the foraging behavior of the parasitoid Apanteles taragamae'. *Journal of Chemical Ecology*, 36, 1083–91.

Dannon, E. A., Tamò, M., van Huis, A. and Dicke, M. (2010b). 'Functional response and life history parameters of Apanteles taragamae, a larval parasitoid of Maruca vitrata'. *BioControl*, 55, 363–78.

Diabate, A., Baldet, T., Chandre, F., Akoobeto, M., Guiguemde, T. R., Darriet, F., Brengues, C., Guillet, P., Hemingway, J. and Small, G. J. (2002). 'The role of agricultural use of insecticides in resistance to pyrethroids in Anopheles gambiae sl in Burkina Faso'. *The American Journal of Tropical Medicine and Hygiene*, 67, 617–22.

Edema, R. and Adipala, E. (1996). 'Effect of crop protection management practice on yield of seven cowpea varieties in Uganda'. *International Journal of Pest Management*, 42, 317–20.

Ekesi, S. (1999). 'Insecticide resistance in field populations of the legume pod-borer, Maruca vitrata Fabricius (Lepidoptera: Pyralidae), on cowpea, Vigna unguiculata (L.), Walp in Nigeria'. *International Journal of Pest Management*, 45, 57–9.

Ellegren, H. (2008). 'Sequencing goes 454 and takes large-scale genomics into the wild'. *Molecular Ecology*, 17, 1629–31.

Fatokun, C. A. (2002). 'Breeding cowpea for resistance to insect pests: attempted 52 crosses between cowpea and Vigna vexillata'. In: Fatokun, C. A., Tarawali, S. A., Singh, B. B., Kormawa, P. M. and Tamò, M. (eds), *Challenges and Opportunities for Enhancing Sustainable Cowpea Production*. Proceedings of the World Cowpea Conference III, 4–8 September 2000,Ibadan, Nigeria. Ibadan, Nigeria: International Institute of Tropical Agriculture (Ibadan), Nigeria, pp. 52–61.

Giga, D. and Biscoe, J. (1989). 'Treating maize grain for storage with registered protectants in Zimbabwe: technical, practical, and economic considerations'. *Zimbabwe Science News*, 23, 101–3.

Harismendy, O., Ng, P. C., Strausberg, R. L., Wang, X., Stockwell, T. B., Beeson, K. Y., Schork, N. J., Murray, S. S., Topol, E. J. and Levy, S. (2009). 'Evaluation of next generation sequencing platforms for population targeted sequencing studies'. *Genome Biology*, 10, R32.

Heong, K., Aquino, G. and Barrion, A. (1992). 'Population dynamics of plant-and leafhoppers and their natural enemies in rice ecosystems in the Philippines'. *Crop Protection*, 11, 371–9.

Hoelmer, K. and Kirk, A. (2005). 'Selecting arthropod biological control agents against arthropod pests: Can the science be improved to decrease the risk of releasing ineffective agents?'. *Biological Control*, 34, 255–64.

Huang, C.-C., Peng, W.-K. and Talekar, N. (2003). 'Parasitoids and other natural enemies of Maruca vitrata feeding on Sesbania cannabina in Taiwan'. *BioControl*, 48, 407–16.

Huse, S. M., Huber, J. A., Morrison, H. G., Sogin, M. L. and Welch, D. M. (2007). 'Accuracy and quality of massively parallel DNA pyrosequencing'. *Genome Biology*, 8, R143.

Jackai, L. and Adalla, C. (1997). 'Pest management practices in cowpea: a review'. In: Singh, B. B., Mohan Raj, D. R., Dashiell, K. E. and Jackai, L. E. N. (eds), *Advances in Cowpea Research*. Ibadan, Nigeria: International Institute of Tropical Agriculture; Japan International Research Center for Agricultural Sciences, pp. 240–58.

Johnson, F. R., Lancsar, E., Marshall, D., Kilambi, V., Mühlbacher, A., Regier, D. A., Bresnahan, B. W., Kanninen, B. and Bridges, J. F. (2013). 'Constructing experimental designs for discrete-choice experiments: report of the ISPOR conjoint analysis experimental design good research practices task force'. *Value in Health*, 16, 3–13.

Kalshoven, L. (1950–1). *De plagen van de culturgewassen in Indonesië.*, 's-Gravenhage, Netherlands, W. van Hoeve.

Kanninen, B., ed. (2007). *Valuing Environmental Amenities using Stated Choice Studies: A Common Sense Approach to Theory and Practice*, Dordrecht: Springer.

Katayama, J. and Suzuki, I. (1984). 'Seasonal prevalence of pod borers [Ostrinia scapulalis, Maruca testulalis and Matsumuraeses sp.] in azuki-beans and injury caused by larval infestation'. Kyoto, Japan: Bulletin of the Kyoto Prefectural Institute of Agriculture, pp. 27–34.

Ke, L., Fang, J.-L. and Li, Z. J. (1985). 'Bionomics and control of the legume pod-borer Maruca testulalis Geyer'. *Acta Entomologica Sinica*, 28, 51–9.

Margam, V. M., Coates, B. S., Ba, M. N., Sun, W., Binso-Dabire, C. L., Baoua, I., Ishiyaku, M. F., Shukle, J. T., Hellmich, R. L. and Covas, F. G. (2011a). 'Geographic distribution of phylogenetically-distinct legume pod borer, Maruca vitrata (Lepidoptera: Pyraloidea: Crambidae)'. *Molecular Biology Reports*, 38, 893–903.

Margam, V. M., Coates, B. S., Bayles, D. O., Hellmich, R. L., Agunbiade, T., Seufferheld, M. J., Sun, W., Kroemer, J. A., Ba, M. N. and Binso-Dabire, C. L. (2011b). 'Transcriptome sequencing, and rapid development and application of SNP markers for the legume pod borer Maruca vitrata (Lepidoptera: Crambidae)'. *PLoS ONE*, 6, e21388.

Margam, V. M., Coates, B. S., Hellmich, R. L., Agunbiade, T., Seufferheld, M. J., Sun, W., Ba, M. N., Sanon, A., Binso-Dabire, C. L. and Baoua, I. (2011c). 'Mitochondrial genome sequence and expression profiling for the legume pod borer Maruca vitrata (Lepidoptera: Crambidae)'. *PLoS ONE*, 6, e16444.

Margulies, M., Egholm, M., Altman, W. E., Attiya, S., Bader, J. S., Bemben, L. A., Berka, J., Braverman, M. S., Chen, Y.-J. and Chen, Z. (2005). 'Genome sequencing in microfabricated high-density picolitre reactors'. *Nature*, 437, 376–80.

McCombie, W. R., Adams, M. D., Kelley, J. M., FitzGerald, M. G., Utterback, T. R., Khan, M., Dubnick, M., Kerlavage, A. R., Venter, J. C. and Fields, C. (1992). 'Caenorhabditis elegans expressed sequence tags identify gene families and potential disease gene homologues'. *Nature Genetics*, 1, 124–31.

Mittapalli, O., Bai, X., Mamidala, P., Rajarapu, S. P., Bonello, P. and Herms, D. A. (2010). 'Tissue-specific transcriptomics of the exotic invasive insect pest emerald ash borer (Agrilus planipennis)'. *PLoS ONE*, 5, e13708.

Moonen, A-C. and Barberi, P. (2008). 'Functional biodiversity: an agroecosystem approach'. *Agriculture, Ecosystems and Environment*, 127, 7–21.

Moore, M. J., Dhingra, A., Soltis, P. S., Shaw, R., Farmerie, W. G., Folta, K. M. and Soltis, D. E. (2006). 'Rapid and accurate pyrosequencing of angiosperm plastid genomes'. *BMC Plant Biology*, 6, 17.

Moran, N. A. (1992). 'The evolution of aphid life cycles'. *Annual Review of Entomology*, 37, 321–48.

Ofuya, T. (1997). 'Control of the cowpea aphid, Aphis craccivora Koch (Homoptera: Aphididae), in cowpea, Vigna unguiculata (L.) Walp'. *Integrated Pest Management Reviews*, 2, 199–207.

Oparaeke, A. M. (2010). 'The potential for controlling Maruca vitrata Fab. and Clavigralla tomentosicollis Stal. using different concentrations and spraying schedules of Syzigium aromaticum (L.) Merr and Perr on cowpea plants'. *Journal of Plant Sciences*, 5, 172–7.

Oyewale, R. and Bamaiyi, L. (2013). 'Management of cowpea insect pests'. *Scholars Academic Journal of Biosciences*, 1, 217–26.

Poelchau, M. F., Coates, B. S., Childers, C. P., de León, A. A. P., Evans, J. D., Hackett, K. and Shoemaker, D. (2016). 'Agricultural applications of insect ecological genomics'. *Current Opinion in Insect Science*, 13, 61–9.

Poelchau, M. F., Reynolds, J. A., Denlinger, D. L., Elsik, C. G. and Armbruster, P. A. (2011). 'A de novo transcriptome of the Asian tiger mosquito, Aedes albopictus, to identify candidate transcripts for diapause preparation'. *BMC Genomics*, 12, 619.

Raheja, A. (1974). 'Report on the insect pests of grain legumes in northern Nigeria. 1st IITA Grain Legume Improvement Workshop (1973)'. Ibadan, Nigeria: International Institute of Tropical Agriculture, pp. 295–9.

Raheja, A. and Leleji, O. (1974). 'An aphid-borne virus disease of irrigated cowpea in Northern Nigeria'. *Plant Disease Reporter*, 58, 1080–4.

Reyes, B., Maredia, M. K., Dabiré, C. B. P. and Ba, M. (2013). 'Economic impacts of bio-control research to manage field insect pests of cowpea in Burkina Faso: baseline survey report (Staff Paper 13–04)'. Michigan State University, Department of Agricultural, Food, and Resource Economics.

Rossel, H. and Thottappilly, G. (1985). 'Virus diseases of important food crops in tropical Africa'. International Institute of Tropical Agriculture.

Rothberg, J. M. and Leamon, J. H. (2008). 'The development and impact of 454 sequencing'. *Nature Biotechnology*, 26, 1117–24.

Rusoke, D. and Rubaihayo, P. (1994). 'The influence of some crop protection management practices on yield stability of cowpeas'. *African Crop Science Journal*, 2, 43–8.

Sharma, H. C., Saxena, K. B. and Bhagwat, V. R. (1999). 'The legume pod borer, Maruca vitrata: bionomics and management. (Information Bulletin no. 55)'. Patancheru, India: International Crops Research Institute for the Semi-arid Tropics (ICRISAT).

Simon, J.-C., Rispe, C. and Sunnucks, P. (2002). 'Ecology and evolution of sex in aphids'. *Trends in Ecology and Evolution,* 17, 34–9.
Singh, S. and Allen, D. (1980). 'Pests, diseases, resistance and protection of Vigna unguiculata (L.) Walp'. In: Summerfield, R. J. and Bunting, A. H. (eds), *Advances in Legume Science*, London. Royal Botanic Gardens; Ministry of Agriculture, Fisheries and Food, pp. 419–43.
Singh, S. and Jackai, L. (1985). 'Insect pests of cowpeas in Africa: their life cycle, economic importance and potential for control'. In: Singh, S. R. and Rachie, K. O. (eds), *Cowpea Research, Production, and Utilization.* Chichester; New York: Wiley, pp. 217–31.
Singh, S. and Jackai, L. (1988). 'The legume pod-borer, Maruca testulalis (Geyer): past, present and future research'. *International Journal of Tropical Insect Science,* 9, 1–5.
Singh, S. and van Emden, H. (1979). 'Insect pests of grain legumes'. *Annual Review of Entomology,* 24, 255–78.
Singh, S. R., Jackai, L. E. N., Dos Santos, J. H. R. and Adalla, C. B. (1990). 'Insect pests of cowpea'. In: Singh, S. R. (ed.), *Insect Pests of Tropical Food Legumes.* Chichester, England: John Wiley & Sons Ltd, pp. 43–90.
Singh, S. R. and Taylor, T. A. (1978). 'Pests of grain legumes and their control in Nigeria'. In: Singh, S. R., Van Emden, H. F. and Taylor, T. A. (eds), *Pests of Grain Legumes: Ecology and Control.* London: Academic Press, pp. 99–111.
Sloan, D. B., Keller, S. R., Berardi, A. E., Sanderson, B. J., Karpovich, J. F. and Taylor, D. R. (2012). 'De novo transcriptome assembly and polymorphism detection in the flowering plant Silene vulgaris (Caryophyllaceae)'. *Molecular Ecology Resources,* 12, 333–43.
Sokame, B. M., Tounou, A. K., Datinon, B., Dannon, E. A., Agboton, C., Srinivasan, R., Pittendrigh, B. R. and Tamò, M. (2015). 'Combined activity of Maruca vitrata multi-nucleopolyhedrovirus, MaviMNPV, and oil from neem, Azadirachta indica Juss and Jatropha curcas L., for the control of cowpea pests'. *Crop Protection,* 72, 150–7.
Souleymane, A., Aken'Ova, M., Fatokun, C. and Alabi, O. (2013). 'Screening for resistance to cowpea aphid (Aphis craccivora Koch) in wild and cultivated cowpea (Vigna unguiculata L. Walp.) accessions'. *International Journal of Science, Environment and Technology,* 2, 611–21.
Suh, J. (1986). 'Entomologie du niébé. Projet conjoint 31 OUA/CSTR: Recherche et developpement des cultures vivrières, rapport annuel SAFGRAD/IITA'. International Institute for Tropical Agriculture, pp. E1–32.
Tamò, M., Ekesi, S., Maniania, N. K. and Cherry, A. (2003). 'Biological control, a non-obvious component of IPM for cowpea'. In: Neuenschwander, P., Borgemeister, C. and Langewald, J. (eds), *Biological Control in IPM Systems in Africa.* Wallingford, Oxon, UK: CAB International, pp. 295–309.
Tamò, M., Srinivasan, R., Dannon, E., Agboton, C., Datinon, B., Dabire, C., Baoua, I., Ba, M., Haruna, B. and Pittendrigh, B. (2012). 'Biological control: a major component for the long-term cowpea pest management strategy'. In: Boukar, O., Coulibaly, O., Fatokun, C., Lopez, K. and Tamò, M. (eds), *Enhancing Cowpea Value Chains Through Research Advances.* Proceedings of the 5th World Cowpea Research Conference, 26 September to 1 October 2010, Saly, Senegal, pp. 249–59.
Taylor, T. A. (1967). 'The Bionomics of Maruca testulalis Gey. (Lepidoptera: Pyralidae), a major pest of cowpea in Nigeria'. *Journal of the West Africa Science Association,* 12, 111–29.
Taylor, T. A. (1978). 'Maruca testulalis: an important pest of tropical grain legumes'. In: Singh, S. R., Van Emden, H. F. and Taylor, T. A. (eds), *Pests of Grain Legumes: Ecology and Control.* London: Academic Press, pp. 193–200.
Tedeschi, R. (2012). 'DNA sequencing and crop protection'. In: Munshi, A. (ed.), *DNA Sequencing – Methods and Applications.* [online]: Intech.
Thottapilly, G., Rossel, H. W., Reddy, D. V. R., Morales, F. J., Green, S. K. and Makkouh, K. M. (1990). 'Vectors of Virus and Mycoplasm diseases: an overview'. In: Singh, S. R. (ed.), *Insect Pests of Tropical Food Legumes.* Chichester, England: John Wiley & Sons Ltd, pp. 323–42.

Traoré, F., Ba, N. M., Dabiré-Binso, C. L., Sanon, A. and Pittendrigh, B. R. (2014). 'Annual cycle of the legume pod borer Maruca vitrata Fabricius (Lepidoptera: Crambidae) in southwestern Burkina Faso'. *Arthropod-Plant Interactions*, 8, 155–62.

Traoré, F., Dabire-Binso, C. L., Ba, N. M., Sanon, A. and Pittendrigh, B. R. (2013). 'Feeding preferences of the legume pod borer Maruca vitrata (Lepidoptera: Crambidae) larvae and suitability of different flower parts for larval development'. *International Journal of Tropical Insect Science*, 33, 107–13.

Trionnaire, G., Hardie, J., Jaubert-Possamai, S., Simon, J. C. and Tagu, D. (2008). 'Shifting from clonal to sexual reproduction in aphids: physiological and developmental aspects'. *Biology of the Cell*, 100, 441–51.

Usua, E. J. and Singh, S. R. (1978). 'Parasites and predators of the cowpea pod borer, Maruca testulalis (Lepidoptera; Pyralidae)'. *Nigerian Journal of Entomology*, 3, 100–2.

Van Alebeek, F. (1996). 'Natural suppression of bruchid pests in stored cowpea (Vigna unguiculata (L.) Walp.) in West Africa'. *International Journal of Pest Management*, 42, 55–60.

Weber, A. P., Weber, K. L., Carr, K., Wilkerson, C. and Ohlrogge, J. B. (2007). 'Sampling the Arabidopsis transcriptome with massively parallel pyrosequencing'. *Plant Physiology*, 144, 32–42.

Whitworth, R. and Ahmad, A. (2009). *Cowpea Aphid*. Manhattan, KS: Kansas State University Agricultural Experiment Station and Cooperative Extension Service.

Wicker, T. Schlagenhauf, E. Graner, A., Close, T. J., Keller, B. and Stein, N. (2006). '454 sequencing put to the test using the complex genome of barley'. *BMC Genomics*, 7, 275.

Wolt, J. D., Wang, K. and Yang, B. (2016). 'The regulatory status of genome-edited crops'. *Plant Biotechnology Journal*, 14, 510–18.

Xue, J., Bao, Y-Y., Li, B-L., Cheng, Y-B., Peng, Z-Y., Liu, H., Xu, H-J., Zhu, Z-R., Lou, Y-G., and Cheng, J.-A. (2010). 'Transcriptome analysis of the brown planthopper Nilaparvata lugens'. *PLoS ONE*, 5, e14233.

Chapter 15

Weed management in grain legume cultivation

Don W. Morishita, University of Idaho, USA

1 Introduction

Weeds are just one of the important groups of pests that can affect grain legume production systems. They have been a pest in grain legumes for as long as grain legumes have been cultivated. This is primarily because grain legumes are not as competitive with weeds as some other crops such as maize or cereal grains (Al-thahabi et al., 1994; Blackshaw and Harker, 1998; Cici et al., 2008; Mesbah et al., 2004; Smitchger et al., 2012; Spies et al., 2011). The general low growth stature of most grain legumes limits their ability to compete with taller weeds. The grain legumes considered in this chapter include dry bean (*Phaseolus vulgaris* L.), chickpea or garbanzo bean (*Cicer arietinum* L.), lentil (*Lens culinaris* Medik.), dry pea (*Pisum sativum* L.) and faba or fava bean (*Vicia faba* L.). Of these species, faba bean typically is the tallest, but is considered a poor competitor with weeds in the early stages of growth.

Liebman et al. (2001) provided a sound basis for the consideration of better understanding the ecological aspects of managing weeds in agro-ecosystems. They identified three principal objectives for weed management. The first is that weed density should be reduced to tolerable levels. In other words, eradication of weeds is not always necessary, but reducing weed populations below a yield-reducing level is necessary. However, this concept argues that waiting until weed populations reach an economic injury level makes us more dependent on the use of herbicides (Norris, 2005). The second objective is that the amount of damage a given density of weeds inflicts on an associated crop should be reduced. They cite evidence to show that delaying weed emergence relative

http://dx.doi.org/10.19103/AS.2017.0023.12

to crop emergence; increasing the proportion of available resources captured by crops; and damaging, but not necessarily killing, weeds with chemical, mechanical or biological agents are enough. The third objective is to shift the composition of weed communities towards less aggressive, easier-to-manage species. The competitiveness of different weed species can differ in how much damage they inflict on different crops. Generally, broadleaf weeds are more competitive with broadleaf crops than with grass species (Everman et al., 2008), although there certainly are exceptions (Harker et al., 2007).

2 Weed interference

Competition between weeds and crops has been studied around the world where grain legumes are grown. In grain legumes, the most studied among the types of competition is understanding the critical period of weed control. One of the reasons for this is that these types of studies helps provide a benchmark for the most important time or crop growth stages that weeds must be controlled to avoid yield loss. In Zimdahl's (2004) review of weed – crop competition, the critical period for weed control is defined as the time between that period after crop seeding or emergence when weed competition does not reduce crop yield and the time in the crop's life after which weed presence does not reduce yield. It would seem obvious that the greater the weed density for any given crop,

Table 1 Duration of competition tolerated before grain legume yield loss was incurred. Sources listed prior to 1996 are credited to Zimdahl (2004)

	Length (weeks) of competition tolerated after				
Crop	Seeding	Emergence	Competing weed(s)	Location	Source
Bean, dry	8		*Echinochloa crus-galli*	Washington, USA	Dawson, 1964
Bean, fava		3–5 after 50% emergence	Mixed annuals	England	Glasgow et al., 1976
Bean, dry	3		Mixed annuals	Chapingo, Mexico	Nieto et al., 1968
Bean, green		4	*Amaranthus retroflexus*	Oregon, USA	Williams, 1971
Bean, fava		4 after 50% emergence	Mixed annuals	England	Hewson et al., 1973
Bean, green	4		*Cyperus rotundus*	Minas Gerais, Brazil	William and Warren, 1975
Bean, lima	6		*Senna obtusifolia*	Georgia, USA	Glaze and Mullinex, 1984
Lentil		4.8–5.8	Mixed annuals and perennials	NE Jordan	Singh et al., 1996

Table 2 Duration of weed-free period required to prevent grain legume yield reduction. Sources listed prior to 1994 are credited to Zimdahl (2004)

	Weed-free period (weeks) required after				
Crop	Seeding	Emergence	Competing weeds	Location	Source
Bean, dry	5		Mixed annuals	Washington, USA	Dawson, 1964
Bean, broad		1–1.5 after 50% emergence	Mixed annuals	England	Hewson et al., 1973
Bean, dry		6–9	*Solanum physalifolium*	Alberta, Canada	Blackshaw, 1991
Bean, dwarf		1–1.5	Unknown	West Indies	Kasasian and Seeyave, 1969
Bean, snap	Unifoliate stage		*Xanthium strumarium*	New Jersey, USA	Neary and Majek, 1990
Lentil		12.1–14.1	Mixed annuals and perennials	NE Jordan	Singh et al., 1996
Chickpea	5–7		Mixed annuals and perennials	NW Jordan	Al-Thahabi et al., 1994
Lentil	7–8		Mixed annuals and perennials	NW Jordan	Al-Thahabi et al., 1994
Chickpea		0.34 to harvest	Mixed annuals and perennials	Van, Turkey	Tepe et al., 2011
Chickpea		3–7	Mixed annuals and perennials	Tabriz, Iran	Mohammadi, 2005
Bean, dry (Black)	V2–V4[a]		Mixed annuals	São Paulo, Brazil	Barroso et al., 2012
Chickpea		7–10	Mixed annuals	Agrigento, Italy	Frenda et al., 2013
Faba bean		4–5	Mixed annuals	Agrigento, Italy	Frenda et al., 2013
Lentil, winter		1–7 after re-growth[b]	Mixed annuals and perennials	Van, Turkey	Tepe et al., 2005
Lentil		3–8	*Anthemis cotula* and *Sonchus asper*	Washington, USA	Smitchger et al., 2012
Lentil		5–10 node growth stage	Mixed annuals	Saskatoon, Canada	Fedoruk et al., 2011
Lentil	2–4		Mixed annuals and perennials	Northern Sudan	Mohamed et al., 1997

[a]V2 and V4 growth stages correspond to approximately 3–5 weeks after planting (Kandel, 2013). [b]Critical period of weed control for winter lentil was 1–7 weeks after growth resumed in the spring.

the smaller the period that the crop can tolerate interference from the weeds. Along with this idea is the fact that the availability of essential resources – soil water, soil nutrients and, to a very limited extent, sunlight – will influence the length of time that the crop can tolerate competition from weeds (Weaver et al., 1992). In most instances, however, these resources tend to become a limiting factor later in the growing season.

Conducting interference studies to determine the critical period that weed control must be imposed depends entirely on the species and the environment in which the crop is grown (Mortimer, 1984). Referring to a single study to use as a guide to the critical period of weed control for any given crop species could lead one to making erroneous assumptions. Zimdahl (2004) compiled the data from a comprehensive list of weed – crop interaction studies that included several grain legume studies that have been conducted and are shown in Tables 1 and 2. Additional data, although not necessarily more recent, from various grain legume studies published after this exhaustive list of interference studies was published in 2004 and are listed in Tables 1 and 2. The duration of competition tolerated and the duration of the weed-free period vary greatly among grain legume species, but also vary greatly within a species, depending apparently on geographical location and the climatic conditions associated with that location and the competing weeds. For example, dry bean tolerated the presence of *Echinochloa crus-galli* [(L.) Beauv] in Washington, USA, for 8 weeks after seeding compared to only 3 weeks after seeding in the presence of mixed annual weeds in Chapingo, Mexico (Dawson, 1964; Nieto et al., 1968). With green bean, the duration of competition with *Amaranthus retroflexus* (L.) in Oregon, USA, before yield loss was 4 weeks after emergence. However, with *Cyperus rotundus* (L.) in Minas Gerais, Brazil, the duration of competition was only 4 weeks after seeding (William and Warren, 1975; Williams, 1971).

The critical weed-free period also varies within species growing in different locations and with different weed species. The weed-free period required for dry bean growing with a mixture of annual weeds was 5 weeks after seeding in Washington, USA, whereas dry bean growing in competition with *Solanum physalifolium* (Rusby) required 6–9 weeks of weed-free conditions after emergence (Blackshaw, 1991; Dawson, 1964). Lentil growing in northeast Jordan had a marked difference in response to a mixture of annual and perennial weeds compared to lentil in northwest Jordan growing in competition with a similar mixture of weeds. The lentil growing in northeast Jordan required 12.1–14.1 weeks of weed-free conditions to avoid yield loss while lentil in northwest Jordan required 7–8 weeks under similar conditions (Al-Thahabi et al., 1994; Singh et al., 1996). Compared to lentil growing in competition with weeds in Washington, USA, and Saskatchewan, Canada, lentil in northeast Jordan were much less competitive (Smitchger et al., 2012; Fedoruk et al., 2011). Winter lentil grown in Van, Turkey, were about as competitive as the United States- and Canadian-grown spring lentil (Tepe et al., 2005).

Evaluating the competitiveness of grain legumes in response to varying weed densities shows a similar variation within species (Table 3). Chickpea growing in competition with *Polygonum aviculare* (L.), a prostrate-growing annual in Viterbo, Italy, and *Euphorbia dracunculoides* (Boiss.), a short erect plant, in Faisalabad, Pakistan, had 14 and 1% yield reductions at 4 and 5 plants m^{-2}, respectively (Radicetti et al., 2012; Tanveer et al., 2015). The same can be said of dry bean. In Wyoming, USA, dry bean yields were affected very differently by *Setaria viridis* [(L.) P. Beauv] and *Helianthus annuus* (L.) (Mesbah et al., 2004). The lowest density that would economically reduce dry bean yield by *S. viridis* was 1.2–2.2 plants m^{-2} compared to *H. annuus*, which economically reduced dry bean yield with densities as low as 0.09–0.15 plants m^{-2}. *Amaranthus retroflexus* reduced dry bean

Table 3 Effect of weed densities on grain legume yield

Crop	Lowest density causing yield reduction (plants m^{-2})	Competing weed(s)	Location	Source
Chickpea	4 plants m^{-2}	*Polygonum aviculare*	Viterbo, Italy	Radicetti et al., 2012
Chickpea	5 plants m^{-2}	*Euphorbia dracunculoides*	Faisalabad, Pakistan	Tanveer et al., 2015
Bean, dry (Red bean)	0.26–0.28 plants m^{-2}	*Amaranthus retroflexus*	Tehran, Iran	Saberali et al., 2012
Bean, dry (Pinto)	1.2–2.2 plants m^{-2}	*Setaria viridis*	Wyoming, USA	Mesbah et al., 2004
Bean, dry (Pinto)	0.09–0.15 plants m^{-2}	*Helianthus annuus*	Wyoming, USA	Mesbah et al., 2004
Bean, dry (Black bean)	0.59–6.53 plants m^{-2}	*Bidens pilosa*	Rio Grande do Sul, Brazil	Galon et al., 2016

yield in Tehran, Iran, with densities slightly higher than *H. annuus* at 0.26–0.28 plants m^{-2} and *Bidens pilosa* (L.) significantly reduced dry bean yield in Rio Grande do Sul, Brazil, at densities of 0.59–6.53 plants m^{-2} (Galon et al., 2016; Saberali et al., 2012).

Regardless of how weed interference in grain legumes is measured, it is clear that all grain legume species are susceptible to weed competition. Within the grain legumes are several species with varying ability to compete with weeds. There are no published studies that compare differences in competitiveness among all the grain legumes discussed in this chapter, although one may argue which species may be better competitors than others, particularly those who have grown or compared at least two grain legumes in the same experiments (Avola1, 2008; Frenda et al., 2013; Mishra et al., 2006; Spies et al., 2011).

Within a species, there can be variations in cultivar plant growth habit and flowering (Kandel, 2013) that can influence their competitiveness with weeds. For example, dry beans have two plant growth habits: 1) determinate growth, in which stem elongation ceases when the terminal flower racemes of the main stem or lateral branches have developed, and 2) indeterminate growth, where flowering and pod filling continue for as long as temperature and moisture availability permits. Beans with determinate growth have a bushy growth habit and beans with indeterminate growth have a vining or trailing growth habit. Thus, dry beans are divided into four plant growth types: Type I beans are described as determinate bush; Type II beans are upright with short vines and a narrow plant profile having three to four branches; Type III beans are indeterminate flowering and the growth habit is a prostrate vine; and Type IV beans are indeterminate with strong climbing tendencies that require a trellis system for optimal production. Dry bean classes are a more important differentiation when considering consuming this legume. The most common dry bean market classes include Black, Cranberry, Great Northern, Red Kidney, Navy, Pinto and Small Red. Each of these classes has specific uses in various parts of the world.

Comparing different market classes of beans is rare, but in 1993 and 1994, a study was conducted in Nebraska, USA (Urwin et al., 1996). The objective of the study was to determine the influence of dry bean canopy type on late season weed emergence of several dry bean market classes. The classes represented a broad range of growth habits and included Great Northern, Pinto, Navy, Black, Small White, Cranberry and Light Red Kidney. Of these market classes, growth habit included viny, short vine, upright and bush. Weed species consisted of *Portulaca oleracea* (L.), *A. retroflexus*, *Setaria glauca* [(L.) P. Beauv.], *Chenopodium album* L. and *S. physalifolium*. Differences in competitiveness with late season weed emergence between market classes were observed, but there were also differences in competitiveness between cultivars within the same market class. The differences, they concluded, were greatly influenced by canopy structure. However, canopy structure varied from one year to the next due to differences in growing conditions. This was especially true when air temperatures were cooler than average.

Vogt et al. (2013) compared the competitive ability of several Black bean genotypes with weeds. Their hypothesis was based on the results of studies conducted in corn, rice, wheat and soya bean that Vogt et al. (2013) reviewed. In all the studies with these crops, growth rate was the key factor in increasing competitiveness of the crop with the weeds. Their aim was to determine if there was variability among Black common beans' genotypes in their competitiveness with weeds and if it was possible to identify plant characteristics that explain the competitive ability. Unfortunately, Black bean growth rate was not the key factor for increasing competitiveness and the response of the Black bean genotypes did not support their hypothesis. They did find, however, that grain yield loss in Black bean genotypes from weed interference ranged from 31% to 55% and the genotypes whose yields were reduced the least also reduced weed growth the most.

Similar comparisons have been made with pea (Spies et al., 2011; Vasilakoglou and Dhima, 2012). The work by Spies et al. (2011) compared ten pea cultivars for their competitive ability against weeds. Two pea types were compared. One type was forage pea cultivars that had a leafy long vine length and the others were grain cultivars that were semi-leafless with short vine length. The forage pea cultivars with leaves and long vines were able to suppress weed growth better without reducing yields compared to the grain cultivars. They concluded that vine length and the leafy characteristic may be important genetic characteristics associated with competition in field pea cultivars. It is worth noting that the forage pea cultivar seed yields were low and this may account for their ability to better compete with the weeds.

In another study conducted in northern Greece, Vasilakoglou and Dhima (2012) compared two field pea cultivars for their competitiveness with *Avena sterilis* spp. *ludoviciana* (L). In this study, one cultivar was leafy with long vine length and the other a semi-leafless, short vine length cultivar. In weed-free conditions, the leafless, short vine length variety seed yield was 37% greater than the leafy, long vine length variety. However, when the two cultivars were exposed to different densities of *A. sterilis*, the leafless, short vine length cultivar seed yield was 24–49% lower than the leafy long vine length cultivar, depending on weed density.

Chickpea genotypes have also been compared for their competitiveness with *P. aviculare* using the competition assessment parameters: relative biomass total, competitive index and competitive balance index (Cb) (Radicetti et al., 2012). The Cb index is used to quantify the intensity of competition. A Cb value >0 means the crop is more competitive than the weed(s). A Cb value equal to or <0 indicates that the crop equal to or less competitive than the weed(s) (Wilson, 1988). Averaged across all chickpea genotypes, *P. aviculare* reduced seed yield by 14, 46, 75 and 88% at densities of 4, 8, 16 and 32 plants m^{-2}, respectively. The two highest-yielding chickpea genotypes in weed-free conditions were very competitive against *P. aviculare*, at intermediate and high weed densities. The lowest yielding genotype in weed-free conditions was more competitive against *P. aviculare* than the two highest yielding genotypes. By averaging over all the genotypes, Radicetti et al. (2012) concluded that *P. aviculare* densities should be <4 plants m^{-2} to avoid seed yield loss and that selecting a highly competitive genotype would only be one component of an integrated approach to weed management.

In Turkey, six winter lentil cultivars were compared for their competitiveness with weeds (Tepe et al., 2005). In this study, there were no cultivars that were more competitive than the other with the weeds. They did identify some varieties that had less yield loss than other varieties in each of the two years they conducted this study. However, they were not the same cultivars from one year to the next, except for one. There was one cultivar that was a consistently poor competitor in both years. Interestingly, this same cultivar ranked among the highest-yielding cultivars grown under weed-free conditions averaged over both years. Tepe et al. (2005) postulated the difference in cultivar response from one year to the next was possibly due to climatic factors. In addition, they concluded that the contribution of cultivar effect to weed interference in lentil was modest.

One point that stands out in more than one of these studies that evaluated the competitiveness of cultivars within a species is that the highest-yielding cultivars grown in weed-free conditions were not the most competitive cultivars when grown with weeds. Perhaps this is telling us that in the pursuit of developing the highest seed-yielding cultivars, competitiveness with other species may be weakened.

3 Weed control

Depending on country and location, weed control methods may differ and certain strategies may be relied upon more than another. These methods may include preventive, biological, cultural, mechanical/physical or chemical control. The critical aspect to remember is that the more diverse weed management strategies used, the greater the likelihood of success.

The value of combining multiple weed management tactics has been shown in several studies with grain legumes (Amador-Ramirez et al., 2001; Ball et al., 1997; Boerboom and Young, 1995; El-Metwally and Abdelhamid, 2008; Fernandez et al., 2012; Johnson and Holm, 2009; Radicetti et al., 2012; Santín-Montanya et al., 2014; Wilson and Sbatella, 2013). Inter-row cultivation used in combination with higher biomass accumulating chickpea genotypes helped reduce the competitive effects of the weeds, but was not totally effective in controlling weeds throughout the growing season (Radicetti et al., 2012).

In dry bean, the combination of overhead irrigation following the application of soil-active herbicides in Nebraska, USA, helped reduce crop injury and weed emergence in no tillage (NT) compared to intensive tillage and no overhead irrigation (Wilson and Sbatella, 2013). With a combination of cultivation and herbicides, Amador-Ramirez et al. (2001) found in Nebraska, USA, that at low weed densities (35 plants m^{-2}), either cultivation or herbicides alone were effective in suppressing weeds. However, with higher weed densities (134 plants m^{-2}), the combination of cultivation and herbicides were needed to effectively control the weeds.

A method for improving weed control in dry pea and lentil using non-chemical weed control methods in Washington, USA, was investigated with increased seeding rates used in combination with post-plant tillage with and without metribuzin herbicide (Boerboom and Young, 1995). Increasing the planting density of either crop increased the potential for reducing weed competition. However, the increased seed cost should be considered when comparing other weed control alternatives such as use of herbicide(s). Post-planting tillage treatments were not detrimental to the pea or lentil, but weed populations also were not reduced under these conditions. Considering the poor competitive ability of lentil and harvest problems resulting from uncontrolled weeds in either crop, they concluded that adequate broadleaf herbicide weed control be used with higher seeding rates.

In a separate study, Johnson and Holm (2009) looked at post-plant tillage used in combination with planting dates, seeding depth and herbicides in Saskatchewan, Canada. Three seeding dates (early May, mid-May and late May) were used in combination with six weed management systems [shallow-seeded (2.5 cm) untreated, shallow-seeded with pre-emergence (PRE) tillage, shallow-seeded herbicide (metribuzin and sethoxydim), deep-seeded (7.5 cm) untreated, deep-seeded with PRE tillage and deep-seeded with herbicide] to optimize pea yield. None of the weed management systems affected pea density. They found that using herbicides was the only treatment that significantly reduced weed density because the herbicide application timing was more synchronous with weed emergence than were the tillage treatments. They concluded that pea producers who choose not to use herbicides would benefit most by delaying seeding until some weeds emerge and to seed 7.5-cm deep followed by two sequential passes with a rod weeder prior to crop emergence. A rod weeder is a tractor-drawn implement with a 2.5 cm square rod that is placed 2.5–5 cm below the soil and rotates in the direction opposite to that of the travel direction (Bowman, 2002). The rod rotation along with its square shape, uplifts small weeds from the soil while at the same time leaving a mulch soil layer that helps prevent new weeds from emerging.

Santín-Montanya et al. (2014) compared the influence of conventional tillage (CT), minimum tillage and NT on weed populations in Comarca de Alcalá, Spain. They found that tillage system had no effect on pea yield or yield components. However, they did find that with below-average precipitation, pea yield, yield components, straw biomass and harvest index was reduced under minimum tillage. *Lolium rigidum* densities increased with less tillage over time similar to other weed species shifts that happens when changing from CT to NT.

The effects of cultivar, delayed sowing and weed removal on yield of spring-planted pea and lentil grown using organic methods was explored in Minnesota, USA (Fernandez et al., 2012). Two lentil cultivars and four yellow field pea cultivars were planted on three dates with and without weed control. Controlling weeds increased pea yields by an average of 63% and lentil yields by an average of 87% compared to the non-weeded treatments. Early planting favoured pea and lentil yield in the weeded and non-weeded treatments. Weed biomass was not influenced by delayed planting. The growing environment influenced pea yield in response to weed control. Cooler, drier climates favoured pea yield as much as or more than weed control. It was concluded that growing pea in Minnesota can be a profitable crop for organic systems when early planted and when weeds are controlled. Lentil yields did not consistently exceed the economic break-even levels in most environments.

The use of small red lentil seeding rates in combination with herbicides was investigated in Oregon and Washington, USA, to determine if lentil seeding rate could enhance weed control with herbicides (Ball et al., 1997). They found that, at Oregon in both years and Washington in one year, they could reduce total weed density when averaged over all the herbicide treatments by increasing lentil seeding rate. Increasing lentil seeding rate reduced weed dry weight more when herbicides were not used or when herbicides were less effective. Herbicides alone generally reduced weed density, but the effectiveness of the herbicides was related to the weed species and environmental conditions present in each experiment location.

The long-term impact of tillage on faba bean productivity, nitrogen fixation and weed incidence has been evaluated in Sicily, Italy (Giambalvo et al., 2012). A more in-depth review of this study is presented near the end of this chapter where two case studies are presented. Basically, what they found was that using NT in a cereal–legume rotation in dryland Mediterranean environments was beneficial because it reduced cost and reduced the impact on physicochemical and biological soil properties. From a weed management standpoint, the success of NT techniques depends on the use of other weed control strategies to complement the effect of using NT in a cereal–legume rotation.

In the central Ethiopian highlands, where faba bean is the most important grain legume, phosphorus is considered to be a critical growth-limiting nutrient and inadequate plus untimely weed control operations is considered one of the other crucial factors causing low faba bean yields there (Agegnehu and Fessehaie, 2006). Most of the crops such as tef, wheat and faba bean are handweeded. Unfortunately, labour for weeding faba bean is usually competing with the timing of weeding other crops, resulting in faba bean being weeded late or not at all. An experiment was conducted to determine the economic dose of phosphorus and weed control for faba bean production. The results showed that the highest marginal rate of return was obtained from weeding once six weeks after crop emergence in combination with applying 20 kg P ha^{-1}, which is economically the most feasible alternative for faba bean production in the central Ethiopian highlands.

In the United States, much of Europe and South America, herbicides such as EPTC and trifluralin were important tools developed in the 1960s for chemically controlling weeds in grain legumes. In the 1970s and 1980s, more and better herbicides became available to selectively control weeds in grain legumes. As the agrichemical industry continued to develop new and better herbicides for controlling weeds, the dependence on chemicals increased. However, in areas around the world where herbicides are intensively used for weed control, the need for utilizing other weed management strategies remains an important part of effectively and economically controlling weeds in grain legumes. The lack of discovery of new herbicide mechanisms of action and the growing list of herbicide-resistant weeds (Heap, 2016) makes it even more important to use integrated weed management practices for controlling weeds not only in grain legumes, but in all crops grown around the world.

It is abundantly clear that reliance on single weed management strategies increases the potential for weed control failure. Integrating crop cultivars that are not necessarily the highest yielding, but most competitive with weeds, along with optimal seeding rates, planting dates, pre-plant tillage operations, in-crop tillage operations and herbicide selection increase the likelihood of successfully controlling weeds in grain legumes.

4 Case study 1

Weed management studies consisting of data compiled from two to three years and possibly from two locations are considered adequate for publication in most refereed journals. It is assumed that this is long enough to determine accurate outcomes of weed interference or weed management under fairly specific guidelines. Long-term experiments lasting over 10 years can provide a more complete picture of what happens or how changes take place in response to various treatments imposed on the experiment. Giambalvo et al. (2012) provided an excellent picture of what happens over an 18-year period with a crop rotation involving different levels of tillage on the outcome of weed interference. These types of studies are rare for many reasons, but among the most important is the cost of conducting such a long-term study.

In the United States, most studies that are funded by the federal government or state commodity groups do not provide funding for multi-year long-term studies or cannot afford to provide funding for long-term studies if it is a state commodity group. There are instances of federally funded long-term studies in the United States, but they are relatively rare (Kok et al., 2009). Private foundations have been known to provide funding for long-term studies, but they too are rare.

The 18-year long-term study that Giambalvo et al. (2012) conducted was to study the effects of continuous tillage system study that compared CT (mouldboard plough 30 cm deep, followed by one or two shallow harrowing operations), reduced tillage [RT] (chisel plough 40 cm deep without inverting soil followed by one shallow harrowing operation) and NT (crop seeded directly into untilled soil) on the performance of crops in a cereal–legume rotation that included wheat (*Triticum aestivum* L.), berseem clover (*Trifolium alexandrinum* L.) and faba bean. This crop rotation was typical of the semi-arid Mediterranean environment in Italy. Faba bean yield in the NT, averaged over the 18-year period, was 23 and 31% higher than RT and CT, respectively. The yield advantage was pronounced when precipitation was <400 mm, due to the soil moisture conserved in NT.

Weeds in the faba beans were controlled by a shallow cultivation when the crop was in the third leaf and if necessary, the seventh leaf growth stages. One of the targeted weeds was the parasitic weed *Orobanche crenata* (Forssk.). *O. crenata* intensity, on average, was lower in NT than in CT or RT. However, this was not constant from year to year. In four years, *O. crenata* densities were higher in CT and RT, but the opposite was observed in both the years. The most abundant non-parasitic weeds were *Lactuca serriola* (L.), *Phalaris* spp., *Papaver rhoeas* (L.), *Ridolfia segetum* (Moris) and *Sinapis arvensis* (L). Other weed species included *Anagallis arvensis* (L.), *Avena* spp., *Chenopodium* sp. *Gladiolus* spp., *Matricaria* spp., *P. aviculare*, *Senecio vulgaris* (L.) and *Triticum durum* (Desf). On average, non-parasitic weed biomass was greater in NT compared to CT and RT. However, not all the weeds responded consistently from year to year. The incidence of *Phalaris* spp., *P. rhoeas* and *L. serriola* decreased with tillage intensity (CT < RT < NT). These differences among tillage systems increases progressively during the study period. In contrast, *S. arvensis*, *P. aviculare* and *R. segetum* were more abundant in CT than in RT or NT. On average, weed diversity increased with tillage intensity (CT > RT > NT), and these differences increased over time among the three tillage systems. Agronomically, faba bean yield was greater in NT compared to CT and RT, especially when annual precipitation was <400 mm. The downside to NT with regard to weed management was that compared to CT, weed control became more challenging because the weed composition in NT shifted to weeds that were more difficult to control. Overall, Giambalvo et al. (2012) concluded that using NT in a cereal–legume rotation in dryland Mediterranean environments was beneficial because it reduced cost and reduced the impact on physicochemical and biological soil properties. From a weed management standpoint, the success of NT techniques depended on the use of other weed control strategies to complement the weed species shift using NT in a cereal–legume rotation.

5 Case study 2

In Idaho, USA, dry bean is the predominant grain legume followed by chickpea (USDA, 2015), where they are also called garbanzo bean. Much of the dry beans are grown in 56 cm rows and because the pods of some market classes hang close to the soil surface, they are harvested in a two-step process. The first step is to cut the bean plants just below the soil surface with an implement equipped with a rod weeder that crosses multiple rows. This essentially uplifts the cut plants from the soil in multiple rows, which are then placed into a windrow as a part of the same cutting operation. After cutting, the second step is to harvest with a combine after the plants have been left for several days to desiccate. There are two drawbacks to this method. The first is that soil adhering to the plant roots is taken into the harvesting combine, which causes additional wear on equipment and some small dirt clods about the size of the beans contaminate the harvested beans. The second drawback is the additional cost associated with undercutting the bean plants. With some market classes of dry bean, the pods remain high enough off the ground to allow either direct harvesting without undercutting first or cutting them near ground level with a swather and leaving the plants cut from multiple rows in a swath. This is gaining interest as a mechanism to reduce cost and wear on harvesting equipment. Direct harvesting or swathing before harvest has led to an increasing interest in growing dry bean in narrow rows similar to solid-seeded soya bean. Several studies evaluating reduced row width, in soya bean and dry bean (Black, Small Red and Navy) have shown that the crop receives a competitive advantage over weeds when the row width is reduced due to faster canopy closure and increased light interception than with wide-row beans (Blackshaw et al., 1999, 2009; Holmes and Sprague, 2013; Rich and Renner, 2009; Yelverton and Coble, 1991; Young et al., 2009). Studies also show that reducing row width is not sufficient to completely reduce the weed population by itself but can be effective when combined with herbicides (Blackshaw et al., 2009; Holmes and Sprague, 2013). In general, these studies show that narrowing the row increases yield versus wide rows (Blackshaw et al., 2009; Cox and Cherney, 2011; Holmes and Sprague, 2013).

Because of the differences in growth habit of the dry bean market classes, it is not known if narrow-row spacing affects the competitiveness of all dry bean market classes. Pinto bean accounts for 35% of the dry bean grown in Idaho, USA, which makes it the most widely grown bean class in the state (USDA, 2015). The objectives of this study were to 1) determine the effect of row spacing and plant architecture of two cultivars on weed control and bean yield and 2) compare the influence of sequential herbicide applications on season-long weed control of hairy nightshade and other weeds.

The study was conducted in Idaho. A total of 24 weed control treatments was established in a 2 by 2 by 6 factorial randomized complete block design with 2 row widths, 2 Pinto bean varieties with different plant architecture and 4 herbicide combinations plus 2 controls. The plots were 2.23 m wide by 7.62 m long with four replications. The narrow-row spacing was 15 cm the first year and 19 cm the second year. Wide-row beans were planted in 56 cm row spacing. Two Pinto bean cultivars were used in the study. Two Pinto bean cultivars, 'Sequoia', which had a Type II upright growth habit, and 'Othello', which has a Type III viny or trailing growth habit, were planted both years. Herbicides evaluated were EPTC in combination with ethalfluralin or dimethenamid-P and pendimethalin in combination with dimethenamid-P as PRE applications. Post-emergence (POST) applications included bentazon alone or in combination with ethalfluralin or dimethenamid-P. Ammonium sulphate and methylated seed oil were added to all POST treatments at a rate of 1.67 kg ai ha^{-1}

and 1% v/v, respectively. The PRE applications were incorporated with approximately 1.25 cm of overhead irrigation one hour after application. Sequential POST herbicides were applied at the dry bean 1–2 trifoliate leaf stage. Clethodim at 0.17 kg ai ha^{-1} was added to all POST applications to control emerged grasses.

C. album, S. viridis, S. physalifolium and *A. retroflexus* generally were controlled better with a POST sequential application compared to a PRE-only application. Weed interference was reduced enough when the dry bean was planted in narrow rows with POST sequential herbicide applications and no cultivation, to have similar weed control and yields as wide-row planted dry bean with POST sequential herbicide applications and cultivation. The viny trailing 'Othello' usually yielded more than the upright growing Sequoia, which led to the conclusion that the Sequoia growth habit did not provide as much competition against weeds as did the viny, trailing Othello. Overall, this study showed that narrow-row spacing can be beneficial in an integrated weed management plan by reducing interference and increasing subsequent Pinto bean yield.

6 Conclusion and future trends

Continued research to explore the interactions of crop rotations, cultivar selection, planting dates, fertilizer rates, application timing and placement, tillage practices between crops and within the cropping year, and careful use of herbicides will be critical for sustainable grain legume production.

It is interesting to note that published studies that have compared cultivar or genotype response to weed interference often show that the most productive or highest-yielding cultivars are not always the most competitive with weeds. Additional research to ascertain what cultivar or market class characteristics are most important for increasing the competitiveness with weeds will be helpful to better understand and develop cultivars that are more competitive with weeds.

Many weed–grain legume interference studies have been conducted around the world over the years. By comparing the results of the relatively narrow number of studies reviewed for this chapter, it is clear that the environment plays a critical role in the outcome of these important field studies. A number of studies showed responses that varied from one year to the next, and even those studies where the same or similar weed species were investigated in the competition studies at different locations, there are often differences in when the critical period of weed control takes place. The environment was considered the critical factor that influences the observed responses. The other important factor is the weed species. While some studies have focused on the interference of individual weed species, it is widely known that one seldom has a cropping environment with only one species. That is not to say that investigating the competitiveness of a single species with a crop is unimportant. On the contrary, these kinds of studies provide crucial information related to the species involved in the study. Comparatively, interference studies conducted with multiple species provides more of a 'real-life' scenario in terms of weeds competing with crops. However, it is very difficult to control the species makeup from year to year or location to location, not to mention the densities of each weed species in a study. This added complexity increases the overall cost of conducting such studies, but these studies provide important information that may help better understand the impact of weeds on crop productivity.

7 Where to look for further information

Information regarding weed competition, interference and weed management studies can be found in journals such as *Weed Science, Weed Technology, Agronomy Journal, Weed Research, Canadian Journal of Plant Science, Crop Protection* and others. Information from US land-grant university cooperative extension systems, Canada and Australia agricultural departments also contain much additional information. Some examples include:

Chickpea Production in the High Plains. University of Nebraska Cooperative Extension EC04-183.

Dry Bean Production Guide. North Dakota State University Extension A1133.

Growing Lentil in North Dakota. North Dakota State University Extension A1636/

Montana Cool-Season Pulse Production Guide. Montana State University Extension EB0210.

Websites

- Saskatchewan Pulse Growers – http://saskpulse.com/
- Pulse Australia – http://pulseaus.com.au/growing-pulses/bmp/faba-and-broad-bean

8 References

Al-thahabi, S. A., Yasin, J. Z., Abu-Irmaileh, B. E., Haddad, N. I. and Saxena, M. C. (1994), 'Effect of weed removal on productivity of chickpea (*Cicer arietinum* L.) and lentil (*Lens culinaris* Med.) in a Mediterranean environment', *J. Agron. Crop Sci.*, 172(5), 333–41.

Amador-Ramirez, M. D., Wilson, R. G. and Martin, A. R. (2001), 'Weed control and dry bean (*Phaseolus vulgaris*) response to in-row cultivation, rotary hoeing, and herbicides', *Weed Technol.*, 15(3), 429–36.

Avola1, G., Tuttobene, R., Gresta, F. and Abbate, V. (2008), 'Weed control strategies for grain legumes', *Agron. Sustain. Dev.*, 28(3), 389–95.

Ball, D. A., Ogg Jr., A. G. and Chevalier, P. M. (1997), 'The influence of seeding rate on weed control in small-red lentil (*Lens culinaris*)', *Weed Sci.*, 45(2), 296–300.

Blackshaw, R. E. (1991), 'Hairy nightshade (*Solanum sarrachoides*) interference in dry beans (*Phaseolus vulgaris*)', *Weed Sci.*, 39(1), 48–53.

Blackshaw, R. E. and Harker, K. N. (1998), '*Erodium cicutarium* density and duration of interference effects on yield of wheat, oilseed rape, pea, and dry bean', *Weed Res*, 38(1), 55–62.

Blackshaw, R. E., Molnar, L. J., Muendel, H. H., Saindon, G. and Li, X. (2009), 'Integration of cropping practices and herbicides improves weed management in dry bean (*Phaseolus vulgaris*)', *Weed Technol.*, 14(2), 327–36.

Blackshaw, R. E., Muendel, H. H. and Saindon, G. (1999), 'Canopy architecture, row spacing and plant density effects on yield of dry bean (*Phaseolus vulgaris*) in the absence and presence of hairy nightshade (*Solanum sarrachoides*)', *Can. J. Plant Sci.*, 79(4), 663–9.

Boerboom, C. M. and Young, F. L. (1995), 'Effect of postplant tillage and crop density on broadleaf weed control in dry pea (*Pisum sativum*) and lentil (*Lens culinaris*)', *Weed Technol.*, 9(1), 99–106.

Cici, S. Z. H., Adkins, S. and Hanan, J. (2008), 'A canopy architectural model to study the competitive ability of chickpea with sowthistle', *Ann. Bot. (London)*, 101, 1311–18.

Cox, W. and Cherney, J. H. (2011), 'Growth and yield responses of soybean to row spacing and seeding rate', *Agron. J.*, 103(1), 123–8.

Dawson, J. H. (1964), 'Competition between irrigated field beans and annual weeds', *Weeds*, 12(3), 206–8.

El-Metwally, I. M. and Abdelhamid, M. T. (2008), 'Control under integrated nutrient management systems in faba bean (*Vicia faba*) production in Egypt', *Planta Daninha, Viçosa-MG*, 26(3), 585–94.

Everman, W. J., Burke, I. C., Clewis, S. B., Thomas, W. E. and Wilcut, J. W. (2008), 'Critical period of grass vs broadleaf weed interference in peanut', *Weed Technol.*, 22(1), 68–73.

Fedoruk, L. K., Johnson, E. N. and Shirtliffe, S. J. (2011), 'The critical period of weed control for lentil in western Canada', *Weed Sci.*, 59(4), 517–26.

Fernandez, A. L., Sheaffer, C. C., Wyse, D. L. and Michaels, T. E. (2012), 'Yield and weed abundance in early- and late-sown field pea and lentil', *Agron. J.*, 104(4), 1056–64.

Frenda, A. S., Ruisi, P., Saia, S., Frangipane, B., Di Miceli, G., Amato, G. and Giambalvo, D. (2013), 'The critical period of weed control in faba bean and chickpea in Mediterranean areas', *Weed Sci.*, 61(3), 452–9.

Galon, L., Forte, C. T., Gabiatti, R. L., Radunz, L. L., Aspiazú, I., Kujawinski, R., David, F. A., Castoldi, C. T., Perin, G. F., Radunz, A. L. and Rossetti, J. (2016), 'Interference and economic threshold level for control of beggartick on bean cultivars', *Planta Daninha*, 34(3), 411–12.

Giambalvo, D., Ruisi, P., Saia, S., Di Miceli, G., Frenda, A. S. and Amato, G. (2012), 'Faba bean grain yield, N_2 fixation, and weed infestation in a long-term tillage experiment under rainfed Mediterranean conditions', *Plant Soil*, 360(1), 215–27.

Glasgow, J. L., Dicks, J. W. and Hodgson, D. R. (1976), 'Competition by, and chemical control of, natural weed populations in spring-sown field beans (*Vicia faba*)', *Ann. Appl. Biol.*, 84(2), 259–69.

Glaze, N. C., and Mullinex Jr., B. G. (1984), 'Competitive effects of sicklepod on lima beans', *Weed Sci.*, 32(1), 1–3.

Harker, K. N., Blackshaw, R. E. and Clayton, G. W. (2007), 'Wild oat (*Avena fatua*) vs. redstem filaree (*Erodium cicutarium*) interference in dry pea', *Weed Technol.*, 21(1), 235–40.

Heap, I. (2016), 'The International Survey of Herbicide Resistant Weeds', www.weedscience.org (accessed 27 December 2016).

Hewson, R. T., Roberts, H. A. and Bond, W. (1973), 'Weed competition in spring sown broad beans', *Hort. Res.*, 13(1), 25–32.

Holmes, R. C. and Sprague, C. L. (2013), 'Row width affects weed management in type II black bean', *Weed Technol.*, 27(3), 538–46.

Johnson, E. N. and Holm, F. A. (2009), 'Pre-emergence mechanical weed control in field pea (*Pisum sativum* L.)', *Can. J. Pl. Sci.*, 90(1), 133–8.

Kandel, H. (2013), *Dry Bean Production Guide*. NDSU Extension Service. Pub. A1133.

Kasasian, L. and Seeyave, J. (1969), 'Critical periods for weed competition', *PANS*, 15(2), 208–12.

Kok, H., Papendick, R. I. and Saxton, K. E. (2009), 'STEEP: Impact of long-term conservation farming research and education in Pacific Northwest wheatlands', *J. Soil Water Cons.*, 64(4), 253–64.

Liebman, M., Mohler, C. L. and Staver, C. P. (2001), *Ecological Management of Agricultural Weeds*, Cambridge, England: Cambridge University Press.

Martelloni, L., Frasconi, C., Fontanelli, M., Raffaelli, M. and Peruzzi, A. (2016) 'Mechanical weed control on small-size dry bean and its response to cross-flaming', *Span. J. Agric. Res.*, 14(1), e0203.

Mesbah, A. O., Miller, S. D. and Koetz, P. J. (2004), 'Common sunflower (*Helianthus annuus*) and green foxtail (*Setaria viridis*) interference in dry bean', *Weed Technol.*, 18(4), 902–7.

Mishra, J. S., Singh, V. P. and Yaduraju, N.T. (2006), 'Wild onion (*Asphodelus tenuifolius* Cav.) interference in lentil and chickpea crops and its management through competitive cropping', *Weed Biol. Manage.*, 6(1), 151–6.

Mohammadi, G., Javanshir, A., Khooie, F. R., Mohammadi, S. A. and Zehtab Salmasi, S. (2005), 'Critical period of weed interference in chickpea', *Weed Res.*, 45(1), 57–63.

Mortimer, A. M. (1984). 'Population ecology and weed science', in *Perspectives on Plant Population Ecology*, eds. Dirzo, R. and Sarukhan, J., 363–88. Sunderland, MA: Sinauer Pub.

Neary, P. E. and Majek, B. A. (1990) 'Common cocklebur (*Xanthium strumarium*) interference in snap beans (*Phaseolus vulgaris*)', *Weed Technol.*, 4(4), 743–8.

Nieto, J., Brando, M. A. and Gonzalez, J. T. (1968), 'Critical periods of the crop growth cycle for competition from weeds', *PANS*, 14(2), 159–66.
Norris, R.F. (2000), 'My view', *Weed Sci*, 48(3), 273.
Radicetti, E., Mancinelli, R. and Campiglia, E. (2012), 'The competitive ability of different chickpea (*Cicer arietinum*) genotypes against *Polygonum aviculare* under field conditions', *Crop Prot.*, 42, 312–19.
Rich, A. M. and Renner, K. A. (2009), 'Row spacing and seeding rate effects on eastern black nightshade (*Solanum ptycanthum*) and soybean', *Weed Technol.*, 21(1), 124–30.
Saberali, S. F., Modarres-Sanavy, S. A. M., Bannayan, M., Baghestani, M. A., Mashhadi, H. R. and Hoogenboom, G. (2012), 'Dry bean competitiveness with redroot pigweed as affected by growth habit and nitrogen rate', *Field Crops Res.*, 135(1), 38–45.
Santín-Montanya, M. I., Zambrana, E., Fernandez-Getino, A. P. and Tenorio, J. L. (2014), 'Dry pea (*Pisum sativum* L.) yielding and weed infestation response, under different tillage conditions', *Crop Prot.*, 65(1), 122–8.
Singh, M., Saxena, M. C., Abu-Irmaileh, B. E., Al-Thahabi, S. A. and Haddad, N. I. (1996), 'Estimation of critical period of weed control', *Weed Sci.*, 44(2), 273–83.
Smitchger, J. A., Burke, I. C. and Yenish, J. P. (2012), 'The critical period of weed control in lentil (*Lens culinaris*) in the Pacific Northwest', *Weed Sci.*, 60(1), 81–5.
Spies, J. M., Warkentin, T. D. and Shirtliffe, S. J. (2011), 'Variation in field pea (*Pisum sativum*) cultivars for basal branching and weed competition', *Weed Sci.*, 59(2), 218–23.
Sustainable Agriculture Network (2002), Steel in the field: a farmer's guide to weed management tools, ed. Bowman, G., Sustainable Agriculture Research and Education Program, CSREES, United States Department of Agriculture.
Swanton C. J. and Weise S. F. (1991), 'Integrated weed management: The rationale and approach', *Weed Technol.*, 5(3), 657–63.
Tanveer, A., Javaid, M. M., Irfan, M., Khaliq, A. and Yaseen, M. (2015), 'Yield losses in chickpea with varying densities of dragon spurge (*Euphorbia dracunculoides*)', *Weed Sci.*, 63(2), 522–8.
Tepe, I., Erman, M., Yergin, R. and Bükün, B. (2011), 'Critical period of weed control in chickpea under non-irrigated conditions', *Turk. J. Agric. For.*, 35(5), 525–34.
Urwin, C. P., Wilson, R. G. and Mortensen, D. A. (1996), 'Late season weed suppression from dry bean (*Phaseolus vulgaris*) cultivars', *Weed Technol.*, 10(4), 699–704.
USDA (2015), *Idaho Annual Statistical Bulletin*, National Agricultural Statistics Service, Northwest Regional Field Office. 64 p.
Vasilakogloua, I. and Dhima, K. (2012), 'Leafy and semi-leafless field pea competition with winter wild oat as affected by weed density', *Field Crops Res.*, 126(1), 130–6.
Vogt, G. A., Balbinot Jr., A. A., Trezzi, M. M. , Backes, R. L. and Nicknich, W. (2013), 'Competitive ability of black common bean genotypes with weeds', *Cienc. Agrotec. Lavras*, 37(5), 397–403.
Weaver, S. E., Kropff, M. J. and Groeneveld, R. M. W. (1992), 'Use of ecophysiological models for crop-weed interference: The critical period of weed interference', *Weed Sci.*, 40(2), 302–7.
Weaver, S. E. and Tan, C. S. (1983), 'Critical period of weed interference in transplanted tomatoes (*Lycopersicon esculentum*): Growth analysis', *Weed Sci.*, 31(4), 476–81.
William, R. D. and Warren, G. F. (1975), 'Competition between purple nutsedge (*Cyperus rotundus*) and vegetables', *Weed Sci.*, 23(4), 317–23.
Williams, C. F. (1971), 'Interaction of crop plant population with weed competition in com (*Zea mays* L.), bush and snap beans (*Phaseolus vulgaris* L.) and onion (*Allium cepa* L.) at differing stages of development', *Diss. Abstr.* Oregon State University. http://hdl.handle.net/1957/25588.
Wilson, J. B. (1988), 'Shoot competition and root competition', *J. Ecol.*, 25(1), 279–96.
Wilson, R. G. and Sbatella, G. M. (2014), 'Integrating irrigation, tillage, and herbicides for weed control in dry bean', *Weed Technol.*, 28(3), 479–85.
Yelverton, F. H. and Coble, H. D. (1991), 'Narrow row spacing and canopy formation reduces weed resurgence in soybeans (*Glycine max*)', *Weed Technol.*, 5(1), 169–74.
Zimdahl, R.L. (2004), '*Weed-Crop Competition, A Review*', 2nd ed. Blackwell Pub., Ames, IA USA.

Chapter 16

Grain legume storage in developing nations

L. L. Murdock and D. Baributsa, Purdue University, USA

1 Introduction

Grain legumes, also called pulses, are legumes cultivated for their seeds. These seeds include familiar peas, lentils and beans of many kinds, as well as unfamiliar ones, such as Bambara groundnuts. Since ancient days all have been cultivated for the nutrition they offer because pulse grain is rich in carbohydrates, B vitamins and proteins. To confer that nutritional benefit the grain must of course be on hand to eat, not only at harvest time but well beyond, and so it must be stored. Storage can last for weeks or many months, and continue until the grain is eventually consumed, sold, bartered away or ruined due to poor storage. Storage exposes pulse grains to multiple threats. These range from hard-to-cook defect that can develop to the depredations of rodents, moulds and perhaps most of all – at least for the resource-poor farmers of the undeveloped world – to insects.

This chapter is an overview only, and focuses on dry pulse storage with emphasis on (1) low-resource farmers in developing nations; (2) the problems they face during storage, mainly from insects; and (3) management tactics devised and applied thus far. Given the broadness of the topic, the limitations of space and the numerous subtopics, we make no attempt to exhaustively cite every published reference nor to review the use of fumigants, insecticides or advances in technologies in use in industrialized nations. References cited here are selected to be illustrative and representative.

http://dx.doi.org/10.19103/AS.2017.0023.13

2 Insect pests of stored grain legumes

It is not surprising that food that is nutritious for humans is also nutritious for insects. The truth is a large cache of nutritious grain in a storage bin can fuel a massive insect population explosion. Millions and millions of low-resource farmers know this all too well from having experienced it.

While there are many different insect pests of stored legume seeds, the most important are 'bruchids' or 'seed beetles'. For a long time these seed beetles were consigned to the family Bruchidae, but in recent years, systematists have reclassified them as a subfamily of the great Chrysomelidae group. There are about 1350 species of bruchids, mostly small beetles that spend the major part of their lives feeding and developing within their host legume seeds. Some bruchids are so closely tied to their major host seeds that their host plant name comprises part of their common name, for example, cowpea bruchid or weevil (*Callosobruchus maculatus* F.), common bean weevil (*Acanthoscelides obtectus* Say.), Mexican bean weevil (*Zabrotes subfasciatus* Boheman), adzuki bean weevil (*C. chinensis* L.) and peanut or groundnut bruchid or weevil (*Caryedon serratus* Olivier). Strictly speaking, bruchids are not weevils at all. They lack the characteristic snouts of weevils and are otherwise taxonomically distinct, but the name weevil has long been in common usage. Despite sometimes being named after its host legume seed, most bruchids can develop on multiple hosts. Thus, the cowpea weevil develops not only on cowpea but on black gram, green gram, lentil, pigeon pea, chickpea and garden pea as well as others.

For many pulse crops infestations begin in the field. Ovipositing females find developing pods of a host plant and lay their eggs on them (Sanon et al., 2005). Field infestations are typically quite small; one per cent or so of the grains brought in at harvest may have an egg on it or a larva in it. One of the characteristics of many bruchids is their rapid developmental times, which, depending on ambient temperature and food quality and quantity, can be as short as three to five weeks from hatched egg to egg-laying adult of the next generation. This, together with high female fecundity and the ability of some species not to diapause but to reproduce continuously, can result in population growth rates that are close to astonishing.

Several factors contribute to these storage insect population explosions. First of course is the fact that there is abundant stored food available to fuel an expanding insect population. Second, generation times for storage insects are short. For example, the cowpea bruchid (*C. maculatus* F.) can take less than a month to complete its life cycle. That is, it may require only about four weeks or less from the time a female deposits her egg on a cowpea seed until the female that developed from that egg begins depositing her eggs to start the next generation. The actual generation time can of course be longer or shorter depending on ambient temperature, grain moisture content and the quality of the grain. Third, individual female storage insect pests can, in a short time, produce large numbers of offspring, for example, up to 100 or more (Cabi, http://www.cabi.org/isc/datasheet/10987) for the cowpea bruchid. Fourth, each individual cowpea seed, the larval food, provides sufficient nutrition to allow as many as eight or ten insects to develop within it. As a result of all this, if the store is a large one, millions of insects can develop in it within a few months. Eventually, every seed in the store has multiple holes through which adult bruchids have emerged. Water and heat produced by the developing insects – metabolic water is their main water supply (Murdock et al., 2012) – lead to mould and decay. In the end the store becomes inedible, has no monetary value and is good, at most, for feeding to chickens or goats. The monetary loss from cowpea bruchid infestations is substantial.

If even one grain out of 100 has a single bruchid emergence hole, the market price is significantly discounted (Langyintuo et al., 2003).

The following calculation illustrates the potential explosive growth of populations. We start with a mated pair of cowpea bruchids that have abundant food available and lives in a favourable temperature and humidity. If we ignore larval mortality – which is often quite low – there will be, after only thirty days, some 100 adults, half of which are females. When yet another month passes, the total number of insects and seeds having emergence holes will amount to about 5000. After a third month the number of emergence holes in seeds approaches nearly a quarter of a million with the number of living individuals being only slightly smaller. The number of insects and emergence holes truly explodes in the following fourth month. By then there are literally millions of insects after only 120 days – but this can only happen, of course, when an adequate food supply is at hand and conditions are otherwise favourable.

The above is just a crude estimate, but it illustrates the potential high reproductive capacity. The practical impact of such high reproductive capacity is approximated in the following, which we do to make a practical point. Suppose you have a 100 L container filled with cowpea grain. This container will hold approximately 350 000 cowpea seeds. If initially clean, insect-free grain is infested with one adult pair of cowpea bruchids and one month passes, the grain will have a total of about 100 holes, assuming no egg or larval mortality. At the end of that first month, you would need to sort through thousands of seeds to find one with an emergence hole in it. Now, after another month passes, the number of holed grains will be about 5000. Even so, at this point only about one grain in 70 will have an emergence hole. Some damaged seeds can be found upon close inspection, but the number is still relatively small, with less than 1.5% of the seeds having emergence holes. Grain containing this level of damage might be discounted in the market, but not by much. Note that vendors in local markets often sort through their seeds to remove the damaged ones, thereby avoiding any discount. Only when the third month has passed and the third generation is completed will most of the seeds in the container have emergence holes, about 5 out of 7 of them. Such seed is well on its way to becoming ill-tasting, odorous and practically worthless, suitable mainly for feeding chickens. When the fourth month has passed, every seed in the store will have multiple holes and the grain will be worthless in the market and good only for animal feed.

The upshot of this is that timely intervention to protect grain that is put into storage is vital. If infested grain coming in from the field is protected against insect population growth before the third generation emerges, that is, during the first two or three months after harvest, damage and loss of value is minor and extensive serious damage is averted. The lesson here is that integrated pest management interventions that stop insect populations from growing should be applied within the first two or three months of harvest but certainly always as soon as possible. This is a time-window convenient for low-resource farmers, who are typically extremely busy at and shortly after harvest time. Subjecting their grain to management practices in the first two or three months or so after harvest can result in excellent control and prevent catastrophic losses.

3 Human need for grain legumes

About 90% of the world's 570 million farms are operated by individual farmers who rely mainly on family labour. About 84% of these family farms are quite small (one to two hectares) or tiny (less than one hectare). Not surprisingly, these diminutive farms have extremely limited financial capital. Likewise, they have little or no access to even minimal technology and they employ few inputs. Nevertheless they produce 80% of the world's food (FAO, 2014). These tough, hardworking but often-struggling farmers – a great many of them women – live precarious lives whose families depend for survival on their success as crop producers. In theory they can increase their food security and cash incomes by producing more food, but this is nearly impossible, given (i) limited access to information and technologies; (ii) their limited available land, which is often not very productive; (iii) their minimal marketing alternatives; (iv) their circumscribed family resources; and (v) lack of adequate government support programmes.

These folks can, however, make more food available for consumption or sale in another way, a way that does not require additional land: by safely storing what they already produce – that is, avoiding the post-harvest losses they inevitably suffer. Low-resource farmers who grow and store pulse crops are the focus of this chapter; to them it is dedicated.

4 The need for storage

Farmers store their grain after harvest for many reasons. The chief one is for food in the future, and of course every farm family needs to eat daily. But people in towns and cities depend upon these farms for food as well. Good storage makes food available for many months following harvest. A second reason for storing is the need for seed to plant the next growing season. Seed is obviously vital for the resource-poor farm family's life. If forced by circumstances to eat its seeds the family door flies open to future gnawing hunger and even eventual starvation. A third reason to store is as insurance, that is, the store serves as a hedge against famine. Famine can happen should the next crop fail for any of many reasons, whether it be family illness or injury, drought, flood, insects, crop disease or civil disorder. A fourth reason for storing is profit. Farmers have the option of obtaining cash by selling their grain. However, grain prices everywhere swing up and down according to supply and demand. At harvest time supply peaks and demand hits rock bottom. Grain prices accordingly drop to the lowest point of the year. As the weeks and months pass, however, the supply of grain dwindles because it is consumed, processed, sold or shipped away to distant towns and cities. Further, when stored poorly, it is often degraded in quality or even rendered worthless by insects, rodents and moulds. On the other hand, farmers and grain merchants who can store grain safely for long periods of time are in a position to profit from seasonal price swings because prices routinely increase markedly, in some years even doubling or more, in the months following harvest (de Beurs and Brown, 2013).

Storing grain safe against loss, whether to insects, moulds, rodents or thieves, is as good as having money in the bank. In one way it is even better: you can eat it.

5 Helping farmers with limited resources

Over the years thousands of scientists have given their time and creative energy trying to devise technologies that low-resource small-scale farmers can use to minimize or even stop losses of their grain to storage insects. The vast majority of their work has been carefully done and responsibly reported. Regrettably, all too rarely those scientists whose hearts were set on helping low-resource farmers have gone about their well-intentioned work unaware of how surprisingly difficult it is to create new technology or new processes that farmers can afford and will actually adopt and use sustainably. We gained our experience at Purdue University while developing practical and useful technologies for post-harvest cowpea storage by low-resource farmers in northern Cameroon (Murdock et al., 2003). Our criterion for success was not the number of research grants we obtained or invited talks we gave or the number of our published papers. Our measure of success was the degree of adoption of our technologies by the low-resource farmers we sought to help. In short, our idea was that if our technologies were adopted, we succeeded. If not, we failed. The greater the adoption the greater our success.

We slowly learnt that to be adopted by resource-poor farmers, a storage technology must have numerous properties. Lacking even one of them, we learnt, increases the chances of failure. Listed here in no particular order are, based on our experience, the required properties of storage technologies that small-scale low-resource farmers in developing nations can and will adopt. They must:

- protect the stored grain effectively;
- be culturally acceptable to the farmers;
- use readily available materials – or materials that can be made available;
- be easy to teach and learn;
- avoid use of chemicals – farmers everywhere dislike handling chemicals;
- be developed with input from end users – the low-resource farmers;
- require minimal labour;
- be low-cost, affordable, accessible and profitable;
- not affect taste or processing qualities of the stored grain;
- be subsidy free or at least not require subsidies to be sustained;
- not affect germination of seeds;
- not promote moulds and mycotoxins;
- address a problem that end users are aware of and want to solve.

It is not easy to create technologies that meet all the above criteria. Furthermore, if successful adoption by large numbers of people over large areas is the impact sought, good technology alone is insufficient. This point is so important that it bears saying again: technology that works and is not yet accepted by end users is NOT enough. It is merely a beginning. Your job is half done, at best, and your technology will remain on the shelf – not reaching and being adopted by end users – until you take additional and indispensable steps.

First, large numbers of would-be users need to be made aware of the technology – awareness raising. This can take the form of focused training events, radio messages, TV spots, posters, public demonstrations, booklets and cell phone videos. Second, the supply chain required to make the technology available to end users has to be created; it does not, as some seem to assume, fall out of the sky just when it is needed. If the technology requires a component to be made and distributed, manufacturers have to be identified, made aware of the potential demand and engaged. Third, a distribution system has to be put in place. Regional distributors have to be identified, and a network of local vendors has to be recruited. Fourth, any required capital – hopefully minimal – has to be found because low-resource farmers rarely have much capital available to them. Fifth, partners have to be engaged, whether with international agriculture research centres, national programmes, NGOs, private charitable foundations, individual businesses or with entrepreneurs. These partners may be motivated by humanitarian goals that can be met by (1) providing funding and by (2) actively serving needy humanity by helping disseminate the new technology in some way. Other partners may be motivated primarily by the prospect of profits. Ultimately, the technology needs to be sustainable – that is, it must continue to be available, distributed and used widely once the outside sources of support disappear, for they will inevitably do so.

The above path to sustainability has been followed by the Purdue Improved Crop Storage (PICS) technology (cf. Baributsa et al., 2010), a storage system using hermetic triple bagging now widely adopted in parts of Africa. PICS seeks to ensure that all of the actors in the value chain – manufacturers, transporters, regional distributors, local vendors and end users – all make a profit from the technology. Profits sustain.

6 Historic grain legume storage methods

6.1 The earliest granaries

The technological capacity to store grain safely after harvest is one of the foundations on which settled agriculture rests. The first purpose-built granaries appeared about 10 000 years ago in the Jordan River Valley. Kuijt and Finlayson (2009) present evidence that these first granaries were used for wild oats and wild barley rather than for domesticated grain. To store grain they gathered, these early people built granaries of stone, mud and clay outside their dwellings. With the passage of time they began to build granaries inside and to use them increasingly to store cultivated grain. As centuries passed from the Natufian period (14 500–11 500 BP) into the Neolithic (12 000–6 000 BP), the granaries not only moved inside, they also became larger. A typical store contained three or four times the food needs of the family or group. Keeping this amount of stored food, which was well in excess of needs, was probably deliberate. It took into account losses due to spoilage and served as insurance against future yield failure (Kuijt, 2015). It may well have served as a sort of grain bank account as well.

6.2 Storage in underground pits

As pointed out earlier, grain coming in from the field often carries small numbers of insects that begin infesting the grain when it is still in the field. Brought into the granary or storage site and now living in the midst of a concentrated source of food in a sheltered situation and thanks to their short reproductive cycles, the initially low number of insects expands exponentially and this population explosion can destroy the store in a matter of a few months. Explosive insect population growth does not occur in airtight sealed pits. This is because insects exposed to low levels of oxygen in the pits stop feeding, growing and reproducing and thus their population growth is arrested before their numbers begin expanding and before serious damage is done. Underground pits have been used throughout the world for grain storage for thousands of years, from Japan and China to New Zealand to Africa, the Middle East and pre-Roman Britain.

Iron Age storage technology already made use of underground storage in pits (Hill et al., 1983; Currid and Navon, 1989). In Celtic Britain these pits were common; they consisted of cylindrical holes in the ground 1 m or so in diameter and 1.8 m deep. They were lined with wicker or stone, filled with grain, then covered with a thick layer of clay or marl (Laing, 1981). Other pit shapes are known from other places; bottle-shaped, beehive-shaped and cylindrical pits have been found on the Jordan River in Palestine (Currid and Navon, 1989). The result was a sealed, substantially airtight container. The composition of the gas in the enclosed airspace began changing soon after the container was sealed. Carbon dioxide levels rose and oxygen levels fell thanks to oxidative metabolism of the grain and the germination of grain that had become wetted by water seeping in at the sides of the container. Added to this – and in pits that remained dry probably supplanting it – was oxidative metabolism of insects initially present.

The Romans of classical times were familiar with grain storage in pits. Indeed, Pliny the Elder (23 AD – 79 AD) remarked on their use: 'Particular care is taken to dig these trenches in a dry soil, and a layer of chaff is then placed at the bottom and the grain, too, is always stored in the ear. In this case, if no air is allowed to penetrate to the corn, we may rest assured that no noxious insects will ever breed in it' (Pliny, 79 AD). A century before Pliny, Marcus Varro (116 BC–27 BC), in his manual of agriculture, pointed out that insects require fresh air to grow and develop 'for the weevil does not breed where air does not reach' (Varro, bef. 27 BC). He pointed out that the air in newly opened pits is dangerous to breathe and can cause death: 'Those who keep their grain under ground in the pits … should remove the grain some time after the pits are opened, as it is dangerous to enter them immediately, some people having been suffocated while doing so.'

6.3 Storage in unthreshed pods

Seeds produced by maturing legume plants are enclosed and nourished by a pod wall. This wall not only helps nourish the seeds, but also protects them against various dangers and enemies like birds, viruses, bacteria, fungi and desiccation. The pod wall likewise helps protect the enclosed seeds against insect attack, not only during seed development in the field but after harvest as well. Therefore, it plays quite an important role. Farmers in northern Cameroon and other areas in Africa often store their cowpeas in pod form for up to several months after harvest because it helps protect the grain against bruchids. When seeds are stored in pod form, the female bruchid cannot deposit her eggs directly on the seeds, but necessarily must place her eggs on the pod wall. When an egg hatches, the tiny larva must penetrate through the wall and somehow bridge the intervening space between the seed inside and the wall and find the leverage it needs to bite its way into the seed. Only about one in three to one in five eggs laid on the pod wall layer – depending on the cultivar – emerge as adults. Mortality, though partly because of egg parasitism, is substantially due to the physical challenge of penetration, and is at this stage quite high (Sanon et al., 2005). Pods are sometimes kept on pole stands in the open (Fig. 1) or held in mud-walled granaries (Fig. 2).

Recognizing the ancient custom of farmers using pod storage after harvest in the cowpea-growing regions of West Africa, Kitch et al. (1991) sought to take advantage of it. Their idea was to breed cowpea lines that had pod wall resistance combined together with the seed resistance of TVu 2027 (see below), thereby creating a cultivar with a high degree of durable seed and pod wall resistance. They examined a wide range of pods from the world cowpea germplasm collection and discovered that the different pods varied widely in their ability to resist penetration of hatchling larvae through the pod wall. The least resistant among them cause 57% failure of hatchling larvae to penetrate and the most resistant had a failure rate of over 99%. A breeding initiative was then launched that led to

Figure 1 In many places in West Africa cowpeas are stored in pod form. The pod wall itself provides some degree of protection against insects by acting as a barrier against access to the seeds inside, both to ovipositing females and to hatchling bruchid larvae. Photo by D. Baributsa.

Figure 2 Mud pot or mud wall granaries for storing family-produced harvested grain in Africa come in many different forms some of them almost artistic, others very functional. The granaries shown here are often compartmentalized and used for storing cowpeas and groundnuts in pod form, as well as sorghum, millet and maize. Photos by D. Baributsa.

the development of cowpea cultivars that were optimally storable in pod form by virtue of expressing both seed and pod wall resistance. Cultivar Lori-Niebe, which was developed and released in 1999 for northern Cameroon farmers, was subsequently widely grown in Chad and Ghana (Olufuwote and Barnes-McConnell, 2002) and has experienced an adoption rate of 18% in Cameroon (Walker and Alwang, 2015).

Storage in pod form is well known to confer a degree of benefit to groundnut seeds as regards protection against insects (Ranga Rao et al., 2010).

7 Other grain legume storage methods

7.1 Storage using botanicals

One storage technology used by the ancients and still often used on many of today's low-resource farms – and one of the most attractive research approaches taken by many who want to help resource-poor farmers – is storing grain mixed in some way with botanical materials (Golob and Webley, 1980). These materials supposedly fumigate the airspaces of the store, repel insects or kill them by contact.

The list of plant materials that have been used by farmers or tried by researchers is extremely long. We here touch only on a few examples. Often, plant parts that have an aromatic quality or are known to be toxic to insects are used. Sometimes the plant material is layered alternatively with the grain, as we (personal observation) have seen farmers doing it with a native mint in northern Cameroon. Other times the plant material is dried, pounded into powder and sprinkled on or mixed with the grain that needs protection against insects.

One variation of this botanical approach involves making extracts of the chosen plant materials and treating the grain with them. Recent work (de Andrade Dutra et al., 2016) demonstrates that citrus peels contain chemicals that are toxic and repellant to the cowpea bruchid (*C. maculatus* F.). The authors' claim that oils derived from citrus peels might be used for control of the bruchids is true, but there are many barriers standing in the way of this becoming a reality. Similar studies with the cowpea bruchid and cowpeas show that essential oils from a variety of African plants might in theory find uses in the control of pulse seed beetles in storage (Ibaoudo et al., 2015). Other botanicals have been shown to provide a degree of protection against insects on various pulses, including neem leaf, garlic and eucalyptus powders mixed with black gram attacked by *C. maculatus* (Suthar and Barpoda, 2015).

Practical questions arise, however – not only about essential oils but about other botanicals as well: Where would low-resource farmers obtain the required essential oils, extracts or botanicals? How would they learn how to use them? Would treating cowpea grain with essential oils change the taste or flavour of foods prepared from grain? What would the monetary cost be? Might the essential oils or botanicals be toxic or allergenic to people? Some farmers have a 50 or 100 kg of pulse grain to store, others hundreds of kgs – how much labour would be involved in treating such large amounts of grain? For that matter, how much does the content of the active principles in the botanical materials vary? Might it not change from location to location, season to season and year to year – and might not the control based on it also vary? Have human toxicity studies been carried out? Some plant chemicals highly toxic to insects are also highly toxic to humans, for example, nicotine. Such studies as cited here and hundreds of others can serve as the starting point for investigations that may in theory lead to the identification of the active principle. Unfortunately, such studies are a moonshot away from becoming practical and useful control technologies that can be adopted by millions of farmers. Appealing and popular as this approach is because it is in some sense 'natural', it is severely limited.

7.2 Storage with ash or sand

Among the more common storage methods used by farmers in developing nations, especially for high-value commodities such as common beans and cowpeas, is co-storage with ash and sand (Golob and Webley, 1980; Ofuya, 1986). Our surveys found that this method was in use most frequently among cowpea farmers in northern Cameroon during the late 1980s (Wolfson et al., 1991). The proportion of ash or sand used relative to the amounts of grain varied widely from farmer to farmer. Some merely dusted their cowpeas, while others mixed roughly equal quantities of ash or dust and grain. Some believed strongly in the efficacy of the procedure, others merely thought it was better than doing nothing. Systematic studies (Wolfson et al., 1991; Baoua et al., 2012) proved that both sand and sieved wood ash have protective effects. Experiments showed that with wood ash the effective ash/grain ratio is three parts ash to four parts grain. The recommendation that emerged from this work was that farmers should store equal parts grain and ash (or sand), and that the mixture be stored in closed containers like the clay jars called canneries that are commonly used for water storage in the region. If the ash or sand/grain mixture in the jar is covered with a 2 or 3 cm layer of the protectant, the store can be kept for long periods of time without further loss to bruchids. One of the negative aspects of this technology is that while it prevents the bruchid population from growing, it does not stop larvae already inside the seeds from developing. These larvae mature and can make pupal cells and even cut emergence holes in the testae of the seeds. The insects usually fail to escape from the seed because they are physically blocked from doing so. The presence of pupal windows and cut emergence holes in seeds that initially showed no such holes lead some farmers to doubt the efficacy of the technology, but in fact the infestation has been arrested.

The mode of action of ash, sand and other granular materials such as dust is that they act as physical barriers. Even if an emerging adult begins to cut its way out of the seed in which it developed, it is physically trapped and so cannot move, find a mate and oviposition sites. Population growth ceases.

Without exception, all technologies have a downside or limitations, often multiple ones. Storage with ash, for example, has more than one serious limitation. Using it is practical if you have only a few kgs of grain to store, but what if you need to store hundreds of kgs? The amount of ash required in that case is impractically large. Using ash and sand has another serious limitation. You have to remove it by sifting and washing when it is time to cook it or sell it. In regions of low rainfall where water is an extremely valuable resource that often must be carried over kilometres to the farmstead, washing grain is a costly and laborious exercise. For storage of small amounts of grain for planting, ash and sand is practical and useful.

7.3 Treatment with oil

A surprisingly simple way to control cowpea bruchids attacking cowpea grain is to mix small amounts of natural oils with the cowpea grain. Coating a kilogram of grain with only 5–8 mL of groundnut oil, castor oil, coconut oil and palm kernel oil stopped an infestation of cowpea bruchids (Mital, 1971; Singh et al., 1978) and this worked well, too, for common beans attacked by *Zabrotes subfasciatus* (Schoonhoven, 1978). Oils from a variety of sources are effective, including the above-mentioned ones as well as neem kernel, karite and Bambara (Pereira, 1983). Oil coating has been shown to work against a variety of bruchids (*Zabrotes subfasciatus, Acanthoscelides obtectus* in addition to *C. maculatus*) and with stored pulses other than cowpeas, including chickpeas, pigeon peas and common beans. Hill and Schoonhoven (1981) mustered evidence that the effective component of vegetable oils was oleic acid and fractions containing triglycerides. Details about the use of oils in grain protection are available (Singh et al., 2012).

While oil treatment of pulse grain is simple and inexpensive, can make use of a large variety of oils and is effective in controlling bruchids, it has limitations. The principal one is that it is suited for only small amounts of grain. A farmer who has hundreds of kg of pulse grain to store safely and who wants to use oil treatments is confronted with the need to thoroughly mix this large amount of grain with oil. This is a huge physical constraint. Out of the hundreds of farmers we surveyed in Cameroon to learn about how they stored their grain, only one person used oil treatment for only a small amount of grain. Oil coating may have disadvantages including the imparting of odours or flavour to the grain, and may affect cooking properties by changing the rate of hydration during food preparation.

7.4 Rolling beans

One particularly clever technique for controlling infestations of the common bean weevil *Acanthoscelides obtectus* is to exploit the peculiar biology of this storage pest. Unlike the cowpea bruchid, which glues its eggs to the testae of cowpea seeds, the ovipositing female *A. obtectus* scatters her eggs loosely among the grain. The tiny newly hatching larvae then crawl around among the beans until they find points of close contact between beans and another fixed surface, be it another bean or the wall of the container. Using the leverage afforded by firmly apposed surfaces, they begin chewing their way through the relatively thick testae of one of the beans. It can take 24 h or more to penetrate into the nutrient-rich cotyledon. A close study of this behaviour led Jim Miller's group at Michigan State University to a bright idea: if the positions of the beans relative to one another were shifted even slightly, the tiny first instar larvae would have to start all over again. If disrupted again and again, the larvae would never succeed in penetrating into the nutritious part of the seed. It would eventually use up its limited energy reserves and die of exhaustion. Indeed, the Michigan State group showed that rolling a variety of containers partially filled with beans and containing plenty of *A. obtectus* eggs reduced bruchid damage by 97% relative to the unrolled control (Spencer et al., 1991; Quentin et al., 1991).

On the basis of this, authors suggested that low-resource farmers in developing nations could use the rolling technique to prevent *A. obtectus* infestations in their beans. All they would need to do was fill cylindrical storage containers, whether jars, jute bags or barrels three-fourths full of beans, and roll them a couple of circumferences regularly, twice a day, morning and evening. They pointed out that this technique required no new investment or materials and would be easy to learn and do. At the same time they wrote: 'It remains to be seen whether this tactic of regular tumbling will prove widely useful in control of storage pests of other pulses and grain' (Quentin et al., 1991). In fact, the technology has not found much application in the 25 years since it was first described. One big reason is that farmers in the bean-growing areas of the world where *A. obtectus* is the primary storage problem simply are not aware of the technology, nor are various extension services that might bring awareness of it to them. Another is that the biology of many bruchids attacking pulse grain is different from that of *A. obtectus*; the females glue their eggs to the surfaces of the seeds. Tumbling does not in those cases prevent the larvae from boring into the seed. It may well be that, simple and clever as the tumbling procedure is, it is too much to expect an often-busy farm family, especially during the harvest season, to twice daily and systematically carry out a control operation when a single neglect of duty could lead to serious damage to their grain.

7.5 Hermetic storage

Hermetic storage in airtight containers is one version of a control approach called modified atmospheres. The term 'modified atmospheres' refers to changing the composition of the gases surrounding the grain, for example, replacing the oxygen in the gas with nitrogen, elevating the level of carbon dioxide at the expense of oxygen or nitrogen and so on. This approach of artificially changing the gas composition has found application in developed nations and has been well reviewed (Navarro, 2012; Jayas and Jeyamcondan, 2002). As Navarro and others recognized long ago, storage of infested grain in hermetic containers is characterized by biologically generated changes in the composition of the gases held within the container and thus is a special form of modified atmospheres.

Storage of grain in airtight containers has a long history reaching back to the use of underground pits (see above). In recent years, there has been a resurgence of interest in airtight storage using modern materials. When 50 tons of barley grain was stored for many weeks in a polyethylene-lined underground pit, oxygen levels fell and carbon dioxide levels rose. Insect population growth was arrested and germination of the grain maintained over many months (Donahaye et al., 1967). The combination of underground storage and the polyethylene liner made the store effectively air tight. Rats chewing through the polyethylene were the only serious problem encountered.

More than 40 years ago, Akpaetok (1974) described the storage of cowpea grain together with desiccants in airtight metal cans, but the purpose of the hermetic container was to prevent moisture gain or loss rather than control insects. Still earlier, Ozburn et al. (1960) found that when sorghum grain at 10% moisture was stored for up to 30 months at Mokwa, Nigeria, in sealed 44 gallon metal drums, germination was unaffected, moisture content was unchanged and insect damage was insignificant. The authors suggested that the drum storage technique – effectively hermetic storage – could be safely used to keep sorghum or maize grain for food or for planting. Such drums indeed proved good for cowpea storage as well. In Senegal, metal drums are widely used for shipping vegetable oils and so are widely available. Widespread use of horse-drawn carts makes it easy for drums to be moved around. Taking advantage of these facts, Seck and Gaspar (1992) demonstrated that sealed drums make excellent hermetic containers for post-harvest storage of cowpea grain. Control of the bruchids was excellent; this fact and the availability of the drums led to widespread adoption of the technology in Senegal. The drum technology came into widespread use for cowpea storage there, thanks to the work of Dogo Seck and Tony Hall of the ISRA/University of California-Riverside Bean/Cowpea Collaborative Research Support Program (cf. Murdock et al., 2003) supported by USAID. Indeed, adoption of drum storage reached 57.5% of Senegalese cowpea production in 1996, although by 2004 the percentage had started to decline (Boys et al., 2007), evidently because the useful life of the drums as hermetic containers had begun to be reached, thanks to rust, cracks and wear. The cost of replacing drums was a negative factor.

Among the earliest to use hermetic plastic bags was O'Dowd (1971) working in northern Nigeria. When cowpea grain was stored in single polyethylene containers, the cowpea bruchids were able to penetrate the plastic, but if the plastic enclosed an inner cotton bag, control was excellent. Oxygen levels fell steeply and quickly, while carbon dioxide levels rose dramatically. Commenting on this work, Caswell (1975) was sceptical about farmers accepting this technology. They were, he said, reluctant to thresh at harvest because it was such a busy time and the use of traditional mud pot granaries to store unthreshed cowpea pods was ingrained. However, he asserted that consumers like the plastic/cotton bag approach because it allowed them to store for long periods after harvest without experiencing loss or degradation.

Storage of legume grain in plastic bags that act as hermetic containers thus has a history going back nearly half a century. Since then, two major versions of this approach have been developed and are now available to various degrees in markets in developing nations. One of these, the PICS bags was developed in the 1980s, which was the output of a collaboration between Purdue University entomologists and Cameroon National Programs collaborators (Murdock et al., 2003). The PICS bags consist of two, separate inner liners of 80-micron thick polyethylene fitted inside a woven polypropylene bag. Developed initially for cowpea storage, the inner bag is filled with the grain and tied shut. The second inner liner is then tied shut, completely surrounding the first bag. Both sealed containers are then enclosed by the woven polypropylene bag, which is also tied shut. Sealed in this way the grain inside the bag suffers virtually no insect damage and can be stored safe from insect damage for months or even years. Control is effective because the insects within the bag initially respire normally and continue feeding and developing. But their respiration depletes the oxygen and raises the level of CO_2 inside the bag. As the oxygen level falls, the insects feed at a steadily lower rate until, when oxygen levels reach a few per cent, they essentially cease feeding. The rise in CO_2 has, by contrast, little effect on the feeding rate, at least during short exposures (Murdock et al., 2012). No longer feeding, the insects do not grow and mature to adulthood and so they do not reproduce. The net result is that the bruchid population fails to grow. Since insects have very high metabolic rates for their size, a relatively small number of insects can use up the oxygen in PICS bags quickly, arresting the infestation quickly. Since the walls of the bag are practically – though not perfectly – impermeable to oxygen, little or no oxygen can pass quickly into the bag fast enough to replace that used by the insects.

In addition to PICS bags, the commercial company GrainPro developed a trademarked single-layer storage bag advertised as owing its effectiveness to an 'Ultra High Barrier Specialty Blend of Polymers and LDPE.' The single multi-laminate layer is extremely impermeable to oxygen and performed well as a hermetic container. However, the high degree of O_2 impermeability of the single layer does not result in better performance than does the triple-layer PICS bag. A side-by-side comparison of GrainPro SuperGrain bags with PICS bags showed that while attained O_2 levels in SuperGrain bags were slightly lower than in PICS bags, preservation of cowpea grain was no better in the high-tech bag than in the low-tech PICS bags (Baoua et al., 2013).

Every legume stored in PICS bags that has been tested thus far showed excellent preservation against insect loss, whether cowpeas (Baoua et al., 2012), common bean (Mutungi et al., 2015), mung bean (Mutungi et al., 2014), pigeon peas (Vales et al., 2014; Mutungi et al., 2014), Bambara groundnuts (Baoua et al., 2014) and peanuts (Sudini et al., 2014). PICS bags have begun to find wide application in Africa, some ten million of them having been sold to date (Fig. 3).

Figure 3 Triple-layer bags now called Purdue Improved Crop Storage (PICS) bags have begun to be widely adopted by low-resource farmers in Africa. They owe their effectiveness to their ability to prevent insects living in the grain put in the bag from obtaining life-essential oxygen. Farmers are enthusiastically adopting them. Some 10 million have been sold thus far. Photos by D. Baributsa.

8 Use of resistant seed

It has long been known that some legume seed accessions are more resistant to attack by storage insects than are others. These differences in resistance have been used by plant scientists in two different ways in attempts to improve the storability of pulse grain. One is to use selective breeding to bring the resistance into cultivars with desirable agronomic properties. The other is to identify a gene or genes conferring resistance and move the responsible gene(s) into desirable genotypes by molecular gene transfer, that is, genetic engineering. Both approaches have had some degree of success. The use of seed-borne resistance to storage pests in legumes has been thoroughly reviewed in the last few years (Keneni et al., 2011; Swapan, 2016), so here we give only a brief sampling.

8.1 Traditional breeding and screening to introduce resistance

Scientists at the International Institute of Tropical Agriculture (IITA) identified landraces of cowpea with substantial resistance to the cowpea bruchid, *C. maculatus* (Singh et al., 1988). Most attention focused on accession TVu 2027. While it was not immune to the bruchid, its seeds caused increased mortality and delayed development of the insects, increasing the storability of grain carrying the resistance. The physical or molecular basis of the resistance has not been conclusively identified. Scientists at the University of Durham in England reported that TVu 2027 had an elevated level of trypsin inhibitor, about twice as much, compared to seeds of more susceptible lines (Gatehouse et al., 1979). They further claimed that (1) protein fractions enriched in trypsin inhibitors were toxic to cowpea weevil larvae when fed in an artificial diet, (2) toxicity was absent when the trypsin inhibitor was removed and (3) feeding purified trypsin inhibitor in artificial diets was toxic to the insects. This was an attractive story indeed, but research by others did not support it. Two major findings militated against the trypsin inhibitor hypothesis. First, when highly purified cowpea trypsin inhibitor was fed to cowpea weevil larvae in an artificial seed system, levels up to 2% by weight had no significant effect on the growth, development and survival of the insects versus control artificial seeds lacking the trypsin inhibitor (Zhu et al., 1994). Second, the cowpea weevil larvae that feed on the cowpea grain surprisingly do not utilize trypsin or chymotrypsin-like serine proteases to digest their dietary protein. They use instead mainly cysteine-type proteases (Gatehouse et al., 1985; Kitch and Murdock, 1986) and to a lesser degree acid proteases. In short, the assumed target of the cowpea trypsin inhibitor, a gut digestive serine protease, is substantially lacking in this insect. Whatever the basis of resistance in TVu 2027, IITA plant breeders have transferred the resistance into elite lines that are less photosensitive, higher-yielding and virus-, disease- and aphid-resistant, and the benefits of additional storability have begun to reach African farmers and consumers.

Many sources of naturally occurring resistance to bruchid storage pests have been described (Swapan, 2016) in other legumes including mung bean/green gram (*Vigna radiata* var. *Sublobata*; Vir et al., 2016), chickpea wild relatives (*Cicer* spp.; Singh et al., 1998), pigeon pea (*Cajanus cajan*; Sommapa et al., 2012), lentils (*Lens culinaris* Medikus; Sarwar, 2013), dry peas (*Pisum sativum*, Nikolova, 2016), common bean (*Phaseolus vulgaris* (L.); Lopes et al., 2016) and rice bean (*Vigna umbellata*, Kashiwaba et al., 2003).

8.2 Directed molecular gene transfer

In the mid-1980s, it became clear that it is possible to impart insect resistance to plants by transferring a resistance gene into the genome of the plant, even using a gene from a different species of organism. Transfer of a gene encoding a *Bacillus thuringiensis* protein into tobacco plants rendered the plants highly toxic to the tobacco hornworm, *Manduca sexta* (Vaeck et al., 1987), and the resistance gene was passed on to the next generation of plant. Nine years later, in 1996, the first commercial transgenic maize plants expressing the Bt toxin were released by Monsanto Co., to combat the European corn borer (cf. Hellmich and Hellmich, 2012). The era of genetically engineered agriculture had begun. This exciting development led researchers to search for other genes that could be used to impart needed insect resistance to agronomically important crops. While the main focus for this effort was on field crops, a group at Purdue University participated in efforts to impart insect resistance to seeds stored after harvest. Initial focus was on cowpea grain attacked by the cowpea bruchid, but in the late 1980s and early 1990s it was impossible to genetically transform cowpea. Purdue scientists working with Maarten Chrispeels at the University of California San Diego and T. J. V. Higgins at CSIRO Plant Industry in Australia managed to transfer a gene imparting cowpea weevil resistance into field peas, *Pisum sativum*. The gene chosen encoded the α-amylase inhibitor from common bean, *P. vulgaris*. This gene came to light because *C. maculatus* larvae cannot survive on common bean, the reason being that the α-amylase inhibitor present at about 0.7% (w/w) in common bean inhibits starch digestion by the larvae and prevents them from obtaining the bulk carbohydrate nutrition they require. *P. sativum* expressing the α-amylase inhibitor from common bean were immune from attack by the cowpea bruchid (Shade et al., 1994). This was the world's first case on the use of gene transfer to impart insect resistance in seeds.

Subsequently, seeds of several legume species have been transformed with the α-amylase inhibitor from common bean and shown to thereby become highly resistant to bruchids (Rubiales et al., 2015). Thus far, pea (Schroeder et al., 1995), azuki bean (Ishimoto et al., 1996), chickpea (Sarmah et al., 2004) and cowpea (Solleti et al., 2008) have been given insect resistance via gene transfer.

8.3 RNAi

There are newer molecular-based alternatives to gene transfer to impart resistance to insects. RNA interference technology, commonly referred to as 'RNAi' and as 'gene silencing', has been developed over the last two decades (cf. Gordon and Waterhouse, 2007). This molecular technique opens a new and promising door to insect management and could lead to a new way of protecting stored products against insects, including stored legume seeds. RNAi technology is built on the concept of exposing an insect to double-stranded RNA (dsRNA) that targets a particular life-important gene in the insect and prevents it from acting by specifically destroying its RNA message. This will eventually kill the insect or at least interfere with its normal health, behaviour, growth, development and reproduction. To effect insect control with RNAi, there are three essential problems. First, one needs to identify an appropriate target gene. Second, an effective dsRNA with the potential to interfere with that key gene action has to be found and third, this dsRNA somehow needs to be brought into the insect body such that it reaches its target cell and RNA. One approach to dsRNA delivery is to transform the plant you want to protect such that it expresses and accumulates the disruptive dsRNA. When the pest insect comes along and consumes the plant, it ingests plant tissues containing the dsRNA, which then enters the insect body and acts to prevent the key gene action from being properly expressed. The consequence is that the insect may be killed or at least becomes incapable of growing and reproducing normally. One dsRNA delivery approach that has been advocated for stored products like pulse seeds would be to spray it on the seeds. Still a third way would be to spray the dsRNA on the insects directly such that insects would absorb or ingest the dsRNA and eventually suffer disruption of the key target gene action. This is a rapidly moving field and to cover it in detail is beyond the scope of this overview of pulse seed storage. However, an overview of the field and its potential is available, which provides an excellent summary of the state of the art, specifically as it relates to stored products (Perkin et al., 2016).

8.4 CRISPR

A still newer approach to insect control by molecular interventions is CRISPR technology (CRISPR stands for Clustered Regularly Interspaced Palindromic Repeats). CRISPR sequences, working with CRISPR-associated protein 9 (Cas9), can eliminate a gene or block it from acting. The potential of the infant CRISPR technology and its applications to stored products' insect management has been outlined recently (Perkin et al., 2016).

It remains to be seen whether transgenic pest-resistant legume seeds or dsRNA- or CRISPR-modified legume seeds will ever find widespread application. On the one hand, it would seem harmless enough to move a gene from one widely consumed food legume, common bean, into another food legume such as cowpea or chickpea. Given that these foods are typically cooked and cooking denatures the proteinaceous α-amylase inhibitor, rendering it inactive, it is hard to imagine that they might be in any way threatening to the health of the consumer. In fact, a continuing controversy has arisen as to whether transgenic bean α-amylase inhibitor-expressing peas, chickpeas and cowpeas are actually allergenic. There is indeed a degree of immunogenicity, but this may be the result of cross-reactivity when plant proteins are fed to experimental mice. It is likely that similar or new objections may be raised against dsRNA and CRISPR-modified plant seeds that resist insects. Every technology has a downside, and there are no exceptions. The question is always, what is the cost and what is the benefit?

9 Conclusion

Given that the population of the earth is likely to increase by nearly three billion souls in just one more generation, it is vital that humanity comes to grip with this challenge of post-harvest storage. Saving food from post-harvest losses is equivalent to increasing food production in the field. Certainly, the people who need more and better food the most are those who live in the poorest regions of the world.

10 Where to look for further information

Numerous sources provide additional information and perspectives on postharvest grain storage, including legumes, in developing nations. Godfray et al. (2010), provide a global perspective on the need to provide food for billions more people in the coming years. Kumar and Kalita (2017) present an up-to-date-overview of the storage problem in developing nations. Detailed information about the use of PICS bags, including extension bulletins, instructional videos, posters, etc. is available: https://picsnetwork.org/. A special issue on hermetic storage of grain in developing nations has been published in the *Journal of Stored Products Research*, Vol. 58, July 2014. FAO *Agricultural Services Bulletin* No. 93 (http://www.fao.org/docrep/t0522e/T0522E04.htm) by M. deLucia and D. Asenato offers an overviewof the storage problem and approaches to dealing with it.

Godfray, H. J. C., Beddington, J. R., Crute, I. R., Hadad, L., Lawrence, D., Muir, J. F., Pretty, J., Robinson, S., Thomas, S. M. and Toulmin, C. (2010). Food security: the challenge of feeding 9 billion people. *Science* 327: 812–18.

Varro, M. (1918; original Latin 47 A.D.). Roman Farm Management: The Treatises of Cato and Varro, In Virginia, F. A. (ed.), p. 172. MacMillan Company, New York. http://penelope.uchicago.edu/Thayer/e/roman/texts/varro/de_re_rustica/1*.html

Pliny the Elder (1856). *The Natural History of Pliny*, vol. 4, Bostock, J. and Riley, H. T. (trans.), Henry J. Bohn, London. https://books.google.com/books/about/The_Natural_History_of_Pliny.html?id=lUoMAAAAIAAJ

11 References

Akpaetok, O. I. (1974). Drying and storage of cowpeas in air-tight containers. *J. Agr. Eng. Res.* 19:279–87.

Andrade Dutra, K., Oliveria, J. V., Amaral Ferraz Navarro, D. M., Silva Barbosa, D. R. and Santos, J. P. O. (2016). Control of *Callosobruchus maculatus* (FABR.) (Coleoptera: Chrysomelidae: Bruchinae) in *Vigna unguiculata* (L.) WALP. with essential oils from four Citrus spp. plants. *J. Stored Prod. Res.* 68:25–32.

Baoua, I., Amadou, L., Margam, V. and Murdock, L. L. (2012). Comparative evaluation of six storage methods for the postharvest storage of cowpea grain. *J. Stored Prod. Res.* 49:171–5.

Baoua, I. B., Amadou, L., Lowenberg-DeBoer, J. D. and Murdock, L. L. (2013). Side by side comparison of GrainPro and PICS bags for postharvest preservation of cowpea grain in Niger. *J. Stored Prod. Res.* 54:13–17.

Baoua, I. B., Margam, V., Amadou, L. and Murdock, L. L. (2012). Performance of triple bagging hermetic technology for postharvest storage of cowpea grain in Niger. *J. Stored Prod. Res.* 51:81–5.

Baoua, I. B., Amadou, L., Batributsa, D. and Murdock, L. L. (2014). Triple bag hermetic technology for post-harvest preservation of Bambara groundnut (*Vigna subterranea* (L.) Verdc.). *J. Stored Prod. Res.* 58:48–52.

Baributsa, D., Lowenberg-DeBoer, J., Murdock, L. and Moussa, B. (2010). Profitable chemical-free cowpea storage for smallholder farmers in Africa: Opportunities and challenges. *Julius Kuehn Archiv* 425:1046–51.

Boys, K., Faye, M., Fulton, J. and Lowenberg-DeBoer, J. (2007). The economic impact of cowpea research in Senegal: an ex-post analysis with disadoption. *Agric. Econo.* 36:363–75.

Caswell, G. H. (1975). The development and extension of non-chemical control techniques for stored cowpeas in Nigeria. In Brady, E. U., Brower, J. H., Hunter, P. E., Jay, E. G., Lum, P. T. M., Lund, H. O., Mullen, M. A. and Davis, R. (Eds), Proceedings of the 1st International Working Conference on Stored-Product Entomology, 7-11 October 1974, Savannah, Georgia, USA. Savannah, Georgia, USA, pp. 63–67.

Currid, J. D. and Navon, A. (1989). Iron age pits and the Lahav (Tell Halif) grain storage project. *Bull. Am. Schools Orient. Res.* 273:67–78.

de Beurs, K. M. and Brown, M. E. (2013). The effect of agricultural growing season change on market prices in Africa. http://dx.doi.org/10.5772/56459, accessed 30 August 2016.

Donahaye, E., Navarro, S. and Calderon, M. (1967). Storage of barley in an underground pit sealed with a polyethylene liner. *J. Stored Prod. Res.* 3:359–64.

FAO. (2014). The State of Food and Agriculture 2014. http://www.fao.org/3/a-i4036e.pdf accessed 17 August 2016.

Gatehouse, A. M. R., Gatehouse, J. A., Dobie, P., Kilminster, A. M. and Boulter, D. (1979). Biochemical basis of insect resistance in *Vigna unguiculata*. *J. Sci. Food Agr.*, 30:948–58.

Gatehouse, A. M. R., Butler, K. J., Fenton, K. A. and Gatehouse, J. A. (1985). Presence and partial characterization of a major proteolytic enzyme in the larval gut of *Callosobruchus maculatus*. *Entomol. Exp. Appl.* 39:279–86.

Golob, P. and Webley, D. J. (1980). *The Use of Plants and Minerals as Traditional Protectants of Stored Products*. Tropical Products Institute publication No. GI38. London: Overseas Development Administration.

Gordon, K. H. J. and Waterhouse, P. M. (2007). RNAi for insect-proof plants. *Nature Biotechnol.* 25:1231–2.

Hill, H. and Schoonhoven, A. V. (1981). Effectiveness of vegetable oil fractions in controlling the Mexican bean weevil Zabrotes subfasciatus on stopred beans Phaseoolus vulgaris. *J. Econ. Entomol.*, 74:478–9.

Hill, R. A., Lacey, J. and Reynolds, P. J. (1983). Storage of barley grain in iron age type underground pits. *J. Stored Prod. Res.* 19:163–71.

Hellmich, R. A. and Hellmich, K. A. (2012). Use and impact of *Bt* maize. *Nat. Educ. Knowledge* 3(10):4.

Ilboudo, Z., Dabire-Binso, C. L., Sankara, F., Nebie, R. C. H. and Sanon, A. (2015). Optimizing the use of essential oils to protect stored cowpeas from *Callosobruchus maculatus* (Coleoptera: Bruchinae) damage. *Afr. Entomol.* 23:94–100.

Ishimoto M., Sato T., Chrispeels M. J. and Kitamura K. (1996). Bruchid resistance of transgenic azuki bean expressing α-amylase inhibitor of common beans. *Entomol. Exp. Appl.* 79:309–15.

Jayas, D. S. and Jeyamcondan, S. (2002). Modified atmosphere storage of grains meats fruits and vegetables. *Biosyst. Eng.* 82:235–51.

Kashiwaba, K., Tomooka, N., Kaga, A., Han, O. K. and Vaughn, D. A. (2003) Characterization of resistance to three bruchid species (*Callosobruchus* spp., Coleoptera, Bruchidae) in cultivated rice bean (*Vigna umbellata*). *J. Econ. Entomol.* 96:207–13.

Keneni, G., Bekele, E., Getu, E., Imtiaz, M., Damte, T., Mulatu, B. and Dagne, K. (2011). Breeding food legumes for resistance to storage insect pests: Potential and Limitations. *Sustainability* 3:1399–415.

Kitch, L. W. and Murdock, L. L. (1986). Partial characterization of a major gut thiol proteinase from larvae of *Callosobruchus maculatus* F. *Arch. Insect Biochem. Physiol.* 3:561–75.

Kuijt, I. and Finlayson, B. (2009). Evidence for food storage and predomestication granaries 11,000 years ago in the Jordan Valley. *Proc. Natl. Acad. Sci. U. S. A.* 106:10966–70.

Kuijt, I. (2015). The Neolithic refrigerator on a Friday night: How many people are coming to dinner and just what should I do with the slimy veggies in the back of the fridge? *Environmental Archaeology* 20:321–36.

Laing, L. (1981). *Celtic Britain – Britain Before the Conquest.* London, Toronto, Sydney and New York: Granada Publishing , 254 pp.

Langyintuo, A., Lowenberg-DeBoer, J., Faye, M., Lambert, D., Ibro, D., Moussa, B., Kergna, A., Kushwaha, S., Musa, S. and Ntoukam, G. (2003). Cowpea supply and demand in West and Central Africa. *Field Crops Res.* 82:215–31.

Lee, R.-Y., Reiner, D., Dekan, G., Moore, A.E., Higgins, T. J. V. and Epstein, M. M. (2013). Genetically Modified α-Amylase Inhibitor Peas Are Not Specifically Allergenic in Mice. *Plos One* 8(1): e52972. doi:10.1371/journal.pone.0052972.

Lopes, L. M., Araujo, A. E. F., Santos, A. C. V., Santos, V. B. and Sousa, A. H. (2016). Population Development of Zabrotes subfasciatus (Coleoptera: Chrysomelidae) in Landrace Bean Varieties Occurring in Southwestern Amazonia. *J. Econ. Entomol.* 109:467–71.

Mital, H. C., 1971. Protection of cowpea from insect infestation with the aid of fixed oils. *J. West African Sci. Assoc.* 16:45–8.

Murdock, L. L., Seck, D., Ntoukam, G. and et al. (2003). Preservation of cowpea grain in sub-Saharan Africa-Bean/Cowpea CRSP contributions. *Field Crops Res.* 82:169–78.

Murdock, L. L., Margam, V., Baoua, I., Balfe, S. and Shade, R. E (2012) Death by desiccation: Effects of hermetic storage on cowpea bruchids. *J. Stored Prod. Res.* 49:166–70.

Mutungi, C. M., Affognon, H., Njoroge, A. W., Baributsa, D. and Murdock, L. L. (2014). Storage of mung bean (Vigna radiata [L.] Wilczek) and pigeonpea grains (Cajanus cajan [L.] Millsp) in hermetic triple-layer bags stops losses caused by Callosobruchus maculatus (F.) (Coleoptera: Bruchidae). *J. Stored Prod. Res.* 58:39–47.

Mutungi, C., Affognon, H. D., Njorge, A. W., Monomo, J., Baributsa, D. and Murdock, L. L. (2015). Triple-layer plastic bags protect dry common beans (*Phaseolus vulgaris* against damage by *Acanthoscelides obtectus* (Coleoptera: Chrysomelidae) during storage. *J. Econ. Entomol.* 108:2479–88.

Navarro, S. (2012). The use of modified and controlled atmospheres for the disinfestation of stored products. *J. Pest Sci.* 85:301–22.

Nikolova, I. (2016). Pea weevil damage and chemical characteristics of pea cultivars determining their resistance to Bruchis pisorum L. *Bull. Entomol. Res.* 106:268–77.

O'Dowd, E. T. (1971). *Hermetic Storage of Cowpea (Vigna unguiculata Walp) in Small Granaries, Silos and Pits in Northern Nigeria.* Samaru Miscellaneous Papers No. 31. Zaria, Nigeria: Institute for Agricultural Research, Ahmadu Bello University.

Ofuya T. I. (1986). Use of wood ash, dry chilli pepper fruits and onion scale leaves for reducing *Callosobruchus maculatus* (Fabricius) damage in cow-pea seeds during storage. *J. Agric. Sci. Camb.* 107:467–8.

Olufuwote, J. and Barnes-McConnell, P. (2002). Cowpea dissemination in West Africa using a collaborative technology transfer model. In Fatokun, C. A., Tarawali, S. A., Singh, B. B., Kormawa, P. M. and Tamo, M. (Eds), Challenges and Opportunities for Enhancing Sustainable Cowpea Production. Proceedings of the World Cowpea Conference III held at the International Institute of Tropical Agriculture (IITA), Ibadan, Nigeria, 4-8 September 2000. IITA, Ibadan, Nigeria.

Ozburn, G. W., Ward, P., George, M., Riley, J. and Cornes, M. (1960). Hermetic storage of guinea corn. Report of the West Africa Stored Products Research Unit. pp. 45–7.

Pereira, J. (1983). The effectiveness of 6 vegetable oils as protectants of cowpeas and Bambara groundnuts against infestation by *Callosobruchus maculatus* (F)) Coleoptera, Bruchidae). *J. Stored Prod. Res.* 19:57–62.

Perkin, L. C., Adrianos, S. L. and Oppert, B. (2016). Gene disruption technologies have the potential to transform stored product insect pest control. *Insects* 7:46–63.

Plinius, Gaius Secundas (Pliny the Elder)(ca. 79 AD) In Natural History, The methods of storing corn, Chapter 73. http://www.perseus.tufts.edu/hopper/text?doc=Perseus%3Atext%3A1999.02.0137%3Abook%3D18%3Achapter%3D73, accessed 10 August 2016.

Quentin, M. E., Spencer, J. L. and Miller, J. R. (1991). Bean tumbling as a control measure for the common bean weevil, *Acanthoscelides obtectus*. *Entomol. Exper. Appl.* 60:105–9.

Ranga Rao, G. V., Rameshwar Rao, V. and Nigam, S. N. 2010. *Post- harvest Insect Pests of Groundnut and Their Management*. Information Bulletin No. 84. Patancheru 502 324, Andhra Pradesh, India: International Crops Research Institute for the Semi-Arid Tropics. ISBN 978-92-9066-528-1. Order code IBE 084, p. 20.

Rubiales, D., Fondevilla, S., Chen, W., Gentzbittel, L., Higgins, T. J. V., Castillejo, M. A., Singh, K. B. and Rispail, N. (2015). Achievements and Challenges in Legume Breeding for Pest and Disease Resistance. *Crit. Rev. Plant Sci.* 34(1–3):195–236.

Sanon, A., Dabire, C., Ouedraogo, A. P. and Huignard, J. (2005). Field occurrence of bruchid pests of cowpea and associated parasitoids in a sub humid zone of Burkina Faso: Importance on the infestation of two cowpea varieties at harvest. *Plant Pathol. J.* 4:14–20.

Sarmah, B. K., Moore, A., Tate, W., Molvig, L., Morton, R. L., Rees, D. P., Chiaiese, P., Chrispeels, M. J., Tabe, L. M. and Higgins, T. J. V. (2004). Transgenic chickpea seeds expressing high levels of a bean alpha-amylase inhibitor. *Mol. Breed.* 14:73–82.

Sarwar, M. (2013). Studies on resistance discrepancy in seeds of local and exotic lentil (*Lens culinaris* Medikus) genotypes against *Callosobruchus analis* L. (Coleoptera: Bruchidae). *Afr. Entomol.* 21:273–9.

Schoonhoven, A. V. (1978). Use of vegetable oils to protect stored beans from bruchid attack. *J. Econ. Entomol.* 71:254–6.

Schroeder, H. E., Gollasch, S., Moore, A., Tabe, L. M., Craig, S., Hardie, D. C., Chrispeels, M. J., Spencer, D. and Higgins, T. J. V. (1995). Bean alpha-amylase inhibitor confers resistance to the pea weevil (*Bruchus pisorum*) intransgenic peas (*Pisum sativum* L.). *Plant Physiol.* 107:1233–9.

Seck, D. and Gaspar, C. (1992). Effectiveness of storing cowpea, *Vigna unguiculata* (L.) Walp, in sealed metal drums as an alternative method for controlling *Callosobruchus maculatus* F (Coleoptera, Bruchidae) in the Sahel of Africa. *Meded. Fac. Landbouwwet., Rijksuniv. Gent* 44:751–8.

Shade, R. E., Schroeder, H. E., Pueyo, J. J., Tabe, L. M., Murdock, L. L., Higgins, T. J. V. and Chrisspeels, M. J. (1994). Transgenic pea-seeds expressing the alpha-amylase inhibitor of the common bean are resistant to bruchid beetles. *Nature Biotechnol.* 12:793–6.

Singh, A., Khare, A. and Singh, A. P. (2012). Use of vegetable oils as biopesticide in grain protection – A review. *J. Fertil. Pestic.* 3:114.

Singh, B. B., Singh, S. R. and Adjadi, O. (1988). Bruchid resistance in cowpea. *Crop Sci.* 25:736–9.

Singh, K. B., Ocampo, B. and Robertson, L. D. (1998). Diversity for abiotic and biotic stress in the wild annual Cicer species. *Genet. Resour. Crop Evol.* 45:9–17.

Singh, S. R., Luse, R. A., Leuschner, K. and Nangdu, D. (1978) Groundnut oil treatment for control of *Callosobruchus maculatus* (F.) during cowpea storage. *J. Stored Prod. Res.* 14:77–80.
Solleti, S. K., Kumar, S., Bakshi, S., Purkayastha, J., Panda, S. K. and Sahoo, L. (2008). Transgeic cowpea (*Vigna unguiculata*) seeds expressing a bean alpha-amylase inhibitor 1 confer resistance to storage pests, bruchid beetles. *Plant Cell Rep.* 27:1841–50.
Sommapa, J., Mohammad, N., Bajpai, G. C. and Joshi, R. (2012). Bioassay studies for bruchid (Callosochinensis) resistance in F7 inter-specific generation of pigeonpea (Cajanus cajan). *Indian J. Agr. Sci.* 82:288–90.
Spencer, J. L., Quentin, M. E. and Miller, J. R. (1991). Tumbling controls bean weevils. *Nature* 352:387.
Suthar, M. D. and Barpoda, T. M. (2015). Evaluation of botanicals against *Callosobruchus chinensis* Linnaeus in black gram under storage condition. *Indian J. Agr. Res.* 50:167–71.
Swapan, K. T. (2016). Bruchid resistance in food legumes – an overview. *Res. J. Biotechnol.* 11:97–105.
Vaeck, M., Reynaerts, A., Hofte, H., Jansens, S., Debueckeleer, M., Dean, C., Zabeau, M., Vanmontagu, M. and Leemans, J. (1987). Transgenic plants protected against insect attack. *Nature* 328:33–7.
Vales, M. I., Ranga Rao, G. V., Sudini, H., Patil, S. B. and Murdock, L. L. 2014. Effective and economic storage of pigeonpea in triple layer plastic bags. *J. Stored Prod. Res.* 58:29–38.
Varro, Marcus *Terentius (bef. 27 BC) on Agriculture*, Vol. 1, Chapter 57 http://penelope.uchicago.edu/Thayer/E/Roman/Texts/Varro/de_Re_Rustica/1*.html, accessed 10 August 2016.
Vir, R., Lakhanpaul, S., Malik, S., Umdale, S. and Bhat, K. V. (2016). Utilization of germplasm for the genetic improvement of mung bean (Vigna radiata (L) Wilczek): The constraints and opportunities. *Gene Pool Diversity and Crop Improvement, Volume 1*, Chapter 12, pp. 367–91.
Walker, T. S. and Alwang, J. (2015). *Crop Improvement, Adoption and Impact of Improved Varieties of Food Crops in Sub-Saharan Africa.* CABI, p. 478. https://books.google.com/books?id=NsgYCwAAQBAJ&dq=Lori+Niebe&source=gbs_navlinks_s, accessed 18 January 2017.
Wolfson, J. L., Shade, R. E., Mentzer, P. E. and Murdock, L. L. (1991). Efficacy of ash for controlling infestations of *Callosobruchus maculatus* (F) (Coleoptera, Bruchidea) in stored cowpeas. *J. Stored Prod. Res.* 27:239–43.
Zhu, K., Huesing, J. E., Shade, R. E. and Murdock, L. L. (1994). Cowpea trypsin inhibitor and resistance to cowpea weevil (Coleoptera, Bruchidae) in cowpea variety TVu-2027. *Environ. Entomol.* 23:987–91.

Chapter 17

Drying, handling, storing and quality monitoring of pulses

C. B. Singh, University of South Australia, Australia; and D. S. Jayas, University of Manitoba, Canada

1 Introduction

Major legume crops, especially pulses, include lentil, chickpea, green pea, yellow pea, beans and lupin. Total annual world production of pulses is around 68.83 million metric tonnes (Mt) (FAOSTAT 2013). India (17.24 million Mt) is the world's largest producer of pulses followed by Canada (5.35 million Mt), Myanmar (4.49 million Mt), China (3.89 million Mt), Nigeria (3.42 million Mt) and Brazil (3.42 million Mt) (FAOSTAT 2013). Pulses are an affordable and vital vegetarian source of protein with low fat and high carbohydrate content (fibre and resistant starch) and are a rich source of micronutrients, amino acids and vitamins (Patterson et al. 2009). Pulses are consumed directly as whole grains as well as in split and milled (flour) form in various food ingredients and snacks. Due to their ability to fix nitrogen in the soil, pulse crops contribute significantly to a sustainable cropping environment.

Legume grains comprise three main parts: the seed coat or hull, two cotyledons and the embryo (Fig. 1). The seed coat is the outer cover, consisting of about 8–10% of the total grain kernel. The two cotyledons, on average, cover about 88–90% of the total grain. The embryo constitutes less than 2% of the total grain. The hull is removed from the seed to obtain split pulses (cotyledons).

http://dx.doi.org/10.19103/AS.2017.0023.15

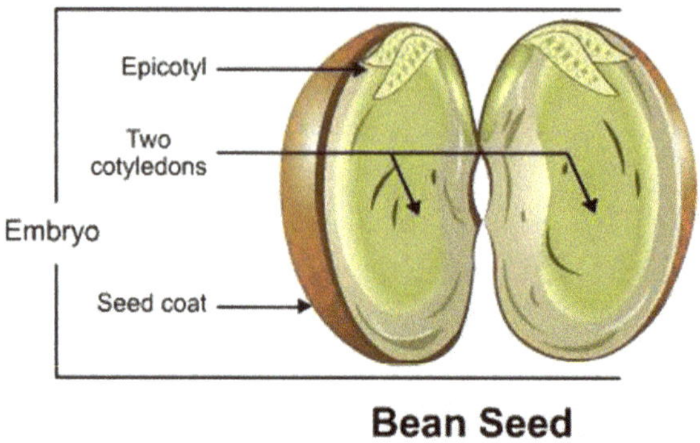

Figure 1 Bean seed structure (Online Labs 2013).

During storage, chemical and physiological properties of pulses change. This can result in significant loss of quality and quantity due to mould growth, development of carcinogenic mycotoxins, insect damage, discolouration and darkening due to oxidation. Other problems can include broken, split or peeled seeds, pulse sprouting, development of unpleasant odours, heat damage and bin burn (i.e. discolouration or damage due to excessive heat during storage) (Singh and Larson 2016). Poor storage, handling and transportation may further downgrade the quality of the pulses. Pulses absorb moisture if exposed to humid air and also tend to produce heat due to respiration inside storage containers.

Pulse quality is assessed throughout the storage and handling process, and optimum-quality pulses, which receive the highest commercial grade, will generate a higher market price. Major pulse-exporting countries have developed standards for commercial grading (CGC 2016; Pulse Australia 2016). The pulses are assessed for quality and are graded based on several physical, processing, compositional and safety-related factors, as outlined in regulator and/or industry-developed standards.

2 Pulse quality

Specifications in standards of commercial grading of pulses are used to assess the quality of pulse grains. Table 1 shows the Australian standard commercial #1 grade requirements for exporting red lentils (Pulse Australia 2016). Pulse quality can be broadly grouped into physical, processing, compositional and safety-related factors, as well as seed quality.

2.1 Physical factors

Physical factors, predominantly appearance, significantly influence the quality of grains and their market price. Grains with uniform shape, size and colour are preferred by consumers. Problems can arise if grains become discoloured, bleached, stained or darkened. Grain quality and commercial grade are also reduced if kernels are physically damaged. This includes bin burnt, cracked/loose, chipped seed coat, peeled, split, shrivelled, immature, sprouted and weather-damaged kernels.

Table 1 Minimum requirement for #1 grade Australian red lentil for export (Pulse Australia 2016)

Parameter	Requirements	Comments/variations
Physical characteristics	The lentils shall be hard and well filled	
Purity	99% min by weight	Whole lentils, defective lentils and seed coats
Moisture	14% max	–
Defective	Container (bulk or bagged): 3% max by weight, bulk vessel hold shipment: 5% max by weight, both include poor seed coat colour and poor kernel colour	Lentils not of the specified variety. Lentil kernels that are black, broken, chipped, diseased, frost damaged, insect damaged, orange tip, sappy, shrivelled, split, sprouted, weather-damaged or wrinkled. Includes pods that contain lentils, whether broken or unbroken, loose seed coat, 3% blondes (Aldinga only) and all red lentil seed material falling through the 2 mm slotted screen (Aldinga 2.2 mm)
Mould	1% by weight	Mould (field and/or storage), caked, bin burnt and heat damaged
Varietal restriction	1% max by weight	Not of the specified variety
Poor seed coat colour	1% max by weight	Discoloured lentils have excessive discolouration of the seed coat as per the GTA Visual Recognition Standards Guide. Includes any disease, frost and water staining
Poor kernel colour	1% max by weight	Discoloured lentils have excessive discolouration of the kernel as per the GTA Visual Recognition Standards Guide. Includes any disease, frost and water staining, and green, brown, black, yellow, bleached and chalky white kernels
Contrasting colour	1% max by weight	Contrasting colour of the seed coat in the same variety. Not classified as defective
Unmillable material	0.1% max by weight	Soil, stones and non-vegetable matter
Snails	Nil tolerance	–
Field insects	One (1) max	Dead per 200 g sample
Foreign seeds	2 cereal seeds max	Cereal seeds exception only
Objectionable material	Nil tolerance	Includes objectionable odour
Ryegrass ergot	2 cm max length	Pieces laid end to end per 200 g sample

Grain moisture, bulk density and test weight are also considered during physical assessment and commercial grading of pulses (CGC 2016). Higher grain moisture increases the risk of spoilage, so there is a maximum moisture limit for straight grade pulses (Table 2) to be traded without any discount in price. Pulses with high density and high test weight are considered healthy and of good quality. Physical properties of pulses are shown in Table 3.

Table 2 Maximum straight grade moisture allowable limits for pulses (CGC 2016)*

Grain	Moisture (% wet basis)
Yellow and green pea	16.0
Chickpea	14.0
Green lentil	14.0
Red lentil	13.0
Faba bean	16.0

*Recommended safe storage moisture for pulses is below 14% (wet basis)

Table 3 Physical properties of pulses

	Physical property							
Grain type	Length (mm)	Width (mm)	Thickness (mm)	Bulk density (kg/m³)	True density (kg/m³)	Porosity (%)	Angle of repose (°)	Reference
Lentil (9–21%)	–	6.15–6.63	2.47–2.8	791–731	1433.2–1393.3*	47.9–50.3	28–26	Szot et al. (2003)
Chick pea 9.21–21.36%	8.27–9.18	6.82–7.73	6.35–7.23	835.55–773.55	1459.73–1449.93	42.75–46.63	–	Shahbazi (2011)
Faba bean 12.6–21.9%	–	–	–	814.8–761.3	1392.8–1373.3	–	28.1–29.3	Irvine et al. (1992)

*11.4–18% moisture (Irvine et al. 1992).

2.2 Processing characteristics

Pulse buyers and processors prefer pulses with improved milling yield and efficiency. Increased percentages of broken grain and fine fractions in milling reduce milling efficiency and result in milling loss. Easy de-hulling is also desired so that splits are clear and uniform in size and colour. Milling yield and efficiency are highly dependent on the physical factors of the grains. However, there are many other variables such as growing conditions, plant disease, harvest conditions, handling methods and storage practices that affect pulse quality and processing characteristics. Since pulses are mainly consumed after cooking (in both whole and split forms), water absorption and cooking time are also used as quality indicators. The taste and texture of cooked pulses and processed products are also considered in their quality evaluation.

2.3 Composition

Pulses are a rich dietary source of proteins and carbohydrates for human consumption and animal feed. Buyers prefer pulses with particular compositions to match their specific needs. They also sometimes blend different varieties and grain types to meet their specific processing and end product requirements. This means that they want to know the specific composition of each shipment. The main pulse constituents considered in quality and grading are protein, starch and fibre content. The composition of grain also affects

Table 4 Composition (g/100 g dry matter) of major pulses (Wang and Daun 2004)

	Pulse crop					
Composition	Field peas	Green lentils	Red lentils	Chickpea, kabuli	Chickpea, desi	Dark red kidney bean
Protein (N × 6.25)	23.7	26.3	25.7	24.4	23.0	28.5
Starch	45.5	45.0	44.0	41.1	36.4	37.0
Amylose*	25.6	25.4	26.5	26.2	23.8	21.2
ADF**	7.0	5.5	5.8	3.7	13.1	8.4
NDF***	9.6	8.1	8.9	5.0	12.8	9.6
Fat	1.3	1.1	1.1	5.9	5.4	1.1
Ash	2.8	2.8	3.0	3.2	3.2	3.9

*(% of total starch).
**Acid detergent fibre.
***Neutral detergent fibre.

processing characteristics such as cooking time and water absorption. The chemical composition of major pulses is shown in Table 4.

2.4 Safety factors

Due to increased consumer awareness of food safety and strict regulations in many countries, grain safety is now considered among the most significant of quality and grading factors in the international grain trade. Buyers not only prefer clean grain with low foreign material, but also want safe grain, which is free from insects, moulds, carcinogenic mycotoxins and chemical or non-chemical residues. Canadian and Australian commercial grades have zero tolerance for live insects. In many countries, there is very strict control on the export and import of grain with moulds and mycotoxins.

Importing countries may apply punitive measures to control the violations of maximum residue limits (MRLs) for pesticide or other chemical residues. Many pulse-trading countries comply with widely accepted Codex MRL standards (FAO 2016). Codex international food safety standards are developed and updated on an on-going basis by the World Health Organization (WHO) and the Food and Agriculture Organization of the United Nations (FAO). Table 5 shows the Codex pesticide MRL for pulses.

2.5 Seed quality

Seed quality of pulses is very important for growers, as crop yield and therefore profitability is highly dependent on the quality of the planted seed. If poor-quality seed is planted, the grower may suffer significant losses due to lower yield and increased input cost. Quality seed should be whole, free from diseases and physical damage and have a high germination rate. Diseased seed is shrunken and discoloured and has a low test weight. Seed germination rates of pulses may be reduced if pulses are not properly stored. Poor storage can result in mould growth, insect infestation, hotspot development, sprouting and physical damage. Effective grain management practices as described in the following sections will ensure that seed quality is preserved during storage and handling.

Table 5 Codex pesticide MRL for pulses (FAO 2016)

Pesticide	MRL (mg/kg)	Year of adoption	Note
Aldrin and Dieldrin	0.05	1997	
Azoxystrobin	0.07	2014	(except soya bean)
Bifenthrin	0.30	2011	
Boscalid	3.00	2010	
Chlorothalonil	1.00	2011	
Clothianidin	0.02	2011	
Cyhalothrin (includes lambda-cyhalothrin)	0.05	2009	
Cypermethrins (including alpha- and zeta- cypermethrin)	0.05	2009	
Deltamethrin	1.00	2004	
Dichlobenil	0.01	2015	
Flubendiamide	1.00	2011	
Imidacloprid	2.00	2013	(except soya bean)
Paraquat	0.50	2006	
Penthiopyrad	0.30	2013	(except soya bean)
Piperonyl Butoxide	0.20	2004	
Pirimicarb	0.20	2007	Except soya bean (dry)
Prothioconazole	1.00	2010	Except soya bean, dry.
Pyrethrins	0.10	2003	
Saflufenacil	0.30	2013	
Sedaxane	0.01	2015	
Spirotetramate	2.00	2012	except soya bean (dry)
Thiamethoxam	0.04	2011	

3 Insects affecting stored pulses

The most common insects affecting stored pulses are the pea weevil (*Bruchus pisorum*) and cowpea weevil (*Callosobruchus* spp.). The genus *Callosobruchus* has several species that are found in tropical and subtropical regions. Other insects that damage stored pulses are the lesser grain borer [*Rhyzopertha dominica* (Fabricius)], red flour beetle [*Tribolium castaneum* (Herbst)], khapra beetle [*Trogoderma granarium* (Everts)], rusty grain beetle [*Cryptolestes ferrugineus* (Stephens)], sawtoothed grain beetle [*Oryzaephilus surinamensis* (Linnaeus)], longheaded flour beetle [*Latheticus oryzae* (Waterhouse)] and cigarette beetle [*Lasioderma serricorne* (Fabricius)]. These insects physically damage the pulses, resulting in quantitative and milling losses, as well as reduced composition of nutrients such as carbohydrates and proteins.

Fumigation with phosphine is common and is the only insect-control method currently being used. However, due to excessive and improper use of this method, insects are developing strong resistance to the phosphine. Chemical sprays are not registered for use in pulses and very tight MRLs are in place for pulse exports. Most of the insects stop reproducing below 15°C so pulses should be aerated to control insects where aeration cooling potential is available. Chilled aeration is worth considering to lower the temperature of pulses in order to control insect pests.

4 Moisture content, drying and aeration

4.1 Equilibrium moisture content (EMC)

Grain EMC is an important factor in the conditioning (drying, aeration and rehydration) of grains. The EMC is the moisture content to which the grain will achieve equilibrium, if exposed to specific air temperature (T) and relative humidity (RH) for a sufficient time. Sorption EMC empirical equations have been developed for different grain types, and the coefficients from these empirical equations are used to calculate EMC at a specific T and RH (Sun 1998). If EMC at a specific air T and RH is lower than the grain moisture, the grain will release its moisture in order to establish equilibrium with the surrounding air, and drying will occur. If EMC at a specified air T and RH is higher than the grain moisture, the grain will absorb the moisture from surrounding air and rehydration will occur. Thus, ambient air conditions (T and RH) are the main factors in determining the drying or rewetting potential of the grain. However, natural air drying or rewetting is a slow process, so a change in grain moisture is not instantaneous and also depends on the quantity and quality of the surrounding air.

4.2 Drying and aeration of pulses: overview

Pulses, like cereal crops, are also often harvested at moisture levels above safe storage limits. There are several benefits of early harvesting including avoiding adverse weather, maintaining quality and minimizing scattering losses and grain damage. However, pulses cannot be safely stored for longer periods without removing the excessive moisture from the grains. Safe grain storage is affected by many factors such as temperature, moisture content, grain maturity, storage condition and storage time. Other factors include intergranular gas composition, insects, microorganisms, mites, rodents, birds, dockage, granary structure and geographical location (Jayas 1995). However, grain moisture and temperature are the two most important factors that affect safe storage of grain without any significant spoilage or quality loss.

Pulses stored at high moisture level and temperature respire due to their metabolic activity. This can result in loss of nutrients and generation of moisture and heat, causing significant qualitative, quantitative and economic losses. Experimental studies showed that the rate of respiration of pulses (pinto bean, chickpea and green lentil) stored below 14% moisture and 20°C temperature was low. In contrast, at high storage temperature (30°C and 40°C) and high moisture content (18% and 20%), the respiration rate increased by 2–3 times (Chidananda et al. 2014). For long-term safe storage, pulses should be dried to 12–14% moisture content. Drying in high-temperature dryers may cause physical damage and grain darkening, so pulses should not be exposed to temperatures above 40°C. In-bin natural air drying using an optional low-temperature heater is considered the most cost-effective method to yield high-quality dried product. After the grain has been dried to safe storage moisture levels, it should be aerated (cooled). Cool and dry grain should be stored in optimal conditions at 14% moisture and 2–3°C temperature in colder climates, and 12–13% moisture and temperature below 15°C (or the lowest possible) in warm climates.

Grain temperature can be monitored in real time using temperature cables, and aeration/cooling can be tracked as it progresses. Automated drying and aeration-fan control can optimize performance, yield better results and deliver efficient drying (Singh et al. 2014). Delay in drying due to poor aeration system design, insufficient airflow rate, poor weather or improper fan control may result in significant grain spoilage. High-moisture pulses stored at high temperature should be dried within four weeks to prevent germination loss and spoilage due to mould growth (Sravanthi et al. 2013).

4.3 In-bin drying system

Figure 2 shows an advanced in-bin drying and aeration system. The details of the in-bin aeration system are provided in Singh et al. (2014). The drying system consists of a freestanding vertical steel structure (commonly known as a silo or bin), which is top-loaded with the grain for drying and/or storage. At the base, a fan(s) is connected to the plenum of the silo through the transition duct.

The silo in Fig. 2 is also equipped with moisture and temperature cables (sensors), plenum temperature and pressure sensors, headspace temperature and RH sensors and a weather station. These devices are linked to grain management software (©OPIsystems

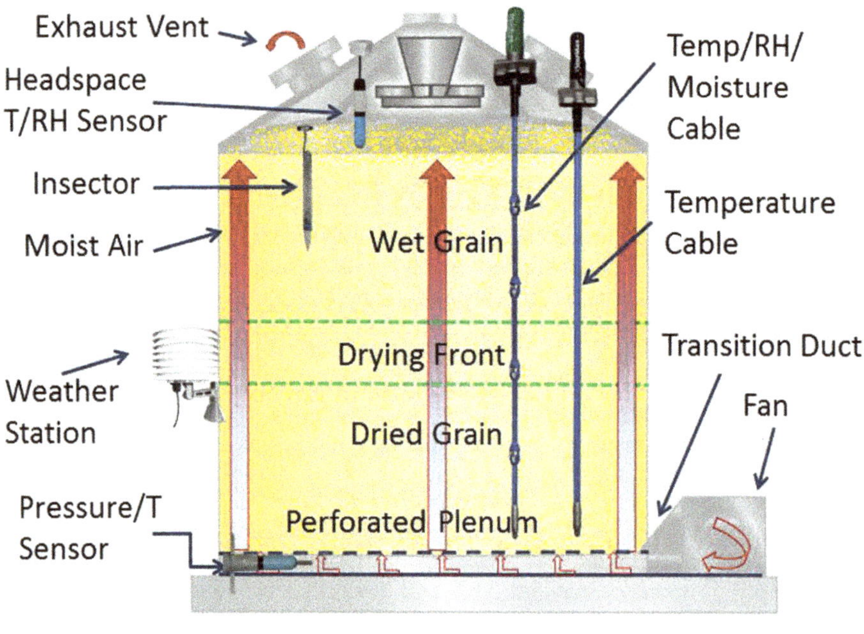

Figure 2 In-bin drying system (Singh et al. 2014).

Inc., Canada) installed in a personal computer through wired or wireless communication. Sometimes a carbon dioxide (CO_2) sensor is also included in the headspace to detect early spoilage by mould and insects. Increased CO_2 concentration indicates mould and insect activity inside the grain bin (Jayas 1995; Maier et al. 2010). The silo can also be equipped with an optional low-temperature heater which is installed inside the transition duct.

A fan pushes the air through the plenum into the grain bin. This dry air either absorbs the moisture from grain, or may add moisture to grain if the air is humid and EMC is low, before moving upwards. Air carrying moisture from grain exits the bins through exhaust vents at the top. As drying progresses, a drying front (a hypothetical layer of 0.5–1.0 m) is established, and once this layer is dried, the front moves to the next layer above it. The bin is fully dried when the drying front reaches the top.

5 Pulse storage and handling

Different types of storage structures, grain handling and management practices have significant impact in preventing storage losses. Storage structures vary from small-scale mud structures to large-scale modern vertical steel structures (grain bins or silos). In developing countries, crops produced in relatively small-sized farms are stored by farmers for their own consumption. Locally grown pulses are stored in indigenous storage structures made of a variety of materials such as mud, paddy straw, wheat straw, wood, bamboo, bricks and concrete. They can be stored in jute or plastic bags and in small plastic and steel containers.

Bulk storage of pulses is usually done in stacked jute or plastic bags kept in warehouses and sheds. However, pulse storage in these indigenous containers and warehouses results in significant grain loss. This is due to an uncontrolled storage environment that provides favourable conditions for insects and mould growth. In developed countries, pulses are mostly stored in steel silos, both at the farms and at grain handling facilities. The steel silos, if equipped with proper aeration fans and grain monitoring systems, ensure the safe long-term storage of grain. The environment inside sealed silos can be manipulated so that pulses can be safely stored for longer duration.

Pulses are prone to oxidation through the combination of sunlight, high moisture and high temperature. Oxidation causes bleaching and discolouration and reduces the pulse quality and premium. Pulses should be stored in bins (corrugated or welded steel silos) to protect them from sunlight. It is important to understand that the safe storage period of grain (including pulses) is cumulative, that is, all the storage and handling conditions after the harvest (such as transportation, moisture, temperature, drying and aeration) should be considered when determining the safe storage period. Table 6 shows the safe storage period for pulses for various moisture and temperature combinations (Sokhansanj and Patil 1995). For safe storage, dried pulses should be cleaned prior to loading into storage structures, and broken or immature grains, fines and foreign material should be removed. Inefficient or uneven drying and aeration, as well as ineffective fumigation, may increase the risk of spoilage, mould growth and hotspot development.

In steel silos, headspace condensation can cause mould growth, hotspot development and spoilage near the top surface of grain, particularly in regions where daily temperature fluctuations are high. In sunny and warm days, headspace temperature may reach up to 15–20°C above the ambient temperature. At night, as ambient air cools, the metal

Table 6 Safe storage period (number of weeks) of pulses (Sokhansanj and Patil 1995)

	Storage moisture (% wet basis)				
Storage temp (°C)	12	14	16	18	21
25	31	16	7	4	2
21	55	28	13	7	4
15	100	50	20	12	6
10	200	38	38	20	20
5	370	175	70	39	21

bin roof also cools but the headspace remains hot. As the warm air hits the cool bin roof, condensation occurs on the inner surface and drops onto the top grain surface. The condensed water is then absorbed by the grain which increases its moisture. This, combined with the grain's high temperature, leads to mould growth, hotspot development, mycotoxins and rapid spoilage (Singh and Larson 2016).

Overdrying (shrink) of pulses below safe storage moisture level should be avoided. Overdried pulses are difficult to handle during loading, transportation, milling and processing with a risk of damaging grain. During handling, dry seed coats chip and often seed coats are fully removed which results in lower commercial grading. Additionally, grain shrinkage causes direct financial loss to the growers and traders as pulses are sold based on their weight.

Free fall of pulses during loading and handling causes significant damage to the grain. Experimental studies have been conducted to assess the physical damage in peas and lentils due to free fall (Bergen et al. 1993). Pulse samples containing various moisture levels (damp, tough, dry and overdry) were dropped onto screed concrete, smooth steel and sanded plywood surfaces from 18.3 m and 9.15 m heights. Damage was assessed based on the percentage of cracked or split grain and by the total damage to the grain. The concrete surface caused the greatest damage (0.60% in peas and 3.39% in lentils) in dry grain when dropped from 18.3 m height. In another study, the effect of moisture content (7–25%) and impact velocity (5–20 m/s) on impact damage to chickpea grain was evaluated (Shahbazi 2011). Increase in impact velocity from 5 m/s to 20 m/s increased the damage from 1.26% to 47.31%. When grain moisture increased from 7.5 to 17.7%, the mechanical damage to grain was reduced by a factor of 2.

These studies suggest that peak loading of dry pulses (especially chickpea and lentils) in tall bins may cause significant physical damage to pulses. Bin ladders should be used in top loading bins to reduce the impact velocity and drop height and therefore minimize grain damage. Augers for pulse handling should be operated at a low speed. Unnecessary movement of pulses from bin-to-bin or during transportation should be avoided.

6 Quality evaluation using colour and hyperspectral imaging

Traditional physical pulse quality inspection by graders and grain inspectors is slow and subjective. Compositional analysis by wet chemical methods is destructive, time-consuming and not suitable for real-time compositional assessment. Colour imaging systems are now commercially used for the identification of external damage in rice (such as chalky, stained and broken kernels). These systems rely on the grain's external features (colour, texture, shape and size) to detect any damage. Colour imaging has also been used to detect fungal damage in wheat (Singh et al. 2012).

Hyperspectral imaging provides the spectral information in a spatially resolved manner (i.e. spectral information from each pixel of the sample image can be obtained). This spectral information, arising from the reflectance or absorbance of grain, could potentially be used for quantification of grain composition and for evaluation of its quality. Singh et al. (2009) successfully evaluated hyperspectral imaging for detecting insect damage in wheat. Colour imaging can be used for sorting the impurities and detecting visible damage in pulses. On the other hand, hyperspectral imaging can be used for compositional analysis and detecting internal damage. Kaliramesh et al. (2014) used hyperspectral imaging for compositional analysis of green gram (also known as mung bean) and for predicting the moisture, protein and starch content with good accuracy.

7 Summary

Pulses are the major vegetarian source of protein. Due to consumer awareness regarding a healthy diet and the range of health benefits attributed to pulses, their global demand is increasing. The high sensitivity to drying, storage and handling methods may change both chemical and physiological properties of the pulses. This can result in a loss of quality. Appropriate and technology-driven pulse management, with moisture and temperature monitoring and automated fan control, can minimize the quality loss throughout the pulse supply chain. Advanced non-destructive colour and hyperspectral imaging techniques have potential for the rapid quality assessment of pulses.

8 Where to look for further information

For detailed further reading, please refer to *Pulse Foods: Processing, Quality and Nutraceutical Applications* (Tiwari et al. 2011). *Stored Grain Ecosystems* (Jayas et al. 1995) is a pre-eminent book for understanding the fundamentals of pulse drying, storage and handling. For detailed information on commercial pulse grading and quality standards, refer to the *Official Grain Grading Guide* of the Canadian Grain Commission (CGC 2016) and the Australian Pulse Standards (Pulse Australia 2016).

9 References

Bergen, G. A., Jayas, D. S. and White, N. D.G. (1993), 'Physical damage to peas and lentils due to free fall', *Can. Agric. Eng.*, 35(2), 151–5.

CGC (2016), '*Official Grain Grading Guide*', Canadian Grain Commission, https://www.grainscanada.gc.ca/oggg-gocg/oggg-gocg-2016-eng.pdf (Accessed 4 December 2016).

Chidananda, K. P., Chelladurai, V., Jayas, D. S., Alagusundaram, K., White, N. D. G. and Fields, P. G. (2014), 'Respiration of pulses stored under different storage conditions', *J. Stored Prod. Res.*, 59, 42–7.

FAO (2016), '*VD 0070-pulseds*', Food and Agriculture Organization of the United Nations, Rome, Italy, http://www.fao.org/fao-who-codexalimentarius/standards/pestres/commodities-detail/en/?c_id=295 (Accessed 15 December 2016).

FAOSTAT (2013), '*FAO Statistical Yearbook 2013*', Food and Agriculture Organization, Rome, Italy.

Irvine, D. A., Jayas, D. S., White, N. D. G. and Britton, M. G. (1992), 'Physical properties of flaxseed, lentils, and fababeans', *Can. Agric. Eng.*, 34(1), 75–81.

Jayas, D. S., White, N. D. G. and Muir, W. E. (1995), *Stored-Grain Ecosystems*, Marcel Dekker Inc., New York, NY.

Jayas, D. S. (1995), 'Mathematical modeling of heat, moisture and gas transfer in stored-grain ecosystems', In: Jayas, D. S., White, N. D. G. and Muir, W. E. (Eds), *Stored-Grain Ecosystems*, Marcel Dekker Inc., New York, NY, pp. 123–67.

Kaliramesh, S., Chelladurai, V., Jayas, D. S. and Alagusundaram, K. (2014), 'Determination of main constituents in green gram using near-infrared hyperspectral imaging', *J. Agri. Eng.*, 51(1), 7–15.

Maier, D. E., Channaiah, L. H., Martinez-Kawas, A., Lawrence, J. S., Chaves, E. V., Coradi, P. C. and Fromme, G. A. (2010), 'Monitoring carbon dioxide concentration for early detection of spoilage in stored grain', In: *Proceedings of 10th International Working Conference on Stored Product Protection*, Rotating conference published by society, Estoril, Portugal, June 27–July 2.

Online Labs (2013), '*Embryo of Dicot Seeds*, Amrita University', http://amrita.olabs.edu.in/?sub=79&brch=16&sim=135&cnt=1 (Accessed 16 January 2017).

Patterson, C. A., Maskus, H. and Dupasquier, C. (2009), 'Pulse crops for health', *Cereal Foods World*, 54(3), 108–12.

Pulse Australia (2016), '*Australian Pulse Standard 2016/17 Season*', Pulse Australia, http://www.pulseaus.com.au/storage/app/media/markets/20160801_Pulse-Standards.pdf (Accessed 4 December 2016).

Shahbazi, F. (2011), 'Impact damage to chickpea seeds as affected by moisture content and impact velocity', *Appl. Eng. Agric.*, 27(5), 771–5.

Singh, C. B. and Larson, R. (2016), '*Advanced pulse storage and management*', In: *Pulses Handbook 2016*, Commodity India, Bangalore, India http://commodityindia.com/publication/pulses/article16.html (Accessed 20 December 2016).

Singh, C. B., Jayas, D. S. and Larson, R. (2014), 'Assessment of fan control strategies for in-bin natural air drying of wheat in Western Canada', *Can. Biosys. Eng.*, 56, 3.25–3.36.

Singh, C. B., Jayas, D. S., Paliwal, J. and White, N. D. G. (2009), 'Detection of insect-damaged wheat kernels using near-infrared hyperspectral imaging', *J. Stored Prod. Res.*, 45(3), 151–8.

Singh, C. B., Jayas, D. S., Paliwal, J. and White, N. D. G. (2012), 'Fungal damage detection in wheat using short-wave near-infrared hyperspectral and digital colour imaging', *Int. J. Food Prop.*, 15(1), 11–24.

Sokhansanj, S. and Patil, R. T. (1995), 'Dehulling and splitting pulses: peas and lentils', In: Chakraverty, A., Mujumdar, A. S., Raghavan, G. S. V. and Ramaswamy, H. S. (Eds), *Handbook of Post Harvest Technology: Cereals, Fruits, Vegetables, Tea, and Spices*, Marcel Dekker Inc, New York, NY, pp. 393–427.

Sravanthi, B., Jayas, D. S., Alagusundaram, K., Chelladurai, V. and White, N. D. G. (2013), 'Effect of storage conditions on red lentils', *J. Stored Prod. Res.*, 53, 48–53.

Sun, D.-W. (1998), 'Selection of EMC/ERH isotherm equations for shelled corn based on fitting to available data', *Drying Tech.*, 16(3–5), 779–97.

Szot, B., Horabik, J. and Rusinek, R. (2003), 'Physical properties characteristic of Polish and Canadian lentil seeds', *Int. Agrophysics*, 17, 123–9.

Tiwari, B. K., Gowen, A. and McKenna, B. (2011), *Pulse Foods: Processing, Quality and Nutraceutical Applications*, Academic Press, London, UK.

Wang, N. and Daun, J. K. (2004), '*The Chemical Composition and Nutritive Value of Canadian Pulses*', Canadian Grain Commission, http://www.pulsecanada.com/uploads/c4/91/c491f652f0cf53390d9a5b86aa63aeea/The-Chemical-Composition-and-Nutritive-Value-of-Canadian-Pulses.pdf. (Accessed 15 December 2016).

Chapter 18

Dietary health benefits, phytochemicals and anti-nutritional factors in grain legumes

Elizabeth Ryan, Colorado State University, USA; Indi Trehan, Kristie Smith and Mark Manary, Washington University, USA

1 Introduction

Grain legumes, also known as pulses, constitute a number of staple foods consumed by humans throughout the world, including, but not limited to, dry beans such as kidney, navy, pinto, lima, adzuki, mung; broad beans; chickpea; lentil; field pea; cowpea and pigeon pea. From the perspective of maintaining health, the first consideration of any staple food is its nutrient composition: How does the consumption of the food contribute to meeting the nutrient requirements? But there are other key health effects of foods, such as:

1. Does the food promote or reduce inflammation?
2. Does the food promote or control carcinogenesis?
3. What is the effect of the food on the gut microbiota and gut health? and
4. What toxins/toxicants might be present in the food and what risks might these pose to the consumer?

This chapter will briefly discuss each of these considerations for some grain legumes that have been investigated in this regard (e.g. dry beans).

http://dx.doi.org/10.19103/AS.2017.0023.17

Legumes are a significant source of essential nutrients including dietary fibres, proteins, dietary minerals and carbohydrates. Abundant in both soluble and insoluble fibre, legumes support colonic and overall health. They are an inexpensive source of good-quality protein and offer a higher protein content than most other plant foods. Grain legumes are low in fat and cholesterol and are rich in energy-giving carbohydrates. They also provide a good source of B-group vitamins, non-haeme iron, calcium, magnesium and zinc.

2 Inflammation and grain legume consumption

The intestinal tract is a major interface between the environment and the human body, and therefore an active site for inflammatory responses. The inflammatory response in the gut must be carefully balanced as to minimize microbial pathogen invasion and also avoid chronic inflammation. This balance is especially important in children whose growth and development are adversely affected by inflammation.

2.1 Effect of grain legumes on gastrointestinal inflammation

Evidence for an effect of dry grain legumes on human gastrointestinal inflammation is limited. Consumption of legumes in general has not been examined for ameliorating or exacerbating inflammatory conditions such as inflammatory bowel disease or celiac disease. In animal models, consumption of legumes and legume extracts were associated with a reduction in histologic findings of inflammatory infiltrates (Duranti, 2006). When a water-soluble, non-soy legume extract of polyphenolic antioxidants was applied to cell culture models of the gastrointestinal epithelia, reductions in apoptosis and attenuation of peroxynitrates were seen, consistent with an anti-inflammatory effect (Sandoval, 1997). In a rodent model, no evidence of increased expression of hepatic pro-inflammatory transcripts was seen with increasing dietary doses of red bean (Daniel et al., 2012). This study further supports the safety of elevated consumption as beans constituted up to 60% of the total dietary intake.

2.2 Effect of grain legumes on systemic inflammation

Low-grade systemic inflammation is associated with several debilitating chronic health conditions, including heart disease, metabolic syndrome, hypertension, obesity and hyperlipidaemia. Serum biomarkers of systemic inflammation, such a C-reactive protein, tumour necrosis factor, interleukin-6 and adhesion molecules, have been found to be routinely elevated in these conditions. Multiple cross-sectional observational studies and clinical trials looking at the association or effect of grain legume consumption on these serum markers of inflammation unequivocally have revealed that legume consumption reduces these biomarkers of inflammation (Salehi-Abargouei, 2015). For example, in Iranian women, a panel of inflammatory biomarkers was tested and a quantitative dietary survey of legume intake was conducted (Esmaillzadeh and Azadbakht, 2012). A 15–40% reduction in inflammation was observed among women consuming 2.4 times as much legumes daily. Likewise, in a crossover intervention trial where overweight diabetic participants consumed controlled diets in which legumes such as navy, pinto, kidney and black beans were substituted for red meat, a greater reduction in a panel of inflammatory biomarkers was seen (Hartman et al., 2010). However, in a crossover intervention trial where participants consumed 250 g/d of common bean or what was described as a 'healthy American diet' without legumes, both groups showed reductions in serum biomarkers of inflammation from the habitual diet, suggesting that legumes may not be uniquely efficacious in reducing systemic inflammation (Hooseinpour-Niazi et al., 2015).

The mechanism of action by which this reduction in inflammation is achieved is believed to occur by dietary fibres and phytochemicals that exist across the different types of legumes, which can bind inflammatory stimuli in the gut, as well as the prebiotic effect of the fibre-promoting populations of microbiota that promote the secretion of anti-inflammatory cytokines (Messina 2014). Additionally, the absorbed phytochemicals (e.g. polyphenols and fatty acids) from the legumes may have a direct effect on the inflammatory signalling pathways. For example, certain flavonoids that have been identified as a small molecule from beans are metabolized by the human colonic microflora into smaller molecules including phenolic acids which may then be absorbed and influence multiple biological signalling pathways including, but not limited to, angiogenesis, oxidative stress, and inflammation (Le Marchand, 2002).

There is no scientific evidence that consuming dry grain legumes affects the severity or incidence of autoimmune diseases.

3 Dietary fibre and health

3.1 Dietary fibre and allergies

Dietary fibre seems to play an important role in gut bacterial ecology, diversity and function. Fibre is a main source of energy for gut bacteria and diets low in fibre have detrimental impact on gut microbiota and the host. It has been hypothesized that fibre protects the host from food allergies through several cellular pathways. For example, short-chain fatty acids (SCFAs) produced through fermentation of fibre can bind to the G protein-coupled receptors that are expressed by both immune and non-immune cells. SCFAs can also prevent histone deacetylase activity in immune cells, which promotes anti-inflammatory cells (Daïen, 2017). While no study has specifically examined the role of dietary fibre in preventing allergies in humans, mice were used to test the relationship between dietary fibre and peanut allergies. High-fibre feeding in mice reshaped gut microbial ecology, increased the release of SCFAs, improved oral tolerance and protected the mice from food allergy. This study suggests that deficiency of dietary fibre by altering gut microbiota and SCFAs may explain the increase in food allergies in Western countries (Tan, 2016).

3.2 Dietary fibre and diabetes

The prevalence of type 2 diabetes is rapidly increasing, with an estimated 422 million cases worldwide (McMacken and Shah, 2017). While the aetiology of the disease is complex, an increasing body of research indicates that dietary factors, including low fibre intake, may increase the risk of developing the disease (Parillo and Riccardi, 2004). Numerous prospective studies found that diets high in fibre have a protective effect against type 2 diabetes, (Meyer, 2000; Weng, 2012; Aune, 2013; Montonen, 2003; de Munter, 2007), though others did not find an association (Hodge et al., 2016; Schulze et al., 2004). One meta-analysis drawing on relevant prospective studies supports an inverse relationship between total dietary fibre, cereal fibre, fruit fibre or insoluble fibre intake and risk of type 2 diabetes (Yao, 2014), while another meta-analysis only found a significant association with cereal fibre (Schulze et al, 2007).

While additional research is needed to explore the association, there are several potential pathways that fibre may protect against type 2 diabetes. Fibre reduces energy density of food, promotes satiety and reduces risk of overweight/obesity, which is a known risk factor for type 2 diabetes (Liu, 2003; McMacken and Shah, 2017). Both soluble fibre, which has viscous properties and slows digestion, and insoluble fibre, which adds bulk to stool and helps food pass quickly through the digestive system, can reduce nutrient absorption. Additionally, SCFAs produced during the fermentation of fibre modulate postprandial glucose response, and can improve glucose signalling and sensitivity (Lattimer and Haub, 2010).

3.3 Dietary fibre and cardiovascular disease

Cardiovascular disease (CVD) is the number one cause of death globally, accounting for nearly 31% of all deaths every year. One of the most important modifiable risk factors is diet, and there is considerable evidence to suggest that diets high in fibre protect against CVD. Systematic reviews and meta-analyses have found an inverse association between dietary fibre and CVD in short, intermediate and long-term follow-up (Erkkilä and Lichtenstein, 2006; Kim et al., 2016; Ning, 2014; Park, 2011; Satija and Hu, 2012; Threapleton, 2013). One dose–response meta-analysis found a 9% decrease in CVD risk for every additional 7g/day of total fibre, while another found a 9% decrease in CVD mortality for every additional 10g/day of total fibre (Kim and Je, 2016). These studies support the American Heart Association's recommendations to increase dietary fibre intake.

Dietary fibre may reduce risk of CVD through several mechanisms. Dietary fibre reduces serum low-density lipoprotein cholesterol, an established risk factor for CVD, through the production of SCFAs which inhibit cholesterol synthesis (Satija and Hu, 2012). Additionally, viscous fibre has a hypocholesterolemic effect by reducing bile acid absorption. When bile acid is released in faeces, more cholesterol is needed to make more bile acid, which increases cholesterol uptake from circulation. High-fibre diets have reported to improve insulin sensitivity as soluble fibre delays gastric emptying and digestion, which slows absorption, therefore reducing rate of glucose absorption and plasma insulin concentrations (Erkkilä and Lichtenstein, 2006). High-fibre diets also seem to be inversely associated with blood pressure and chronic inflammation, though the mechanism is not fully understood (Satija and Hu, 2012).

4 Carcinogenesis and dry grain legume consumption

Cancer is a condition where mutated cells of a specific tissue type multiply uncontrollably such that they damage or destroy normal tissue architecture and function. Cancer is a leading cause of death worldwide, accounting for 13% of all deaths worldwide. The process of carcinogenesis is prolonged, with multiple, sequential genetic mutations required before clinically important cancer is appreciated. Increased consumption of fibre from grain legumes was compared to fibre consumed from other fruits and vegetables, and showed stronger associations with reduced risk of cancer in a large prospective cohort study (Park, 2011).

4.1 Grain legumes and colorectal cancer

In terms of grain legumes and cancer, most is known about human colorectal cancers. Two very large meta-analyses with data from about 4 million subjects show that higher legume consumption is associated with a relative risk of colorectal cancer of 0.90–0.91 for consumption of 10 g/d of fibre (Zhu, 2015; Aune, 2011). Fibre is one essential anti-carcinogenic component of all legumes. A dose-dependent relationship with fibre consumption was shown whereby increasing fibre consumption to 30 g/d was associated with a relative risk reduction of colorectal cancer (0.83).

Animal models of colon carcinogenesis showed that legume consumption prevents tumour function with ingested carcinogens. The mechanism of action is in part prevention of exposure of the tissue to dietary carcinogens through binding of the carcinogens to intraluminal fibre (Sanchez-Chino, 2015) and possibly hepatic detoxification as shown by Daniel et al., 2012. It was also shown that dietary chickpea reduced the carcinogenicity of an injected carcinogen on colon cancer, suggesting the mechanism is something more than non-specific complexing of compounds in the intestinal lumen (Murillo, 2004). Stachyose, an oligosaccharide commonly found in legume fibre, has been shown *in vitro* with Caco-2 cell cultures to induce apoptosis in a dose-dependent manner, mediated by the caspase-dependent mitochrondrial pathway (Huang, 2015). The mechanism by which legume fibre prevents colorectal carcinogenesis is likely multifactorial, and merits continued evaluation in people at high risk (Mentor-Marcel, 2009). Feasibility and tolerability of increased consumption of cooked navy beans to doses that were reported to reduce colon cancer formation in animals were recently demonstrated in colorectal cancer survivors (Borresen, 2016).

4.2 Dry grain legumes and other cancers

Given the evidence for legume consumption reducing colorectal cancers through direct contact with potentially dysplastic tissues, it is plausible that legume consumption alters the process of carcinogenesis in other sections of the gastrointestinal tract. Epidemiologic evidence from South America shows that consuming more legumes reduces the risk of pharynx, larynx, oesophageal, stomach and renal carcinoma, but does not reduce the risk of more common cancers such as prostrate and breast cancers (Aune, 2009). In a rat model, six distinct market classes of common dry beans were compared for reducing mammary cancer incidence and multiplicity (Thompson, 2009), and showed common bean consumption prevents the absorption of breast cancer carcinogens (Thompson, 2008). Further interrogation of mechanisms of action for dry beans appears to be preventing absorption of carcinogenic nitrogen adducts and modulation of tissue lipid metabolism (Mensack, 2012; Thompson, 2012).

5 Gut health and grain legume consumption

Gut health can be broadly described as the ability of the gastrointestinal tract to efficiently absorb essential and nonessential nutrients present in the diet, while maintaining appropriate antimicrobial defences and barriers to the many microbes in the lumen of the gastrointestinal tract. The community of all gut microbes, known as the microbiota, plays a key role in maintaining gut health. Emerging evidence supports that fibre types, such as the indigestible polysaccharides in the diet, may be primary determinants of the members of the microbiota (Simpson and Campbell, 2015). Fibre is a mixture of hundreds of compounds, some of which are biologically inert, while others are fully consumed by the gut microbes (Graf, 2015). Still other fibre components may exert transient effects on the microbial environment as they pass through the lumen prior to their degradation. At this time, it is not clearly understood which types and combinations of dietary fibre are optimal for gut health across the lifespan, or how the intricate interplay of microbes and host in various segments of the gastrointestinal tract determines optimal gut immunity and barrier function. While diets enriched with cranberry beans favourably altered gut microbiota to mitigate colitis in mice (Monk, 2016), pinto beans and navy beans were in separate studies shown to influence gut bacterial metabolism of adults (Finley, 2007; Sheflin, 2017), and these microbial metabolic mechanistic investigations for grain legumes to enhance gut health merits continued research attention in larger cohorts globally.

Legumes are rich in indigestible carbohydrates often called fibre. Legume fibre is a conglomerate of pectic substances, lignin, cellulose, hemi-cellulose, non-cellulose glycans and α-galactosides (Campos-Vega, 2009; Han and Baik, 2006; Khan, 2007). α-Galactosidase is not secreted into the human digestive tract, and thus α-galactosides are not available for human nutriture. Table 1 lists the most commonly consumed grain legumes, their fibre content and the types of α-galactosides they contain. It is thought that the dietary fibre provided by legumes is the source of the positive effects on gut health (DeFilippis et al., 2015). The fibre content of cooked dry beans is also associated with decreasing serum

Table 1 Fibre content of commonly consumed legumes[1]

Legume	Total fibre	Raffinose	Stachyose	Verbascose	Ciceritol	Total α-galactosides
Chickpea	3.8	0.4–1.2	2.0–3.6	0.6–4.2	2.5	7.4–7.5
Pea	10.4	0.3–1.6	1.3–5.5	1.6–4.2	0%	5.1–8.7
Lentil	6.8	0.3–1.0	1.7–3.1	0.6–3.1	2.8	3.0–7.1
Broad bean	31.3	0.1–0.3	0.7–3.1	1.7–3.1		3.1–4.2
Common bean	18.4	< 0.05–0.93	0.5–4.1	0.06–4.0		2.6–6.6
Lupin bean	9.7	0.5–1.1	0.9–7.4	0.6–3.4		7.5–9.5

[1] All components expressed as % of dry weight.

cholesterol levels (Duane 1997), increasing satiety and improving weight management (McCrory, 2010).

Some compounds in fibre interfere with epithelial adhesion and invasion of microbes from the gut lumen into the intestinal mucosa *in vitro* and *in vivo* models. These effects can be exerted by the microbe changing its cell surface in response to the fibre, rendering it less virulent or more easily displaced. This may also be referred to as a contrabiotic effect with the potential to act in a directed manner and benefitting human health (Dinito et al., 2006; van Zanten, 2012). Legumes are thought to be promising candidates as contrabiotics.

α-Galactosides can include raffinose, stachyose and verbacose that exhibit prebiotic properties with biological effects in favour of gut health. Bacteria assemble in an extracellular matrix, a biofilm, which enhances their ability to persist in an environment and infect a human host. Raffinose has recently been shown to disrupt the biofilms from *Pseudomonas aeruginosa* (Kim et al., 2016) via raffinose binding to LecA, one of the extracellular proteins secreted to create the biofilm.

Fermentation of foods and the resulting by-products largely destroy α-galactosides. Microbial fermentation of fibre in the human intestine also results in the production of substantial amounts of SCFAs (Morrison and Preston, 2016). SCFAs are a necessary microbial waste product, required for balancing redox reactions in an anaerobic environment. Acetate, proprionate and butyrate comprise more than 95% of the SCFAs in the gut (den Besten, 2013). These SCFAs are largely absorbed by the human, and contribute 10% of the energy in the human diet. They largely feed the cells of the colon. In the liver, endogenous lipid metabolism is modulated by the amounts of SCFAs in the circulation. SCFAs also modulate the intestinal immune response, inhibiting neutrophil migration and reducing chemokine secretion, and promote apoptosis of lymphocytes, macrophages and neutrophils (Vinolo, 2011). Thus, SCFAs reduce gut inflammation. SCFA production is increased by legume consumption, as anaerobic metabolism is common in the gut lumen. It is estimated the increase in SCFAs is 10–100 fold.

One manifestation of poor gut health is environmental enteric dysfunction, widespread T-cell infiltration of the mucosa associated with disruption of paracellular junctions and reduction in the absorptive surface area of the small intestine (Trehan, 2016). Environmental enteric dysfunction affects hundreds of millions of sub-Saharan Africans and South Asians. By increasing SCFA production, and its consequent reduction in inflammation, legumes are considered to be good candidates to improve gut health of the people in the developing world.

A large clinical trial is ongoing in young children in Malawi, where the effect of common bean and cowpea consumption as complementary foods on gut health and the gut microbiota is being examined (Trehan, 2015). If legumes are found to enhance gut health, this could have profound implications for childhood stunting.

6 Proteins, antinutrients and other classes of phytochemicals in legumes

The term antinutrient has been complicated for a clear understanding by nutritionists because cooking substantially reduces the levels of many components that are potentially of concern to human health. Legumes provide a staple source of digestible protein for multiple nations across the lifespan and are a complete complement to whole grains for essential amino acids such as lysine. Common dry beans comprise up to 30% protein, including albumins and globulins. Albumin fractions consist of lectins and enzyme inhibitors that are present in high amounts of raw beans. The α-amylase inhibitors, as well as trypsin and chymotrypsin inhibitors, are present at negligible levels after proper cooking (Admassu Shimelis and Kumar Rakshit, 2005; Barampama and Simard, 1995). Independently, lectins, α-amylase inhibitors and trypsin inhibitors have demonstrated modulation of gut hormone secretion (Lajolo and Genovese, 2002), interference with tumour cell growth and metabolism (Yau, 2015), influence on the gut microbiome, decreases in inflammation and modulation of lipid metabolism. The relevant doses for these components found in dry beans to elicit these effects as part of a complete diet with other foods merit further investigation from human studies emphasizing consumption of whole cooked beans at appreciable, increased levels. Given the observations for these effects in clinical trials and epidemiological studies, cooked bean proteins may retain enough inhibitory activity to confer a health benefit, while minimizing any potential nutritional deficit.

For example, α-amylase inhibition activity may be associated with slowing carbohydrate digestibility in the small intestine, while reducing dietary energy extraction has potential benefit for overweight/obese body conditions. While bean-derived α-amylase inhibitors have been used to improve human weight loss, the compounds that have been isolated from raw bean extracts have had variable results and further support the need for investigations with whole cooked bean consumption. There exists a wide range of small, bioactive molecules in legumes that may provide additional health benefits outside of the traditional carbohydrate, fibre, protein and micronutrient contents. These other non-nutrient compounds, also called phytochemicals that are found in legumes, include, but are not limited to, polyphenolics , lectins, phytates, tannins, oxalates, α-amylase inhibitors and saponins. In laboratory studies, saponins have anticarcinogenic activity. Whether these compounds have beneficial clinical effects is an active area of current investigation in human trials.

6.1 Phenolics

Cooked legumes contain a wide range of polyphenols that include phenolic acids, flavonoids and triterpenic acids. Total phenolic acids reported in cooked bean cultivars average around 1,600 μg/100 g dry matter (as measured by gallic acid equivilents) and some abundant acids include ferulic acid, o-coumaric and p coumaric acid, sinapic acid and caffeic acid (Kalogeropoulos, 2010). As expected, legumes with coloured seed coats tend to have the highest concentration of anthocyandins and flavonoids including catechin, low levels of genistein and epicatechin and kaempferol and even some of the polyphenol resveratrol. In general, polyphenols have been shown to reduce inflammation, cardiometabolic disease, oxidative stress, and inhibit the cell cycle of neoplastic cells (Pandey and Rizvi, 2009; Cruz-Bravo, 2014; Amarowicz et al., 2010). However, emerging evidence supports that these phenolic compounds are associated with improved weight loss and dyslipidaemia by altering fat metabolism and improving glucose homeostatsis (Forester et al., 2012). Polyphenol metabolism occurs via common pathways as the aglycones can be absorbed from the small intestine, yet many polyphenols may also be modified with esters, glycosides or polymers that cannot be readily absorbed in native forms. Hydrolysis by liver and intestinal enzymes or by the colonic microflora assists in polyphenols, absorptive capacity. Efficiency of absorption may be reduced because the microflora degrade aglycones and a series of metabolic events promotes hydrophilicity with the goal of targeted excretion in urine. Given the conjugation and mechanisms for excretion, there are low to undetectable concentrations of polyphenols in the blood after consumption. Systemic polyphenols are conjugated derivatives that are heavily bound to albumin, and while they may penetrate tissues, limited information is available regarding tissue accumulation and should be further investigated across organ systems.

6.2 Phytosterols

Cooked beans also contain the phytosterols such as β-sitosterol, stigmasterol and camesterol – plant-derived lipids that exhibit structural similarity to cholesterol. Phytosterols have been shown to reduce hyperlipidaemia by competitively inhibiting cholesterol incorporation into micelles in the intestinal lumen. There is also increased expression of reverse transporters in enterocytes that in turn will increase the rate at which lipids are excreted back into the lumen after initial absorption, and finally competitively inhibiting the incorporation of cholesterol into chylomicrons (Gylling, 2014).

6.3 Phytates

While there is emerging interest and investigations on the health effects of bean phytochemicals, as mentioned above, some compounds have become strongly associated with antinutrient properties. For example, the antinutrient terminology applied to phytates stems from the ability of phyates to potentially reduce selected micronutrient absorption. This observation is supported by *in vitro* studies using pure phytic acid or from raw legume extracts. As such, phytates were shown to lower carbohydrate digestibility, and in high concentrations, chelate calcium, iron, zinc and magnesium (Broughton, 2003). Phytate is the primary storage molecule of phosphorus in plants and while it is reduced with both soaking and cooking (Helbig, 2003), bean seeds selected for low phytate concentration were correlated with improved iron absorption (Petry, 2013). The possible trade-off with total protein content of the low phytate cultivars warrants attention as other plants have shown a direct correlation between phytate and protein content (Broughton, 2003). Furthermore, phytates demonstrate strong cancer chemopreventive properties (Jariwalla, 1999) and serum lipid-lowering activity (McCrory, 2010), indicating that in low concentrations, they may confer more of a health benefit than a nutritional liability in the context of consuming a complete, balanced diet. Phytates not only are found in legumes, but are also components from other nutrient-dense foods such as whole grains and nuts. Emerging literature supports no effects of dietary phytate on micronutrient status in children (Miller, 2015). Additional studies are warranted in populations showing geographical and dietary diversity to clarify the concern regarding mineral absorption.

6.4 Saponins

Saponins are another class of phytochemicals found in legumes and many other plant foods that have shown health benefits for lipid lowering and cancer protection, but have also been referred to as an antinutrient based on their ability to mimic actions of some endogenous steroids. The effects of food processing and cooking (e.g. soaking) have been examined for influences on saponin contents from various types of beans (Shi, 2004), yet limited data exist for the health effects of saponins from bean consumption.

6.5 Lectins

Lectins are natural insecticides to many plant foods, and are destroyed with proper cooking of legumes. Lectins are at low levels in the edible parts of many fruits and vegetables, and have varied concentrations across types of legumes including soybeans, lentils, lima beans and kidney beans. Also referred to as haemagglutinins, these components may lead to toxicities in the form of nausea and vomiting, and possibly diarrhoea, though only if consumed in large amounts by people. The symptoms that have been most commonly associated with eating improperly cooked red kidney beans occurs due to elevated levels of phytohaemagglutinin relative to other types of legumes. Low levels of lectins consumption do not present a safety concern to humans.

7 Possible chemical toxicant exposures from grain legume consumption

The use of pesticides (e.g. insecticides, fungicides and herbicides) applied to legume crops is subject to regulations in some countries for the maximum allowable residue limits on the raw, uncooked product, but pesticides have received limited research and regulatory attention with regard to their levels in the cooked form. Analysis of foods for chemical toxicant exposures and the levels found in people for this important staple food crop is an emerging global human health concern. Little is known regarding the levels of these agrochemicals that reach the intestinal tract and whether there is substantial biotransformation into toxic intermediates or less toxic compounds by the gut microflora. The relationship between agrochemical exposures from legume intake and intestinal health is an area of toxicologic and public health importance. The combinations of low-dose agrochemical exposures from legume consumption should be addressed alongside complex efforts aimed towards enhancing food and nutritional security for enhanced health across the lifespan. Some of the reported agrochemicals commonly used in legume production are listed in Tables 2 and 3.

8 Conclusion

Legume consumption is clearly associated with a lower risk of colorectal cancer, and basic science studies support this association demonstrating direct anti-carcinogenic activities of compounds found in legumes, as well an effect of the dietary fibre complexing carcinogens in the intestinal lumen. The oligosaccharides plentiful in legumes may reduce gut inflammation and improve gut health, and this effect is likely mediated by the microbiota. The multitude of biologically active metabolites in grain legumes on the whole are safe and likely to promote human health beyond that from staple nutritional support. Consideration should be given to the multitude of agrichemicals used in grain legume production, as they may adversely affect human health.

Table 2 Agrochemicals applied during legume production

Agrochemical	Legumes	Type	URL
Cypermethrin	Soybeans, mung beans and navy beans; field peas and field beans	Insecticide	http://www.herbiguide.com.au/Labels/CYPE200_52025-0500.PDF.PDF
Karate	Lentils, guar, lablab bean, lupines, soybeans and navy beans	Insecticide	http://www.syngentacropprotection.com/pdf/labels/scp1097al1j0109.pdf
Imidacloprid	Adzuki bean, asparagus bean, broad bean (succulent or dry), catjang bean, chinese longbean, field bean, guar bean, jackbean, kidney bean, lablab bean, lima bean (succulent or dry), moth bean (succulent or dry), mung bean, navy bean, pinto bean, rice bean, runner bean, snap bean, sword bean, tepary bean, urd bean, wax bean, yardlong bean, black-eyed pea (succulent or dry), chickpea, cowpea (succulent or dry), crowder pea, dwarf pea, edible pod pea, english pea, field pea, garden pea, green pea, pigeon pea (succulent or dry), snow pea, southern pea succulent or dry), sugar snap, pea gram, lupin, sweet lupin, white lupin, white sweet lupin, lentil	Insecticide	http://www.agrian.com/pdfs/Imidacloprid_4ST_Label.pdf
Malathion	Beans and lupine	Insecticide/Miticide	http://www.kellysolutions.com/erenewals/documentsubmit/KellyData%5CGA%5Cpesticide%5CProduct%20Label%5C66330%5C66330-228%5C66330-228_MALATHION_TECHNICAL_12_15_2011_11_22_27_AM.pdf
Arrow 100 EC	Navy beans, faba beans, field peas, lupins and lucerne	Insecticide/Miticide	http://www.conquestag.com.au/documents%5Cproductlabels%5CConquest%20Arrow%20100%20Label.pdf
Switch	Fresh beans (green, runner, broad), dried beans (dwarf French bean, navy bean [Phaseolus], kidney beans, haricot beans, dried broad beans), vining pea and edible podded pea, dried peas (combining peas including marrowfat)	Fungicide	http://www.fargro.co.uk/products/agrochemicals/switch/switch.asp

(*Continued*)

Table 2 (*Continued*)

Agrochemical	Legumes	Type	URL
Clothianidin	Common bean (dry) (navy bean), Mung bean (dry)	Insecticide	http://apvma.gov.au/sites/default/files/gazette/food-standards/gazette-08032016-pages_23-26.pdf
Fludioxonil	Black bean, cranberry bean, field bean, great Northern bean, kidney bean, lima bean, navy bean, pinto bean, runnerbean, snap bean, small red bean, tepary bean, wax bean, yellow bean	Fungicide	http://www.agrian.com/pdfs/Seed_Shield_Beans_Label.pdf
Mefenoxam		Fungicide	http://www.agrian.com/pdfs/Seed_Shield_Beans_Label.pdf
Thiamethoxam		Fungicide	http://www.agrian.com/pdfs/Seed_Shield_Beans_Label.pdf
Dimethoate	Adzuki bean, cowpea, mung bean, navy bean, pigeon pea, lentils, bean and soy bean	Insecticide	http://www.nufarm.com/assets/32931/1/Dimethoate104879180116.pdf
Azadirachtin	Adzuki beans, field beans, kidney beans, lima beans, moth beans, mung beans, navy beans, pinto beans, rice beans, runner beans, snap beans, tepary beans, urd beans, wax beans, asparagus beans, black-eyed peas, catjang, chinese longbeans, cowpeas, chowder peas, southern peas, yard-longbeans, broad beans, (fava beans), chick peas (garbanzo beans), guar, jackbean (sword bean), lablab bean (hyacinth bean), lentils, peas (garden peas, field peas, sugar snap peas, english pea, snow pea), pigeon peas, soybeans, sweet lupin beans, white lupin beans, white sweet lupin, sword bean	Insecticide	https://www.valent.com/Data/Labels/Azera-Spec-Label-8.13.13.pdf
Pyrethrins		Insecticide	https://www.valent.com/Data/Labels/Azera-Spec-Label-8.13.13.pdf
Quintozene	Beans (navy)	Fungicide	http://www.inchem.org/documents/jmpr/jmpmono/v071pr18.htm

Table 3 Herbicides applied to legumes during production

Agrochemical	Type	URL
Glyphosphate	Herbicide	http://www.nrcresearchpress.com/doi/pdf/10.4141/cjps-2015-098
Carfentrazone-ethyl	Herbicide	http://www.nrcresearchpress.com/doi/pdf/10.4141/cjps-2015-098
Glufosinate ammonium	Herbicide	http://www.nrcresearchpress.com/doi/pdf/10.4141/cjps-2015-098
Diquat	Herbicide	http://www.nrcresearchpress.com/doi/pdf/10.4141/cjps-2015-098
3-amino-1,2,4-triazole	Herbicide	PMID: 2307069
2-sec-butyl-4,6-dinitrophenyl acetate	Herbicide	doi:10.1007/BF01193435

9 Where to look for further information

Research on the effect of grain legumes is of growing interest, given the increasing incidence and prevalence of metabolic syndrome and chronic illnesses. Key issues to delineate with regard to legume consumption include the adequacy of nutrient availability from grain legumes during life when growth and development is accelerated, such as during early childhood and pregnancy. It is largely unknown what health effects may be mediated through the gut microbiota, and grain legumes are a likely to be key in this regard, because of the large fraction of fiber they contain. Leaders in this regard include national research agencies, such as the National Institute of Health in the United States or the Medical Research Council in the United Kingdom, as well as private disease focused foundations.

10 References

Admassu Shimelis, E. and Kumar Rakshit, S. (2005), 'Antinutritional factors and in vitro protein digestibility of improved haricot bean (Phaseolus vulgaris L.) varieties grown in Ethiopia', *Int. J. Food Sci. Nutr.*, 56, 377–87.

Amarowic, R., Estrella, I., Hernández, T., et al. (2010), 'Free radical-scavenging capacity, antioxidant activity, and phenolic composition of green lentil (Lens culinaris)', *Food Chem.*, 121(3), 705–11.

Aune, D., Chan, D. S., Lau, R, et al. (2011), 'Dietary fibre, whole grains, and risk of colorectal cancer, systematic review and dose-response meta-analysis of prospective studies', *BMJ*, 343, d6617.

Aune, D., de Stefani, E., Ronco, A., et al. (2009), 'Legume intake and the risk of cancer, a multisite case-control study in Uruguay', *Cancer Causes Control*, 20, 1605–15.

Aune, D., Norat, T., Romundstad, P., et al. (2013), 'Whole grain and refined grain consumption and the risk of type 2 diabetes: A systematic review and dose-response meta-analysis of cohort studies', *Eur. J. Epidemiol.*, 28(11), 845–58.

Barampama, Z. and Simard, R. E. (1995), 'Effects of soaking, cooking and fermentation on composition, in-vitro starch digestibility and nutritive value of common beans', *Plant Foods Hum. Nutr.*, 48, 349–65.

Borresen, E. C., Brown, D. G., Harbison, G., et al. (2016), 'A randomized controlled trial to increase navy bean or rice bran consumption in colorectal cancer survivors', *Nutr. Cancer*, 68(8), 1269–80.

Broughton, W. J., Hernandez, G., Blair, M., et al. (2003), 'Beans (Phaseolus spp.)–model food legumes', *Plant Soil*, 252(1), 55–128.

Campos-Vega, R., Reynoso-Camacho, R., Pedraza-Aboytes, G., et al. (2009), 'Chemical composition and in vitro polysaccharide fermentation of different beans (Phaseolus vulgaris L.)', *J. Food Sci.*, 74, T59–65.

Cruz-Bravo, R. K., Guevara-Gonzalez, R. G., Ramos-Gomez, M., et al. (2014), 'The fermented non-digestible fraction of common bean (Phaseolus vulgaris L.) triggers cell cycle arrest and apoptosis in human colon adenocarcinoma cells', *Genes. Nutr.*, 9, 359.

Daïen, C. I., Pinget, G. V., Tan, J. K., et al. (2017), 'Detrimental impact of microbiota-accessible carbohydrate-deprived diet on gut and immune homeostasis: An overview', *Front. Immunol.*, 8, 548.

Daniell E. L., Ryan E. P., Brick M. A., et al. (2012), 'Dietary dry bean effects on hepatic expression of stress and toxicity-related genes in rats', *Br. J. Nutr.*, 108(Suppl 1), S37–45.

de Filippis, F., Pellegrini, N., Vannini, L., et al. (2015), 'High-level adherence to a Mediterranean diet beneficially impacts the gut microbiota and associated metabolome', *GUT*, 65, 1812–21.

de Munter, J. S., Hu, F. B., Spiegelman, D, et al. (2007), 'Whole grain, bran, and germ intake and risk of diabetes: A prospective cohort study and systematic review', *PLoS Med.*, 4, e261.

den Besten, G., van Eunen, K., Groen, A. K., et al. (2013), 'The role of short-chain fatty acids in the interplay between diet, gut microbiota, and host energy metabolism', *J. Lipid Res.*, 54, 2325–40.

Dinoto, A., Suksomcheep, A., Ishizuka, S., et al. (2006), 'Modulation of rat cecal microbiota by administration of raffinose and encapsulated Bifidobacterium breve', *Appl. Environ. Microbiol.*, 72, 784–92.

Duane, W. C. (1997), 'Effects of legume consumption on serum cholesterol, biliary lipids, and sterol metabolism in humans', *J. Lipid. Res.*, 38, 1120–28.

Duranti, M. (2006), 'Grain legume proteins and nutraceutical properties', *Fitoterapia*, 77, 67–82.

Erkkilä, A. T. and Lichtenstein, A. H. (2006), 'Fiber and cardiovascular disease risk: How strong is the evidence?', *J. Cardiovasc. Nurs.*, 21(1), 3–8.

Esmaillzadeh, A. and Azadbakht, L. (2012), 'Legume consumption is inversely associated with serum concentrations of adhesion molecules and inflammatory biomarkers among Iranian women', *J. Nutr.*, 142, 334–9.

Finley, J. W., Burrell, J. B. and Reeves, P. G. (2007), 'Pinto bean consumption changes SCFA profiles in fecal fermentations, bacterial populations of the lower bowel, and lipid profiles in blood of humans', *J. Nutr.*, 137(11), 2391–8.

Forester, G., Ollila, C., Burton, J. H., et al. (2012), 'Nutritional weight loss therapy with cooked bean powders regulates serum lipids and biochemical analytes in overweight and obese dogs', *J. Obes. Wt. Loss.*, 2(8), 1–8.

Graf, D., di Cagno, R., Fak, F., et al. (2015), 'Contribution of diet to the composition of the human gut microbiota', *Microb. Ecol. Health Dis.*, 26, 26164.

Gylling, H., Plat, J., Turley, S., et al. (2014), 'Plant sterols and plant stanols in the management of dyslipidaemia and prevention of cardiovascular disease', *Atherosclerosis*, 232, 346–60.

Han, I. H. and Baik, B. K. (2006), 'Oligosaccharide content and composition of legumes and their reduction by soaking, cooking, ultrasound, and high hydrostatic pressure', *Cereal Chem.*, 83(4),428–33.

Hartman, T. J., Albert, P. S., Zhang, Z., et al. (2010), 'Consumption of a legume-enriched, low-glycemic index diet is associated with biomarkers of insulin resistance and inflammation among men at risk for colorectal cancer', *J. Nutr.*, 140, 60–7.

Helbig, E., de Oliveira, A. C., Queiroz Kda, S., et al. (2003), 'Effect of soaking prior to cooking on the levels of phytate and tannin of the common bean (Phaseolus vulgaris, L.) and the protein value', *J. Nutr. Sci. Vitaminol. (Tokyo)*, 49, 81–6.

Hodge, A. M., English, D. R., O'Dea, K., et al. (2004), 'Glycemic Index and Dietary Fiber and the Risk of Type 2 Diabetes', *Diabetes Care*, 27(11), 2701–6.

Hosseinpour-Niazi, S., Mirmiran, P., Fallah-Ghohroudi, A., et al. (2015), 'Non-soya legume-based therapeutic lifestyle change diet reduces inflammatory status in diabetic patients, a randomised cross-over clinical trial', *Br. J. Nutr.*, 114, 213–19.

Huang, G., Mao, J., Ji, Z., et al. (2015), 'Stachyose-induced apoptosis of Caco-2 cells via the caspase-dependent mitochondrial pathway', *Food Funct.*, 6, 765–71.
Jariwalla, R. J. (1999), 'Inositol hexaphosphate (ip6) as an antineoplastic and lipid-lowering agent', *Anticancer Res.*, 19, 3699–702.
Kalogeropoulos, N., Chiou, A., Ioannou, M., et al. (2010), 'Nutritional evaluation and bioactive microconstituents (phytosterols, tocopherols, polyphenols, triterpenic acids) in cooked dry legumes usually consumed in the Mediterranean countries', *Food Chem.*, 121(3), 682–90.
Khan, A. R., Alam, S., Ali, S., et al. (2007), 'Dietary fiber profile of food legumes', *Sarhad J. Agric.*, 23(3), 763.
Kim, H. S., Cha, E., Kim, Y., et al. (2016), 'Raffinose, a plant galactoside, inhibits Pseudomonas aeruginosa biofilm formation via binding to LecA and decreasing cellular cyclic diguanylate levels', *Sci. Rep.*, 6, 25318.
Kim, Y. and Je, Y. 2016. Dietary fibre intake and mortality from cardiovascular disease and all cancers: A meta-analysis of prospective cohort studies. *Arch. Cardiovasc. Dis.*, 109(1), 39–54.
Lajolo, F. M. and Genovese, M. I. (2002), 'Nutritional significance of lectins and enzyme inhibitors from legumes', *J. Agric. Food Chem.*, 50, 6592–98.
Lattimer, J. M. and Haub, M. D. (2010), Effects of dietary fiber and its components on metabolic health. *Nutrients*, 2, 1266–89.
Le Marchand, L. (2002), Cancer preventive effects of flavonoids–a review. *Biomed. Pharmacother.*, 56(6), 296–301.
Liu, S., Willett, W. C., Manson, J. E., et al. 2003, Relation between changes in intakes of dietary fiber and grain products and changes in weight and development of obesity among middle-aged women. *Am. J. Clin. Nutr.*, 78(5), 920–7.
McCrory, M. A., Hamaker, B. R., Lovejoy, J. C., et al. (2010), 'Pulse consumption, satiety, and weight management', *Adv. Nutr.*, 1, 17–30.
McMacken, M. and Shah, S. (2017), 'A plant-based diet for the prevention and treatment of type 2 diabetes', *J. Geriatr. Cardiol.*, 14(5), 342–54.
Mensack, M. M., McGinley, J. N. and Thompson H. J. (2012), 'Metabolomic analysis of the effects of edible dry beans (Phaseolus vulgaris L.) on tissue lipid metabolism and carcinogenesis in rats', *Br. J. Nutr.*, 108(Suppl 1), S155–65.
Mentor-Marcel, R. A., Bobe, G., Barrett, K. G., et al. (2009), 'Inflammation-associated serum and colon markers as indicators of dietary attenuation of colon carcinogenesis in ob/ob mice', *Cancer Prev. Res. (Phila)*, 2(1):60–9.
Messina, V. (2014), 'Nutritional and health benefits of dried beans', *Am. J. Clin. Nutr.*, 100(Suppl 1), 437s–42s.
Meyer, K. A., Kushi, L. H., Jacobs Jr., D. R., et al. (2000), 'Carbohydrates, dietary fiber, and incident type 2 diabetes in older women', *Am. J. Clin. Nutr.*, 71(4), 921–30.
Miller, L.V., Hambidge, K.M. and Krebs, N.F. (2015), 'Zinc absorption is not related to dietary phytate intake in infants and young children based on modeling combined data from multiple studies', *J. Nutr.* 145(8), 1763–9.
Monk J. M., Lepp, D., Zhang, C. P., et al. (2016), Diets enriched with cranberry beans alter the microbiota and mitigate colitis severity and associated inflammation. *J. Nutr. Biochem.*, 28, 129–39.
Montonen, J., Knekt, P., Järvinen, R., et al. (2003), 'Whole-grain and fiber intake and the incidence of diabetes', *Am. J. Clin. Nutr.*, 77, 622–9.
Morrison, D. J. and Preston, T. (2016), 'Formation of short chain fatty acids by the gut microbiota and their impact on human metabolism', *Gut Microbes.*, 7(3), 189–200.
Murillo, G., Choi, J. K., Pan, O., et al. (2004), 'Efficacy of garbanzo and soybean flour in suppression of aberrant crypt foci in the colons of CF-1 mice', *Anticancer Res.*, 24, 3049–55.
Ning, H., van Horn, L., Shay, C. M., et al. (2014), 'Associations of dietary fiber intake with long-term predicted cardiovascular disease risk and C-reactive protein levels (from the National Health and Nutrition Examination Survey Data [2005–2010])', *Am. J. Cardiol.*, 113(2), 287–91.

Pandey, K. B. and Rizvi, S. I. (2009), 'Plant polyphenols as dietary antioxidants in human health and disease', *Oxid. Med. Cell Longev.*, 2, 270–8.
Parillo, M. and Riccardi, G. (2004), 'Diet composition and the risk of type 2 diabetes: Epidemiological and clinical evidence', *Br. J. Nutr.*, 92(1), 7–19.
Park, Y., Subar, A. F., Hollenbeck, A., et al. 2011, 'Dietary fiber intake and mortality in the NIH-AARP diet and health study', *Arch. Intern. Med.*, 171(12), 1061–8.
Petry, N., Egli, I., Campion, B., et al. (2013), 'Genetic reduction of phytate in common bean (Phaseolus vulgaris L.) seeds increases iron absorption in young women', *J. Nutr.*, 143, 1219–24.
Salehi-Abargouei, A., Saraf-Bank, S., Bellissimo, N., et al. (2015), 'Effects of non-soy legume consumption on C-reactive protein, a systematic review and meta-analysis', *Nutrition*, 31, 631–9.
Sanchez-Chino, X., Jimenez-Martinez, C., Davila-Ortiz, G., et al. (2015), 'Nutrient and nonnutrient components of legumes, and its chemopreventive activity, a review', *Nutr. Cancer*, 67, 401–10.
Sandoval, M., Ronzio, R. A., Muanza, D. N., et al. (1997), 'Peroxynitrite-induced apoptosis in epithelial (T84) and macrophage (RAW 264.7) cell lines, effect of legume-derived polyphenols (phytolens)', *Nitric Oxide*, 1, 476–83.
Satija, A. and Hu, F. B. (2012), 'Cardiovascular benefits of dietary fiber', *Curr. Atheroscler. Rep.*, 14(6), 505–14.
Schulze, M. B., Liu, S., Rimm, E. B., et al. (2004), 'Glycemic index, glycemic load, and dietary fiber intake and incidence of type 2 diabetes in younger and middle-aged women', *Am. J. Clin. Nutr.*, 80(2), 348–56.
Schulze, M. B., Schulz, M., Heidemann, C., et al. (2007), 'Fiber and magnesium intake and incidence of type 2 diabetes: A prospective study and meta-analysis', *Archives of Internal Medicine*, 167(9), 956–65.
Sheflin, A. M., Borresen, E. C., Kirkwood, J. S., et al. 2017, 'Dietary supplementation with rice bran or navy bean alters gut bacterial metabolism in colorectal cancer survivors', *Mol. Nutr. Food Res.*, 61(1).
Shi, J., Arunasalam, K., Yeung, D., et al. (2004), 'Saponins from edible legumes, chemistry, processing, and health benefits', *J. Med. Food*, 7, 67–78.
Simpson, H. L. and Campbell, B. J. (2015), 'Review article, dietary fibre-microbiota interactions', *Aliment Pharmacol. Ther.*, 42, 158–79.
Tan, J., McKenzie, C., Vuillermin, P. J., et al. (2016), 'Dietary fiber and bacterial SCFA enhance oral tolerance and protect against food allergy through diverse cellular pathways', *Cell Rep.*, 15(12), 2809–24.
Thompson, M. D., Brick, M. A., McGinley, J. N., et al. (2009), 'Chemical composition and mammary cancer inhibitory activity of dry bean', *Crop Sci.*, 49, 179–86. https://dl.sciencesocieties.org/publications/cs/abstracts/49/1/179.
Thompson, M. D., Mensack, M. M., Jiang, W., et al. (2012), 'Cell signaling pathways associated with a reduction in mammary cancer burden by dietary common bean (Phaseolus vulgaris L.)', *Carcinogenesis*, 33(1), 226–32.
Thompson, M. D., Thompson, H. J., Brick, M. A., et al. (2008), 'Mechanisms associated with dose-dependent inhibition of rat mammary carcinogenesis by dry bean (Phaseolus vulgaris, L.)', *J. Nutr.* 138, 2091–7.
Threapleton, D. E., Greenwood, D. C., Evans, C. E., et al. (2013), 'Dietary fibre intake and risk of cardiovascular disease: Systematic review and meta-analysis', *BMJ*, 347, f6879.
Trehan, I., Benzoni, N. S., Wang, A. Z., et al. (2015), 'Common beans and cowpeas as complementary foods to reduce environmental enteric dysfunction and stunting in Malawian children, study protocol for two randomized controlled trials', *Trials*, 16, 520.
Trehan, I., Kelly, P., Shaikh, N., et al. (2016), 'New insights into environmental enteric dysfunction', *Arch. Dis. Child*, 101, 741–4.
van Zanten, G. C., Knudsen, A., Roytio, H., et al. (2012), 'The effect of selected synbiotics on microbial composition and short-chain fatty acid production in a model system of the human colon', *PLoS One*, 7, e47212.

Vinolo, M. A., Rodrigues, H. G., Nachbar, R. T., et al. (2011), 'Regulation of inflammation by short chain fatty acids', *Nutrients* 3, 858–76.

Weng, L. C., Lee, N. J., Yeh, W. T., (2012), 'Lower intake of magnesium and dietary fiber increases the incidence of type 2 diabetes in Taiwnese. *J. Formos Med. Assoc.*, 111(11), 651–9.

Yao, B., Fang, H., Xu, W., et al. (2014), 'Dietary fiber intake and risk of type 2 diabetes: A dose-response analysis of prospective studies', *Eur. J. Epidemiol.*, 29(2), 79–88.

Yau, T., Dan, X., Ng, C. C., et al. (2015), 'Lectins with potential for anti-cancer therapy', *Molecules*, 20, 3791–3810.

Zhu, B., Sun, Y., Qi, L., et al. (2015), 'Dietary legume consumption reduces risk of colorectal cancer, evidence from a meta-analysis of cohort studies', *Sci Rep.*, 5, 87–97.

Chapter 19

The nutritional potential of grain legumes: an economic perspective

Alan de Brauw, International Food Policy Research Institute, USA

1 Introduction

The second sustainable development goal (SDG), ratified by the United Nations in 2015, explicitly lays out a challenge linking agriculture to nutrition, and therefore health, while also promoting sustainable agriculture. At the heart of this challenge is the observation that while the global food system may produce enough food, it does not produce the right type of food for the global population to consume healthy diets. Indeed, evidence shows that consumers at most income levels may find it challenging to obtain a nutritious diet (Alston, MacEwan and Okrent, 2016).

Consequently, while measures of overall food security have been improving over time, as many as 1 in 3 people worldwide are malnourished. There are three major types of malnutrition: undernutrition, in which individuals do not obtain enough calories; micronutrient malnutrition, in which individuals may obtain enough calories but not enough micronutrients; and overnutrition, in which individuals consume too many calories. At present, even within countries, multiple forms of malnutrition are observed concurrently (IFPRI, 2016).

Grain legumes can play an important role in reducing at least two of the three types of malnutrition. First, pulses and legumes are typically rich in several vitamins and minerals than the three major grains (rice, wheat and maize) (USDA, 2015). Consequently, they can potentially play an increased role in reducing micronutrient malnutrition among those suffering from micronutrient deficiencies, if grain legumes are added to the diet. Second, grain legumes or pulses can also potentially play an important role in reducing

http://dx.doi.org/10.19103/AS.2017.0023.16

overnutrition, if grains or processed foods are substituted for in the diet (Kim et al., 2016). When unprocessed, cooked pulses are included in the diet, they create a feeling of satiety, reducing hunger and at least theoretically reducing calorie intake (McCrory et al., 2010).

If pulses and legumes are so potentially beneficial from a nutritional standpoint, an important question is why they do not play a more prominent role in either agricultural production systems or consumers' diets around the world. In fact, considering all available calories globally since 1961, Khoury et al. (2014) show that the share represented by pulses has been declining. The decline can partially be explained by changes in production patterns; productivity gains have been higher among other crops. Whereas pulse production has grown by 1.3% per annum between 1980 and 2013 (Joshi and Rao, 2017), total grain production has grown by 1.9% per annum (FAO, 2016). As populations are also increasing, pulse availability per capita has actually stagnated over the more recent period, and has fallen substantially since 1961 (Joshi and Rao, 2017). So pulses have become less available over time, increasing their scarcity.

This chapter discusses the gap between the nutritional and economic values of grain legumes in the diet, and proposes some potential ways through which these values could be better aligned. The chapter generally takes the perspective of less-developed countries, where pulses can potentially play an important role in reducing micronutrient malnutrition and the rate of increase in overnutrition. First, it discusses the nutritional value of pulses from several perspectives, as well as biofortification, which aims to further improve the nutritional value of specific pulses. Secondly, the reasons why a gap exists between the nutritional and economic values of pulses are discussed. The chapter concludes with a discussion of ways through which the economic value could be aligned better with nutritional value in the future.

2 Pulses and their nutritional value

Relatively, pulses tend to be higher in selected macronutrients and micronutrients than in the three main grain crops (Table 1). Pulses typically have 66% more protein than durum wheat, the grain with the most protein content, and 46% more fibre. When compared to brown rice or maize, the differences are even higher. Pulses are also rich in specific minerals than major grains. According to the USDA Standard Reference the average pulse has 6.6 mg of iron per 100 g, whereas major grains have between 1 and 3 mg of iron. Pulses are a major source of potassium, as the average pulse has three times the iron of durum wheat, the best source of iron among major grains. However, iron content of beans varies slightly, ranging from 4.3 to 10.4 mg per 100 g. The zinc content of pulses is also 50% higher than that of brown rice or maize on average; the minimum content in the database (black beans) is comparable to maize or brown rice.[1] Finally, pulses are a good source of some B vitamins; specifically, they are quite high in folate, averaging 469 μg per 100 g, which exceeds the US-recommended daily allowance for all groups except pregnant and lactating women.

Given their nutritional attributes, there are substantive reasons why an increase in pulse consumption might lead to a reduction in malnutrition. First, as shown in Table 1,

1. Moreover, wheat and maize in particular must be milled before consumption, which can reduce mineral and micronutrient content.

Table 1 Nutritional attributes of pulses relative to major grains

Crop	Protein (g)	Fibre (g)	Iron (mg)	Zinc (mg)	Potassium (mg)	Magnesium (mg)	Folate (mcg)	Niacin (mg)
Average value, major pulses	22.8	17.8	6.65	3.14	1237	166	469	2.09
Minimum value, major pulses	20.5	10.7	4.31	2.20	677	47	274	0.48
Maximum value, major pulses	26.1	25.5	10.44	6.11	1795	333	639	2.97
Brown rice	9.4	7.3	2.71	2.21	287	127	19	3.63
Maize	7.5	3.4	1.80	2.02	268	143	20	4.31
Durum wheat	13.7	12.2	3.52	4.16	431	144	43	6.74

Notes: Values for pulses listed above are average, maximum and minimum values among black turtle beans, roman beans, great northern beans, kidney beans, navy beans, pink beans, pinto beans, yellow beans, white beans, fava beans, chickpeas, cowpeas, lentils, mung beans, pigeon peas and split green peas. *Source:* US Department of Agriculture, Agricultural Research Service, Nutrient Data Laboratory. *USDA National Nutrient Database for Standard Reference, Release 28*. Version Current: September 2015.

substituting pulses for some coarse grain consumption would lead to increased intakes of some vitamins and minerals. Among major micronutrient deficiencies, pulses are perhaps best placed to help alleviate iron deficiency and iron-deficiency anaemia, the most common micronutrient deficiency in the world (WHO, 2014). Iron-deficiency anaemia causes fatigue, lethargy and slower cognitive development in young children, and maternal anaemia increases the risk of miscarriages, prematurity and low birth weight. Horton and Ross (2003) estimated that combined physical and cognitive losses are equivalent to 4% of GDP across ten developing countries, with potential impacts on child survival being ignored. So reducing iron-deficiency anaemia could have important implications for economic growth.

However, it is well known that iron in plant-based foods is not as bioavailable as in meat, because of the presence of phytic acids that hinder absorption (e.g. Hunt, 2003).[2] Phytic acids bind with iron in foods and inhibit iron absorption. An important question, therefore, is whether increasing intakes of pulses or other grain legumes would be bioavailable. Several studies have examined questions about the relative bioavailability of iron from pulses or other grain legumes. Specifically, Murray-Kolb et al. (2003) demonstrated that women with low iron stores are able to absorb iron from soybeans. Lonnerdal (2009) suggested that by breeding soybeans for ferritin, iron absorption can increase as iron binds to ferritin molecules and then can be absorbed by the body. Normally, beans have low concentrations of ferritin. In a more recent study, Petry et al. (2016) found that low-phytic acid beans improve iron absorption relative to control

2. The presence of calcium in foods can also inhibit iron absorption; this is less of an issue for grain legumes (Roughead, Zito and Hunt, 2005).

beans among Rwandan women with low iron status, but low-phytic acid beans also appear to cause adverse gastrointestinal symptoms, specifically nausea, abdominal pain and diarrhoea.

Rather than reducing phytic acids in beans, an alternative strategy is to increase the iron content through conventional breeding, or biofortification (Bouis et al., 2011). High-iron beans have been released in 15 countries, both in Central and South America and in East and Central Africa (HarvestPlus, 2016). Evidence exists for the following: iron status improves among Mexican primary school–age children when consuming high-iron beans (Haas, 2014), and improves among Rwandan women in a feeding trial, relative to beans with normal iron levels (Haas et al., 2016). Moreover, Petry et al. (2016) also showed that biofortified beans demonstrate improvements in iron absorption similar to a finding for low-phytic acid beans, without the adverse gastrointestinal symptoms. So adding biofortified beans to the diet is certainly promising from the perspective of reducing one type of micronutrient malnutrition, iron-deficiency anaemia.

Pulses may also play a role in reducing overnutrition when added to the diet. In a systematic review, Kim et al. (2016) found that participants in clinical trials, comparing the effects of diets including whole pulses to diets without pulses, demonstrated a weight reduction of 0.34 kg over a median duration of 6 weeks, without any caloric restrictions. Subjects in these trials tended to already be overweight or obese. Although no mechanism was identified, they suggested increased satiety after eating pulses as one mechanism, and the reduced bioavailability of calories as another potential mechanism. Though limited to nine studies with small sample sizes, Li et al. (2014) found that satiety, defined as the reduction in appetite after a meal, increases when pulses are consumed. Reduced caloric bioavailability would occur through the relatively high fibre content of pulses (e.g. Howarth, Saltzman and Roberts, 2001). More research is necessary to confirm both weight reductions from diets including pulses and the potential mechanisms by which such a reduction occurs.

Including plant proteins such as pulses in diets can also improve glycaemic control among individuals with diabetes, relative to animal proteins (Viguiliouk et al., 2015). As with the other systematic reviews cited earlier, the total sample size for reported trials is small, and improvements are modest but statistically significant. Similarly, systematic reviews of small numbers of trials have shown that pulse interventions can reduce systolic blood pressure (Jayalath et al., 2013) and low-density lipoprotein cholesterol (Ha et al., 2014). As all these conditions are positively related to obesity status, this combined evidence at least modestly suggests that including pulses in the diet can lead to weight reductions and the incidence of some disease related to obesity.

In addition, pulses have important nutritional value over coarse grains that provide the bulk of calories in the diets of most developing countries. In places where micronutrient malnutrition is prevalent, an increased role of pulses in the diet could lead to reductions in iron deficiency, particularly if biofortified beans have been introduced, with no other adverse health effects. Systematic reviews also point to potential benefits of including pulses in diets to help reduce weight among the overweight and obese or to reduce the probability of an individual becoming overweight or obese. As the prevalence of overweight and obesity status is increasing literally in every country of the world (IFPRI, 2015), there is potential urgency for larger trials to pinpoint the dietary benefits of pulses suggested by the small trials combined in the reviews cited earlier.

3 The economic value of pulses: overview

Given that pulses have a relatively high nutritional value, or value within diets, next it is worth exploring how they are valued economically. To address this question, we need to first discuss briefly how economic values (or prices) are generated, and then we reflect upon why the economic value may not reflect the nutritional value.

Prices reflect an equilibrium between the demand and supply for a good. If a market is perfectly competitive, equilibrium prices also reflect the cost of producing one more unit of that good. Therefore, prices tell us something about how much demand exists for specific goods as well as the cost of producing them. In agricultural markets, in particular for bulky storable crops such as coarse grains and pulses, it is also important to understand transaction costs for getting crops to market; if transaction costs are too high, farmers in remote areas may not trade in specific goods because transaction costs lead the net price for the producer to be too low (e.g. Key, de Janvry and Sadoulet, 2000).

That said, prices do not reflect externalities, which are factors for which markets do not adjust. For pulses, the important externality is in production, as pulses (and other grain legumes) fix nitrogen in the soil. As such, there is a value in producing pulses in a crop rotation that is not reflected in the crop's market value. Such value may be considered by farmers who plant pulses, but they may not account for such value in making the decision to plant pulses, especially if they are unaware of this benefit.

From the perspective of production, there is strong evidence that at least the relative cost of producing pulses has been increasing. Figure 1 illustrates the worldwide average change in yields of coarse grains, cereals and pulses over a longer period (1961–2014);

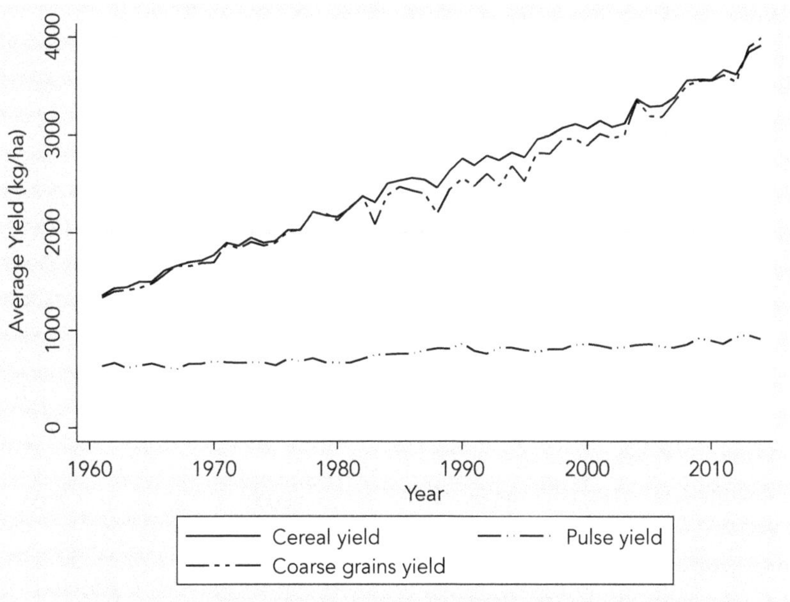

Figure 1 Average yield growth of cereals, coarse grains and pulses, World, 1960–2011. *Source:* FAOSTAT database.

note that whereas coarse grain and cereal yields were double those of pulses in 1961, they are approximately four times higher by 2014. From a producer's perspective, if other costs of production are held constant, it costs about twice as much to grow pulses on average in 2014 than it had in 1961. Under an assumption of competitive markets for major grains and pulses, it is safe to assume that pulse prices have roughly doubled since 1961 relative to major grains in real terms.

Of course, this worldwide pattern masks a great deal of heterogeneity, as pulses grow well in some conditions but not in others. Figure 2 shows a plot of average cereal yields and average pulse yields, measured in 2010. Countries with higher yields in general tend to have higher GDPs; what we find is that there is an approximately linear and increasing relationship between major grain yields and pulse yields, but some countries are better at producing pulses relative to grains than others; those that fall below the linear fit are relatively good at producing pulses. It is notable that the bulk of countries with low pulse yields also have low grain yields, suggesting that those are countries in which average yields are falling below agronomic potential.

Therefore, increasing pulse and legume production need not require increasing their sown area; it would seem there is plenty of potential for improvement in yields. Farmers may not choose production processes that maximize yields because either they face binding constraints that hinder technology adoption, or maximum yielding varieties and production processes are not profitable for them. Farmers in less-developed countries face a number of constraints that hinder technology adoption (Jack, 2011). Constraints relevant to legume production beyond the availability of improved varieties include credit, risk and information constraints. Infrastructure problems might also increase transaction

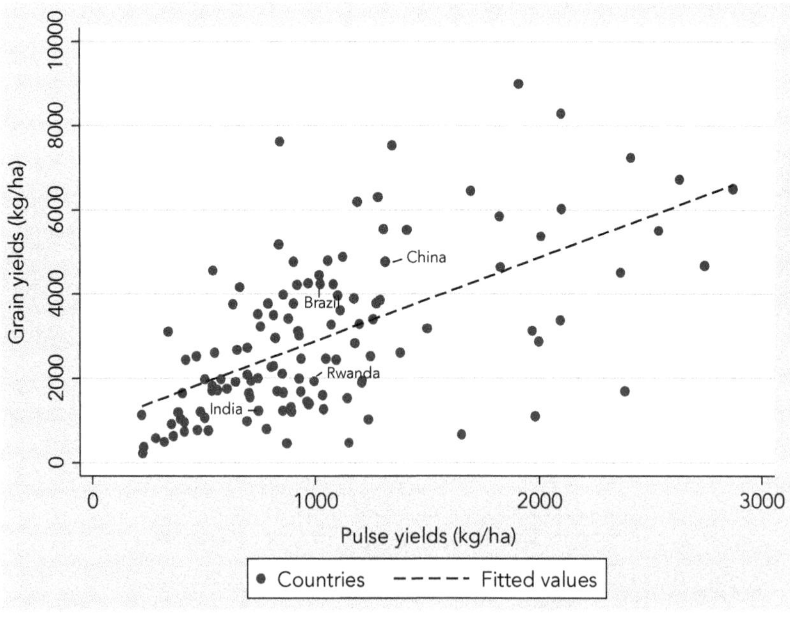

Figure 2 Average grain yields versus pulse yields, by country, 2010. *Source:* FAO (2016).

costs between markets, but this constraint would also be faced by farmers growing major grains, as pulses are similar in that they are not perishable commodities.

If such constraints can be alleviated, presumably farmers can plant the same amount of sown area and produce more, increasing either self-consumption or the amount they can provide to the market. Note that if these constraints can be alleviated, increasing the adoption of improved beans could both increase sown area, because growing beans becomes more profitable, and improve the productivity of area already being allocated to beans. We discuss each in turn in the following sections.

4 Constraints on legume production

4.1 Credit constraints

Smallholder farmers often lack liquidity around planting time and when they lack credit, they cannot purchase improved inputs, such as improved seeds, fertilizers that would enhance production or inoculants for soy production. In general, credit-constrained farmers are less likely to use modern varieties (Lipton, 1976), implying that overcoming constraints in credit markets could have important implications for inclusive growth. In pulse growing areas, this constraint can severely limit private sector participation in the market, especially if they sense a severe lack of demand for inputs. For example, in a 2016 survey conducted among mung bean growers in southern Bangladesh, we found that only 13.4% of farmers purchased mung bean seed in the past 12 months.

Credit constraints appear to be quite difficult to overcome without cash transfers or direct input transfers. For example, evidence from randomized control trials demonstrates that microcredit has low take-up rates (Banerjee, Karlan and Zinman, 2015). Perhaps a more effective device is commitment savings, which requires participants to commit savings at harvest for planting time (e.g. Brune et al., 2016). However, such programmes would only work well for increasing legume production if part of a project attempts to specifically boost output of legume yields, or else choices about the most profitable crops to grow continue to dominate.

Perhaps a more promising approach to alleviating credit constraints specifically targeted for grain legume production would be to provide farmers with one-time cash grants to begin participating in a legume value chain. Specifically, such a programme should include a market for the crop as well as access to seeds in growing seasons beyond the initial season. One-time cash grants have been quite successful at boosting farmer productivity (Karlan et al., 2014; Ambler, de Brauw and Godlonton, 2016a), but would not necessarily lead to sustainable changes in productivity without a method to maintain increased output. A variant of this model was tested by Ambler, de Brauw and Godlonton (2016b), which guaranteed access to seeds placed in the community seed bank for the following planting season.

4.2 Risk constraints

Farmers in developing countries face several sources of risk in the agricultural production process. For crops such as pulses, the main risks faced by farmers relate to input quality, weather and output prices. Farmers may not believe that inputs being sold in local markets are of high quality, potentially for good reason (Ashour et al., 2016). Indeed, many farmers purchase beans for seed from grain markets, which can decrease yields relative to using fresh seed (e.g. Murekezi, 2013). Weather risks largely result from rainfall that is either too high or too low, decreasing yields or high temperatures, which can damage plants while in the field. Finally, farmers also face output price risk; many legume prices are affected by global commodity prices, and as a result, individual farmer's profits can be quite different than expected when the crop is actually sold.

Smallholder response to weather risk, in terms of growing pulses or legumes, will depend upon the perception of the risk relative to growing other crops, as well as farmer's perceptions of future weather. Risk perceptions might be positive or negative for producing beans, depending upon whether or not farmers perceive they help hedge against *ex ante* risk. From the perspective of prices, if farmers perceive prices for beans as more volatile than prices for major grains, risk-averse farmers would be less likely to plant those crops, since their price expectations would not be as stable. Indeed, farmers make *ex ante* decisions that reduce their potential productivity (Dercon, 1996). So if planting beans is perceived as risky, farmers may reduce their planting even if it is more profitable without other ways to deal with risk.

To alleviate this constraint, a number of recent projects have attempted to introduce index-based insurance to smallholders in developing countries in order to alleviate the risk constraint (Carter et al., 2014). To improve demand for weather-based insurance, it is important to understand the factors that determine demand, which include price, basis risk and risk aversion (Hill, Robles and Ceballos, 2016). An innovation suggested by Dercon et al. (2014) was to focus on selling insurance among groups that already share risk among one another. A perhaps more promising alternative is to bundle insurance with other agricultural inputs in high demand. For example, Karlan et al. (2010) bundled crop price insurance with loans to purchase fertilizers, thereby dealing with an alternative risk market inefficiency. This example could be applied to bean production if weather or price risk was considered as a main constraint to increasing pulse productivity.

4.3 Information constraints

In deciding what crops to grow, farmers face several information constraints. They may know that they can plant pulses, but they may either not know how to appropriately time inputs on pulses (e.g. Conley and Udry, 2001) or may not know about specific aspects of pulses that might make them attractive to grow.

The standard type of intervention to overcome information constraints is to provide agricultural extension. However, agricultural extension services are often an important part of the adoption process, and in many countries, extension systems have become underfunded and unresponsive to needs of farmers (Bientema and Stads, 2011). A major reason that they are often unresponsive is that they tend to be organized in a top-down manner, rather than being decentralized and responsive to farmers.[3]

In particular, there are several information constraints that likely affect pulses in various ways, and likely reduce both the productivity of pulses and other crops. Specifically, smallholders may not know how to appropriately save seed from one season to the next or how to decide when to replace saved seed with the seed purchased from the market.

3. Swanson and Rajalahti (2010) provided a detailed guide that discusses ways to modernize agricultural extension systems and make them more responsive to the needs of farmers in developing countries.

Smallholders may also not know how to utilize pulses in a crop rotation to enhance production of other crops, as discussed earlier.

Two types of interventions could work to overcome informational inefficiencies: improvements in or variants on agricultural extension and the introduction of innovative new crops. There are several ongoing interventions attempting to improve information available to farmers through information and communication technologies (ICTs). Such interventions include video-based extension or phone in extension lines, and could potentially include text messages (e.g. Aker, 2011; Cole and Fernando, 2014). At present, the most successful ICT interventions have occurred through price dissemination (e.g. Goyal, 2010), but in densely populated places, internet kiosks could potentially provide information for growing crops as well.

4.4 Summary

To increase pulse production and therefore availability, additional production is needed. Many countries demonstrate unrealized potential in average pulse yields, suggesting that producers face socio-economic binding constraints against improving their production. Such constraints may include credit, risk and information constraints and can be overcome through interventions, particularly promising are interventions that provide cash grants to farmers to catalyse participation in pulse value chains, or interventions that attempt to deliver information to farmers in new ways.

5 Increasing demand for pulses

5.1 Consumers and pulses

To this point, in this chapter, we have described why pulses should be added to diets, and how production increases could potentially be stimulated to improve the availability of pulses. Whereas a production increase is necessary to make pulses more widely consumed, it would not be sufficient. It is important also to understand whether consumers will purchase and eat pulses if they are abundantly available. In this subsection, we describe requirements from the demand side to make pulses a larger portion of the diet.

According to consumer theory, demand for pulses is affected by prices and income levels, as well as preferences. Considering income, it is clear that incomes have risen substantially in real terms since 1961 in almost all parts of the world. The question, then, is how pulse demand would respond to income. Very little rigorous evidence exists on income elasticities for pulses, with the exception of recent scholarship in India, a major consumer of pulses, in part due to the large number of vegetarians there (Joshi, Kishore and Roy, 2016).[4] A major problem with current estimates is that we do not know whether they vary by income level. Anecdotal evidence suggests that in many places pulses are 'poor people's food', and get substituted out of the diet once incomes rise (e.g. Jayasena and Abbas, 2016). For example, meat may replace pulses in the diet at higher income

4. Note that there is little current research on either price or income elasticities of demand, in part because of increased emphasis in the literature on finding exogenous sources of variation to measure economic parameters. Since prices rarely vary randomly, it is difficult to find such sources of variation, and consequently with almost any estimates in the literature. Moreover, in developing countries where farmers sometimes consume part or all of their own production, responsiveness to prices is even more difficult to measure.

levels, except in cases when religious restrictions against eating meat apply. Additional research on how demand for pulses varies by income would help tailor interventions meant to increase consumption of pulses.

Prices are also particularly important to pulse demand, and in fact in India, pulses appear to be more responsive to prices than major grains (Joshi, Kishore and Roy, 2016). As such, note that if production were to increase substantially, prices would drop, theoretically increasing consumer demand substantially so long as transaction costs were not high. In contrast, as India imports a substantial share of pulses consumed, if world prices rose, one might expect demand to quickly equilibrate at a lower level. From the perspective of increasing pulses consumed, it is crucial to increase production, at least in India.

That said, a particular challenge in the pulse sector is the existence of a wide variety of pulses, and focussing crop research on one type of pulses might not lead to major gains in terms of consumption for the following reason. From the consumer perspective, different pulses may not be considered substitutes for one another. Among a sample in southern India, Kumar and Joshi (2016) found little evidence of cross-price elasticities among different pulses. In other words, even if the price of lentils decreased, consumers would demand more lentils, but they would not also switch from chickpeas (or other pulses) to lentils. Therefore, research into productivity of pulses may need to cover a wide variety of pulses, rather than just one or two pulses, since demand may be quite pulse-specific. However, evidence on this point is only available for India, so it might not hold elsewhere.

5.2 Behaviour change

The third channel through which demand for specific products can be enhanced is broadly through preferences. More broadly, one can think of attempting to increase demand for pulses through behaviour change communication (BCC) programmes, which work to change consumption behaviour among targeted groups. From the perspective of pulses, the most important groups are those suffering from iron-deficiency anaemia – in particular children, girls near the age of menarche and women of child-bearing age; or those at risk of overnutrition. In general, many integrated agriculture–nutrition interventions conducted in the past decade have included BCC programmes, though none of them have focussed specifically on pulses. BCC could be important to reducing the apparent stigma of consuming pulses as incomes rise.

Yet evidence from other interventions suggests that inducing behaviour change may be complex and potentially expensive. For example, a randomized controlled trial in Burkina Faso evaluated the impact of an integrated agriculture and nutrition BCC programme using bimonthly home visits to impart information on nutrition; the programme witnessed positive impacts to the nutrition status of both children and women from the intervention but found that the impacts varied based upon who delivered the information (Olney et al., 2015). However, de Brauw et al. (2015) found little contribution of key BCC messages to the adoption of vitamin A-rich orange flesh sweet potato in Mozambique and Uganda. Furthermore, multiple strategies may be necessary to increase demand for more nutritious food. For example, studies conducted on porridge made with orange (vitamin A-fortified) maize versus white maize showed different reactions to the same information; in Zambia, consumers preferred the orange maize when told it was more nutritious, while in Ghana, the preference for orange maize decreased once this information was provided (Birol et al., 2016).

From the research that exists on BCC, it is clear that experience as well as information is needed, and often, through multiple channels. Innovative experiments are occurring with both radio and video. Murray et al. (2015) tested a package radio intervention in Burkina Faso including high exposure to messages related to survival of children under five, through creative storytelling. Sarassat et al. (2015) found that as a result of Murray et al.'s intervention, there was an increase in seeking care for diarrhoea as well as fast or irregular breathing, but not on habitual behaviours such as hand-washing, suggesting complementary interventions might be required to change habitual behaviours. An experiment in Ethiopia used video screenings of successful individuals to indirectly influence attitudes towards child education and found positive results, but expressed concern that these positive impacts would not be sustained over the longer term (Bernard et al., 2015).

An alternative way to stimulate demand through desirability of products is to provide consumers with multiple, public signals that pulses or lightly processed pulse products are relatively healthy or nutritious. Such signals could come either through the public health system or through nutritional labelling. These strategies are commonly used in developed countries to increase demand for specific foods (e.g. Alston et al., 2007), so their effectiveness in developing countries is not well known. Further research is necessary to understand how effective public promotion or labelling can be in developing countries.

6 Conclusion

This chapter discusses some reasons why there exists a gap between the nutritional and economic values of pulses. Relative to major grains, pulses have a much higher nutritional value and demonstrate characteristics that induce satiety after they are eaten, and as such they are quite high in nutritional value to the diet in relative terms. Moreover, in many countries, biofortified varieties are either available or becoming available, increasing the nutritional value of pulses.

Whereas pulses have a great deal of nutritional value, in economic terms, they are not as valued, particularly from the producer perspective; sown area of pulses has been slowly declining over time, in part because pulse productivity has not increased as fast as coarse grain productivity. However, the production of pulses can have positive externalities as they fix nitrogen in the soil, so farmers already undervalue the potential role of pulses for their soil, and in many developing countries, a yield gap exists between potential and actual yields. Interventions might be required to help farmers overcome constraints that hinder them from realizing potential yields, which include credit, risk and information constraints. Transaction costs due to poor infrastructure may also reduce the amount of pulses that make it to market.

Finally, simply increasing pulse production is sufficient to stimulate the consumption of pulses among those who might benefit the most; that is, individuals at risk of iron-deficiency anaemia or of overweight status. BCC programmes may help, but they appear to require multiple strategies and fairly high cost to be effective. Public health campaigns with wide dissemination techniques, or nutritional labelling, might be more effective ways to stimulate pulse consumption in developing countries, but more research is required to understand how effective they might be.

7 Where to look for further information

For more information on agricultural production constraints that affect pulses, see Jack (2011) or the ATAI website (www.atai-research.org). For more information on biofortification of pulses among other crops, see the HarvestPlus website (www.harvestplus.org).

8 References

Aker, J. 2011. 'Dial "A" for agriculture: A review of information and communication technologies for agricultural extension in developing countries', *Agricultural Economics* 42(6): 631–47.

Alston, J.M., J. M. Crespi, H. M. Kaiser and R. Sexton. 2007. 'An evaluation of California's mandated commodity promotion programs', *Review of Agricultural Economics* 29(1): 40–63.

Alston, J. M., J. P. MacEwen and A. M. Okrent. 2016. 'The economics of obesity and related policy', *Annual Review of Resource Economics* 8: 443–65.

Ambler, K., A. de Brauw and S. Godlonton. 2016a. 'Cash Transfers and Crop Production in Senegal', working paper.

Ambler, K., A. de Brauw and S. Godlonton. 2016b. 'Relaxing Constraints for Family Farmers: Providing Capital and Information in Malawi', working paper.

Ashour, M., L. Billings, D. Gilligan, J. Hoel and N. Karachiwalla. 2016. 'Do Beliefs about Agricultural Inputs Counterfeiting Correspond with Actual Rates of Counterfeiting? Evidence from Uganda', IFPRI Discussion Paper no. 1552.

Banerjee, A., D. Karlan and J. Zinman. 2015. 'Six randomized evaluations of microcredit: Introduction and further steps', *American Economic Journal: Applied Economics* 7(1): 1–21.

Bernard, T., S. Dercon, K. Orkin and A. S. Taffesse. 2015. 'Will video kill the radio star? Assessing the potential of targeted exposure to role models through video', *World Bank Economic Review* 29(S1): S226–37.

Bientema, N. and G.-J. Stads, 2011, *African Agricultural R&D in the New Millennium. Progress for Some, Challenges for Many*. Washington, DC: IFPRI.

Birol, E., A. Saltzman, A. M. Ball, E. Boy, B. Mudyahoto, E. Simpungwe, A. Ubomba-Jaswa and M. Zeller. 2016. 'The role of biofortification as part of more diverse diets in Africa: Progress, challenges, and opportunities'. In N. Covic and S. L. Hendriks (Eds) *Achieving a Nutrition Revolution for Africa: The Road to Healthier Diets and Optimal Nutrition*, ReSAKSS Annual Trends and Outlook Report 2015. Washington: International Food Policy Research Institute (IFPRI).

Bouis Howarth, E., C. Hotz, B. McClafferty, J. V. Meenakshi and W. H. Pfeiffer, 2011, 'Biofortification: A new tool to reduce micronutrient malnutrition', *Food and Nutrition Bulletin* 32(suppl 1): S31–40.

Brune, L., X. Giné, J. Goldberg and D. Yang, 2016, 'Facilitiating savings for agriculture: Field experimental evidence from Malawi', *Economic Development and Cultural Change* 64(2): 187–220.

Carter, M., A. de Janvry, E. Sadoulet and A. Sarris, 2014, 'Index-based Weather Insurance for Developing Countries: A Review of the Evidence and a Set of Propositions for Up-scaling', working paper, University of California at Davis Department of Agricultural and Resource Economics.

Cole, S. and N. Fernando, 2014, 'The Value of Advice: Evidence from the Adoption of Agricultural Practices', Harvard Business School Working Group paper 1(1.3).

Conley, T. and C. Udry, 2001, 'Social learning through networks: The adoption of new agricultural technologies in Ghana,' *American Journal of Agricultural Economics* 83(3): 668–73.

de Brauw, A., P. Eozenou, D. Gilligan, C. Hotz, N. Kumar and J. V. Meenakshi. 2015. 'Biofortification, Crop Adoption and Health Information: Impact Pathways in Mozambique and Uganda', HarvestPlus Discussion Paper no. 21.

Dercon, S. 1996. 'Risk, crop choice, and savings: Evidence from Tanzania', *Economic Development and Cultural Change* 44(3): 485–513.

Dercon, S., R. Vargas Hill, I. Outes-Leon, A. Bayrou, D. Clarke and A. Seyoum Taffesse. 2014. 'Rainfall insurance and informal groups: Evidence from a field experiment in Ethiopia', *Journal of Development Economics* 106: 132–43.

Food and Agriculture Organization. 2016. FAOSTAT database, available at http://www.fao.org/faostat/en/#home.

Goyal, A. 2010. 'Information, direct access to farmers and rural market performance in Central India', *American Economic Journal: Applied Economics* 2(3): 22–45.

Ha, V., J. L. Sievenpiper, R. J. de Souza, V. H. Jayalath, A. Mirrahimi, A. Agarwal, L. Chiavaroli, S. B. Mejia, F. M. Sacks, M. Di Buono, A. M. Bernstein, L. A. Leiter, P. M. Kris-Etherton, V. Vuksan, R. P. Bazinet, R. G. Josse, J. Beyene, C. W. C. Kendall and D. J. A. Jenkins. 2014. 'Effect of dietary pulse intake on established therapeutic lipid targets for cardiovascular risk reduction: A systematic review and meta-analysis of randomized controlled trials', *Canadian Medical Association Journal* 186(8): E252–62.

Haas, J. 2014. 'Efficacy and other nutrition evidence for iron crops', In The 2nd Global Conference on Biofortification: Getting Nutritious Foods to People Conference Brief number 12.

Haas, J. D., S. V. Luna, M. G. Lung'aho, M. J. Wenger, L. E. Murray-Kolb, S. Beebe, J.-B. Gahutu and I. M. Egli. 2016. 'Consuming iron biofortified beans increases iron status in rwandan women after 128 days in a randomized controlled feeding trial', *Journal of Nutrition* 146:1586–92.

HarvestPlus. 2016. Biofortified Staple Crops: Who is Growing What? International Food Policy Research Institute crop map, http://www.harvestplus.org/knowledge-market/publications (accessed 21 December 2016).

Hill, R., M. Robles and F. Ceballos. 2016. 'Demand for a simple weather insurance product in India: Theory and evidence', *American Journal of Agricultural Economics* 98(4): 1250–70.

Horton, S. and J. Ross. 2003. 'The economics of iron deficiency,' *Food Policy*, 28, 51–75.

Howarth, N. C., E. Saltzman and S. B. Roberts. 2001. 'Dietary fiber and weight regulation', *Nutrition Review* 59: 129–39.

Hunt, J. 2003. 'Bioavailability of iron, zinc, and other trace minerals from vegetarian diets', *American Journal of Clinical Nutrition* 78(3): 633S–9S.

International Food Policy Research Institute. 2015. *Global Nutrition Report 2015: Actions and Accountability to Advance Nutrition and Sustainable Development.* Washington, DC: International Food Policy Research Institute.

International Food Policy Research Institute. 2016. *Global Nutrition Report 2016: From Promise to Impact: Ending Malnutrition by 2030.* Washington, DC: International Food Policy Research Institute.

Jack, K. 2011, 'Constraints on the Adoption of Agricultural Technologies in Developing Countries', White paper, Agricultural Technology Adoption Initiative, J-PAL (MIT) and CEGA (UC Berkeley).

Jayalath, V. H., R. J. de Souza, J. L. Sievenpiper, V. Ha, L. Chiavaroli, A. Mirrahimi, M. Di Buono, A. M. Bernstein, L. A. Leiter, P. M. Kris-Etherton, V. Vuksan, J. Beyene, C. W. C. Kendall and D. J. A. Jenkins. 2013. 'Effect of dietary pulses on blood pressure: A systematic review and meta-analysis of controlled feeding trials', *American Journal of Hypertension* 27: 56–64.

Jayasena, V. and S. Abbas. 2016. 'Pulses: Poor man's food or future super food?' *Food Australia* 68(3): 32.

Joshi, P. K., A. Kishore and D. Roy. 2016. 'Making Pulses Affordable Again: Policy Options from the Farm to Retail in India', IFPRI Discussion Paper no. 1555.

Joshi, P. K. and P. Parthasarathy Rao. 2017. 'Global Pulses Scenario: Status and Outlook', *Annals of the New York Academy of Sciences*, 1392: 6–17.

Karlan, D., E. Kutosati, M. McMillan and C. Udry, 2010, 'Crop price indemnified loans for farmers: A pilot experiment in rural Ghana', *Journal of Risk and Uncertainty* 78(1): 37–55.

Karlan, D., R. Osei, I. Osei-Akoto and C. Udry. 2014. 'Agricultural decisions after relaxing credit and risk constraints', *The Quarterly Journal of Economics* 129(2): 597–652.

Key, N., A. de Janvry and E. Sadoulet, 2000, 'Transaction costs and agricultural household supply response', *American Journal of Agricultural Economics* 82(2): 245–59.

Khoury, C. K., A. D. Bjorkman, H. Dempewolf, J. Ramirez-Villegas, L. Guarino, A. Jarvis, L. H. Rieseberg and P. C. Struik. 2014. 'Increasing homogeneity in global food supplies and the implications for food security,' *Proceedings of the National Academy of Sciences* 111, 4001–6.

Kim, S. J., R. J. de Souza, V. L. Choo, V. Ha, A. I. Cozma, L. Chiavaroli, A. Mirrahimi, S. Blanco Mejia, M. Di Buono, A. M. Bernstein, L. A. Leiter, P. M. Kris-Etherton, V. Vuksan, J. Beyene, C. W. C. Kendall, D. J. A. Jenkins and J. L. Sievenpiper. 2016. 'Effects of dietary pulse consumption on body weight: A systematic review and meta-analysis of randomized controlled trials', *American Journal of Clinical Nutrition* 103: 1213–23.

Kumar, P. and P. K. Joshi. 2016. 'Changing Demand Patterns of Pulses and the Projections to 2030', Paper presented at the Conference on Pulses for Sustainable Agriculture and Human Health, New Delhi, 31 May – 1 June.

Li, S. S., C. W. C. Kendall, R. J. de Souza, V. H. Jayalath, A. I. Cozma, V. Ha, A. Mirrahimi, L. Chiavaroli, L. S. A. Augustin, S. B. Mejia, L. A. Leiter, J. Beyene, D. J. A. Jenkins and J. L. Sievenpiper. 2014. 'Dietary pulses, satiety and food intake: A systematic review and meta-analysis of acute feeding trials', *Obesity* 22: 1773–80.

Lipton, M. 1976. 'Agricultural finance and rural credit in poor countries,' *World Development* 4(2): 543–54.

Lonnerdal, B. 2009. 'Soybean ferritin: Implications for iron status of vegetarians', *American Journal of Clinical Nutrition* 89(5): 1680S–5S.

McCrory, M. A., B. R. Hamaker, J. C. Lovejoy and P. E. Eichelsdoerfer. 2010. 'Pulse consumption, satiety, and weight management', *Advances in Nutrition* 1: 17–30.

Murekezi, A. 2013. 'Bean Seed and Grain Value Chains in Rwanda: What Are the Entry Points for High Iron Bean Varieties?', Presented at 4th African Association of Agricultural Economists Conference, Hammamet, Tunisia, 22–25 September.

Murray, J., P. Remes, R. Ilboudo, M. Belem, S. Salouka, W. Snell, C. Wood, M. Lavoie, L. Deboise and R. Head. 2015. 'The *saturation +* approach to behavior change: Case study of a child survival radio campaign in Burkina Faso', *Global Health: Science and Practice* 3(4): 541–56.

Murray-Kolb, L. E., R. Welch, E. C. Theil and J. L. Beard. 2003. 'Women with low iron stores absorb iron from soybeans', *American Journal of Clinical Nutrition* 77(1): 180–4.

Olney, D. K., A. Pedehombga, M. T. Ruel and A. Dillon. 2015. 'A 2-year integrated agriculture and nutrition and health behavior change communication program targeted to women in Burkina Faso reduces anemia, wasting, and diarrhea in children 3–12.9 months of age at baseline: A cluster-randomized controlled trial', *The Journal of Nutrition* 145(6): 1317–24.

Petry, N., F. Rohner, J. Bosco Gahutu, B. Campion, E. Boy, P. L. Tugirimana, M. B. Zimmerman, C. Zwahlen, J. P. Wirth and D. Moretti, 2016, 'In rwandese women with low iron status, iron absorption from low-phytic acid beans and biofortified beans is comparable, but low-phytic acid beans cause adverse gastrointestinal symptoms', *Journal of Nutrition* 146: 970–5.

Roughead, Z. K. (Fariba), C. A. Zito and J. R. Hunt, 2005, 'Inhibitory effects of dietary calcium on the initial uptake and subsequent retention of heme and non-heme iron in humans: Comparisons using an intestinal lavage method', *American Journal of Clinical Nutrition* 82(3): 589–97.

Sarrassat, S., N. Meda, M. Ouedraogo, H. Some, R. Bambara, R. Head, J. Murray, P. Remes and S. Cousens. 2015. 'Behavior change after 20 months of a radio campaign addressing key lifesaving family behaviors for child survival: Midline results from a cluster randomized trial in Burkina Faso', *Global Health: Science and Practice* 3(4): 557–76.

Swanson, B. and R. Rajalahti. 2010. Strengthening Agricultural Extension and Advisory Systems: Procedures for Assessing, Transforming, and Evaluating Extension Systems, World Bank, Agriculture and Rural Development working paper no. 45.

US Department of Agriculture (USDA), Agricultural Research Service, Nutrient Data Laboratory. 2015. *USDA National Nutrient Database for Standard Reference, Release 28*. Version Current: September 2015. Washington, DC: USDA.

Viguiliouk, E., S. E. Stewart, V. H. Jayalath, A. Praneet Ng, A. Mirrahimi, R. J. de Souza, A. J. Hanley, R. P. Bazinet, S. Blanco Mejia, L. A. Leiter, R. G. Josse, C. W. C. Kendall, D. J. A. Jenkins and J. L. Sievenpiper. 2015. 'Effect of replacing animal protein with plant protein on glycemic control in diabetes: A systematic review and meta-analysis of randomized controlled trials', *Nutrients* 7: 9804–24.

World Health Organization. 2014. *Global Nutrition Targets 2025: Anaemia Policy Brief* (WHO/NMH/NHD/14.4). Geneva: World Health Organization.

Index

Lightning Source UK Ltd.
Milton Keynes UK
UKHW02n0105200318
319686UK00001B/4/P

9 781786 761361